Alternating Current Machines

M. G. SAY

Emeritus Professor of Electrical Engineering
Heriot-Watt University Edinburgh

Alternating Current Machines

Fifth Edition

A Halsted Press Book

JOHN WILEY & SONS

New York

First published in Great Britain in 1983
by Pitman Publishing Ltd

Published in the U.S.A.
by Halsted Press
a division of John Wiley & Sons, Inc.
New York

Library of Congress Cataloging in Publication Data
Say, M.G. (Maurice George), 1902 —
Alternating Current machines.
Includes bibliographical references and index.
    1. Electric machinery — Alternating current.
    2. Electric transformers. I. Title.
TK2711.S3 1984   621.31'33   83—10719
ISBN 0-470-27451-4

Reproduced and printed by photolithography and bound in Great Britain

Contents

Contents

13 Special Machines

Preface

The forerunner of this book was *The Performance and Design of A.C. Machines*. First published in 1936, its combination of design and performance proved gratifyingly acceptable. But times have changed. Design, revolutionized by the digital computer, is now an esoteric art practised by the few. In contrast, performance concerns the many, as a result of the advent of the big generator, the interest in the transient dynamics of starting, braking and speed control, and the development of power electronics.

Much attention has recently been given to the 'generalized machine' in which 'types' become variants of a common principle expressed in electric-circuit equations of behaviour. Though powerful, the method can model a machine only in a restricted way. Real machines have real magnetic, mechanical and thermal attributes as well, and although their basic electro-magnetic principles are certainly the same, the several types do, in fact, differ markedly in performance.

The present text expresses my belief that machine performance is grasped more readily through flux-current interaction than by the manipulation of matrix equations, which are meaningful only if their physical basis is understood. The treatment therefore begins with the simplified electromagnetic principles of transformers and rotating machines, followed by material common to both in magnetic circuits, windings and loss dissipation. For each type, the ideal case is modified by those departures from the ideal that affect actual performance. Design is discussed briefly, and some constructional details are illustrated. Most of the text covers transformers and 'standard' polyphase machines. The two final chapters respectively discuss recent developments, and 'special' machines including single-phase types.

All formulae are couched in SI units, but decimal multiples and sub-multiples are employed for convenience in numerical examples. Guidance in SI usage can be found in a booklet (*Symbols and Abbreviations*) published by the Institution of Electrical Engineers.

My thanks are due to good friends in the industry and in the universities for their ready help. I am particularly indebted to the critics (unknown to me) who read the manuscript: in only a few cases have I failed to take their expert advice.

Edinburgh 1983   M. G. Say

1
Introductory

1.1 *Alternating-current machines*

Electric and magnetic fields both store energy, and useful mechanical forces can be derived from them. In air (or other gas at normal pressure) the dielectric strength of the medium restricts the working *electric* field intensity to about 3 MV/m and the energy density in consequence to about 40 J/m^3; this corresponds to a force between displaced charges of 40 N/m^2. There is no comparable restriction on *magnetic* fields, but the saturation of ferromagnetic media required to complete the magnetic circuit limits the working flux density to about 1.6 T, for which the energy density in air is about 1 MJ/m^3. As this is 25 000 times as much as for the electric field, almost all industrial electric machines are magnetic in principle; and they are inherently a.c. machines because of the occurrence of magnetic poles in pairs, leading to a natural N–S sequence. The movement of a conductor through such a sequence induces an e.m.f. which changes direction in accordance with the magnetic polarity: in brief an *alternating* e.m.f.

At present, the chief source of electrical energy is the mechanical form, direct (as in hydraulic storage reservoirs or rivers), or indirect from chemical or nuclear energy via heat. Conversion of mechanical to or from electrical energy is by use of the mechanical force manifested when a current-carrying conductor lies in a magnetic field.

The wide use of electrical energy has been fostered by the development of systems of transmission and distribution, so that interconnected networks enable generating plant to be concentrated, and a reliable, economic supply to be achieved. A high voltage is required to limit loss in the transmission of large powers; a much lower one is necessary for distribution for reasons connected with safety and convenience. The transformer makes this economically possible.

Thus both generation and transformation are inherently a.c. phenomena, and the collateral development of a.c. generators and motors has been the result. The a.c. system is not, however, a finality: in some applications the d.c. machine is superior, and d.c. transmission is technically possible. But a.c. machines and transformers seem likely to remain important components in the pattern of electrical energy generation and utilization.

1.2 *Performance*

Generators, motors and transformers form part of an electro-mechanical energy network. The various parts of such a system are interconnected and react upon each other, sometimes considerably and on occasion catastrophically. No machine can be studied in isolation, for its behaviour is influenced by the electrical supply at one end and the mechanical attachments at the other, between which it forms a dynamic link. The performance of a machine is therefore not just its steady-state characteristics, but more importantly its dynamic and transient behaviour, in which the rest of the system plays a possibly paramount role.

Consider an underground coal-cutter motor at the far end of a long mine cable. When the machine is started on load there will be a considerable volt-drop in the cable impedance, the motor terminal voltage will be abnormally low, and the starting torque will be seriously reduced. The 'normal' steady-state characteristic of the motor is not applicable: the performance must be taken in relation to the system to which the machine is connected. Again, although a large supply system is not likely to be significantly affected by a 50 W hand-tool motor, the starting of a 25 MW industrial machine may require previous notification to the supply authority. A machine that develops substantial harmonic voltages and currents can be an embarrassment to other consumers' plant.

The performance of a machine has a great many facets, among them the mechanical characteristics of prime-movers or fluctuating loads, frequent starting and stopping duty-cycles, stored kinetic energy, ambient conditions (particularly of cooling), and the like. The machine is a link in a *system*.

System

Systems engineering is concerned with the planning, design, construction and operation of complex interconnected systems. A *system* is regarded as an assembly of *sub-systems*: in an electric supply network these could be fuel-handling equipments, boiler plant, turbines, generators, transformers, transmission lines, distributors, motors and mechanical loads. The operation of the system as a whole is a function of the subsystem characteristics. The overall performance is studied in systems engineering in accordance with an objective such as optimized cost, overall reliability, or stability under conditions of fault.

Model

To evaluate system performance it is necessary to have a quantitative model of each subsystem in the form of a set of equations, a simulation or a compilation of graphical data. The subsystem models are then interconnected with due regard to their individual constraints. The model of an electrical machine can take many forms, from an elementary description in terms of current sheets and fluxes to sets of sophisticated equations in terms of voltage, current, speed, time, inertia, friction and electrical parameters. Much depends on the use to which a model is to be put. A transformer, for example, can for normal operating conditions be represented by a simple T-network of impedances; but this will not serve for problems such as surge distribution, short-circuit force or temperature-rise. Performance must be interpreted in terms of the significant conditions of operation. The more comprehensive the study, the more complex must be the model. Eventually, the only possible model of a machine is the machine itself.

1.3 *Design*

Design, the creative translation of a concept into a reality, is (in the present context) the application of science, technology and invention to the realization of a machine to perform specified functions with optimum economy and efficiency. In a large-scale manufacturing process the design activities include a balanced consideration of the following.

Basics. Previous experience and new ideas; research into extension of technology of materials, stresses and limits; manufacturing and transport facilities; ergonomics, environmental effects, aesthetics, safety, reliability, maintenance; economics and cost analysis; optimization.

Communication. Transfer of design to factory floor (drawings, instructions, processes, numerical and automatic methods); job-flow, critical path, timing, delivery date.

Specification. Customer's requirements; standards, codes; interpretation of specification, subdivision into specialist design areas, co-ordination.

Information. Data sources (previous contracts, technical journals, research reports); *ad hoc* research and development.

Design Team. Allocation of tasks; technical design analyses; group design, computation; synthesis of group designs, matching and compatibility, prediction of performance, achievement of specification and guarantees, optimization.

Optimization is not easy to define. An optimum solution can be so only in relation to specific objectives, but the choice of these depends on what is meant by optimization. It may not be merely 'cost reduction', but may involve a quantitative appreciation of all the criteria that identify the best design for a given duty. A single criterion, such as minimum mass or minimum manufacturing cost, might be readily definable; more often the weighted combination of several conflicting requirements is the determining factor. Such formulations are inherently less amenable to quantization than are analyses of technical performance.

Machines and Transformers

As an example, Fig. 1.1 lists the main areas concerned in the design of a motor. Many of the items are interrelated; the choice of slot dimensions affects leakage flux and therefore reactance, starting performance, torque/speed relation, heating, ventilation . . . , but the analytical relations between them are indirect and not easy to formulate.

Given the specification that a machine has to meet, the main design areas are: the magnetic circuit, the windings, the insulation, the heat dissipation and the mechanical construction. There is an economic conflict for space

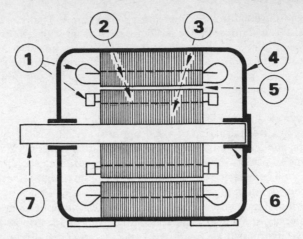

1.1 Design and constructional features of a rotating machine.

Specification: rating, standards, guarantees, loadings, frame size, starting, efficiency, temperature-rise.

Performance: starting torque, transient behaviour, system considerations, short-circuit forces, load loss, optimization.

Manufacture: materials handling, works practice, works processes, available skills, transportation to site.

(1) *Windings*: type, connection, conductors, turns, transposition, bracing, current density, resistance, eddy-currents, I^2R loss, cooling, leakage inductance, insulation, connections, commutator/slip-rings, mass.

(2) *Slots and Teeth*: dimensions, insulation, flux density, core loss, mechanical forces, wedges.

(3) *Cores*: material, stamping, clamping, insulation, mounting, dimensions, flux density, core loss, mass.

(4) *Frame*: size, stiffness, vibration, fixing, enclosure, distortion, expansion, brushgear, mass.

(5) *Airgap*: length, gap factor, flux distribution, space harmonics. clearance, magnetic pull, noise.

(6) *Bearings*: type, loading, loss, cooling, lubrication.

(7) *Shaft*: size, stiffness, deflection, critical speed.

between constructional steel, ferromagnetic material, copper (or aluminium), insulation and coolant. Limitations arise from the inherent material properties such as elastic limit, magnetic saturation, temperature-rise and efficiency — or, rather, loss as affecting the dissipation of heat. Many questions have to be answered: What form of insulation will have adequate life when subject to given electrical, thermal and mechanical stresses? How much coolant can be induced to flow along paths of given size, and how effective will it be in picking up and carrying away the heat? What size of shaft with a given space between bearings will give acceptable limits of deflection, with the consequent effects on out-of-balance magnetic pull and on critical speed?

Computer-aided design

Supposing that the functional relationships between many machine constructional variables can be established, the evolution of a design to meet some specified optimum criterion is a matter of long and tedious iteration. As such, it is clearly in the field of the digital computer. The fully *synthetic* design of a machine is possible in principle, but so far the great complexity of setting up the array of functional relationships has not been mastered. On the other hand, the 'slave' function of the computer has been widely developed to carry out difficult and complicated (but still routine) calculations. This is *computer-aided* design.

Programs have been devised for calculating the leakage inductance of induction machines, and refined by comparison with test results; for accepting details of machine rating, frame size and windings and evaluating performance; and for the prediction of transformer surge-voltage distribution. On occasion a program is written for rapid 'first-shot' design for size, mass and cost as an aid to tendering. Subsystem programs specifying performance and design parameters are then run, and subsequently matched to form the complete design. Some of the design parameters are continuously variable, others are discontinuous and can have only certain discrete values. There will also be absolute limits, such as temperature-rise or dimensions or mechanical stress, that must not be exceeded, and the subsystem programs must take account of them. After integrating the subsystems the overall behaviour is investigated for conformity with the master specification. In this process the computer is used as a decision-making device.

The computer accelerates the design process enormously, makes possible more 'trial' designs, and enables sophisticated calculations to be made without intolerable tedium and excessive time. It can reduce empiricism, readily handle non-linearities, and incorporate much of the designer's 'know-how'. A valuable feature is the man/computer symbiosis, in which the designer can 'talk' to the computer, ask questions and get quick informative answers, change his mind and ask again. The designer is released from numerical drudgery to grapple with physical and logical ideas.

In brief, computer-aided design gives the advantages of eliminating tedious hand-calculations, making data checks at every stage, facilitating trial designs preliminary to a final decision, permitting much more detailed and precise functional relations (provided that they can be formulated), and giving the possibility of developing new and more comprehensive design procedures.

This book does not deal specifically with computer design. It discusses the electromagnetic principles upon which the parameters depend, and the effect on performance.

1.4 *Methods of approach*

In design, a machine is considered a structure in which the interaction of magnetic fields and electric currents is the basic physical principle. It gives rise to calculable forces, torques, e.m.f.s, saturation effects, heat production, mechanical stresses, vibration, noise and several other phenomena of practical importance. In the assessment of performance, various secondary effects, both internal and external to the machine, have to be studied in terms of flux-current interactions, machine structure and system considerations. When the machine is analysed as a member connected into a complex electromechanical system, it is almost essential to treat it as a 'black box' possessing a set of terminal characteristics of behaviour.

The method of treatment depends upon the problem to be solved. The approach chosen is that which appears to 'model' the machine in the most appropriate way. It may be in terms of flux densities and current sheets, or multiple circuit equations, or voltage—current/torque—speed relations; it may be simplified by restriction to the steady state or by ignoring all purely mechanical considerations. The method of approach unavoidably shifts its ground to suit. Even a basic electromagnetic approach can be undertaken in several ways: we choose a flux-current interaction approach because it is most readily pictured in physical terms.

1.5 *Basic principles*

All electromagnetic machines can be related to three principles, namely those of *induction, interaction* and *alignment*.

Induction

The essentials for the production of an electromotive force by magnetic means are electric and magnetic circuits, mutually interlinked. Summation of all products of magnetic flux with complete turns of the electric circuit gives the total flux linkage ψ. If ψ is made to change, an e.m.f. is induced in the electric circuit. The e.m.f. exists only while the change is taking place, and the magnitude of its time-integral is equal to the change of linkage, so that

$$\int e \cdot dt = -\psi \quad \text{or} \quad e = -d\psi/dt$$

The direction of the e.m.f. is such as to oppose the change. If the electric circuit is closed and the linkage with an externally produced magnetic field is reduced, then the e.m.f. generates a current in the closed circuit, developing a self-flux the linkage of which tends to make up the deficiency. Comparable conditions obtain when the change of linkage is the result of a change of current in the electric circuit concerned: then the self linkage may be written as $\psi = Li$ where L is the inductance coefficient. The self-induced e.m.f. is

$$e = -d\psi/dt = -d(Li)/dt$$

acting in the direction of positive current. For reasons of simple convention, set out in Sect. 1.7, it is more convenient to consider the e.m.f. as directed in *opposition* to positive current, giving

$$e = +d\psi/dt = +d(Li)/dt \tag{1.1}$$

For engineering purposes this is used in the form

$$e = N(d\Phi/dt) \tag{1.2}$$

where all N turns link the flux Φ and result in a linkage $\psi = N\Phi$. The flux Φ

may be resolved into (i) a *useful* or working component, and (ii) a non-useful or *leakage* component.

Change of linkage in a coil may occur in three ways:

(i) Supposing the flux constant, the coil may move through it.
(ii) Supposing the coil stationary with respect to the flux, the flux may vary in magnitude.
(iii) Both changes may occur together: i.e. the coil may move through a time-varying flux.

In (i) the flux-cutting rule can be applied, and the e.m.f. in a single conductor of length l calculated from the rate at which it cuts across a magnetic field of uniform density B when moving at speed u in a direction at right-angles to the direction of the flux: then

$$e = Blu \tag{1.3}$$

This is the *motional* e.m.f., always associated with conversion of energy between the mechanical and electrical forms. In (ii) the e.m.f. is found by applying eq. (1.1) to give the *pulsational* (or *transformer*) e.m.f.; no motion is involved and there is no energy conversion. In (iii) both e.m.f. components are concerned.

Interaction

When a current i lies in and perpendicular to a magnetic field of density B over an active length l, a mechanical force

$$f_e = Bli \tag{1.4}$$

is developed on it in a direction perpendicular to both current and field. In (*a*) of Fig. 1.2, B represents the flux density of an undisturbed field. Introduction of a conductor carrying a current imposes a corresponding field component, developing the resultant in (*b*). In the neighbourhood of the conductor the resultant density is greater than B on one side and less than B on the other; provided that the increase on one side has the same magnitude as the reduction on the other, the force on the conductor is given by eq. (1.4).

Alignment

Pieces of high-permeability material (e.g. iron) in an ambient low-permeability

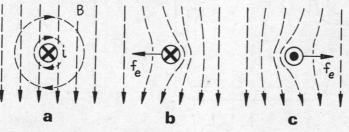

1.2 Interaction law.

medium (e.g. air) in which a magnetic field is established, experience mechanical forces tending to align them with the field direction in such a way as to minimize the reluctance of the system.

1.6 *Basic forms*
Electromagnetic transformers and machines all utilize the induction principle. Most machines exploit interaction, a few are based on alignment, and some utilize both principles. The devices can be built in a wide variety of forms, of which the most elementary are illustrated in Fig. 1.3.

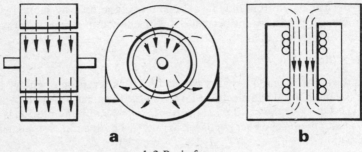

a   **b**

1.3 Basic forms

Rotary Machines
Two magnetic elements (*a*), one fixed (stator) and the other (rotor) capable of rotation, are separated by a narrow annular airgap. For mechanical convenience the stator is normally the outer member. Each element carries windings arranged to lead currents axially along the airgap surface, and a mutual magnetic flux crosses the gap to link them. Rotation of the rotor results in e.m.f. induction and in electromechanical energy conversion through interaction and alignment torques.

Because of the heteropolar nature of the magnetic field, it is convenient in a 2-pole machine to take pairs of gap-surface conductors and join them in series to make turns, as in Fig. 1.4. Then the two opposing forces combine to form

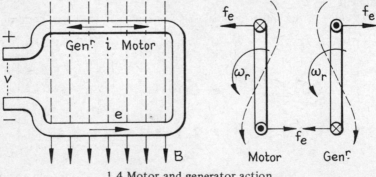

1.4 Motor and generator action.

a torque. Directions of flux, current and force are shown, together with the directions of current for *generator* and *motor* action. This reversibility of mode is a valuable feature of electromechanical devices.

Generator. An e.m.f. is produced by movement of a coil in a magnetic field. The current in an external circuit produced by the e.m.f. interacts with the field to develop a mechanical force opposing the movement, so that a mechanical driving torque must be provided. The generated electrical power is converted from the mechanical power input.

Motor. If current is supplied to the machine from an external source, interaction develops mechanical forces to move the rotor. The motion causes an e.m.f. to be induced in opposition to the current. Thus the motor requires electrical power to produce a corresponding mechanical power.

Despite considerable constructional variety, the electrical differences between machines are only secondary: they arise from (i) the kind of electrical system, a.c. or d.c., to which the windings of the machine are connected; and (ii) the kind of connection made between the windings and the electric system — i.e. tapped phase windings, or commutator windings with switched connections.

Transformers

Transformers do not convert energy to or from the mechanical form and in this sense are not 'machines'. For the *static* transformer, Fig. 1.3(*b*), no gap is needed as there is no movement, so that only pulsational e.m.f.s are concerned. Interaction forces are certainly developed, but the windings are braced against them. In the induction *regulator* transformer, in which the relative position of the two elements has to be adjustable, the constructional form is identical with that in (*a*) of Fig. 1.3. Again there is an interaction effect, and the torque produced makes it necessary to hold the rotor mechanically against rotation.

1.7 *Conventions*

Voltage, current and power

An electrical device, connected to a source of voltage v of polarity as indicated in Fig. 1.5(*a*), takes a current i of positive direction (i.e., in at the positive terminal) and a positive input power $p_e = vi$. If the current reverses (to become $-i$ in the positive direction) then the power also reverses, becoming the negative input $-p_e$ corresponding to an output.

The terminal voltage is taken as the voltage *applied* and the current as that in the positive input direction, so permitting a direct distinction between the action of the load or machine as an absorber of electrical input power (*motor* mode) or as a producer of electrical output power (*generator* mode) corresponding to negative input.

Electromotive force

In that part of a machine where energy storage or conversion takes place magnetically, an e.m.f. given by eq. (1.1) appears. Its direction is in opposition to the direction of positive current. The active part of the machine, enclosed in

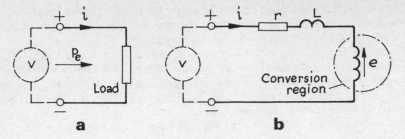

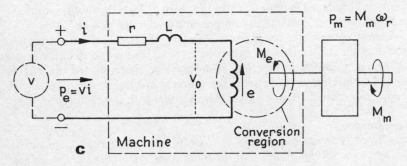

1.5 Machine conventions.

a circle in Fig. 1.5(*b*), is the seat of the e.m.f. $e = d\psi/dt$ which in general can be regarded as comprising two components:

Motional E.M.F.: This e.m.f., e_r, in a winding results from the movement at angular speed ω_r of its conductors through the flux densities in the gap.

Pulsational E.M.F.: The pulsational or 'transformer' e.m.f. e_p appears in the winding if the flux in which it lies has a rate of change with time.

The motional e.m.f. is concerned with electrical/mechanical energy conversion, while the pulsational e.m.f. provides a means of electrical energy transfer between magnetically coupled windings, as from primary to secondary winding in a transformer or from stator to rotor winding in a rotary machine. Then, if a winding carries a current i, the product $e_r i$ represents the rate of energy conversion, and $e_p i$ the rate of energy transfer.

General form: With its two components stated explicitly, the e.m.f. $e = d\psi/dt$ can be written

$$e = \frac{\partial \psi}{\partial \theta} \cdot \frac{d\theta}{dt} + \frac{\partial \psi}{\partial t} = \frac{\partial \psi}{\partial \theta} \omega_r + \frac{\partial \psi}{\partial t} = e_r + e_p \tag{1.5}$$

The term $\partial \psi/\partial \theta$ is a measure of the distribution in the gap of the flux density B.

Machine Winding
The conventions for a machine winding are set out in Fig. 1.5(*c*). The effective resistance of the winding is r, and the loss in it is ri^2. A further quality of

the winding is its leakage inductance L: not all of the flux linkage is usefully employed in linking the primary and secondary windings of a transformer or the stator and rotor windings of a machine. The leakage flux-linkage is considered as separated from the active region of the device.

The voltage and input power relations at any instant are

$$v = ri + L(\mathrm{d}i/\mathrm{d}t) + e_r + e_p \quad \text{and} \quad p_e = ri^2 + L(\mathrm{d}i/\mathrm{d}t)i + e_r i + e_p i$$

Energy-rate balance: A machine is an electro-mechanical device linking an electric source, of voltage v supplying current i, to a mechanical source of angular speed ω_r and torque M_m (or moving at linear speed u under a driving force f_m). Both vi and $M_m \omega_r$ are treated as *input* powers p_e and p_m. The total input to the machine is $p_e + p_m$, and within the machine there are three general rates of power absorption, namely

p: the rate of loss in $I^2 R$, in the core, and in mechanical friction and windage;
$\mathrm{d}w_f/\mathrm{d}t$: the rate of change of stored magnetic-field energy;
$\mathrm{d}w_s/\mathrm{d}t$: the rate of change of mechanical energy stored in inertia, deformation, vibration, shaft strain, etc.

The energy-rate balance is consequently

$$p_e + p_m = \mathrm{d}w_f/\mathrm{d}t + \mathrm{d}w_s/\mathrm{d}t + p \tag{1.6}$$

For simplicity, let the machine run at constant speed in a steady-state condition and with no change of field or mechanical stored energy. Relative motion of stator and rotor generates a motional e.m.f. e_r in opposition to the applied voltage v. The behaviour can be summarized as follows. With the machine acting as a *motor*, the applied voltage v drives an input current $+i$ against e_r to give an electrical input power $p_e = v(+i)$, of which the part $e_r i$ is converted. The conversion process means the development of an electromagnetic torque M_e which drives the rotor against the mechanical input torque M_m at speed ω_r, to produce the negative mechanical input $p_m = (-M_m)\omega_r$, corresponding in effect to a positive mechanical output. With the machine as a *generator* driven at ω_r by a positive mechanical input torque M_m and so receiving the positive input $p_m = (+M_m)\omega_r$, the e.m.f. e_r now exceeds v. The current reverses to provide the negative electrical input (or positive output) $p_e = v(-i)$. The sum of the positive inputs is the rate of energy dissipation within the machine.

At the points of attachment of the machine to its associated electrical and mechanical system (i.e. at the terminals and the shaft), the convention is that input powers are taken as positive. In the motor mode, positive current enters the machine from the positive terminal and there is a positive electrical power input (but the mechanical input is negative). In the generator mode, the current direction is reversed and the electrical input is negative (but the mechanical input is positive).

Conversion Region: That region within the chain-dotted line in Fig. 1.5(c) contains only the essential electromagnetic/mechanical conversion quantities, respectively e and i, and ω_r and M_e. Losses, magnetic leakage and mechanical storage are excluded so that attention can be directed to the essential physical process of energy conversion. This process is intrinsically 'perfect', making it possible to consider electromagnetic machines as ideal devices with respect to their conversion regions, with other effects taken into account externally.

Transformers

A two-winding transformer is basically a pair of coils (of which either can be the primary) embracing a common magnetic circuit of high-permeability

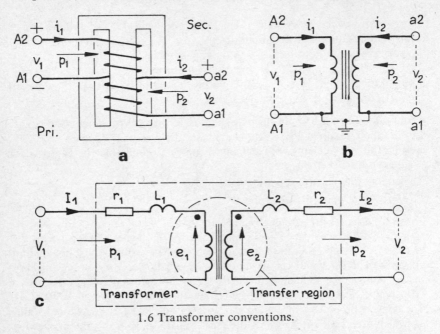

1.6 Transformer conventions.

steel, Fig. 1.6(*a*). Both windings encircle the common flux in the same *sense*: then positive polarity at terminal A2 of the primary pair A1A2 corresponds to positive polarity at a2 of the secondary pair a1a2. In the schematic diagram (*b*) this information is conveyed by placing a dot at the coil ends A2 and a2. If, in conformity with Fig.1.5(*c*), both primary and secondary currents were flowing into a dotted terminal, their m.m.f.s. would be additive on the magnetic core. Such a condition is not a practical one. The convention adopted is therefore that in Fig. 1.6(*c*), in which the basic primary/secondary m.m.f. balance is emphasized. This further relates more directly to the usual equivalent circuit, in which the transformer is considered as a two-port network with input at the primary and output at the secondary terminals.

The windings possess resistance, and not all of the flux due to one m.m.f. links the other winding. Leakage fluxes not contributing to the mutual coupling (by means of which energy can be transferred from one winding to the other) therefore exist. By treating the resistance and the leakage inductance as separate entities, it is possible to set up an elementary equivalent circuit, Fig. 1.6(*c*).

Transfer Region: The region within the chain-dotted line in (*c*) is that in which the essential energy transfer takes place. The primary and secondary powers are $e_1 i_1$ and $e_2 i_2$ at any instant. Again the process is intrinsically 'perfect', making it is possible to consider a transformer as an ideal device with external defects.

1.8 *Dynamic circuit theory*

The study of electrical machines is devoted (*a*) to understanding how they work, and (*b*) to predicting their performance. The former is the concern of the applications engineer and the student: the latter chiefly of the designer.

Ideally, it ought to be possible to express the behaviour of electrical machines in terms of Maxwellian electromagnetic field equations. Such an approach is exceedingly difficult, and the normal method is through the simplified concepts of induction and interaction. Various techniques (e.g. complexor and locus diagrams, equivalent circuits), mostly of particular rather than of general application are employed. The disadvantage of this approach is its practical limitation to the steady state. But if one more step is taken from 'reality' into the abstract, machines can be dealt with in terms of *dynamic circuits*. A truly equivalent circuit applies to its prototype machine under all conditions, steady and transient, balanced and unbalanced. To such a circuit can be applied all the powerful techniques of circuit analysis — circuit laws, network theorems, operational and matrix methods, network analysers and computers. The circuit can be used to obtain the three primary characteristics (voltage, current and torque) in advanced studies of generator faults, changes of load, or automatic control, all so important in modern engineering.

There are some sacrifices. Manipulation of the equations is only tractable on the assumption of linear parameters. Non-linear effects, such as saturation, must be the subject of subsequent adjustment in the light of experience by semi-empirical methods. Space-harmonics of flux, brush-contact phenomena and commutation effects must be ignored. These effects are far from negligible in practical machines; nevertheless to ignore them is to assume conditions differing only in degree from those of normal circuit theory. Finally, the circuits are still not truly equivalent because propagation phenomena are disregarded, so that surge-voltage effects need separate consideration.

Circuit Elements: Fig. 1.7 shows the basic elements of dynamic circuits with their equations of behaviour, in terms of the operators p (representing d/dt) and $1/p$ (representing time-integral).

1.9 *Prototype ideal transformer*

Power transformer action under steady-state sinusoidal conditions can be established in terms of an *ideal transformer* embodying the transfer region. The magnetic circuit of an ideal transformer has infinite permeability and zero reluctance, so that a vanishingly small m.m.f. is needed to excite the magnetic flux; all the flux completely links both primary and secondary windings (there is no magnetic leakage); there are no losses in the windings or in the core.

As the practical power transformer approximates to these conditions, it can be taken as equivalent to an ideal transformer with the following external 'defects': a small but finite m.m.f. to magnetize the core, small but finite leak-

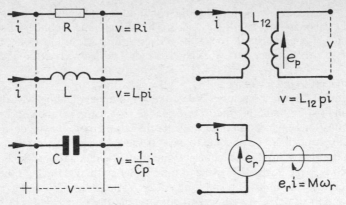

1.7 Dynamic circuit elements.

age inductances to account for the slightly imperfect coupling, and resistances to simulate loss effects in the windings and the core.

Ideal Transformer under Sinusoidal Conditions

When a sine voltage v_1 of angular frequency $\omega = 2\pi f$ is applied to the primary winding with the secondary open-circuited, the voltage is balanced by an e.m.f. induced by the rate of change of flux linkage $\psi_1 = N_1\Phi$, and the core flux must therefore vary with time. Let it be $\Phi = \Phi_m \cos \omega t$, a variation at angular frequency ω between oppositely directed peaks of Φ_m: then from eq. (1.1) the corresponding primary voltage is

$$v_1 = d\psi_1/dt = N_1\omega\Phi_m \cos \omega t$$

which has the peak value $v_{1m} = N_1\omega\Phi_m$ and the r.m.s. value

$$V_1 = (1/\sqrt{2})\omega N_1\Phi_m = 4.44fN_1\Phi_m$$

As the same flux links each turn, the voltage per turn is

$$V_t = (1/\sqrt{2})\omega\Phi_m = 4.44f\Phi_m \tag{1.7}$$

Thus the primary terminal voltage is $V_1 = N_1 V_t$ and that of the secondary is $V_2 = N_2 V_t$. The primary/secondary voltage ratio V_1/V_2 is equal to the primary/secondary turns ratio N_1/N_2.

With the secondary an open circuit the primary input current is vanishingly small because the core is infinitely permeable and the flux is generated in it by negligible m.m.f. The primary winding therefore presents between terminals an infinite magnetizing inductance. But if the secondary terminals are connected to a load impedance such that the instantaneous current is i_2, this develops an m.m.f. $N_2 i_2$ acting on the magnetic circuit. To preserve unaffected the flux Φ whose e.m.f. balances v_1 in the primary winding, the primary must take a balancing current i_1 such that $N_1 i_1 = N_2 i_2$ at any instant. In r.m.s. terms, $N_1 I_1 = -N_2 I_2$. The current ratio is thus the inverse of the turns ratio, with $I_1/I_2 = N_2/N_1$ and the m.m.f.s in opposition.

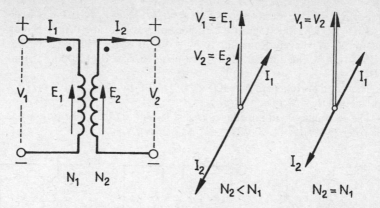

1.8 Ideal transformer.

Phasor Diagram. With the 'sense' of the windings as in Fig. 1.8, the primary and secondary voltage phasors are coincident in direction and of length proportional to the respective number of turns. The requirement of m.m.f. balance is satisfied by opposition of the current phasors. If I_1 enters the primary at the positive terminal, then I_2 leaves the secondary at the positive terminal, giving the condition $N_1 I_1 = -N_2 I_2$. Thus a clear distinction is made between primary input and secondary output conditions. In machine phraseology, the primary acts in a 'motor' mode, the secondary in a 'generator' mode. The primary input apparent power is $S_1 = V_1 I_1$, the secondary *input* is $V_2(-I_2) = -S_2$. These are numerically equal:

$$S_1 = V_1 I_1 = V_2(N_1/N_2) \cdot (-I_2) (N_2/N_1) = -V_2 I_2 = -S_2$$

The apparent power rating is therefore the volt–ampere product of either primary or secondary.

Impedance Transformation. If the secondary terminals are connected to a load of impedance Z, then $V_2/I_2 = Z$. The corresponding ratio V_1/I_1 is the effective input impedance of the primary:

$$V_1/I_1 = (N_1/N_2)^2 (V_2/I_2) = (N_1/N_2)^2 Z = k_n^2 Z$$

Thus Z across the secondary terminals appears in effect across primary terminals as Z multiplied by the square of the turns ratio $k_n = N_1/N_2$.

Summary. For sine conditions we have the following relations for the ideal transformer:

Ratios: $V_1/V_2 = N_1/N_2$ and $I_1/I_2 = N_2/N_1$
Rating: $S = V_1 I_1$ or $V_2 I_2$ M.M.F. balance: $I_1 N_1 + I_2 N_2 = 0$ (1.8)

EXAMPLE 1.1: The core of an ideal 1-ph 11 000/415V, 50 Hz, 300 kVA transformer has a net cross-section of area 0.032 m^2. Find (i) the number of primary and secondary turns, (ii) the rated currents, (iii) the secondary load and primary input impedances, for a peak core flux density of about 1.5 T.

(i) For $B_m = 1.5$ T the voltage per turn, eq. (1.7), is $V_t = 10.7$ V, whence $N_2 = V_2/V_t = 38.8$. Fractions of a turn are not possible, and it may be convenient for N_2 to be even. Rounding to $N_2 = 40$ turns, for which $B_m = 1.46$ T and $V_t = 10.4$ V, the number of primary turns is $N_1 = 40(11\,000/415) = 1060$.

(ii) The rated currents are $I_1 = 300 \times 10^3/11\,000 = 27$ A, and $I_2 = 720$ A. The m.m.f.s are $N_1 I_1 = N_2 I_2 = 28\,800$ A-t.

(iii) The secondary load impedance is $Z = V_2/I_2 = 0.58$ Ω. The primary input impedance is $V_1/I_1 = 11\,000/27 = 415$ Ω $= Z(N_1/N_2)^2$.

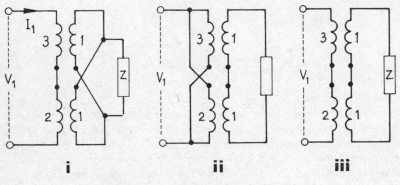

1.9 Concerning m.m.f. balance.

M.M.F. balance

This is a concept useful where connections are not straightforward. Fig. 1.9 shows ideal transformers with ratios 3/1 and 2/1 interconnected to a common load Z. In (i) the primaries are in series and the secondaries in parallel. If the voltage across Z is V_0, then the primary voltages must be $3V_0$ and $2V_0$, giving $V_1 = 5V_0$. As the current is I_1 in each primary, for m.m.f. balance the secondary currents must be $3I_1$ and $2I_1$. The current in Z is $5I_1$ (or I_1 depending on the 'sense' of the secondaries). As $V_0 = 5I_1 Z = V_1/5$, then $I_1 = V_1/25Z$ (or $I_1 = V_1/5Z$). The terminal input impedance V_1/I_1 is consequently $25Z$ (or $5Z$).

In (ii) the secondary voltages are respectively $V_1/3$ and $V_1/2$, and $V_0 = 5V_1/6$ (or $V_1/6$). From m.m.f. balance, the two primary currents are $5V_1/18Z$ and $5V_1/12Z$ summing to $I_1 = 25V_1/36Z$ (or $V_1/18Z$ and $V_1/12Z$, summing to $5V_1/36Z$).

In (iii) m.m.f. balance is impossible with any finite current. The network therefore draws no current and the input impedance is infinite (equivalent to an open circuit). With both primaries and secondaries in parallel, the input terminals are short-circuited.

1.10 *Prototype machines*

If the ferromagnetic parts of the magnetic circuit of a machine can be considered as infinitely permeable, then the whole action is concerned with electromagnetic conditions in the airgap. The energy-rate balance of eq. (1.6) can be expressed in terms of the action of windings on the gap surface, which are subject to e.m.f.s induced by the gap flux and give rise to distributed m.m.f.s resulting from the winding currents. By ignoring the ends of the machine, the problem becomes *two-dimensional*. In practical machines it is not, of course, justifiable to ignore the end regions, which present leakage-flux problems of some complexity.

1.11 *Prototype two-pole uniform-gap machine*

The 2-pole machine of Fig. 1.3(a) is shown in detail in Fig. 1.10. The stator

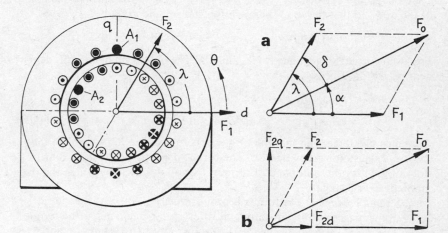

1.10 Prototype two-pole uniform-gap machine.

and rotor members have each a conductor arrangement so fed as to produce a *sine-distributed current-sheet,* the respective peak linear densities (in amperes per metre of gap periphery) being A_1 and A_2. The magnetic axis of the stator, called the *direct-* or *d*-axis, is taken as horizontal; that of the rotor is displaced counterclockwise from the *d*-axis by an angle λ. The distribution around the gap of the stator current-sheet density is $A_1 \sin \theta$, whilst for the rotor it is $A_2 \sin (\theta - \lambda)$. Interaction of the two current-sheets develops a torque tending to reduce λ to zero and so to align the rotor axis with that of the stator.

In an ideal machine, attention is concentrated on the active gap region, the dimensions of which are given in Fig. 1.14. It is assumed that the iron parts of the magnetic circuit are highly permeable, and that in consequence the gap length l_g accounts for the whole of the magnetic circuit reluctance.

The total current in one-half of the stator current-sheet is $A_1(2/\pi)\frac{1}{2}\pi D = A_1 D$. This is the total stator m.m.f., acting horizontally across two gaps in series, so that the m.m.f. *per pole* is $F_1 = \frac{1}{2}A_1 D$. Across a gap of length l_g it produces the flux density of peak value $B_1 = F_1 (\mu_0/l_g) = \frac{1}{2}A_1 D(\mu_0/l_g)$ on the d-axis. At any angle θ the m.m.f. is $F_{1\theta} = F_1 \cos \theta$, and the corresponding gap flux density is $B_{1\theta} = B_1 \cos \theta$. In a similar way the rotor, acting alone, produces a total m.m.f. per pole $F_2 = \frac{1}{2}A_2 D$ and a peak density $B_2 = \frac{1}{2}A_2 D(\mu_0/l_g)$ on the axis at the angle λ. For any angle θ the rotor m.m.f. is $F_2 \cos(\theta - \lambda)$ and the corresponding gap flux density is $B_2 \cos(\theta - \lambda)$.

With both stator and rotor excited, their combined m.m.f. at angle θ is $F_{1\theta} + F_{2\theta}$ producing $B_{1\theta} + B_{2\theta}$ in the gap. This is a new sine-distributed flux density of peak B_0 corresponding in magnitude and axis to the resultant F_0 of F_1 and F_2, which can be found by treating F_1 and F_2 as if they were space vectors and using the relation

$$F_0{}^2 = F_1{}^2 + F_2{}^2 + 2F_1F_2 \cos \lambda \tag{1.9}$$

The peak gap energy density is $\frac{1}{2}B_0{}^2/\mu_0 = \frac{1}{2}F_0{}^2 (\mu_0/l_g{}^2)$, the mean is one-half of the peak, so that the total energy in the gap volume $\pi Dllg$ is

$$w_f = \tfrac{1}{4} F_0{}^2 \, \pi Dl(\mu_0/l_g)$$

For given currents the torque is $M = dw_f/d\lambda$. The only term in eq.(1.9) dependent on λ is $2F_1F_2 \cos \lambda$, so that

$$M = -\tfrac{1}{2} \pi Dl(\mu_0/l_g)F_1F_2 \sin \lambda$$
$$= -\tfrac{1}{2} \pi DlB_1 A_2 \sin \lambda = -\tfrac{1}{2} \pi DlB_2 A_1 \sin \lambda \tag{1.10}$$

the minus sign indicating that the torque acts to decrease λ, i.e. to align the rotor and the stator magnetic axes. The torque depends on the main dimensions of the gap, the stator and rotor m.m.f.s, and the angle by which the two magnetic axes diverge.

Torque and load angles: The torque is proportional to sine of the angle between the magnetic axes of stator and rotor, and λ is defined as the *torque angle*. The m.m.f. diagram in Fig. 1.10(a) shows λ to be the sum of α and δ, the respective angular displacements between F_1 and F_2 and their resultant F_0; then α and δ are defined as respectively the stator and rotor *load angles*. In the analysis of certain machines, one or other of the load angles can be more readily identified from outside than can the torque angle λ. Suitably applying eq. (1.9) it is possible to write the torque in the alternative forms

$$M = -kF_1F_2 \sin \lambda = -kF_1F_0 \sin \alpha = -kF_2F_0 \sin \delta \tag{1.11}$$

where $k = \frac{1}{2}\pi Dl(\mu_0/l_g)$.

Torque distribution: Eq.(1.10) shows that we could consider the torque as arising from the interaction of a rotor current-sheet of peak density A_2 lying in the gap density B_1 of the stator acting alone; alternatively of A_1 lying in B_2. Fig. 1.11 illustrates the former in a developed diagram. In an element of angle $d\theta$ there lies a rotor current element $A_2 \sin(\theta - \lambda) \cdot d\theta$ in a stator-produced flux density $B_1 \cos \theta$. From eq. (1.4) the torque on the current element is

$$dM = \tfrac{1}{2}DlB_1 \cos \theta \cdot A_2 \sin(\theta - \lambda) \cdot d\theta$$

Integrating over the range $0 < \theta < 2\pi$ gives for the total torque

$$M = -\tfrac{1}{2} \pi DlB_1 A_2 \sin \lambda$$

as in eq. (1.10). The curve of torque distribution over the periphery shows

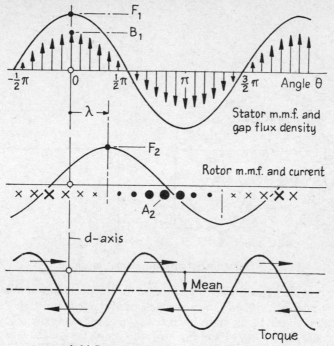

1.11 Interaction torque distribution.

there to be regions of oppositely directed torque. These become equal and the total resultant torque vanishes when, for $\lambda = 0$ or π rad, the rotor and stator m.m.f. axes are in alignment. The distribution is completely undirectional and the mean torque is greatest when the magnetic axes are at right-angles and $\lambda = \pm\frac{1}{2}\pi$ rad.

Two-axis concept: The first expression in eq.(1.11) can be written

$$M = -kF_1 (F_2 \sin \lambda) = -kF_1 F_{2q}$$

where $F_{2q} = F_2 \sin \lambda$ is the component of F_2 directed in the *quadrature* axis or q-axis, Fig. 1.10(*b*). The effect is as if the rotor had two separate current sheets, one of peak linear density $A_2 \sin \lambda$ magnetizing along the q-axis and producing torque, the other $A_2 \cos \lambda$ magnetizing along the d-axis and coupled with the stator (d-axis) winding. It would be equally valid to fix the d-axis on the rotor and resolve the stator m.m.f. into components $F_1 \sin \lambda$ and $F_1 \cos \lambda$; and the stator would now carry the two equivalent axis windings with their sine-distributed m.m.f.s.

1.12 *Elementary two-pole salient machine*

If one member, e.g. the rotor as in Fig. 1.12, is structurally polarized (salient), its m.m.f. F_2 cannot be taken as sine-distributed, nor does the resultant m.m.f. F_0 act everywhere across a uniform gap. The simple m.m.f. summation in (*a*) does not apply.

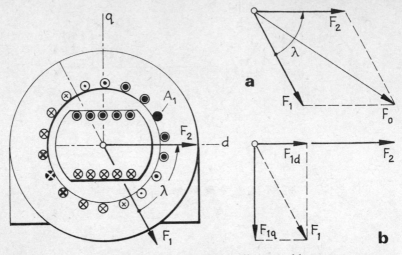

1.12 Elementary two-pole salient machine.

Two-axis Concept: This is particularly useful for the treatment of saliency effects. The d-axis is chosen to be the pole-axis of the *salient* member. In the case shown in Fig. 1.12 the stator m.m.f. F_1 is resolved along the rotor d- and q-axes to give $F_{1d} = F_1 \cos \lambda$ and $F_{1q} = F_1 \sin \lambda$ respectively. The d-axis m.m.f.s F_2 and F_{1d} act mainly across the short gap l_g, but the q-axis m.m.f F_{1q} is presented with a much longer gap of variable length. It is still possible to write the torque as

$$M = -\tfrac{1}{2} \pi D l B_2 (A_1 \sin \lambda) = -\tfrac{1}{2} \pi D l B_2 A_{1q} \tag{1.12}$$

where B_2 is the peak of the *fundamental* component of the gap-flux density produced by F_2, and A_{1q} is the peak of the equivalent sine-distributed stator q-axis current sheet.

1.13 *Multipolar machines*

A machine with p pole-pairs has p successive sets of N-S poles, each considered as spread over an electrical angle of 2π rad, so that although the physical (or mechanical) angle around the periphery is 2π (mech) rad, the total electrical angle is $2\pi p$ (elec) rad. An electrical torque angle λ therefore corresponds to a mechanical displacement angle $\lambda_m = \lambda/p$. The torque, being a mechanical quantity measured externally, is given now by $M = dw_f/d\lambda_m = p(dw_f/d\lambda)$. This does not mean that a machine of given gap dimensions produces more torque the greater the number of poles, because with given peak linear current densities A_1 and A_2 (which are determined by heat dissipation) the m.m.f.s per pole are reduced by the factor $(1/p)$.

EXAMPLE 1.2: The main dimensions of a uniform-gap machine are $D = 0.50$ m, $l = 0.20$ m and $l_g = 0.005$ m (5 mm). The stator has a winding excited to produce a sine-distributed current-sheet of peak density $A_1 = 12\ 800$

A/m; the rotor has a corresponding density $A_2 = 9\ 600$ A/m, with its magnetic axis displaced by a torque angle $\lambda = \pi/3$ (elec) rad from the stator d-axis. Neglecting the m.m.f. required for the iron parts of the magnetic circuit, find the torque with the machine windings arranged to give (i) a 2-pole field, (ii) an 8-pole field.

(i) The m.m.f.s per pole are $F_1 = \frac{1}{2}A_1 D = \frac{1}{2}(12\ 800 \times 0.5) = 3\ 200$ A-t, and $F_2 = 2\ 400$ A-t. Applying eq. (1.7) with $\sin \lambda = 0.87$ and $\mu_0 = 4\pi \times 10^{-7} \cong 1/800\ 000$ gives

$$M = -\tfrac{1}{2}\pi \times 0.50 \times 0.20(1/800\ 000 \times 0.005)3\ 200 \times \ 2\ 400 \times 0.87$$

$$= -260 \text{ N-m}$$

Alternatively, F_1 produces in the gap a peak flux density $B_1 = F_1 \mu_0 / l_g = 0.80$ T. From the appropriate expression in eq. (1.10),

$$M = -\tfrac{1}{2}\pi \times 0.50 \times 0.20 \times 0.80 \times 2\ 400 \times 0.87 = -260 \text{ N-m}$$

(ii) Each 2-pole unit covers an arc $\pi D/p$, reducing the stator m.m.f. per pole to $F_1 = 800$ A-t and the corresponding peak gap flux density to $B_1 = 0.20$ T. Similarly $F_2 = 600$ A-t. The rate of change of gap energy with electrical angle has been reduced by the factor $1/p^2 = 1/16$; but as $\lambda_m = p\lambda = 4$, the reduction of torque is given by the factor $(1/p)$. By either of the forms of eq. (1.7) the torque is therefore

$$M = -260/4 = -65 \text{ N-m}$$

The average *flux* density $\bar{B}_1$ in the gap due to the stator excitation has been reduced by the factor $1/p = 1/4$, but the average rotor linear *current* density is unchanged.

1.14   Main dimensions
Although the thermal and mechanical design of transformers and machines is often complicated, there is a simple design relation between the electrical rating and the main dimensions.

Transformer

The electromagnetic quantities concerned are the mutual flux in the magnetic circuit and the current linkage in the electric circuit. The basic dimensions, Fig. 1.13(a), are the cross-section of the magnetic core and the 'window' area enclosed by it. The active ferromagnetic area A_i of the core (which is an assembly of steel plates thinly coated with insulation to reduce eddy-current loss) is a roughly constant fraction (about 0.9) of the gross core area. The peak mutual flux is $\Phi_m = B_m A_i$, the peak density B_m being chosen with due regard to core loss and saturation. The voltage per turn of the primary winding, from eq. (1.7), has the r.m.s. value

$$V_t = V_1/N_1 = 4.44\ f B_m A_i \tag{1.13}$$

The primary and secondary windings must be accommodated in the window area A_w. Insulation and clearance reduce the available area for actual conduc-

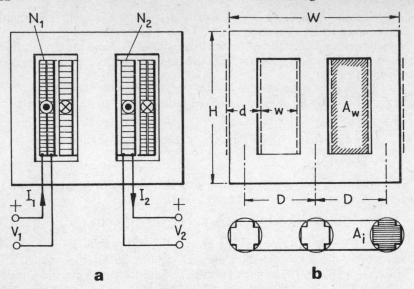

a   **b**

1.13 Transformer: main dimensions.

tor cross-section to $k_w A_w$: the space factor k_w may be quite small if the working voltage is high. Because of the m.m.f. balance ($N_1 I_1 = N_2 I_2$) the primary and secondary windings each occupy about one-half of the available area so that the primary current linkage is $N_1 I_1 = \frac{1}{2} k_w A_w J$. The r.m.s. current density J is limited by $I^2 R$ loss and heat dissipation. The total rating $S = V_1 I_1$ of the 2-limbed 1-phase transformer in Fig. 1.13(*a*) is

$$S = (V_1/N_1) N_1 I_1 = 2.22 f B_m J A_i k_w A_w \tag{1.14}$$

This can be considered as $\frac{1}{2}S$ per limb, so that for the 3-phase transformer in (*b*), with the overall dimensions defining its 'frame size', the total rating is

$$S = 3.33 f B_m J A_i k_w A_w \tag{1.15}$$

directly relating the rating to the basic dimensions through B_m and J, the flux and current densities.

EXAMPLE 1.3: A 50 Hz 1-ph transformer has a core of net section $A_i = 0.032$ m^2 and a window area $A_w = 0.07$ m^2. The density loadings are $B_m = 1.5$ T and $J = 2.7$ MA/m$^2 = 2.7$ A/mm^2. Insulation and clearance take up 0.7 of the window area (i.e. $k_w = 0.3$). Find the rating of the transformer.
    From eq. (1.14),

$$S = 2.22 \times 50 \times 1.5 \times 2.7 \times 10^6 \times 0.032 \times 0.3 \times 0.07 = 300\,000 \text{ VA}$$

so that the rating is 300 kVA.

Machines
The gap dimensions of a cylindrical machine, Fig. 1.14(*a*), are the diameter

D and the axial length l. Let the density of the flux crossing the gap be B_1 and the linear density of the axial rotor current sheet be A_2. Their interaction develops on the rotor a tangential force $B_1 A_2$ per unit of gap surface, or $B_1 A_2 \pi D l$ for the whole machine. The torque with radius $\frac{1}{2}D$ is $M = B_1 A_2 \frac{1}{2} \pi D^2 l$; and at a rotor angular speed $\omega_r = 2\pi n$ the power is

$$M\omega_r = B_1 A_2 \tfrac{1}{2}\pi D^2 l \omega_r = B_1 A_2 \pi^2 D^2 l n$$

For mechanical reasons the current sheet is normally formed by discrete conductors, sunk into slots of pitch y_s and carrying current $I_c = b h_1 J_2$ as in Fig. 1.14(b), the current density being chosen on a heat-dissipation basis. Then $A_2 = I_c/y_s = (b h_1/y_s) J_2 = c J_2$, and the power can be expressed as

$$M\omega_r = B_1 J_2 c \pi^2 D^2 l n$$

which directly relates developed power to the electric and magnetic densities J and B_1, the speed n and the basic dimensions. In this idealized approach it is assumed that the interaction forces have, all at points on the rotor periphery, the same direction. A relation specific to a.c. machines requires further data.
Rating: Number of phases m, phase voltage V, phase current I, frequency f, rotational speed n, rating $S = mVI$.
Magnetic Circuit: Flux per pole Φ, number of pole-pairs p, mean gap-flux density $\bar{B} = 2p\Phi/\pi Dl$. The specific magnetic loading $\bar{B}$ is chosen in relation to excitation and saturation conditions, core loss and power factor.
Electric Circuit: Active conductors Z per phase, each carrying current I_c, and the specific electric loading (linear r.m.s. current density) $A = mZI_c/\pi D$, chosen with regard to m.m.f., I^2R loss and cooling.

The product of the *specific magnetic* and *electric loadings* is, as discussed above, the specific force $\bar{B}A = 2p\Phi m Z I_c/\pi^2 D^2 l$. Now $2(pn)\Phi Z = 2f\Phi Z$ is proportional to the phase e.m.f. and so to the applied phase voltage V, while I_c is (or is a fraction of) the phase current I. Hence the rating of the machine is

$$S = mVI = k\bar{B}A \cdot D^2 l n \tag{1.16}$$

where k is a coefficient related to the technical details of the phase windings.

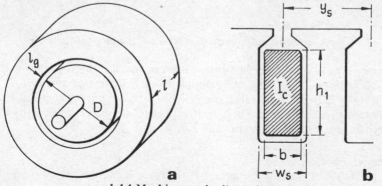

1.14 Machine: main dimensions.

EXAMPLE 1.4: In a typical rotating machine the mean gap flux density is
$\bar{B} = 0.50$ T. The slots are spaced $y_s = 40$ mm apart and each contains a conductor of section 35 mm x 12 mm. The current density is $J = 3.3$ MA/m^2 =
3.3 A/mm^2. Find the tangential force per unit length of gap periphery and per
unit axial length of the machine.

The specific electric loading is $A = (35 \times 12)3.3 \times 1\,000/40 = 35\,000$ A/m,
and consequently the specific force is

$$f_e = \bar{B}A = 0.5 \times 35\,000 = 17\,500 \text{ N/m}^2$$

which is typical of a medium-rated machine. The torque depends on the diameter D and axial length l of the rotor member.

1.15   *Torque maintenance*

To maintain torque production when the machine rotates, the relative current-
sheet patterns and the angle λ between their magnetic axes must be maintained.
The manner in which this happens is dependent on the form (d.c. or a.c.) of
the stator and rotor currents and on the windings in which they circulate.
Only polyphase synchronous and induction machines are considered here.

Synchronous machine
Fig. 1.15(*a*). The stator phase windings carry currents of frequency f_1 and
the rotor is d.c. excited, i.e. its poles are 'fixed' with respect to the rotor.

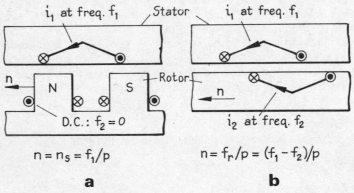

$$n = n_s = f_1/p \qquad\qquad n = f_r/p = (f_1 - f_2)/p$$

a   **b**

1.15 Synchronous and induction machines: torque maintenance.

Stator current reversal every half-period must be matched by rotor movement.
In a 2-pole machine the rotor must complete one revolution in one period
of the stator current, so that its speed is $n = f_1$, or $\omega_r = 2\pi n = 2\pi f_1 = \omega_s$. In a
machine with p pole-pairs the argument applies to successive 2-pole units and
the synchronous speed is f_1/p, or $\omega_s = 2\pi f_1/p$.

Induction machine
Fig. 1.15(*b*). The stator is the same, but the rotor is excited wth currents of
frequency f_2 so that its poles move with respect to its windings. In a 2-pole

machine the torque is maintained only if rotation at $n = \omega_r/2\pi = f_r$ together with current variation at f_2 combine to give $f_r + f_2 = f_1$. For let the speed be a little less than synchronous: the rotor slips back from the position it would occupy if running synchronously, but its poles move forward at f_2 with respect to its windings so that the field in effect moves synchronously with respect to the stator. In a machine with p pole-pairs the only difference is that $n = f_r/p$.

1.16 *Classification*

Machine windings are excited so as to develop particular current-sheet and m.m.f. patterns. A winding for the d.c. excitation of a working flux is often concentrated. For an a.c. flux it is generally distributed to reduce the leakage reactance. Polyphase windings can be arranged to give, to a close approximation, the sine-distributed current sheets of Fig. 1.10. A discussion of windings in general is given in Chapter 3.

Transformers and machines can be classified in accordance with (i) the kind of electrical supply, d.c. or a.c., to which each of its windings is connected; and (ii) the kind of connections made between the winding and its supply, i.e. by phase or tapped windings, or by switched connections (as in a commutator winding).

The Table lists a number of possible combinations, giving the names by which they are generally known.

Gap flux	Supply and Windings		Name
	Member 1	Member 2	
Fixed-axis			
Constant	d.c. concentrated	d.c. commutator	D.C. commutator A
Constant	a.c. phase	d.c. concentrated	1-ph synchronous B
Alternating	a.c. distributed	a.c. commutator	1-ph commutator C
Alternating	a.c. phase	a.c. phase	1-ph induction D
Alternating	a.c. concentrated	a.c. concentrated	Transformer E
Travelling-wave			
Constant	a.c. polyphase	d.c. concentrated or distributed	3-ph synchronous F
Constant	a.c. polyphase	a.c. polyphase	3-ph induction G

This book is devoted mainly to cases E, F and G. Types F and G, particularly the latter, can take a linear as well as a rotary form.

2
Magnetic Circuits

2.1 *Magnetic circuit properties*

The engineering approach to magnetic circuit design is to consider the magnetic flux Φ in a circuit of uniform cross-sectional area a, length l and absolute permeability μ established by the m.m.f. $F = Ni$ of a linked electric circuit of N turns carrying a current i. This relation is

$$\Phi = F\mu(a/l) = F\Lambda = F/S \tag{2.1}$$

where $\Lambda = \mu a/l$ is the permeance (in Wb per A-t) and $S = l/a\mu$ is the reluctance (in A-t per Wb). The flux Φ is assumed to be uniformly distributed over the area to give a flux density $\bar{B} = \Phi/a$. Practical magnetic circuits are usually divisible into component parts of uniform cross-section, and assignable length and permeability: for each the m.m.f. is $(B/\mu)l = Hl$. For several such parts in series the total m.m.f. is

$$F = Ni = H_1 l_1 + H_2 l_2 + \ldots \tag{2.2}$$

For air and for nonmagnetic materials in general, the absolute permeability is $\mu = \mu_0 = 4\pi \times 10^{-7}$ (in T per A-t/m, or H/m, approximating to $H = 800\,000\,B$. For ferromagnetic materials the absolute permeability is $\mu = \mu_r \mu_0$, where the relative permeability μ_r (a numeric) may lie between a few hundred and one-half million, and is highly dependent upon the level of B, i.e. upon the saturation condition. Thus H, the m.m.f. per metre length of path, may be between $2B$ and $2\,000\,B$, and it is necessary to employ graphical magnetization curves relating B to H. Strictly, each specimen has a unique B/H relation, but it is customary to employ 'average' curves. Typical comparative curves are given in Fig. 2.1, more detailed curves for structural, transformer and machine-core materials in Figs. 2.2–2.4.

For digital computation some analytic relation between B and H may be more convenient. Some typical expressions (from many) are

$$B = \frac{aH}{1 + bH} \quad \text{and} \quad B = \frac{a_0 + a_1 H + a_2 H^2 + \ldots}{1 + b_1 H + b_2 H^2 + \ldots} \tag{2.3}$$

The former gives reasonable approximation for higher values of B, the latter is a better overall fit. If the flux is to reverse, only odd powers of H must be included.

26

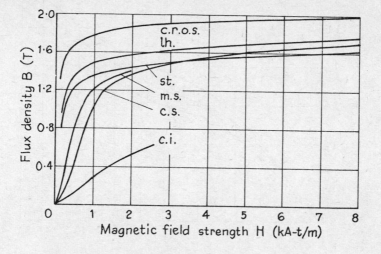

2.1 Magnetization curves: comparative.
Cold-rolled grain-oriented steel (c.r.o.s.) Lohys (lh.)
Stalloy (st.) Mild steel (m.s.) Cast steel (c.s.) Cast iron (c.i.)

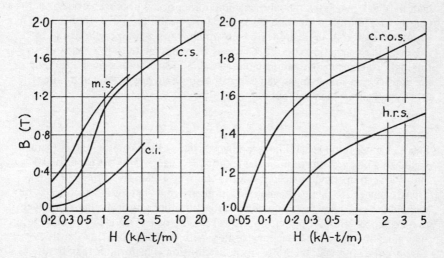

2.2 Magnetization curves:
structural materials.

2.3 Magnetization curves:
transformer steel.

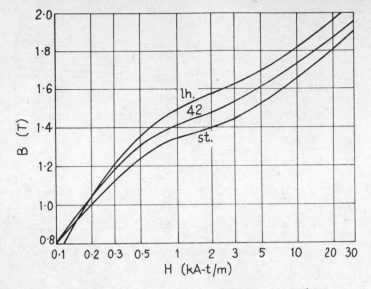

2.4 Magnetization curves: machine core steel.

Flux Distribution

The total flux in a ferromagnetic part of a magnetic circuit subject to a sine-varying excitation voltage varies sinusoidally with time. It might be expected that the flux density at every point in the cross-section would be Φ/a, directed parallel to the section and sine-varying. This is not so: practical magnetic materials lack magnetic, metallurgical and dimensional homogeneity, so that the density at a point is a function of the local permeability and structure. In laminated materials there are regions in which B has a component at right-angles to the magnetic axis as flux crosses from one plate to the next ('cross-fluxing'). In transformer and machine design a macroscopic view is taken of the density as a simple mean value, but the consequences of the non-homogeneity appear in the core loss.

Magnetostriction

Small changes in physical dimensions occur in a ferromagnetic material subjected to a magnetizing m.m.f. The lengthwise change is of the order of 10^{-5} and it is accompanied by corresponding transverse changes of opposite sign. These changes are due to magnetostriction, are of a hysteretic nature (i.e. they dissipate energy) and are intimately related to other physical conditions such as mechanical stress. A practical result of magnetostriction is the audible noise that emanates from a power transformer. With the high core flux densities commonly employed, the noise from distribution transformers may be the subject of limiting terms in the specification, and offending transformers may require sound-baffle structures to reduce the nuisance level.

Core loss

A ferromagnetic core carrying an alternating flux is the seat of magnetization loss that appears as heat. Traditionally the phenomenon is regarded as arising (i) from *hysteresis* loss due to the cycling of the material through its hysteresis loop, and (ii) from *eddy-current* loss resulting from the induction of currents circulating within the material. For sinusoidal flux-density variation of peak value B_m and r.m.s. value B, the specific loss (in W/kg) on a simple theoretical basis has the components

$$(i)\ p_h = k_h f B_m{}^x \qquad (ii)\ p_e = k_e d^2 f^2 B^2 / \rho \qquad (2.4)$$

The factor k_h depends on the molecular structure of the material; the exponent x lies between 0.8 and 2.3, and for the usual flux density levels may be about 2. The thickness d of the material takes account of the fact that cores carrying varying flux are sub-divided, usually into sheet form, and the appearance of ρ shows that to limit eddy-current loss the material should have a high electrical resistivity.

The idealized conditions on which eq.(2.4) is based do not obtain in machines and transformers, nor is it profitable in other than a metallurgical context to 'separate' the components. The loss is better regarded as a mass effect, distinguished only by the manner in which the flux variation occurs: (i) as a cyclic reversal on the same axis (i.e. a true alternating flux), or (ii) as a roughly constant magnitude but an axis that varies cyclically or rotates. In an actual core both processes occur. In a transformer, for example, there would seem to be only true alternating flux, but in fact rotational core-loss effects occur at the joints between limbs and yokes. The core loss may therefore be regarded as comprising components proportional respectively to the first and second powers of the frequency, and these indeed can be separated.

Attempts to optimize core loss must be based on tests, as some anomalies

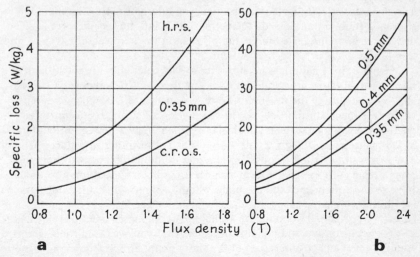

2.5 Core loss at 50 Hz: (*a*) Transformers, (*b*) Machines.

occur. The actual loss in a built core is increased by contact between plate edges, imperfect insulation between sheets, hardening and burring by shearing and punching processes, and the non-uniformity of flux density distribution. Designers employ curves such as those in Fig. 2.5 for the core loss in constructed plant, using test results as a basis.

Core clamping
Cores must be kept tight by compression. A loose core vibrates under magnetic repulsion forces and is noisy, and narrow parts such as slot lips may break under fatigue; but excessive pressure may distort the core and impair its magnetic properties.

Magnetic sheet and strip material received from the steelmakers has small surface asperities, waviness, and variations of thickness of as much as ± 0.05 mm. Cores may shrink in service, due in part to progressive bedding of asperities, in part to the compression of intersheet insulation. Tests show that the average pressure in a machine core should not be less than about 300 kN/m^2; that stiffener bars should not be more than 25 mm apart; and that long through-bolted stacks should have high-tensile bolts stressed during core building to about 0.75 of the elastic limit to take up shrinkage in service. Such bolts develop eddy-current loss, which must be limited by insulating them from the surrounding core.

2.2 *Magnetic materials*
Materials may be classified in accordance with the flux, pulsating or steady, that they normally carry. The former must be laminated; the latter may be, if constructionally convenient. The mechanical properties of a built core are also of importance.

Laminated materials
The characteristics important in producing an alternating flux in minimum section with minimum loss are the permeability and the specific core loss. The designer is more interested in the m.m.f. than in the permeability as such, and uses 'static' B/H curves such as those in Figs. 2.1–2.4.

Magnetic sheet materials are alloy steels. The chief alloying constituent is silicon, which raises the permeability at low flux densities, reduces hysteresis, and, by augmenting the resistivity, eddy-current losses. The silicon content affects the mechanical properties, increasing tensile strength but impairing ductility. A 5% silicon steel is difficult to punch or shear, but addition of nickel makes *cold-rolling* possible with important resulting advantages. Where core loss is not a primary consideration, and in induction motors where the magnetizing current must be kept small, a low silicon content may be used. In large and totally-enclosed machines where cooling is difficult, a high-grade low-loss steel may be called for.

Transformer steels. Two kinds are available: (i) high-resistance steel (h.r.s.) of 4–5% silicon content and nominal thickness 0.35 mm; (ii) cold-rolled grain-oriented sheet and strip (c.r.o.s.) of 0.33 mm nominal, for which the magnetic properties in the rolling direction are far superior to those on any other axis.

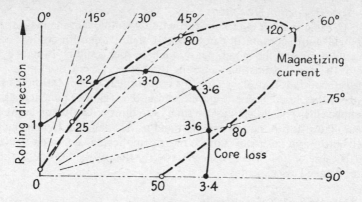

2.6 Directional properties of cold-rolled grain-oriented transformer steel plate.

Typical directional variations are shown in Fig. 2.6, where the loss and magnetizing current in the rolling direction for a given peak flux density are each taken as unity. For example, the relative loss and m.m.f. are respectively 3 and 80 for magnetization on an axis at 45° to the rolling direction. If the directional properties can be exploited, c.r.o.s. is highly advantageous and can be worked to higher flux densities than h.r.s. It is therefore used in all large transformers and in most of the lower ratings. It may even be applied in large machines in spite of the absence of a fixed flux axis. Both c.r.o.s. and h.r.s. are available in several grades, with specific losses down to 1.0 and 2.4 W/kg respectively at 1.5 T, 50 Hz.

Solid materials

Where the flux is normally constant, the magnetic criterion is the B/H curve. Mechanical considerations may, however, entail alloying that also decreases the permeability, as in steel castings or forgings for high peripheral speeds.
Castings. Low-carbon steels are used for stationary parts not subject to large mechanical stressing. The carbon content is increased for rotating parts, with some sacrifice in permeability. A 10% nickel, 5% manganese cast iron is nearly non-magnetic, and may be employed when it is necessary to avoid shunting a magnetic field. For small frames the use of cast iron is less common as cast aluminium is light and convenient.
Forgings. Very large steel forgings are used for turbo-generator rotors where high mechanical strength is essential. The overhang windings are retained by forgings of non-magnetic alloy.

Working flux densities

The maximum flux densities (in T) normally employed are:
Core plates. Transformers (c.r.o.s.), 1.3—1.6 (up to 2.0 on 20% overvoltage).
Machines: teeth, 1.8—2.2; cores, 0.8—1.2.
Cast steel. Poles, 2.0; yokes, 1.5.

2.3   *Magnetic circuit calculation*

Having available the *B/H* curves for the materials employed, the magnetic circuit is divided into convenient parts in series, the flux densities estimated for given conditions, and the m.m.f. per unit length found and multiplied by the length of the part. Summation yields the total magnetic circuit m.m.f.

The estimation is basically simple for transformers except for the corners or joints of the closed magnetic circuit, and in the region of punched clamping or cooling vents. The design must be such as to limit cross-fluxing between adjacent plates, as this increases the loss and m.m.f. demand. In machines the shape of the magnetic circuit is much more complicated, and in particular the airgap, because of its low permeance relative to that of the ferromagnetic parts, requires detailed assessment.

2.4   *Transformer magnetic circuits*

Rolls of c.r.o.s., up to 2 Mg in mass, are slit into widths by gang-operated slitters. The strips are then cut to length. With h.r.s. the material is sheared to size in power guillotines and punched in multiple presses. The working impairs the properties of c.r.o.s., so that sheets (or complete small cores) are given a stress-relief anneal at about 800 °C, in an inert atmosphere to avoid oxidation and carbon contamination. Surface insulation to reduce eddy currents is a thin coating of kaolin or varnish for h.r.s., but c.r.o.s. usually comes from the makers with a phosphate-base coating capable of withstanding the annealing process. Small cores need no further insulation, but kaolin or varnish is applied to laminations intended for transformers of 10 MVA upward.

Magnetic circuit design

Transformer cores up to about 5 kVA rating can be formed by winding continuous strip through pre-formed coils using special machinery, eliminating joints and using c.r.o.s. exclusively in its rolling direction. Larger sizes, Fig. 2.7, may have strip-wound cores, impregnated and then cut across to form 'C' cores which can be butt-jointed into pre-formed coils; core thicknesses up to 30 mm in widths up to 60 mm can be wound to give window dimensions of 100 mm square. By far the most common construction, however, is by stacking cut strips with interleaved joints to form core- or shell-type magnetic circuits, Fig. 2.8.

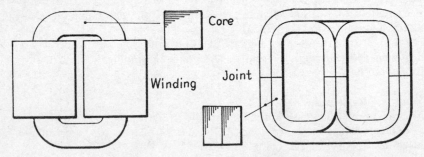

2.7 Small cores for grain-oriented steel.

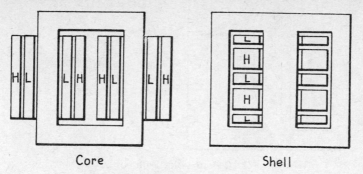

Core   Shell

2.8 Basic 1-ph coil and core arrangements.

Most 1-ph and 3-ph transformers are of the core type, Fig. 2.9. If transport loading limits are likely to be reached, a 5-limbed core may be used to reduce the overall height. For very large units it may be necessary for transport to use three 1-ph transformers, or some form of demountable construction. Fig. 2.10 shows an example of four yokes fitted round a central limb for height limitation.

The basic parameters of a 3-ph core (Fig. 1.13b) are the core-circle diameter d, the limb spacing between centres D, and the window length l. From these the net core area A_i and the overall core width w give the available window area A_w. The overall height H and width W are of importance in tank design

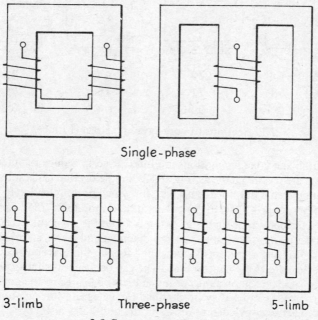

Single-phase

3-limb   Three-phase   5-limb

2.9 Core arrangements.

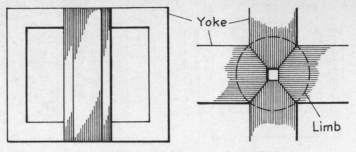

2.10 Four-yoke 1-ph core.

and transport. For a given type and shape there is a relation $d \propto \sqrt{S}$ between the core-circle diameter and the rating S. For economic design it is essential to restrict d by using as much as possible of the core-circle area. The shape of the core-packet assembly must therefore approximate to a circle. Typical optimum proportions are shown in Fig. 2.11 with the gross core area as a percentage of the circumscribing circle area, and with the net area A_i based on a stacking factor of 0.9 (which on large units can be raised). Seven or more steps may be used for larger sections, but if too massive to allow adequate cooling, ducts are arranged between sections of the core packets.

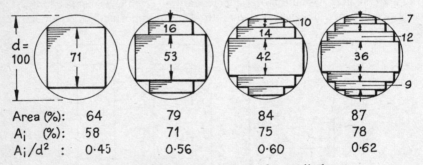

Area (%):	64	79	84	87
A_i (%):	58	71	75	78
A_i/d^2 :	0·45	0·56	0·60	0·62

2.11 Sections of core-type transformer limbs.

Joints. Limb and yoke laminations are interleaved to reduce reluctance and loss. There are several methods, Fig. 2.12. Simple overlap (*a*) is suitable for h.r.s. plate material but not for c.r.o.s., for which mitred overlap (*b*) is preferred, either in the 45° form (*c*) or the more elaborate 35°/55° variant in (*d*). When the core-circle diameter exceeds about 0.8 m, commercially available strip is too narrow to span the full diameter and a core split becomes necessary. Bridging sections (*e*) are needed for the flux passing between the halves of the core at the central limb. The gap between the half-cores aids cooling because it exposes twice as much plate-edge area to the coolant.
Clamping. Limb sections may be clamped by plates and through-bolts; both may be subject to eddy-current loss, and the bolt-holes distort the flux path. A preferred method is to hold the limb plates together by bands e.g. of woven

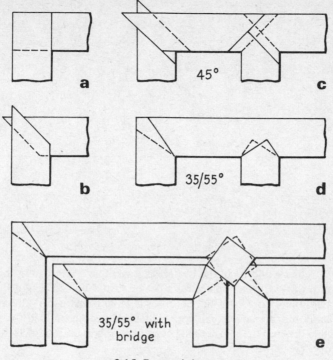

2.12 Corner joints.

glass-fibre tape impregnated with a polyester resin. After banding with the core under pressure, the bands are heated to cure the resin. To spread the pressure more evenly, the outer packets of core plates may be bonded by a suitable adhesive such as Araldite.

Magnetizing current and power

The exciting m.m.f. is calculated with adjustment for gap and joint effects and for divergence of the flux path from the rolling direction, on the basis of the peak density B_m. The r.m.s. value of the m.m.f. is obtained by dividing the result by $\sqrt{2}$. This takes no account of saturation nonlinearity nor of the asymmetry introduced by hysteresis. A plot of the magnetizing current waveform for a particular case is shown in Fig. 2.13. An ascending flux density corresponding to point P on the hysteresis loop and to Q on the sinewave of time variation requires 8 units of magnetizing current. Proceeding in this way, the current waveform is seen to contain harmonics, which in this case (in terms of a fundamental of 100) analyse to

$$100 \sin(x + 18°) - 39 \sin(3x + 7°) + 18 \sin(5x + 9°) - 8 \sin(7x + 10°)$$

The displacement of the fundamental implies that it lags the voltage applied by $90 - 18 = 71°$, and that there is consequently an active power component

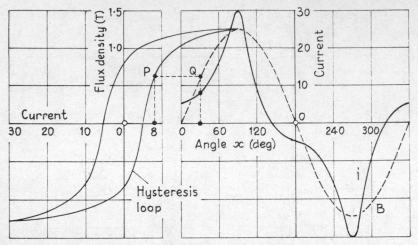

2.13 Magnetizing current waveform.

100 cos 71° corresponding to the hysteresis loss. (The eddy-current loss is additional to this). The example is for a peak density of 1.25 T, and there is a 40% 3rd harmonic. For the higher densities in large transformers the 3rd, 5th and 7th harmonics are considerably intensified. However, on load the harmonics are negligible because the magnetizing current is only 0.05 p.u. of full-load current, or thereabouts,

Magnetizing reactive power. Writing the total m.m.f. for a peak density B_m as $H_m l_m$, where l_m is the mean length of the flux path around the magnetic circuit, the magnetizing current in the N_1 turns of the primary winding is $I_{or} = H_m l_m / \sqrt{2} N_1$, ignoring harmonics. Using eq.(1.7) for the primary voltage, the magnetizing reactive power (in var) is

$$Q_0 = V_1 I_{or} = 4.44 f B_m A_i (H_m l_m)/\sqrt{2}$$

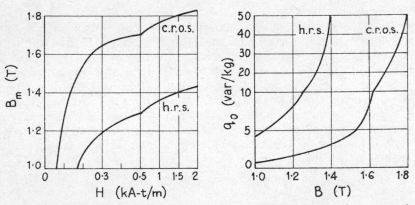

2.14 Magnetizing m.m.f. and reactive power for 0.35 mm transformer steel cores.

As $A_i l_m$ is the core volume and the density of the steel is 7 500 kg/m^3, the specific magnetizing reactive power is

$$q_0 = 4.2 f B_m H_m \times 10^{-4} \text{ var/kg} \qquad (2.5)$$

Curves relating q_0 and B_m for given steels may be used for a rapid estimate of the requirements of a preliminary design. Fig. 2.14 gives curves for h.r.s. and c.r.o.s. materials.

No-load current. With a total core loss P_i, the active component of the no-load current I_0 is $I_{0a} = P_i/V_1$, and the total no-load current is the phasor sum of the active and reactive current components.

Asymmetry. In 3-ph core-type transformers the magnetizing current of the central limb is somewhat less than that of either outer limb. The unbalance is small because the limbs are relatively long in comparison with the yokes.

2.5 Machine magnetic circuits

In a machine, the intricacy of the shapes of the various parts, particularly of the airgap boundaries, makes a precise determination of the flux distribution very difficult. Fortunately it is possible to employ certain simplifications without undue loss of practical accuracy. It is usual to consider the end-winding as separate from the gap region; in fact, most of the individual parts are given an *ad hoc* treatment because, to be tractable, the boundaries must be redefined for each. The problem is aggravated by the several states in which the machine may be magnetized. The simplest is that for which only one member (stator or rotor) is excited. Less simply, but very important, is the estimate of leakage in the condition when both members have current-carrying windings. Most difficult of all is the transient condition, when induced currents in normally quiescent windings or in core masses affect the flux distribution and therefore the saturation condition from point to point. In this Section attention is confined to one of single excitation in the steady state.

Airgap

The gap is bounded on either side by iron surfaces, either or both slotted and sectionalized by cooling ducts. The principal quantities concerned are

l	total axial length of core	y_s	slot pitch
l_i	net iron length of core	w_o	slot opening
l_g	radial length of gap	w_d	duct width

The gap reluctance is solved using the analytical results given by Carter [2]. In Fig. 2.15, (a) shows a possible gap flux distribution in the presence of openings in the gap boundaries. The permeance of the gap is less than if the openings were ignored (b), but greater than if they were deemed to carry no flux at all (c). The effective 'contracted' width of the slot pitch is

$$y_s' = y_s - k_o w_o \qquad (2.6)$$

where k_o is a function of the ratio (w_o/l_g) from Fig. 2.16, a distinction being made as to the nature, semi-closed or open, of the boundary opening. Radial

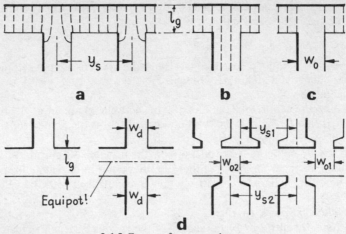

2.15 Gap surface openings.

ducts can be handled in a similar way by contracting the axial length l of the gap surface to l' where

$$l' = l - k_d n_d w_d \tag{2.7}$$

for a total of n_d ducts, with k_d a function of (w_d/l_g). If stator and rotor ducts coincide, their central plane is considered as an equipotential and eq.(2.7) applied, but with k_d now a function of $(w_d/\frac{1}{2}l_g)$.

When both gap surfaces are slotted, a more convenient method is to consider the gap length l_g to be extended to l_g', i.e.

$$l_g' = k_g l_g \tag{2.8}$$

where $k_g = k_{g1} \cdot k_{g2} = \dfrac{y_{s1}}{y_{s1} - k_{o1}w_{o1}} \cdot \dfrac{y_{s2}}{y_{s2} - k_{o2}w_{o2}}$

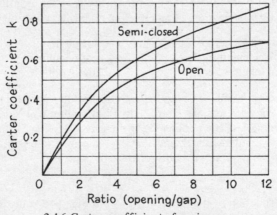

2.16 Carter coefficients for airgaps.

The several quantities are shown in Fig. 2.15: the factors k_{o1} and k_{o2} are functions respectively of (w_{o1}/l_g) and (w_{o2}/l_g) from Fig. 2.16.

For an airgap between a dentated surface on one side and a salient pole on the other, the gap m.m.f. is

$$F_g = 800\ 000\ K_g B_g l_g \qquad (2.9)$$

where l_g is the gap length at the pole centre, B_g is the peak gap-flux density (at the same point) and $K_g = (l/l')(y_s/y_s')$ takes account of slot-openings and ducts.

Gap-density distribution

A method for evaluating the gap-density waveform for odd-order harmonics up to the 7th is illustrated in Fig. 2.17. The outline of the half pole-arc is laid out and the half pole-pitch divided into six equal intervals. From each division mark is sketched a flux line, leaving and entering the iron surfaces in a normal direction, using as aid the technique for two-dimensional flux plotting by curvilinear squares. At the pole centre where the gap length is l_g, the flux density is assumed to be 100 units. At other points it is $100(l_g/l)$, where l is the length of the sketched flux line. A plot (b) of the result shows a positive value at position 0. This should in fact be zero, so a straight line is drawn from the ordinate of B at position 0 down to base at a point where the pole bevel begins. The actual flux density is taken as the intercept between the straight line and the B-plot, on the assumption that the omitted flux is leakage direct to the adjacent pole. The net curve, with ordinates $b_0 = 0, b_1, b_2 \ldots b_5, b_6 = 100$ can

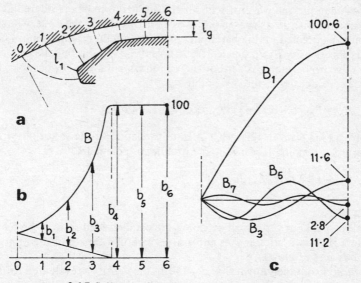

2.17 Salient-pole gap-density distribution.

now be analysed for its space harmonic content by use of the expression
$B_n = k_1 b_1 + k_2 b_2 + \ldots + k_6 b_6$ with the following factors k:

Harmonic density	k_1	k_2	k_3	k_4	k_5	k_6
B_1	0.086	0.167	0.236	0.289	0.323	0.167
B_3	0.236	0.333	0.236	0	−0.236	−0.167
B_5	0.323	0.167	−0.236	−0.289	0.086	0.167
B_7	0.323	−0.167	−0.236	0.289	0.086	−0.167

Then $B = B_1 \sin \theta + B_3 \sin 3\theta + B_5 \sin 5\theta + B_7 \sin 7\theta$.

EXAMPLE 2.1: A flux plot, Fig. 2.17, yields the values $b_1 = 10, b_2 = 31, b_3 = 68$, $b_4 = b_5 = b_6 = 100$ units. Derive the harmonic series for the flux density B, and find its mean and r.m.s. values.
The harmonic components, shown in Fig. 2.17(c), are:

$B_1 = 0.86 + 5.16 + 16.72 + 28.9 + 32.3 + 16.7 = 100.6$
$B_3 = 2.36 + 10.32 + 16.04 + 0 - 23.6 - 16.7 = -11.6$
$B_5 = 3.23 + 5.16 - 16.04 - 28.9 + 8.6 + 16.7 = -11.2$
$B_7 = 3.23 - 5.16 - 16.04 + 28.9 + 8.6 - 16.7 = 2.8$

The mean gap density and its r.m.s. value are

$\bar{B} = (2/\pi) [100.6 - \frac{1}{3}11.6 - \frac{1}{5}11.2 + \frac{1}{7}2.8] = 60.4$
$B = \sqrt{[\frac{1}{2}(100.6^2 + 11.6^2 + 11.2^2 + 2.8^2)]} = 72.0$

The amplitude of the fundamental density usually approximates to the peak gap density. The effect of the harmonics on the r.m.s. value is minimal.
Induction machines. These have very short airgaps, and tooth saturation flattens the peak density, the waveform of which can therefore be considered as a fundamental with a superposed positive 3rd harmonic. As the r.m.s. winding e.m.f. is almost unaffected by the harmonic, magnetic circuit calculation can be based on the flux density B_{30} at an electrical angle of $30°$ to the flux axis, a point where the 3rd harmonic is zero. If then a fundamental flux Φ_1 occupies a path of area a, its density at the $30°$ point is

$$B_{30} = B_{m1} \cos 30° = \frac{1}{2}\pi(\Phi_1/a)0.87 = 1.36(\Phi_1/a)$$

The total m.m.f. on this basis is F_{30}, and its relation to the stator phase current I_1, turns N_1, winding factor K_{w1} and number of pole-pairs p is

$$F_{30} = 1.17 I_1 N_1 K_{w1}/p \qquad (2.10)$$

Teeth
The slots of all but very small machines are parallel-sided so that the teeth are tapered. High saturation makes necessary a careful estimate of the m.m.f. The methods are as follows.
Graphical. From the known flux per tooth, the density is evaluated for a number of sections from tip to root, correcting for the true density $B_t{}'$ if necessary

by the method described later. From a B/H curve the m.m.f. summation is made.

Three-ordinate. For simple trapezoidal teeth of moderate taper, Simpson's rule can be applied to the values H_1, H_2 and H_3 relating to three equidistant sections. The mean value is then $H = (H_1 + 4H_2 + H_3)/6$.

One-third density. For slight taper and low densities it is sufficient to find the density $B_{t1/3}$ at a section one-third of the tooth length from the narrower end, and to assume that this obtains over the whole tooth length.

Integration. Accurate practical methods have been worked out for a number of tooth profiles and presented as curves for the usual core materials.

True and apparent tooth density. If the tooth flux density is high, the m.m.f. is sufficient to produce appreciable flux in the parallel paths in slots and ducts. Consequently an estimate of tooth density based on the tooth area alone is erroneous. Calling this the apparent tooth density B_t' and the true density $\overset{\circ}{B}_t$, the two are related by

$$B_t' = B_t + \mu_0 H(K - 1) \tag{2.11}$$

the second term being the flux density in the non-magnetic paths. The factor K is obtained from the ratio

$$K = \text{(gross area of iron and air)/(net iron area)}$$

which has to be found for each section taken. When reckoning the iron area the stacking factor (about 0.9) of the laminated core must be incorporated. K may have values between 1.5 for a machine without cooling ducts and 4 for the roots of turbo-generator teeth.

EXAMPLE 2.2: A machine with a gross axial length $l = 0.17$ m and active length $l_i = 0.153$ m has a flux $\Phi_t = 3.16$ mWb in one tooth pitch $y_s = 25.6$ mm. Data for points 1 ... 6 of the tooth profile, Fig. 2.18, are:

Section	Tooth width mm	Tooth area mm^2	B_t' T	H A-t/m	K	H A-t/m
1	13.6	2 080	1.52	2 300	–	–
2	11.4	1 740	1.82	11 000	–	–
3	9.2	1 410	2.24	80 000	3.0	50 000
4	13.2	2 020	1.56	2 800	–	–
5	11.5	1 760	1.79	9 500	–	–
6	9.8	1 500	2.10	40 000	2.5	32 000

Find the tooth excitation F_t by (i) a graphical method, (ii) the density at one-third of the tooth length.

(i) Plots of B_t and H are given in Fig. 2.18, taking account of the factor K for sections 3 and 6. Summation gives $F_t = 390$ A-t.

(ii) At 10 mm from the root, the uncorrected density $B_{t1/3} = 1.86$ T requires $H = 13\,000$ and $F_t = 390$ A-t. The method is adequate for this moderate saturation level.

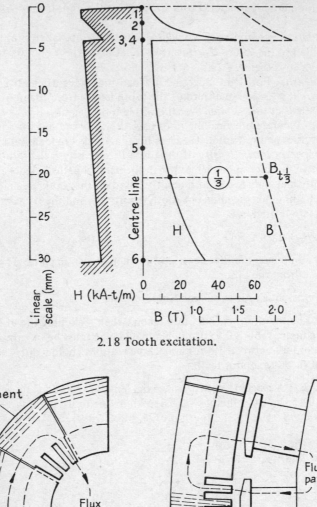

2.18 Tooth excitation.

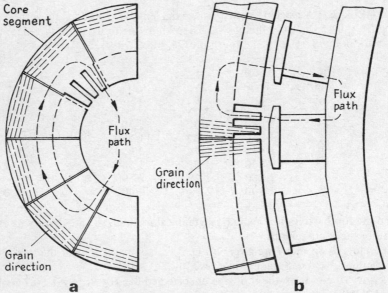

2.19 Application of grain-oriented steel.

Cores

Excitation for poles, cores and similar sections is based on the choice of suitable regions for the calculation of area and the estimation of flux density. For salient poles it may be necessary to take account of the change of pole flux resulting from leakage, Sect. 2.6.

If it is feasible to choose a favourable magnetizing direction, the use of c.r.o.s. has possibilities in machines. In 2-pole turbo-generators the core-flux direction is mainly circumferential in the parts external to the slotting, Fig. 2.19(a). The core-plates are cut with the preferred direction at right-angles to a central radius. In large hydro-generators it is the teeth rather than the core that may benefit, and the arrangement (b) may give some reduction in both excitation and core loss.

Magnetizing current

Summation of the m.m.f.s required for the several series parts of the magnetic circuit gives the total to be provided by the exciting winding. The magnetizing current depends upon the number of turns and how they are distributed. For a concentrated coil of N turns such as that on a salient pole, the exciting current is $I = F/N$. For 3-ph uniformly distributed windings with the common 60° phase-spread and carrying sinusoidally varying currents, the m.m.f. distribution has a peak F_a related (as shown in Sect. 3.14) to the r.m.s. phase current by

$$F_a = 1.35\, I N K_w / p \tag{2.12}$$

where p is the number of pole-pairs and K_w is the winding factor of the N turns per phase in series.

2.6   *Total flux*

The total flux in a machine is a combination of the working, main or mutual flux crossing the gap, and the leakage flux in slots and end-windings. These may be summed by superposition on the assumption that the presence of one does not affect the other. The assumption is unwarranted because of the effects of saturation, but it has to be made for simple analysis. Saturation can be taken into account only by iterative digital computation, and even then the program is difficult to formulate.

On the total flux and its complicated distribution depends the estimation of the self and mutual inductance of windings, the conditions that give loss-generating flux pulsations, and the development of magnetic forces giving rise to the production of noise, the need for braced end-windings, harmonic torque effects and a number of other practical matters. Binns [2] has attacked the flux problem (particularly of airgaps with both iron surfaces slotted) by conformal transformation, the resulting equations being solved numerically by computation. Here we shall deal with leakage flux basically, in order to illustrate simple estimates which (though sometimes based on questionable assumptions) do give reasonable approximate results.

Leakage flux

The flux that traverses paths outside the active transfer or conversion regions (e.g. the core of a transformer or the airgap of a rotating machine) affects the excitation demands of salient poles, the leakage inductance of windings, forces in and between windings (especially under conditions of short circuit), load (stray) losses, circulating currents in transformer tank walls and metallic clamps, voltage regulation, and several more phenomena of practical import.

From eq.(2.1), the flux per unit current in a coil of N turns linked with a magnetic path of permeance Λ is $\Phi = N\Lambda$. The leakage inductance is therefore $L = N\Phi = N^2\Lambda$. If the current in the coil alternates at angular frequency $\omega = 2\pi f$ its leakage reactance is

$$X = \omega L = 2\pi f N^2 \Lambda \tag{2.13}$$

Because of the complex geometry of leakage-flux paths, it is not easy to estimate Λ. Normally it is assumed that those parts of the path within iron require negligible m.m.f.; that is, they are infinitely permeable, so that the iron surfaces from which leakage flux emerges can be taken as equipotentials or to have some predetermined magnetic potential distribution. Attention is then concentrated on the non-magnetic ('air') parts of the flux path. Estimating the flux pattern by a flux plot or some similar method, the path can be divided into series or parallel sections for each of which a *permeance coefficient* λ is estimated from the ratio area/length to give $\lambda = a/l$. The permeance of a series of x sections is then written

$$\Lambda = \mu_0 \Sigma(a_x/l_x) = \mu_0 \Sigma(\lambda_x) \tag{2.14}$$

and the problem reduced to determination of λ. In practical cases the exciting circuits comprise conductors of finite area, and some of the leakage flux penetrates the area.

2.7   *Leakage flux effects in transformers*

The most common arrangements of the core and the windings are those of the core- and shell-types, Fig. 2.8. The flux plots in Fig. 2.20, based on calculations by Hague [3] show typical leakage field patterns for these. In (*a*), for concentric cylindrical primary and secondary windings of equal length, the leakage flux runs parallel with the axis for nearly the full winding length. Where, as in (*b*), the windings are of unequal lengths, the field is greatly modified. Plot (*c*) is for a shell-type transformer with sandwich coils. Detailed analyses of these field patterns have been given by Billig [4]. Here we attack the problem by simplifying the field geometry to obtain straightforward approximate estimates. The leakage flux is considered as a superposition of straight components in one or in two directions, Fig. 2.21(*a*).

Concentric cylindrical coils

Equal length. The leakage field is taken to comprise an axial flux (i) of uniform and constant value in the interspace between primary and secondary, and (ii) of value decreasing linearly to zero at the outer and inner surfaces of the windings and threading through them, as in Fig. 2.21(*a*). Further, the permeance of

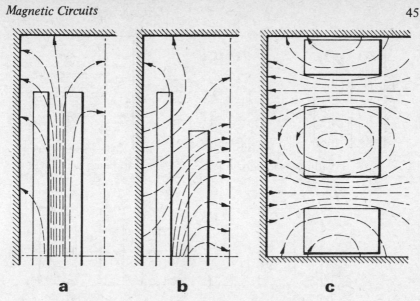

a   **b**   **c**

2.20 Transformer leakage flux.

the leakage path external to the winding length L_c is taken to be so large as to require the expenditure of a negligible m.m.f.; i.e. the whole of the combined winding m.m.f. is expended on the path length L_c. It is further assumed that there is primary and secondary m.m.f. balance $N_1I_1 = N_2I_2 = F$. The flux density in the duct of width a is $B_a = \mu_0 F/L_c$, and the duct flux is $\Phi_a = B_a a L_{mt}$, where L_{mt} is the mean circumference of the duct, nearly the same as the length of a primary or secondary turn. One-half of the duct flux is taken as linking the N_1 primary turns to give a linkage $\frac{1}{2}N_i \, \Phi_a$. Within the primary winding at radial distance x from the surface, Fig. 2.21(b), the flux density is $B_a(x/b_1)$ and the flux in an elemental annulus of width dx and (approximate) circumference L_{mt} is $d\Phi_b = B_a(x/b_1)L_{mt} \cdot dx$. Now $d\Phi_b$ links only the fraction (x/b_1) of the turns: hence integrating $N_1(x/b_1) \cdot d\Phi_b$ over the width b_1 gives the internal linkage as $B_a N_1 L_{mt}(b_1/3)$.

The primary leakage inductance is the linkage per ampere, for which $F = N_1I_1 = N_1$. Summing the two components gives

$$L_1 = \mu_0 N_1{}^2 \frac{L_{mt}}{L_c}\left[\frac{a}{2} + \frac{b_1}{3}\right]$$

Substituting N_2 for N_1 and b_2 for b_1 gives the secondary leakage inductance. The *total* equivalent inductance in primary terms (from the impedance transformation in Sect. 1.9) is $L = L_1 + L_2(N_1/N_2)^2$. The corresponding leakage reactance for frequency f is therefore

$$X_1 = \mu_0 2\pi f N_1{}^2 \frac{L_{mt}}{L_c}\left[a + \frac{b_1 + b_2}{3}\right] \tag{2.15}$$

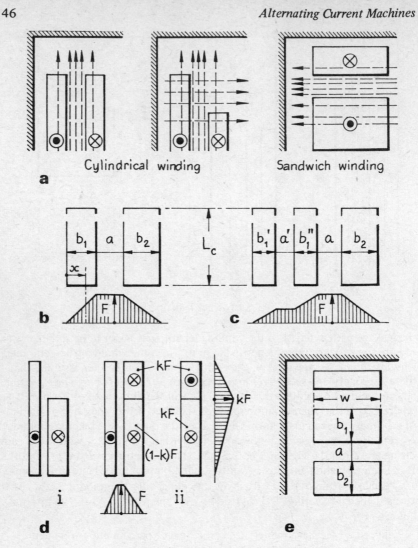

Cylindrical winding   Sandwich winding

2.21 Calculation of leakage flux.

Alternatively, with F as the rated m.m.f. per limb of either primary or secondary, and with the e.m.f. per turn $E_t = V_1/N_1$, the *per-unit* leakage reactance of the transformer is

$$\epsilon_x = \mu_0 \, 2\pi f \, \frac{F}{E_t} \, \frac{L_{mt}}{L_c} \left[a + \frac{b_1 + b_2}{3} \right] \qquad (2.16)$$

If the h.v. winding is split, as in Fig. 2.21(c), into equal parts the square-bracketed terms in the two equations above are replaced by the expression $[a + \frac{1}{3}(b_1' + b_1'' + b_2) + \frac{1}{4}a']$.

Unequal length. This common case may be the result of the manufacturing process, non-uniform turn insulation, or the provision of tapped sections for voltage regulation. One approach is to resolve the leakage flux into axial and radial components, as in Fig. 2.21(*a*) for a difference of length at one end.

The concentric coils, each of m.m.f. *F*, are shown at (i) in Fig. 2.21(*d*). The arrangement is resolved as in (ii) into two superposed m.m.f. distributions

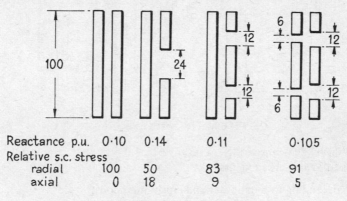

Reactance p.u.	0·10	0·14	0·11	0·105
Relative s.c. stress				
radial	100	50	83	91
axial	0	18	9	5

2.22 Cylindrical coil asymmetry.

in accordance with the lengthwise unbalance *k*. The first is dealt with as if it comprised balanced coils of equal length, the latter in a corresponding way but with different orientation and appropriate physical dimensions. Typical cases are shown in Fig. 2.22, with figures of percentage reactance compared with the case of symmetrical coils of equal length.

Sandwich coils

Fig. 2.21(*a*) shows the simplifying assumption made. It is usual to arrange the coils as in (*e*), where the end coils have one-half of the turns of the remainder. There are *n* h.v. coils, (*n* − 1) l.v. coils plus two half-coils. Taking two adjacent half-coils as the unit, the reactance is obtained by analogy with the cylindrical coil arrangement. For the *n* sections and 2*n* units the total leakage reactance referred to the primary is

$$X_1 = \mu_0 2\pi f \frac{N_1^2}{2n} \frac{L_{mt}}{w} \left[a + \frac{b_1 + b_2}{6} \right] \qquad (2.17)$$

where N_1 is the total number of primary turns. The per-unit reactance is given by

$$\epsilon_x = \mu_0 2\pi f \frac{F}{E_t} \frac{L_{mt}}{2nw} \left[a + \frac{b_1 + b_2}{6} \right] \qquad (2.18)$$

with *F* the rated phase m.m.f. of either primary or secondary.

Mechanical forces

The primary and secondary currents of a loaded transformer are in magnetic opposition with respect to the common core, but cumulative with respect to the space between them, so exciting leakage flux. One important consequence is the production of mechanical forces of mutual repulsion. Consideration of

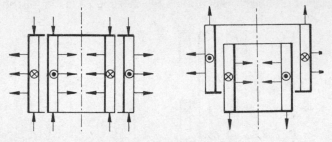

2.23 Mechanical force directions.

the simplified diagrams in Fig. 2.23 shows that these forces can be analysed into the following components:

  (i) *radial*, tending to burst outer and crush inner coils;
 (ii) *axial*, tending to compress coils lengthwise; and
(iii) *axial*, tending to force apart axially displaced coils.

For rated currents the forces are comparatively small. Under the worst fault conditions, however, instantaneous currents may reach a level of nearly $2/\epsilon_x$ per-unit because of the doubling effect. In a 100 kVA transformer with $\epsilon_x = 0.04$ p.u. the radial force may reach 400 kN. In a 100 MVA transformer with the same per-unit reactance the force might be 180 MN; and even with 0.10 p.u. reactance the force is 30 MN.

Radial force. Circular coils have a shape best suited to withstand radial forces. For an instantaneous current i in the N turns of one winding the flux density in the annular duct is $B_a = \mu_0 F/L_c$ and the mean value in the windings is $\frac{1}{2}B_a$, where $F = Ni$. The mean force per unit length of perimeter and per unit axial length is therefore $\frac{1}{2}\mu_0(F/L_c)(F/L_c)$. For the whole surface area $L_{mt}L_c$ the force is

$$f_r = \tfrac{1}{2}\mu_0 F^2 (L_{mt}/L_c) \tag{2.19}$$

Axial force. Radial components of leakage flux originate axial compressive forces tending to squeeze the windings. When the coil arrangement is symmetrical the resulting stresses are unimportant even under fault conditions, and can normally be supported by the winding material. In shell types the outer coils experience repulsion forces which are resisted by the core in the buried length, but those parts of the coils external to the core must be braced.

Axial force of asymmetry. The radially directed leakage field produces coil forces illustrated by three typical cases in Fig. 2.24. Case (*a*) is the worst condition, and if as a result of a short circuit the coils are displaced by the forces, the forces themselves are further intensified. In (*b*) there is (ideally) less ten-

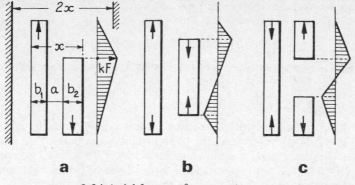

2.24 Axial forces of asymmetry.

dency to produce such an effect. Cases (*a*) and (*c*) are typical respectively of end-tapped and centre-tapped coils. The mechanical superiority of (*c*) over (*a*) is obvious. The forces of asymmetry can be estimated from the cross-flux developed by the out-of-balance m.m.f. *kF*, shown for example in case (*a*). The windings are deemed to be enclosed between iron surfaces separated by a distance $2x = 2(a + b_1 + b_2)$: the cross-flux density for a given instantaneous m.m.f. $F = Ni$ has the maximum value $\mu_0 kF/2x$. The axial force is that produced by the mean cross-flux density (one-half of the maximum) acting on N turns, i.e.

$$f_a = \tfrac{1}{2}\mu_0 kF^2 L_{mt}/2(a + b_1 + b_2) \tag{2.20}$$

In the comparative cases in Fig. 2.22 the short-circuit currents reduce with subdivision because of increased leakage flux. The forces due to asymmetry depend upon how closely the primary and secondary discontinuities are 'matched'.
Bracing. In so far as the stresses are inwards axially, they are taken by the windings themselves. Inward radial stresses are passed to the formers, packing pieces and cores. Outward axial stresses must be withstood by the end insulation. A well-constructed transformer will have a suitable choice of conductor dimensions and interturn insulation, and coils well supported and braced, with the compressive stresses kept in view. The end supports are not easily arranged to give good insulation and at the same time great mechanical strength, so that as far as possible the need for the latter quality must be avoided by maintaining symmetry, or by a suitable distribution of the unpreventable asymmetry caused by tappings, connections, etc.

EXAMPLE 2.3: The windings of a 600 kVA, 7.5/0.435 kV, 50 Hz, 1-ph, 2-limb core-type transformer have the following dimensions with reference to Fig. 2.21(*c*):

$$a = 15.75 \text{ mm} \quad b_1' = b_1'' = 9.0 \text{ mm} \quad L_{mt} = 1.26 \text{ m} \quad N_1 = 378$$
$$a' = 8.5 \text{ mm} \quad b_2 = 23.0 \text{ mm} \quad L_c = 0.35 \text{ m} \quad N_2 = 22$$

Then $a + (b_1' + b_1'' + b_2)/3 + a'/4 = 31.5 \text{ mm} = 0.0315 \text{ m}$. Each limb has $L_c = 0.35$m and $N_1 = 378/2 = 189$ primary turns, and an equivalent total reactance (referred to the primary) given by eq.(2.15). The reactance of the two limbs forming the 1-ph winding is therefore

$$X_1 = 2[\mu_0 2\pi 50 \times 189^2 \times (1.26/0.35) \times 0.0315] = 3.2\ \Omega$$

The full-load primary current is $600/7.5 = 80$ A, so that the per-unit reactance is

$$\epsilon_x = 3.2(80/7\ 500) = 0.034\ \text{p.u.}$$

Neglecting winding resistance, the r.m.s. symmetrical short-circuit current is $1/0.034 = 29.4$ p.u., corresponding to an instantaneous peak of $\sqrt{2} \times 29.4 = 42$ p.u. With asymmetry this can be taken as, say, $1.8 \times 42 = 75$ p.u. or $75 \times 80 = 6\ 000$ A. Applying eq.(2.19) with 189 turns per limb, the peak instantaneous radial outward force on the h.v. winding is

$$f_r = \tfrac{1}{2}\mu_0 (6\ 000 \times 189)^2 \times (1.26/0.35) = 3\ 100\ \text{kN}$$

Suppose that the h.v. winding is 5% shorter than the l.v. winding and that the difference is all at one end. The result is an unbalance between the h.v. and l.v. distributions of m.m.f., with $k = 0.05$. From eq.(2.20), with $a + a' + b_1' + b_1'' + b_2 = 0.065$ m, the axial force of asymmetry is

$$f_a = \tfrac{1}{2}\mu_0 0.05 \times (6\ 000 \times 189)^2 \times 1.26/2 \times 0.065 = 4\ 200\ \text{kN}$$

If this force causes further divergence of the windings, a subsequent short-circuit current of the same value will produce an even greater force.

2.8 *Leakage flux effects in machines*

Estimation of leakage flux is more difficult than for transformers. Flux-plot

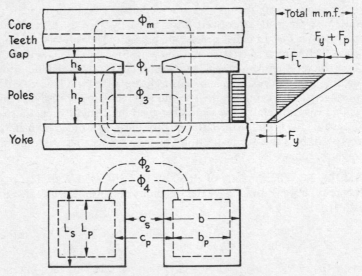

2.25 Leakage flux of salient poles.

methods include the fitting of a pattern of curvilinear squares by freehand drawing, electrolytic tanks, membranes, resistance networks, etc. With the advent of digital computers it has become possible to predetermine flux patterns accurately without having recourse to analogues. Mamak and Laithwaite [5] have described a method of deriving the relevant finite-difference equations for a two-dimensional problem using the concept of vector potential, enabling the flux pattern to be computed. A simpler approach through m.m.f. and permeance is given here, indicating the order of the results rather than their precise values.

Salient poles

Most low-speed salient-pole machines have many poles on a large diameter, so that the pole-axes can be assumed to be parallel. Components of the leakage flux are set out as in Fig. 2.25. The m.m.f. F_l producing leakage on no load comprises that for the airgap, teeth and core. Approximately, F_l produces Φ_1 and Φ_2, while $\frac{1}{2}F_l$ produces Φ_3 and Φ_4. Taking the leakage flux between facing sides to be rectilinear, and that between parallel faces (Φ_2 and Φ_4) to comprise straight lines plus two quarter-circles, then the estimations of leakage flux per pole are

(i) between pole-shoes, $\Phi_{sl} = 2\Phi_1 + 4\Phi_2$:

$$\Phi_{sl} = \mu_0 F_l \left[2\frac{L_s h_s}{c_s} + 2.9 h_s \log_{10}\left\{1 + \frac{\frac{1}{2}\pi b}{c_s}\right\}\right] \tag{2.21a}$$

(ii) between poles, $\Phi_{pl} = 2\Phi_3 + 4\Phi_4$:

$$\Phi_{pl} = \mu_0 F_l \left[\frac{L_p h_p}{c_p} + 1.5 h_p \log_{10}\left\{1 + \frac{\frac{1}{2}\pi b_p}{c_p}\right\}\right] \tag{2.21b}$$

Φ_{sl} is generally of the order $20F/10^8$, while Φ_{pl} approximates to $80F/10^8$. The flux at the back of the pole-shoes is $\Phi + \Phi_{sl}$, and that at the root of the pole (where it joins the yoke) is $\Phi + \Phi_{sl} + \Phi_{pl}$, where Φ is the active gap flux.

Non-salient poles

The non-salient rotor has considerable leakage if the retaining rings clamping the overhang are not non-magnetic; the amount is dependent on the saturation flux density of the material, for the density will reach this level in all normal cases. A disadvantage of such leakage is that it raises the m.m.f. required for the core and teeth of the rotor. Non-magnetic steel is generally preferred for the retaining rings for this reason. Estimation of the leakage of non-salient rotors has to be taken step by step with that of the working flux, and cannot be treated separately.

Armature

By 'armature' is meant the member carrying distributed a.c. windings. In the gap region the leakage flux is a component of the total flux; only in the end

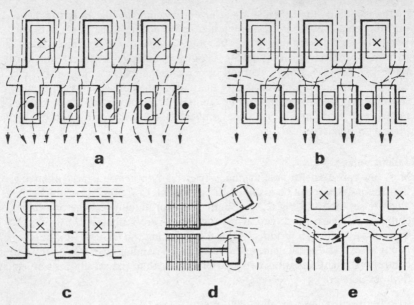

2.26 Leakage flux of slotted members.

regions can it be recognized as a separate entity. Fig. 2.26 shows at (*a*) the total flux in the gap region with both members excited, and (*b*) its arbitrary components. The working flux links both windings while the leakage flux either links one winding only, or is so directed as not to contribute to useful energy transfer between stator and rotor.

Summarizing the leakage flux components with reference to Fig. 2.26, there are:

Overhang leakage (*d*), produced by the end-windings and of value depending on the winding arrangement and on the proximity thereto of metal masses, such as core-stiffeners and end-covers, having conducting and magnetic properties.

Slot leakage (*c*), crossing the slot conductors from tooth to tooth.

Zig-zag leakage (*e*), which depends on the gap length and on the relative position of the stator and rotor tooth-tips; it tends to 'zig-zag' across and along the gap, utilizing the relatively high tooth-tip permeances.

Differential leakage, Fig. 2.27, the result of a difference between the current distributions on opposite sides of the gap. In (*a*), where two windings with identical distributions have a turn-ratio of 2/1, m.m.f. balance both in magnitude and in distribution is obtained if $I_2 = 2I_1$. But with the same turns-ratio and a dissimilar distribution as in (*b*), the condition for zero secondary linkage is that $(72yI_1 - 34yI_2) = 0$, giving $I_2 = (72/34)I_1$, a ratio differing from 2. However, the primary linkage is $(160yI_1 - 72yI_2) = 7\frac{1}{2}yI_1$. Thus the m.m.f.s cannot precisely balance. The net flux acts as leakage and is called the differential or harmonic leakage, the latter term because of an alternative approach by the

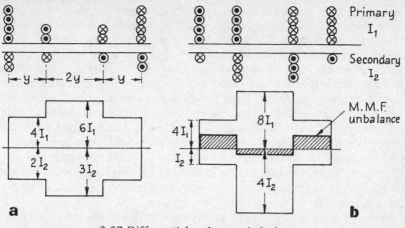

2.27 Differential or harmonic leakage.

harmonic resolution of the two m.m.f. distributions. The differential leakage is mainly the result of the higher space harmonics, and for normal windings may often be small enough to ignore. In a cage rotor, the currents can balance the stator currents at every point without restraint, so that differential leakage vanishes except for those space harmonics whose wavelength is comparable with the slot-pitch.

Slots

Fig. 2.28 shows the very common parallel-sided slot shape, accommodating one or more rectangular conductors. When the conductors carry current, they set up a slot-leakage flux pattern of complex shape. Typical patterns have been obtained for steady-state conditions by Hague [3], by applying the condition that the magnetic vector potential within the slot must satisfy the

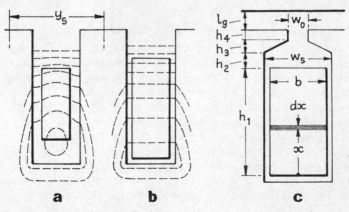

2.28 Slot leakage.

Poisson equation within the section of the conductors, and the Laplace equation elsewhere. If there is appreciable space below the bottom conductor and its width is substantially less than that of the slot, the field pattern is that shown in (*a*). However, apart from the case of high-voltage machines requiring thick slot insulation, the proportions are more usually those in (*b*), for which the field pattern is very nearly in the form of straight lines across the slot. There is thus some justification for the classical assumption that an elemental leakage path is completed around the bottom of the slot as a consequence of the assumption that the ferromagnetic material has infinite permeability. The pattern is readily analysed, and is adequate for steady currents or those of low frequency, but fails (unless the conductors are subdivided and transposed) for frequencies approaching normal industrial supply values (e.g. 20 Hz upward). We adopt the straight-line assumption here, leaving more complicated cases for Chapter 3.

Semi-closed rectangular slot: The permeance coefficient per unit of axial slot length is the ratio (area/length) of the air path considered. A number of component parallel paths can be taken. Referring to the dimensions in Fig. 2.28(*c*) we have: (h_2/w_o) for the region immediately above the conductor; $2h_3/(w_s + w_o)$ for the wedge portion taken as an average width; and (h_4/w_o) for the lip.

The slot height h_1 occupied by the conductor has to be treated differently. All the leakage flux *above* the conductor links it wholly, but the flux within the height x links only the lower part of the conductor and contributes only partial linkage. Also, such flux has a lower density as only the fraction (x/h_1) of the total conductor current is available to provide the m.m.f. For 1 A in the conductor, uniformly distributed, the flux in the elemental path dx is $d\Phi = \mu_0(x/h_1 w_s) \cdot dx$ and it links the fraction (x/h_1) of the conductor. The permeance coefficient for this part is the summation from $x = 0$ to $x = h_1$ of $(x^2/h_1^2 w_s) \cdot dx$, giving $h_1/3w_s$. Adding the several parallel coefficients, the specific slot permeance is obtainable from

$$\lambda_s = \frac{h_1}{3w_s} + \frac{h_2}{w_s} + \frac{2h_3}{w_s + w_o} + \frac{h_4}{w_o} \tag{2.22}$$

The total slot permeance of a net core-length L_s is the product $\mu_0 \lambda_s L_s$.

Other slot shapes: Fig. 2.29 shows some of the several forms that slots can take. The *semi-closed rectangular* has already been discussed. The *circular* slot is often used for the cage rotors of induction machines, as is also the *T-bar* slot. Two *deep-bar* slots are shown, one trapezoidal and the other parallel-sided. The *open* slot shape permits of the easy insertion of pre-formed coils. The *closed* slot is sometimes used for induction machine rotors with cast-in aluminium cages, the 'bridge' mechanically retaining the molten metal during casting.

Round slot: With the assumption that flux crosses the slot in straight lines (an even less tenable concept than with rectangular slots) the result is independent of the slot diameter. It is found to be 0.62, usually raised to 0.66 to compensate for the actual flux pattern. The permeance coefficient is then

$$\lambda_s = 0.66 + (h_4/w_o) \tag{2.23}$$

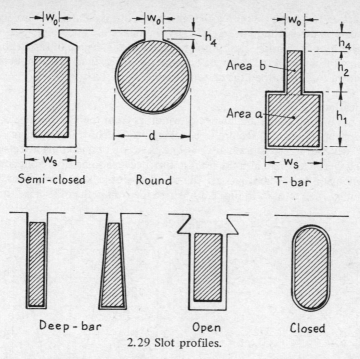

Semi-closed   Round   T-bar

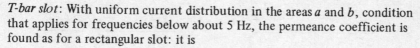

Deep-bar   Open   Closed

2.29 Slot profiles.

T-bar slot: With uniform current distribution in the areas *a* and *b*, condition that applies for frequencies below about 5 Hz, the permeance coefficient is found as for a rectangular slot: it is

$$\lambda_s = \frac{1}{(a+b)^2}\left[\frac{h_1}{w_s}\cdot\frac{a^2}{3} + \frac{h_2}{w_o}\left\{a^2 + ab + \frac{b^2}{3}\right\}\right] + \frac{h_4}{w_o} \qquad (2.24a)$$

The total conductor section is $(a + b)$, the two parts sharing the current in proportion to their area. For currents of industrial frequency the current is substantially confined to *b*, and the permeance coefficient approximates to

$$\lambda_s = \frac{h_1}{2w_s} + \frac{h_2}{3w_o} + \frac{h_4}{w_o} \qquad (2.24b)$$

Values intermediate between these two extremes may be of significance during starting and speed change. An elegant solution of the T-bar and similar shaped conductors is given by Bruges [6].

Deep slot: This is a variant of the foregoing, exploiting the effect of leakage flux in disturbing the current distribution by making the conductor deep and narrow. Eqs.(2.24) apply, with the obvious modifications.

Open slot. Facility in winding is counterbalanced by the aggravation of high-frequency no-load tooth loss and the increase of gap reluctance, particularly in induction machines whose gaps are very short. It is possible to ameliorate such effects by use of semi-magnetic slot-wedges.

Closed slot: The 'bridge' reduces gap reluctance, but increases slot leakage flux by increasing the last term in eq.(2.22) by the factor μ_r, the effective relative permeability of the bridge. As μ_r is dependent on the degree of saturation its assessment is difficult: its value may change e.g. from 25 at low conductor currents down to 3 or 4 when the current is large, as at starting.

Modifications

In the calculation of the inductance of windings from their permeance coefficients, certain effects may require to be taken into account. The influence of slot *skewing* is discussed in Chapter 3. The *chording* of a winding affects its winding factor K_w; and also the leakage flux because some slots hold conductors belonging to different phases. The effective slot-leakage permeance is reduced by the factor k_s in Fig. 2.30. Where the iron paths of leakage flux are

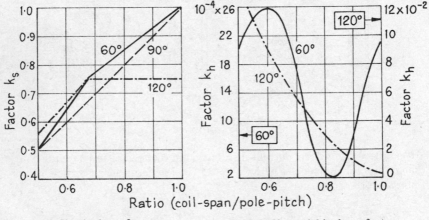

2.30 Slot leakage factor.   2.31 Differential leakage factor.

saturated, the flux per ampere is reduced. It is common to base the leakage on the assumption of infinite permeability and then to apply saturation factors. Agarwal and Alger [7] produced a simple method in which the iron is assumed to have a B/H relation in which the permeability is infinite ($H = 0$) for any flux density up to 2.0 T, and thereafter a fixed saturation flux density of 2.0 T regardless of the value of H. The method was developed to deal with combined tooth-tip and zigzag leakage components, and has been extended by Chalmers and Dodgson [8].

Zigzag

One expression (of many) gives the zigzag leakage reactance of a phase winding in terms of the reactance offered to the main flux by the winding, i.e. the magnetizing reactance x_m, in the form

$$x_z = \frac{5}{6}x_m\left[\frac{1}{g_1{}^2} + \frac{1}{g_2{}^2}\right]$$

(2.25)

where $g_1 = S_1/2p$ and $g_2 = S_2/2p$ are the numbers of slots per pole of the two members.

Differential

An expression for the differential or harmonic leakage reactance is

$$x_h = x_m(k_{h1} + k_{h2}) \qquad (2.26)$$

where values of the coefficients k_h are obtained from Fig. 2.31 due to Alger [9]. This reactance is ignored for induction machines with cage rotor windings.

Overhang (end-winding)

One simplified expression for small machines is in terms of the length L_o of an overhang conductor and the pole-pitch Y:

$$L_o \lambda_o = k_s Y^2/\pi y_s \qquad (2.27)$$

where y_s is the slot-pitch and k_s is given by Fig. 2.30.

Simple direct methods fail completely to deal with the overhand leakage problem in cases where its assessment is very important, as in turbo-generators. The leakage must be related to the length of the end-connectors and their shapes, the interleaving of phase conductors, the spacing between the stator and rotor overhangs, the proximity and configuration of neighbouring magnetic or conducting parts (cores, clamping and stiffening plates, housings, retaining rings, etc.) and the level of saturation in ferromagnetic parts. Not only is the leakage reactance required, but also the consequent 'load' loss and the magnetically generated mechanical forces.

Approximate analyses based on a solution of the Laplace equation for the scalar magnetic potential have been advanced by Reece and Pramanik [10]. A 2-dimensional field-plot method for estimating the leakage of a given machine has been described by Hawley, Edwards, Heaton and Stoll [11]. Carpenter [12]

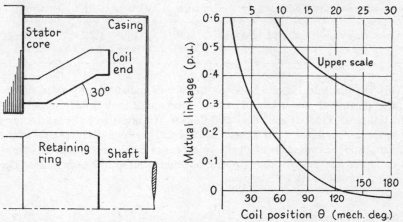

2.32 Overhang leakage in turbo-generator.

has extended the principle of images to cases of currents partly free and partly embedded in one of the reflecting surfaces, and in particular to end-winding problems in which this surface is divided by an airgap. Lawrenson [13] has used the method of images for dealing with *magnetic* and *eddy-current-carrying* boundaries. Eddy-current boundaries are assumed to be non-permeable and to reduce leakage-flux linkage, while magnetic boundaries increase it. Lawrenson's method is (i) to obtain the basic linkage of a coil-end, (ii) to modify the value by factors introduced by the boundary surfaces, and (iii) to superpose the linkages of all the coils in a phase band in accordance with their relative position.

The shape of a single end-coil, and the significant boundary surfaces in the neighbourhood of the end winding, are shown in Fig. 2.32. The steps in the estimation of the linkage per ampere are as follows:

(i) Using the length L_0 of the overhang of a coil of cross-section $b \times c$ and comprising N_c conductors all in series, the basic linkage of an isolated coil-end is obtained from

$$\psi = 1.5L_0(1 + 0.0075L_o)N_c{}^2\left[\ln \frac{2L_0}{b+c} + 0.5\right]10^{-7}$$

(ii) Taking ψ as 1.0 p.u., the boundaries make proportional contributions of 0.18 for the cylindrical casing, 0.07 for the end casing, 0.08 for the stator core-end, and 0.11 for a magnetic retaining ring on the rotor. These contributions must be made with the appropriate sign, positive for a magnetic boundary and negative for an eddy-current boundary. A non-magnetic retaining ring has no steady-state effect, but appears as an eddy-current boundary under transient conditions. These contributions result in the factors k_b by which ψ has to be multiplied to give the total linkage of an isolated end-coil:

VALUES OF FACTOR k_b

| Type of rotor retaining ring: | magnetic | | non-magnetic | |
With or without core screen:	with	without	with	without
Operating condition: steady-state	1.30	1.44	1.19	1.35
transient	0.56	0.56	0.56	0.56

(iii) The graph in Fig. 2.32 shows the mutual inductance between one coil-end at position $0°$ and a second at some mechanical angle θ to it, in per-unit of the linkage of the first coil. The total linkage of a phase group is (a) the summation for each coil in turn of its own linkage and that due to other coils in the group, and (b) summation of all the totals in (a). The result is then doubled (in a 2-pole machine) for the opposite coil-end group of the same phase. The graph is for coils with a $150°$ span.

2.9   *Phase reactance*

The reactance of a phase winding of $g'z_s$ turns per pole-pair, with a conductor length L_c and a permeance coefficient λ is, from eq.(2.13),

$$X = 2\pi f N^2 \lambda = 2\pi f (g' z_s)^2 \mu_0 2 L_c (\lambda/g')$$

because the turn length includes two conductors and the flux has to traverse g' slots. If the turns in all pole-pairs are in series, the phase reactance is

$$X = 4\pi f \mu_0 N_{ph}^2 L_c (\lambda/pg') \tag{2.28}$$

Induction machine
The total phase reactance referred to the stator is

$$X_1 = x_1 + x_2' = x_{s1} + x_{s2}' + x_o + x_z + x_h \tag{2.29}$$

where x_{s1} is the stator slot reactance using the net core-length L_s in eq.(2.28) with the appropriate permeance coefficient for the slot shape; x_{s2} for the rotor is calculated in a similar way and then multiplied by the square of the effective turns ratio, i.e. by $K_{w1}^2 S_1 / K_{w2}^2 S_2$; and x_o, x_z and x_h are the overhang, zig-zag and harmonic (differential) reactances.

For calculation of performance from an equivalent circuit it is sufficient to take $x_1 = x_2' = \frac{1}{2}X_1$. As (apart from slot leakage) the leakage flux is a result of the stator and rotor currents in combination, it is not strictly possible to separate it completely into individual stator and rotor components.

Synchronous machine
The leakage flux must be known to assess regulation under steady-state conditions; but it enters also into many complex problems such as short-circuit behaviour and damping. For the calculation of regulation, the leakage flux may be taken as the sum of the slot and overhang leakage fluxes for normal full-load r.m.s. current I_c per conductor. With eq.(2.1) as basis, the peak leakage flux associated with the slots for a coil of N_c turns is $\sqrt{2}I_c N_c \Lambda_s$, where Λ_s is $\mu_0 2 L_s \lambda_s$. For a group of g' coils of the same phase, the turns producing the flux are $g' N_c$, but the slot flux now has to cross g' slots and its air path is g' times as long. This gives the total slot leakage flux per phase as

$$\Phi_s = 2\sqrt{2}\mu_0 I_c N_c (L_s \lambda_s) \tag{2.30}$$

An identical expression, except that $(L_o \lambda_o)$ is substituted for $(L_s \lambda_s)$, gives the overhang leakage flux. Zigzag leakage does not occur for a plain salient-pole machine, but may be present (though reduced by the comparatively long air-gap) if the pole shoes carry damper windings.

2.10 *Unbalanced magnetic pull*
In an excited rotary machine, a magnetic flux crosses the gap between stator and rotor, and these members exert on each other a strong magnetic attraction, considerably more intense than the useful force of interaction between the flux and the gap-surface currents. In a completely symmetrical machine the forces of attraction balance out with no resultant; but if there is asymmetry the balance is incomplete, leaving a resultant force called the *unbalanced magnetic pull* (u.m.p.).

A simplified introduction to the analysis of u.m.p. can be based on the well-known force of attraction between infinitely permeable ferromagnetic surfaces with a flux of density B crossing the gap between them. Per unit area this force is $\frac{1}{2}B^2/\mu_0$.

Symmetry. In the ideal 2-pole machine of Fig. 1.10 the rotor is set symmetrically within the stator bore, both gap surfaces are purely cylindrical, and the gap length is everywhere l_g. Taking the gap flux as distributed sinusoidally with its peak B_m on the x-axis, then at angle θ to the axis the radial density is $B = B_m \cos \theta$, and the radial attractive force over an elemental angle $d\theta$ is $[\frac{1}{2}(B_m \cos \theta)^2/\mu_0]\frac{1}{2}Dl \cdot d\theta$, where D is the diameter of the gap and l is the axial length. This quantity multiplied by $\cos \theta$ is the component of force along the x-axis, and integration over the whole periphery $(0 < \theta < 2\pi)$ gives the resultant x-axis magnetic pull as zero, because the force $(Dl/3\mu_0)B_m^2$ for *one* pole is balanced by an equal and opposite force on the other.

Asymmetry. Let the rotor axis be displaced by a small distance e in the direction of the polar x-axis. At angle θ the radial gap length is now $l_g' = l_g(1 - \epsilon \cos \theta)$, where $\epsilon = e/lg$ is the eccentricity expressed as a fraction of the symmetrical gap length. Assuming the flux density to be inversely proportional to gap length, then at θ, where with no eccentricity the flux density was B, it is now $B' = B/(1 - \epsilon \cos \theta) \simeq B(1 + \epsilon \cos \theta)$. Integrating the x-axis force components around the whole rotor gives the resultant $(\pi Dl/4\mu_0) B_m^2 \epsilon$, which is directly proportional to the eccentricity.

EXAMPLE 2.4: For the 2-pole machine of Example 1.2, find the unbalanced magnetic pull on the rotor when its axis is displaced by $e = 0.001$ m = 1 mm. The machine has $D = 0.50$ m, $l = 0.20$ m, $l_g = 0.005$ m = 5 mm; the stator and rotor peak current-sheet densities are $A_1 = 12\ 800$ A/m and $A_2 = 9\ 600$ A/m; and the torque angle is $\lambda = \pi/3$ rad. The m.m.f.s per pole are

$$F_1 = \frac{1}{2}A_1 D = 3\ 200 \text{ A-t} \quad \text{and} \quad F_2 = \frac{1}{2}A_2 D = 2\ 400 \text{ A-t}$$

and applying eq.(1.9), the resultant gap m.m.f./pole is $F_0 = 4\ 450$ A-t, which produces in the gap a sine-distributed flux density of peak value $B_m = F_0\mu_0/l_g$ = 1.11 T. Then the magnetic pull per pole is

$$(Dl/3\mu_0)B_m^2 = 33.0 \text{ kN}$$

which, balanced by the opposition of polar attractive forces, gives zero u.m.p. as resultant if the rotor is *symmetrically* centred within the stator bore. With the displacement e giving an *eccentricity* $\epsilon = 0.001/0.005 = 0.2$, then there is a resultant u.m.p. of

$$(\pi Dl/4\mu_0) B_m^2 \epsilon = 15.5 \text{ kN}$$

The useful torque of the machine is $M = 260$ N-m, corresponding to a peripheral force of $260/0.25 = 1.04$ kN, so that the u.m.p. is fifteen times as great as the useful force.

The simplified analysis given above assumes that the stator and rotor axes, though displaced, remain parallel; this is not necessarily the case if the bearing alignments differ. It is also assumed that the peak flux density is unaltered by the eccentricity, which may not be true. The u.m.p. is calculated for the worst condition, i.e. coincidence of eccentricity with the axis of maximum flux density, but this condition may change with speed and load. A form of eccentricity is produced by non-circularity: it is possible to grind a cage rotor, but much more difficult to machine the stator bore, so that asymmetry may arise from this cause.

Binns and Dye [110] have summarized much of the work on u.m.p. with reference to induction motors, in which the short airgap length intensifies the effect of small eccentricities. Experiment shows that the u.m.p. is proportional to ϵ for eccentricities not greater than 0.1, but that between 0.1 and 0.3 the increase is small as a result of saturation (especially in the tooth-tips) imposing a limit on the peak flux density. Leakage flux, which increases with load, plays a significant role in saturating tooth-tips so that u.m.p. effects are likely to be more prominent under light- or no-load conditions.

Practical effects

In manufacturing electrical machines it is necessary to allow dimensional tolerances on core-plate stampings, core assemblies, frames, bearings and endshields. Wide tolerances cheapen a machine, but may impair its performance; in particular, there may be variations in gap length which are significant in short-gap machines such as induction motors in giving rise to u.m.p. and vibration. Stiffer shafts and bigger bearings are made necessary, and the machines may be more noisy than is tolerable. Reckoned as an effective reduction in the stiffness of the shaft, u.m.p. reduces the critical speed, possibly into a region too close to the normal running speed.

In 2-pole machines appreciable homopolar flux is produced by the gap asymmetry, and the flux path may be completed through the shaft and frame. In effect there is a modulation of the gap length, inducing currents in cage-rotor bars and in parallel-connected parts of the stator winding.

The use of parallel paths in machine windings allows counteracting currents to flow and mitigates u.m.p. Cage windings provide multiple parallel paths which have maximum effect on no load, less on load and almost none at starting. Parallel paths in stator windings are most effective when equalizer connections are fitted.

Noise production in machines can be strongly supported by u.m.p., especially if its variation corresponds to a natural frequency of the stator structure. During the start of an induction machine, the u.m.p. has an alternating component at the rotor slot frequency and may set up vibration at speeds given by the natural resonance frequency divided by the number of rotor slots, possibly producing noise.

2.11 *Open magnetic circuits*

The rotary machines so far discussed have magnetic circuits that are substantially symmetrical, at least apart from constructional imperfections. The development of 'linear' machines (Sect. 13.10) has made practical use of magnetic circuits of marked asymmetry and of an 'open' nature. The inter-linked electrical and magnetic circuits that are characteristic of electromagnetic machines have preferably minimum possible resistance and reluctance, the latter achieved by use of a short working airgap and high-permeability steel cores. Further, the symmetry leads (ideally) to a properly balanced magnetic pull. If the magnetic flux is made to pass over a long air path, some compensation for the high reluctance can be obtained by arranging for the path area to be large. But it is now a far more complex path in which the assessment of flux distribution must be based on three dimensions, and not on two as has been possible for the annular airgap of a conventional machine. Again, it is not possible to neutralize the unbalanced magnetic pull which is inherent in the structural form, with the result that mechanical means must be provided to sustain it.

The methods used in this Chapter are therefore unsuitable for the linear machine, and it is necessary to change radically the analytical processes for magnetic circuit design. The problem is discussed in detail by Laithwaite [82, 96].

3
Windings

3.1 *Materials*

The winding of a machine or transformer conveys electrical energy to or from the working region, and is concerned with e.m.f. induction and the development of magneto-mechanical force. Windings are formed from suitably insulated conductors.

Conductors

If a current of density J flows in a conductor of resistivity ρ and density δ, the loss per unit cube is $J^2\rho$ and the loss per unit mass (the specific loss) is $p = J^2\rho/\delta$. Copper is the most common conducting metal, but there is an increasing use of aluminium.

Copper: The International Annealed Copper Standard (IACS) has at 20°C a resistivity 0.017 241 $\mu\Omega$-m, a resistance-temperature coefficient 0.003 93 per °C, and a tensile strength 220–250 MN/m². Hot and cold working (such as wire-drawing) raises the mechanical strength at a small sacrifice in conductivity. Most machines employ windings of annealed high-conductivity copper, but it is necessary in large transformers to use a worked copper of controlled proof-strength, in turbo-generators a silver-bearing copper with resistance to thermal softening and creep. Cadmium-copper is adapted to cage windings because it can be flame-brazed without deterioration.

Aluminium. Aluminium-alloy conductors to BS 3242 have a resistivity at 20 °C of 0.0325 $\mu\Omega$-m (i.e. a conductivity of 53% IACS). Sheet material is often used for the low-voltage windings of small- and medium-rated transformers. Cage windings can be fabricated by brazing slot bars to the end-rings, or by integral casting with silicon-aluminium alloy (6–12% Si) of resistivity 0.04–0.05 $\mu\Omega$-m.

Typical characteristics at 20 °C for resistivity ρ, temperature coefficient α per °C, and density δ, are:

	$\rho(\mu\Omega\text{-m})$	α (per °C)	$\delta(\text{kg/m}^3)$
Copper: annealed	0.0172	0.003 93	8 900
hard-drawn	0.0178	0.003 90	
Aluminium: cast	0.045	0.003 90	2 700
hard-drawn	0.0325	0.003 90	

I^2R *Loss*: This is the significant electrical property. As an example of magnitude, the loss in hard-drawn material at 75 °C carrying a current of uniform density 5 MA/m^2 (or 5 A/mm^2) is
    Copper: 540 kW/m^3, 61 W/kg   Aluminium: 990 kW/m^3, 365 W/kg.

Insulants

Insulating materials, essentially non-metallic, are organic or inorganic, uniform or heterogeneous in composition, natural or synthetic. An ideal insulant would have: (i) high dielectric strength, sustained at elevated temperatures; (ii) good thermal conductivity; (iii) permanence, non-deteriorating at high temperatures; (iv) good mechanical properties such as ease of working and application, non-hygroscopic, and resistant to vibration, abrasion and bending. Almost all usable materials are subject to temperature limitation, and are classified in accordance with agreed limits of operating temperature:

Class Y. Cotton, silk, paper, wood, cellulose, fibre, etc., without impregnation or oil-immersion.

Class A. The materials of Class Y impregnated with natural resins, cellulose esters, insulating oils, etc.; also laminated wood, varnished paper, cellulose-acetate film, etc.

Class E. Synthetic-resin enamels, cotton and paper laminates with formaldehyde bonding, etc.

Class B. Mica, glass fibre, asbestos, etc. with suitable bonding substances; built-up mica, glass-fibre and asbestos laminates.

Class F. The materials of Class B with more thermally-resistant bonding materials.

Class H. Glass-fibre and asbestos materials, and built-up mica, with appropriate silicone resins.

Class C. Mica, ceramics, glass, quartz and asbestos without binders or with silicone resins of superior thermal stability.

The classification is based on the following maximum permitted temperatures:

Insulation class:	Y	A	E	B	F	H	C
Maximum temp., °C.:	90	105	120	130	155	180	> 180

The figures are based on a 20-year working life under average conditions. The life of an insulating material is closely related to the "hot-spot" temperature within the winding it covers. Some notes follow on typical machine insulation.

Mica in the virgin or sheet state is difficult to work, and it is used in the form of sheets of splittings with shellac, bitumen or synthetic polyester bonding.

Micafolium is a wrapping composed of mica splittings bonded to paper and air dried. It may be wound on to conductors, then rolled and compressed between heated plates to solidify the material and exclude air.

Fibrous glass is made from material free from alkali metal oxides (soda or potash) that might form a surface coating that would attack the glass silicates. Glass absorbs no moisture volumetrically, but may attract it by capillary action between the fine filaments. Tapes and cloths woven from continuous-filament yarns have a high resistivity, thermal conductivity and tensile strength, and form a good Class B insulation. The space factor is good, but the material is susceptible to abrasive damage. Thin glass-silk coverings are available for wires for field coils or mush windings: varnishing is necessary to resist abrasion.

Asbestos is mechanically weak, even when woven with cotton fibres, and can only occasionally compete with fibre glass. Laminates of asbestos with synthetic resins have good mechanical strength and thermal resistance. Asbestos, wire- and strip-coverings have resilience and abrasion resistance, but the space factor is low.

Cotton fibre tapes woven from acetylated cotton, recently developed, have remarkable resistance to heat "tendering", and are much less hygroscopic than ordinary cotton materials.

Polyamides such as nylon make tapes of high mechanical strength and effect a saving in space by their thinness. Nylon film is one of the few plastic films having adequate resistance to temperature and opposition to tearing.

Synthetic-resin enamels of the vinyl-acetate or nylon types give an excellently smooth finish and have been applied for mush windings, with considerable improvement in winding times and in length of mean turn. Varnishes of the same basic materials give good bonding of windings.

Slot-lining materials have in the past been various mica-composites, but the mica content is easily damaged in forming. With small motors a two-ply varnished cotton cloth bonded to pressboard has been found satisfactory; while three-ply material may serve for heavier windings.

Wood, in the form of synthetic-resin-impregnated compressed laminations, has proved a robust and accurate material for packing blocks, coil supports and spacers. If the electrical properties are not adequate, phenolic paper laminates will be preferred although their cost is greater.

Silicones are semi-inorganic materials with a basic structure of alternate silicon and oxygen atoms. They are remarkably resistant to heat, and as binders in Class H insulation permit of continuous operation at 180 °C. Even when disintegrated by excessive temperatures, the residue is the insulator silica. Silicones are water-repellent and anti-corrosive: they have been successfully used in dry (oil-less) transformers, traction motors, mill motors and miniature aircraft machines operating over a winding temperature range of 200 to −40 °C. An additional advantage is the superior thermal conductivity, improving heat-transfer from conductors and facilitating dissipation.

Epoxide thermosetting resins have become important in casting, potting, laminating-adhesive and varnishing applications, and in the encapsulation of small transformers.

Synthetic resin bonded paper, cotton and glass-fibre laminates have good electrical and mechanical properties as sheets, large cylinders and tubes.

Petroleum-based mineral oils are commonly employed in the cooling and insulation of immersed transformers. The characteristics of importance are chemical stability, expansion coefficient, resistance to sludging by oxidation, and viscosity. When clean and moisture-free their electric strength is good.

Askarels are synthetic non-flammable insulating liquids which, when decomposed by an electric arc, evolve only non-explosive gases. The commonest

askarel is a 60/40 mixture of hexachlorodiphenyl/trichlorobenzine giving a
low pour point and a satisfactory viscosity/temperature characteristic.

3.2   Eddy currents

The main or working flux in a transformer or machine induces in a winding
the useful e.m.f. concerned in energy transfer or conversion. The non-useful
leakage flux also produces an e.m.f., which may introduce a loss arising from
the disturbance of the current distribution in the conductors of the winding.

The parasitic eddy-currents in an *isolated* conductor due to its own field
are called the *skin effect*. They arise on account of the inductance of the central parts of the conductor exceeding that of the outer parts. The reactance
of the centre is therefore greater, and the current flows consequently more
readily in the outer layers of the conductor. But any departure from uniform
current density increases the $I^2 R$ loss over its d.c. value. The greater induced
e.m.f. of self-induction in the middle parts of the conductor causes circulating
currents which, superimposed on the main current, increases the $I^2 R$ loss.

In machines and transformers the conductors forming the windings are not,
however, isolated, and the effects of alternating leakage fields are intensified
by the proximity of ferromagnetic material. There are losses — usually undesirable — in the conductors themselves, and in neighbouring permeable or conducting masses (such as transformer tanks, and the stiffeners and housings of
machines) resulting from currents induced therein by the conductors. The
boundary shapes differ so radically that it is necessary to deal piecemeal with
a variety of combinations of conductors and neighbouring masses.

Slot conductors

As the core is laminated, the teeth flanking a slot can be taken as having high
permeability and infinite resistance, so that the leakage flux path is the same
as that assumed in Fig. 2.28. Consider the single rectangular conductor of
Fig. 3.1(a), set into a parallel-sided open slot. Let the conductor carry a uniformly distributed current of peak i_m alternating at angular frequency ω. At
the instant of peak current the m.m.f. F acting across the slot from wall to
wall increases from zero at the bottom to i_m at the top.

At height h the cross-flux density is $\mu_0 i_m / w_s$, so that the total flux crossing
the conductor per unit axial length of slot is $\Phi_h = \frac{1}{2}\mu_0 i_m (h/w_s)$. Up to the lower
height x the corresponding flux is $\Phi_x = \Phi_h (x/h)^2$. The flux linking the lower
part x is the flux $\Phi = \Phi_h - \Phi_x$ crossing the slot *above* it. The peak eddy e.m.f.
induced in the elemental lamina dx is $e_{dx} = \omega\Phi$:

$$e_{dx} = \tfrac{1}{2}\mu_0 \omega i_m \frac{h}{w_s}\left[1 - \frac{x^2}{h^2}\right] = k i_m \left[h - \frac{x^2}{h}\right]$$

where $k = \frac{1}{2}\mu_0 \omega/w_s$. The eddy e.m.f. is plotted in Fig. 3.1(b) together with the
eddy current that it produces. The average e.m.f. of $k i_m \frac{2}{3}h$ occurs at $x =$
$h/\sqrt{3} = 0.58h$: at this height there is no circulating eddy current, but above it
an eddy current i_d flows axially in one direction, to return in the lower part of

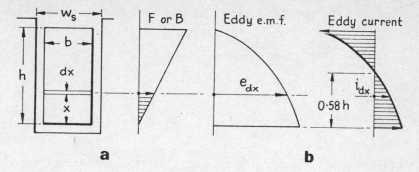

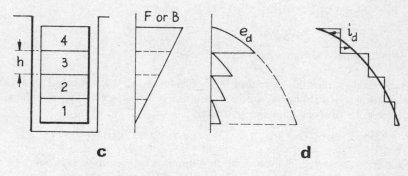

3.1 Eddy currents in slot conductors.

the conductor. As i_d lags i_m by $\frac{1}{2}\pi$ rad in time phase, the I^2R loss due to each can be calculated separately.

This *first-order* eddy current disturbs the originally uniform distribution of current density, superposing a *second-order* m.m.f., flux, e.m.f. and eddy current, which in turn develops a third-order component and so on. Not only do the eddy-current components cause further I^2R loss, but also they disturb the field distribution and affect the slot leakage flux and linkage. At this point it is necessary to distinguish between the eddy-current effects: A, in conductors in which the eddy currents are minimized; and B, in conductors in which the effect is encouraged for a useful purpose, such as the suppression of hunting in synchronous machines or the development of high values of starting torque in cage induction motors.

CASE A: This is the most common, and for it we can ignore all eddy-current components other than that of the first order, and neglect all disturbance to the effective leakage inductance.

The resistance r_x per unit length of an elemental lamina in a conductor of resistivity ρ is $\rho/b \cdot dx$. The effective eddy e.m.f. acting on it is $e_x = ki_m \times [h - (x^2/h) - \frac{2}{3}h]$. The laminar eddy current is therefore $i_{dx} = e_x/r_x$. In the same lamina the main current component is $i_m \cdot dx/h$. If the main and eddy components are squared and multiplied by the resistance they give the corresponding I^2R losses. Integrating their ratio over the conductor height gives the result

$$\frac{\text{eddy-current } I^2R}{\text{main-current } I^2R} = \frac{4}{45}\frac{k^2b^2}{\rho}h^4 = \frac{4}{45}(\alpha h)^4$$

where

$$\alpha = \sqrt{\frac{kb}{\rho}} = \sqrt{\frac{\frac{1}{2}\mu_0\omega}{\rho}}\cdot\sqrt{\frac{b}{w_s}}\qquad\qquad (3.1)$$

The first term always occurs in the analysis of skin effect. The second is the square root of the ratio (conductor-width/slot-width).

It is now possible to define the *eddy-loss ratio*

$$K_d = \frac{\text{total loss}}{\text{loss due to main current}} = 1 + \frac{4}{45}(\alpha h)^4 \qquad\qquad (3.2)$$

which makes it possible to deduce the actual loss from the I^2R based on a d.c. calculation. For a frequency of 50 Hz and the resistivities (hot) given in Sect. 3.1 the value of α becomes approximately

$$\text{Copper: } \alpha = 100\sqrt{(b/w_s)} \qquad \text{Aluminium: } \alpha = 150\sqrt{(b/w_s)} \qquad (3.3)$$

For a solid copper conductor of width 4 mm in a slot 5 mm wide, then $\alpha = 100\sqrt{(4/5)} = 90$. If the conductor height is $h = 5$ mm $= 0.005$ m, we have $\alpha h = 0.45$, and $K_d = 1 + (4/45)(0.45)^4 = 1.0035$; and the d.c. loss is increased by eddy currents only to the extent of under 1%. But if the conductor is 30 mm deep, with $\alpha h = 2.7$, then $K_d = 1 + 4.7 = 5.7$, and the I^2R loss is increased by a factor of nearly six. Such an excessive loss could not be tolerated. It is unusual to find $\alpha h > 0.7$ for a single bar.

Where the cross-section of a conductor has to be large, it is essential to sectionalize it as indicated in Fig. 3.1(*c*), with the sections lightly insulated from each other. The reduction in the eddy current is made clear in (*d*). With *m* sections or layers, the loss in the *p*th layer ($p = 1$ at the bottom of the slot) is

$$K_{dp} = 1 + \frac{p(p-1)}{3}(\alpha h)^4 \qquad\qquad (3.4)$$

and the average loss ratio for the *m* layers, each of height *h*, is

$$K_{dav} = 1 + \frac{m^2}{9}(\alpha h)^4 \qquad\qquad (3.5)$$

The original solution for K_d was given by Field [14], who published his results in the form of curves.

The formulae for the loss factors given for sectioned conductors are valid so long as the subdivisions are not connected in parallel in such a way as to permit a path between laminae for eddy currents. Otherwise the utility of sub-division is almost entirely lost. Of course, a parallel connection is essential for distributing the conductor current, so that the connection in parallel must be made so as to secure complete balance between the individual subdivisions. This may be obtained by *twisting* or *transposition.*

Twisted slot-conductors have their layers transposed or twisted in the slot so as to obtain symmetrical lengths such that each sub-division occupies all possible layer positions for the same length of slot. In a long-cored turbo-generator, the twisting may be carried out two or three times in a single slot. The effect is to equalize the eddy e.m.f.s in all laminae, and to allow the layers to be paralleled at the ends without producing eddy circulating currents

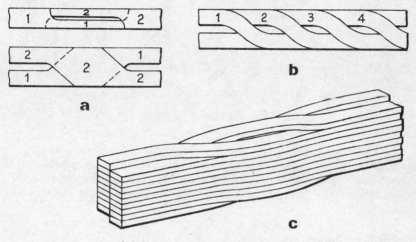

3.2 Slot conductor transposition.

between the layers. Considerable ingenuity is shown in the design of the twist, which must be accommodated in the slot space. Fig. 3.2 shows typical slot transpositions. For (*a*) and (*b*) the crossover is obtained by special shaping of the conductor. Method (*c*) is more compact but involves some reduction of the conductor area.

CASE B: The simplified expressions for Case A do not apply to deep con-ductors in which αh is large, because the eddy-currents of higher order are much too great. Fig. 3.3 shows a simple deep conductor with three instantan-eous current distributions in a half-period of the alternating current carried. There is a distinct impression of downward *propagation*, and in fact the case is one of wave motion. The treatment, based on the Maxwell equations, gives the effective resistance and inductive reactance of a slot conductor in terms of its d.c. resistance r_d in the following form:

$$z = r + jx$$
$$= r_d(\alpha h)\left[\frac{(\sinh 2\alpha h + \sin 2\alpha h) + j(\sinh 2\alpha h - \sin 2\alpha h)}{\cosh 2\alpha h - \cos 2\alpha h}\right] \qquad (3.6)$$

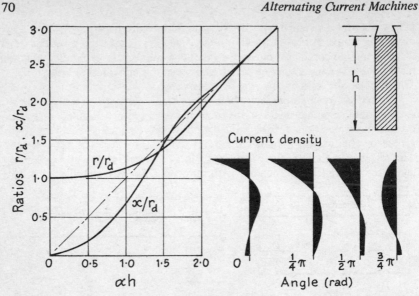

3.3 Eddy currents in deep bars.

This is plotted in Fig. 3.3. The case of a slot with a lip has been solved by Swann and Salmon [15], and an expression based on a ladder network (i.e., an equivalent transmission line) by Bruges [6]. All assume that the core has infinite permeability; the effects of saturation are evaluated by Chalmers and Dodgson [8].

Overhang conductors
The problem of the estimation of eddy current losses in overhang conductors is very difficult, as the leakage field is not readily estimated either in magnitude or direction. Consider Fig. 3.4, in which a typical case is illustrated. The actual field distribution may, for analysis, be resolved into approximate fields parallel to and across the conductors, (*a*) and (*b*). Since the eddy loss factor is proportional to h^4, the parallel field may frequently be neglected. The length of path of the leakage flux is assumed to be $b_1 + h$, and the

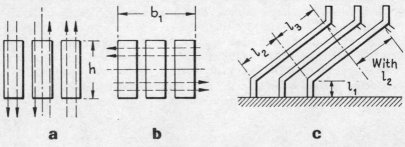

3.4 Overhang leakage.

overhang considered as contained in an imaginary slot $b_1 + h$ wide and $h/2$ deep. With a basket winding (c), the overhang length might be taken in three parts l_1, l_2 and l_3; with $n = 1$ for l_1, $n =$ two-thirds of the number in the group for l_2, and $n =$ total number in a phase group for l_3. Whence for the overhang

$$K_d = \frac{K_{d1}l_1 + K_{d2}l_2 + K_{d3}l_3}{l_1 + l_2 + l_3}$$

Transposition in the overhang connectors is a method of making a layer occupy all possible slot positions by twisting. A relation between the number of layers and the number of slots per phase is essential. This transposition is obtained automatically in diamond and basket windings, in that the upper conductors of one coil side become the lower conductors of another. This may be sufficient for all but the largest machines, where twisting in slots has to be combined with overhang transposition to keep the eddy loss within bounds.

Transformer coils

For normal transformers, Fig. 3.5(a), the eddy-current loss factor may be found by considering the effective conductor length as the coil length L_c contained in a 'slot' of width L so that the ratio b/w_s is represented by L_c/L. The

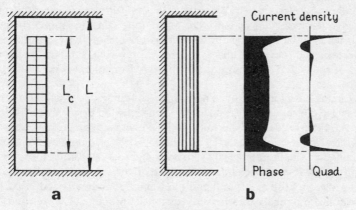

3.5 Eddy currents in transformer windings.

conductor height h is replaced by b, and eqs.(3.1) – (3.5) applied.

In sheet-wound transformers, where the l.v. winding is a spiral (usually of aluminium) of the full usable length of the limb, the current density is much higher at the ends than elsewhere. Fig. 3.5(b) shows typical distributions of the l.v. current, one for the component opposing the h.v. current, the other in quadrature. The loss factor is not likely to exceed 1.1, and will generally be less. This case is examined by Mullineux, Reed and Whyte [16].

Permeable and conducting masses

Conductors (apart from those in slots) carrying alternating currents may lie

Alternating Current Machines

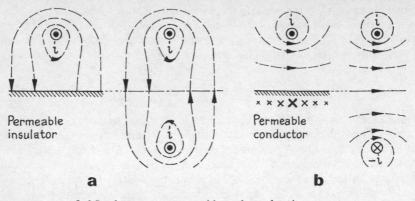

3.6 Leakage near permeable and conducting masses.

near metal masses having permeable and/or conducting properties. The leakage field may penetrate such masses, inducing eddy currents which give rise to I^2R loss and which react to modify the leakage field pattern.

The conductor carrying current i in (a) of Fig. 3.6 lies parallel to a semi-infinite iron mass. If the iron is assumed to have *infinite* permeability and *zero* conductivity, its effect on the external field is as if it were replaced by an *image conductor* carrying an equal current i in the same direction. The mass in (b) is taken to be non-magnetic and of *infinite* conductivity, and its external effect is again that of an image current, but reversed, summing the contributions of the actual eddy-current distribution. In neither (a) nor (b) is there any eddy loss.

In practice both kinds of mass have finite conductivity, which impairs the simple image concept and introduces loss. Further, the permeable mass has saturation. If these 'defects' are to be included, an analysis of the leakage field and of the eddy-current loss in the mass becomes exceedingly complicated, and in any case the initiating leakage field is due to a distribution of current-carrying conductors and not to a single isolated current filament. Several attempts, e.g. by Stoll and Hammond [17] and by Mullineux and Reed [18], have been made to analyse various aspects of the problem. The solutions are complex, but can be computed without much difficulty for specific cases. As might be expected, the results depend on the relation between the permeability of the mass and the eddy-current skin depth (a function of the same variables as α above), and may approach closely to the two idealized extreme cases in Fig. 3.6.

The general conclusion is that the eddy loss in a permeable mass can be significantly reduced by facing it with a relatively thin layer of material of good conductivity, such as copper or aluminium.

3.3 *Electromotive force*

The useful e.m.f. in the windings of a transformer or machine are obtained by application of the Faraday–Lenz law, eq.(1.1). Because of the two ways in which linkage can change, we can write

$$N\frac{d\Phi}{dt} = N\left[\frac{\partial\Phi}{\partial\alpha} \cdot \frac{d\alpha}{dt} + \frac{\partial\Phi}{\partial t}\right] = e_r + e_p \tag{3.7}$$

a form which allows for variation of linkage due to both causes occurring together. The first term refers to the change of flux linking a coil moving at angular speed $\omega_r = d\alpha/dt$ relative to the flux, the distribution of the flux at the instant considered being expressed as $\partial\Phi/\partial\alpha$. The second term refers to flux pulsation.

Although the turns of a winding are arranged in coils having a variety of shapes and (in machines) are located in slots in the airgap periphery, it is initially helpful to idealize the conditions. The assumptions made here are:
 (i) the structure is 2-polar;
 (ii) the distribution of the flux around the airgap is sinusoidal;
(iii) the flux alternates sinusoidally in time;
(iv) the turns of a winding, although finite in number, are considered to be uniformly distributed and of full diametral pitch;
 (v) there is complete magnetic and electric symmetry.
The effects of slotting and of non-sinusoidal flux distribution, which are matters of practical importance, are considered in detail in Sect. 3.12.
Flux. The stator can be cylindrical (*a*) or salient (*b*), Fig. 3.7. The maximum

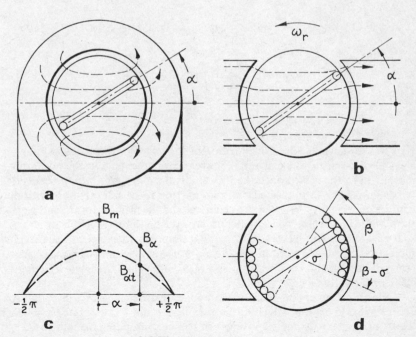

3.7 Electromotive force.

instantaneous flux per pole is Φ_m, which corresponds to a peak gap flux density $B_m = \Phi_m/Dl$ at the pole-centre ($\alpha = 0$) and a density $B_\alpha = B_m \cos \alpha$ at an angle α to the pole centre. The flux alternates at angular frequency ω and is taken as $\Phi = \Phi_m \sin \omega t$. The instantaneous gap flux density at angle α is therefore

$$B_{\alpha t} = B_\alpha \sin \omega t = B_m \cos \alpha \cdot \sin \omega t$$

as shown in Fig. 3.7(c). The instantaneous flux linking a single-turn full-pitch coil in a plane making the angle α to the pole centre is

$$\Phi_\alpha = \int_{-\alpha}^{+\alpha} B_{\alpha t} \cdot \tfrac{1}{2}Dl \cdot d\alpha = Dl B_m \sin \omega t \cdot \sin \alpha = \Phi_m \sin \omega t \cdot \sin \alpha$$

Coil E.M.F. Eq.(3.7) can now be applied to obtain the e.m.f. in a concentrated coil of N_c turns rotating at angular velocity ω_r and momentarily situated at angle α: it is

$$e_c = \omega_r N_c \Phi_m \sin \omega t \cdot \cos \alpha + \omega N_c \Phi_m \cos \omega t \cdot \sin \alpha \qquad (3.8)$$

Winding E.M.F. A winding, Fig. 3.7(d), is made up of m coils in series, uniformly spread over an angle σ and making $mN_c = N$ turns in all. The winding e.m.f. is the sum of the e.m.f.s of coils making various angles α:

$$e = \Sigma(e_c) = \left[\omega_r N \Phi_m \sin \omega t \cdot \frac{1}{\sigma} \int_{\beta-\sigma}^{\beta} \cos \alpha \cdot d\alpha \right.$$

$$\left. + \omega N \Phi_m \cos \omega t \cdot \frac{1}{\sigma} \int_{\beta-\sigma}^{\beta} \sin \alpha \cdot d\alpha \right] \qquad (3.9)$$

$$= e_r + e_p$$

Here β is the angle between the position of the 'start' of the phase winding and the pole-centre. The coil e.m.f. has two components: e_r is the result of movement of the coil through the flux and is the *e.m.f. of rotation*, associated (when current flows in the coil) with interaction force and therefore with the conversion of energy between the electrical and mechanical forms; and e_p is the *e.m.f. of pulsation*, concerned with energy transfer between rotor and stator windings. Eq.(3.9) is general, in that it applies to any electrical machine subject to the stated idealizations. Clearly e_r is a maximum when the phase-*centre* lies in the flux axis, while e_p is a maximum when the phase-*axis* coincides with the flux axis.

A classification of common flux and winding arrangements is given in Sect. 1.17. As it is immaterial which of the members 1 and 2 is the stator and which the rotor, we take the more convenient form in considering Cases A to G, Fig. 3.8.

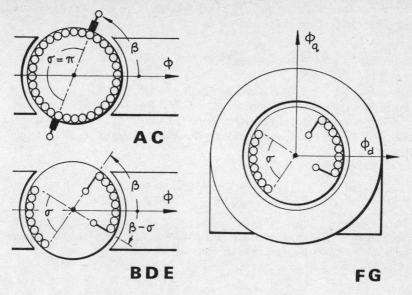

3.8 Basic winding arrangements in rotating machines.

Fixed-axis flux

CASE A: *Constant stator flux, rotor with commutator winding.* In a commutator winding there is, between brushes, a constant number of turns occupying always the same position with respect to the flux axis, making the brush e.m.f. have the same frequency as that of flux pulsation, in this case zero frequency. The e.m.f. is therefore unidirectional and due only to rotation. The N turns of the rotor are divided into two equal parallel sets each of $\frac{1}{2}N$ in series. The angle of spread is $\sigma = \pi$, so that eq.(3.9) reduces to

$$e = e_r = \omega_r \tfrac{1}{2} N \Phi \frac{1}{\pi} \int_{\beta-\pi}^{\beta} \cos \alpha \cdot d\alpha = \omega_r N \Phi \frac{1}{\pi} \sin \beta$$

With the rotor brush axis set at $\beta = \frac{1}{2}\pi$, i.e. at right-angles to the flux axis, then $\sin \beta = 1$. The e.m.f. can be written

$$e = E_r = 2nN\Phi$$

where $n = \omega_r/2\pi$ is the rotor speed (in rev/s) This case represents that of the common d.c. machine.

CASE B: *Constant stator flux, rotor with phase winding.* Again the flux is constant and there is no pulsational e.m.f. From eq.(3.9)

$$e = e_r = \omega_r N \Phi \frac{1}{\sigma} \int_{\beta-\sigma}^{\beta} \cos \alpha \cdot d\alpha$$

The integral is $[\sin \beta - \sin(\beta - \sigma)]$. It is a maximum for $\beta = \frac{1}{2}\sigma$ and the corresponding peak e.m.f. of the phase winding is

$$e_{rm} = \omega_r N\Phi \left[\frac{2}{\sigma} \sin \frac{\sigma}{2} \right] = \omega_r N\Phi k_d$$

where $k_d = (2/\sigma) \sin (\sigma/2)$ is the distribution factor. The turns, being distributed over the angle σ, cannot all link the same flux at a given instant, and the distribution factor is a measure of the reduction of the average linkage. Values of k_d for a number of common phase-spread angles are given in the Table:

Phase-spread, σ	$\pi/3 = 60°$	$\pi/2 = 90°$	$2\pi/3 = 120°$	$\pi = 180°$
Distribution factor k_d	0.955	0.905	0.827	0.636

The phase e.m.f. alternates at angular frequency $\omega_r = 2\pi f_r$. Its instantaneous value is $e = \omega_r k_d N\Phi \cos \omega_r t$ and its r.m.s. value

$$E = \sqrt{2} \pi f_r k_d N\Phi \qquad (3.10)$$

This case is applicable to the synchronous machine.

CASE C: *Alternating stator flux, rotor with commutator winding.* This is the condition for the a.c. commutator machine. Both modes of e.m.f. induction are present and it is necessary to include both terms of eq.(3.9). Remembering that the number of series-connected turns is $\frac{1}{2}N$ and that the angular spread of the winding is $\sigma = \pi$, the e.m.f. for a flux $\Phi = \Phi_m \sin \omega t$ is

$$e = \omega_r \tfrac{1}{2}N \Phi_m \sin \omega t \cdot \sin \beta - \omega \tfrac{1}{2}N \Phi_m \cos \omega t \cdot \cos \beta = e_r + e_p$$

The e_r component has the flux frequency ω and is in phase with the flux; its magnitude is proportional to speed ω_r. The presence of e_p means that, in a closed rotor winding, energy could be transferred to the rotor from the stator through the common magnetic field by transformer action.

CASE D: *Alternating stator flux, rotor with phase winding.* The e.m.f. is given by eq.(3.9) as it stands. Using the distribution factor given for case B, the rotor-phase-winding e.m.f. can be written

$$e = k_d N\Phi_m [\omega_r \cdot \sin \omega t \cdot \cos \omega_r t + \omega \cdot \cos \omega t \cdot \sin \omega_r t]$$

We can express the difference of the angular frequencies of flux pulsation and of rotor rotation as $\omega - \omega_r = \omega_2$, and define $\omega_2/\omega = s$ as the fractional *slip*.

Thus $\omega_2 = s\omega$ and $\omega_r = (1 - s)\omega$. With these substitutions, the e.m.f. of the rotor phase winding is

$$e = \tfrac{1}{2}k_d N \Phi_m \; [(2\omega - \omega_2) \sin(2\omega - \omega_2)t - \omega_2 \sin \omega_2 t]$$

$$= \tfrac{1}{2}k_d N \Phi_m \omega [(2 - s) \sin (2 - s)\omega t - s \sin s\omega t] \qquad (3.11)$$

The terms in the brackets involve sinusoidal variations at frequencies defined by $2\omega - \omega_2 = \omega + \omega_r$ and $\omega_2 = \omega - \omega_r$. The effect is as if the single *pulsating* flux Φ_m were divided into equal parts $\tfrac{1}{2}\Phi_m$ *rotating* in opposite directions at synchronous speed ω. The armature runs at an angular speed ω_r differing by ω_2 from one of these rotating fluxes, and therefore by $2\omega - \omega_2$ from the other. The induced e.m.f. can be considered as the combination of the e.m.f.s induced by the two fluxes acting separately. This equation forms the basis of one theory of the single-phase induction machine.

CASE E: *Alternating primary flux, stationary secondary with phase winding.* The 1-ph induction regulator takes the form shown in Fig. 3.8E; as there is no sustained rotation the motional e.m.f. component is zero and only the pulsational term e_p of eq.(3.9) appears. Its magnitude is adjustable by changing the axis of the rotor phase winding. Writing $\gamma = \sigma - \tfrac{1}{2}\beta$ for the position of the phase centre, the instantaneous rotor e.m.f. is

$$e = e_p = \omega k_d N \Phi_m \cos \omega t \cdot \sin \gamma$$

and the r.m.s. value is

$$E = \sqrt{2}\pi f k_d N \Phi_m \sin \gamma$$

The static transformer belongs to this class of machine. The secondary coils ideally link the whole primary flux, the airgap is no longer necessary, and the e.m.f. is

$$e = e_p = \omega N \Phi_m \cos \omega t$$

with the r.m.s. value

$$E = \sqrt{2}\pi f N \Phi_m = 4.44 f N \Phi_m \qquad (3.12)$$

Travelling-wave flux
CASES F and G: A travelling-wave field is produced by a balanced m-phase excitation, where m is normally 2, 3 or 6. Under steady-state ideal conditions such a field can be represented by direct- and quadrature-axis fluxes respectively given by

$$\Phi_d = \Phi_m \sin \omega t \quad \text{and} \quad \Phi_q = - \Phi_m \cos \omega t$$

The e.m.f. of a winding can then be obtained by summing the results of applying eq.(3.9) for Φ_d and a comparable equation for Φ_q, with due regard to the relative orientations of winding and flux axis. The expressions obtained are naturally lengthy, but they all reduce to the r.m.s. form

$$E = (1/\sqrt{2})\omega' k_d N \Phi_m \tag{3.13}$$

where ω' depends on the angular speed of the phase considered relative to the angular speed of the travelling wave (Ref.[1]).

3.4 *Transformer windings*
The basic coil forms are shown in Fig. 2.8, where H and L refer to the high- and low-voltage sides respectively. The constructional details depend on rating and voltage, the forms in common use for the normal core-type transformer being given in the Table:

Transformer Windings					
Service	Rating MVA	H.V. winding kV	type	L.V. winding kV	type
Distribution	up to 1	11–33	foil, cross-over or multilayer	0.43	helix
System	1–30	33–66	disc	11	disc or helix
Trans-mission	30 and upward	132–500	disc or multilayer	11, 33, 66	disc or disc-helix
Generator	30 and upward	132–500	disc or multilayer	11–22	disc-helix

Various coil designs are illustrated in Fig. 3.9.

Helix. This simple coil consists of a single layer over the full axial winding length, the conductor being a single section or a number of strands in parallel. At each end of the coil a wedge-shaped packing is required.

Disc-helix. When the current rating demands many parallel conductors the individual strips can be assembled in a radial pack, either as a single column or as two columns in parallel.

Multilayer helix. This can be used for the h.v. windings of large transformers. The augmented capacitance between layers gives advantage in surge-voltage distribution, but it is difficult to provide the long, thin layers of conductors with adequate strength to sustain axial forces.

Cross-over. A cross-over or bobbin coil is wound in sections on a former with side cheeks. It is similar to a multi-layer winding except that each section is short. Separate sections are connected in series, with horizontal cooling ducts provided by spacers. Interconnection is more usually made back-to-back and front-to-front with successive sections reversed.

Disc. The disc coil differs from other forms in that adjacent turns, consisting of strip conductors, are wound in a spiral from the centre outward, rather like a cross-over coil with only one turn per layer. To achieve the required disposition of turns the coils are formed in pairs, the requisite turns for one disc being loosely wound so that the conductor finishes in a position to provide the start of the inside turn of the adjacent disc, which is then wound from inside outwards. The first disc is then rearranged in such a mannner that the start is located as an outside turn: it can then be tightened.

Continuous disc. Following the formation of one disc, the procedure is repeated without cutting the conductor, saving jointing and joint space — an advantage with an inner winding. For large currents the coils can be wound with multiple conductors, or constructed as separate parallel-connected pairs. Alternatively two half-windings of reversed direction can be stacked, with the line connection taken from the centre of the stack.

Foil. Aluminium could be used for any of the forms mentioned above, but is uniquely employed in foil windings because it can be rolled to thinner and more flexible sheet than copper. The sheet is used in bobbin-type coils of one turn per layer.

Insulation

Classes A, B, C, F and H are all used in "dry" (i.e. non-oil-immersed) transformers, the silicone-treated materials being advantageous because of their water-repellent property. For oil-immersed transformers the coil insulation is

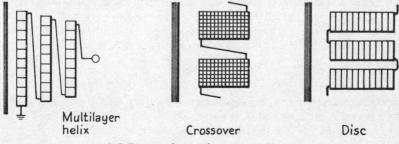

Helix   Disc-helix

Multilayer
helix   Crossover   Disc

3.9 Types of transformer windings.

a Class A material, generally paper or a derivative thereof. Cotton is confined to small units, but the use of synthetic enamels for interturn insulation is increasing, and may be extended to the l.v. windings (up to about 17.5 kV) of larger units.

3.5   D.C. field windings

Synchronous machines require d.c.-excited field windings, which may be placed on salient-pole cores or in the distributed slots of a cylindrical rotor. In synchronous-induction motors a 3-ph winding on the rotor is arranged for d.c. excitation in the synchronous mode.

Salient-pole windings

The features of coil design are the voltage, required m.m.f., coil dimensions, conductor material, and thermal dissipation. For a given coil voltage V and current I the resistance of a winding of N turns each of mean length L_{mt} is given by $R = V/I = \rho L_{mt} N/a$; whence the cross-sectional area a of the conductor is

$$a = \frac{\rho L_{mt} IN}{V} = \frac{\rho L_{mt} F}{V} \tag{3.14}$$

for a given m.m.f. $F = NI$. The current cannot be settled until the cooling conditions are known. Assuming a current density J, the current is $I = Ja$ and the number of turns is $N = F/I$. The conductor area totals aN so that the gross cross-section of the coil is $A_w = aN/k_s$ where k_s is the space factor. Thus for a given coil voltage, mean length of turn and conductor resistivity the conductor area is directly proportional to the m.m.f. required.

If the thermally emitting surface S of the coil can dissipate a power c per unit area and per degree rise of temperature above ambient, the loss $P = VI$ of the coil under steady sustained conditions gives a surface temperature-rise $\theta_m = cP/S$. On intermittent or short-time rating the rise is a function of the time-constants (heating and cooling).

As the field voltage of a salient-pole a.c. machine can usually be chosen to suit the requirements, a value permitting strip-on-edge conductors, Fig. 3.10, is selected to give mechanical strength and effective cooling.

EXAMPLE 3.1: Each rotor pole of a 20-pole synchronous machine has the dimensions given in Fig. 3.10 and is required to develop an m.m.f. $F = 9\ 000$ A-t. The field working voltage is 80 V (i.e. 4 V per pole); the mean length of turn is estimated to be $L_{mt} = 1.25$ m. For a copper winding at 75 °C the conductor area is

$$a = 0.021 \times 10^{-6} \times 1.25 \times 9000/4 = 60 \times 10^{-6}\ \text{m}^2 = 60\ \text{mm}^2$$

The outside surface area is approximately $S = L_{mt} h_p = 1.25 \times 0.15$ m; the cooling coefficient is estimated to be $c = 0.019$ °C m^2/W; whence for a temperature-rise of 65 °C the field coil can dissipate

$$p = \theta_m S/c = 65 \times 1.25 \times 0.15/0.019 = 620\ \text{W}$$

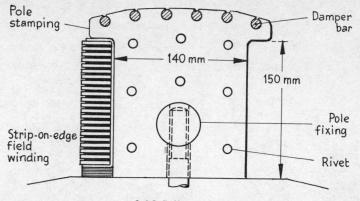

3.10 Salient pole.

The field current is therefore $I = 620/4 = 155$ A and the number of turns per pole is $N = 9\,000/155 \simeq 60$. With 60 turns of 2 mm x 30 mm strip wound on edge, insulation of 0.3 mm between turns and 12 mm at the coil ends, the total coil depth is 150 mm. The current density is $J = 155/60 = 2.6$ A/mm^2.

Non-salient-pole windings

Centrifugal force in turbogenerator rotors inhibits the use of salient poles. The field winding is therefore fixed in slots milled axially into the surface of a solid steel forging. Figure 3.11(a) shows one pole-pitch of a 4-pole rotor and a slot detail. For small medium-speed generators the form (b) may be adapted from the rotor stampings of a slip-ring induction motor, the pole-centre slots being left empty except for the bars of a damper winding. The arrangement provides a more nearly sinusoidal no-load gap flux distribution, an improved on-load waveform, and a better transient response in respect of 'voltage dip' and recovery time.

3.6   *A.C. armature windings*

The a.c. armature winding develops e.m.f.s in a number of *phases* when associated with a heteropolar magnetic field. The e.m.f.s are normally equal in magnitude and correctly displaced in time-phase relationship. The winding is composed of conductors in slots spaced around the periphery of the airgap, connected together at the ends, and grouped to form separate phase windings. When excited the phase windings in combination should produce an m.m.f. that is, as far as possible, sinusoidally distributed in the gap.

Conductors, turns and coils

Consider the armature conductors moving in the constant-flux field systems in Fig. 3.12. Each conductor, of length l, moving at peripheral speed u in a flux density B has an induced e.m.f. $e = Blu$. As the conductor passes N and

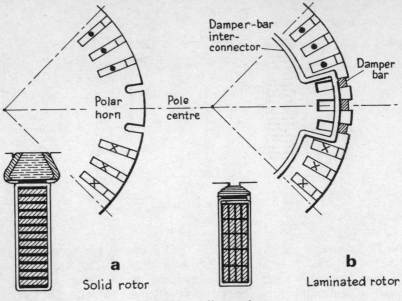

a
Solid rotor

b
Laminated rotor

3.11 Non-salient poles

S poles successively, its e.m.f. alternates. Diagram (*a*) shows a double pole-pitch of a machine in which the spatial distribution of the gap flux density over the circumference is sinusoidal. A conductor, *1*, in the position shown has an e.m.f. proportional to the density *B* in which it is moving. A second conductor, *2*, displaced a pole-pitch from *1*, moves in a flux density of the same magnitude but opposite polarity. The instantaneous conductor e.m.f.s e_1 and e_2 are therefore equal but oppositely directed. Let the conductors be joined together at one end to form a single turn: then the e.m.f.s e_1 and e_2 are additive round the turn and sum to the turn-e.m.f. e_t plotted in (*a*) in accordance with the position of conductor *1*. The flux linkage of the turn is also shown: comparison with e_t confirms eq.(1.1).

Fig. 3.12(*b*) is for a rectangular distribution of the airgap flux density. The turn-e.m.f. has a similar form. In an actual distributed phase winding the summation of turn-e.m.f.s gives a phase-e.m.f. more nearly sinusoidal.

Connections. The two conductors are series connected to form a turn. A coil may have more than one turn. The connections joining the conductors form the end-connectors or, in the mass, the *overhang* or *end-winding*. When, as in Fig. 3.12, the coil-sides are spaced one pole-pitch apart they are said to be of *full pitch*. The coil pitch may be less than a pole-pitch, in which case it is described as *short-pitched* or *chorded*.

Windings
The slots in which the coils lie are normally punched uniformly around the gap surface. The coils are then arranged and connected to form phase groups.

In a *single-layer* winding each coil-side occupies a whole slot, whereas in a *double-layer* winding one coil-side lies in the upper position in a first slot with the other occupying the lower position in a second slot spaced one pole-pitch (or less) from the first.

Phases

The requirements of an a.c. winding are that it shall produce a symmetrical *m*-phase system of e.m.f.s of identical magnitude, frequency and waveform,

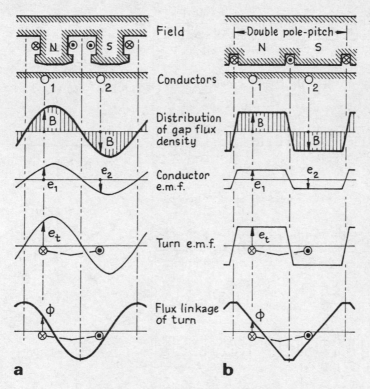

3.12 Rotational e.m.f. in a single turn.

displaced in time-phase by $\beta = 2\pi/m$ elec. rad. This is secured by providing identical phase groups arranged with their axes displaced by $2\pi/m$. The most usual arrangement is that in which the spread of each phase is $\sigma = \pi/m$, i.e. $1/m$ of a pole-pitch. The result is a sequence of phase bands A C B A C B in the double pole-pitch of a 3-ph machine.

In Fig. 3.13, an array of conductors in twelve slots is set out at (*a*) in a diagrammatic development of a double pole-pitch. The individual conductor e.m.f.s are shown on the right as a symmetrical group of phasors. At (*b*) a 3-ph grouping into phase-bands is made. Each band has a 120° spread, the summation of conductor e.m.f.s yielding three phase-e.m.f.s with a time-phase displacement of $2\pi/3$ rad. It will be seen that the phase e.m.f.s are less than the arithmetic sum of the conductor e.m.f.s as a result of spreading the winding. In (c) is shown a grouping for six phases with 60° spread. Grouping (b) could not be used with a single-layer winding for three phases, nor could (c) be used for six, because the subdivision of the winding leaves no further free conductors a pole-pitch apart to form the return conductors. But (c) could be used for *three* phases if D, E and F were employed as the completion of

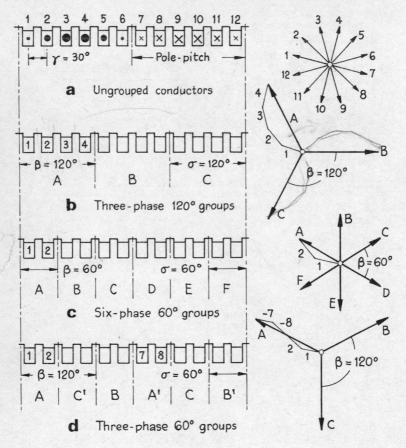

3.13 Phase grouping.

A, B and C. This is shown in (d), where the phase-bands are arranged in this way, and the phase sequence A, B, C obtained by re-lettering. The *phase-bands* follow in the order A, C′, B, A′, C, B′.

Connections between coil groups

The overhang comprises the end-connections which form the conductors into turns. In addition, the links made at the connection end of a winding comprise

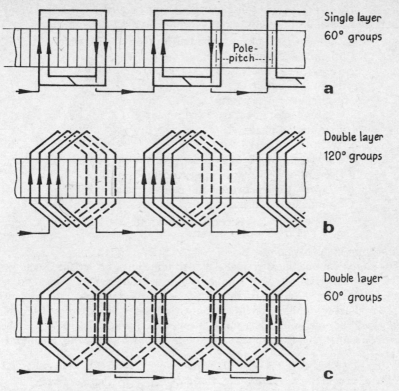

Single layer
60° groups

a

Double layer
120° groups

b

Double layer
60° groups

c

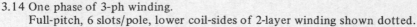

3.14 One phase of 3-ph winding.
Full-pitch, 6 slots/pole, lower coil-sides of 2-layer winding shown dotted.

(*a*) coil-to-coil connectors, (*b*) phase-group connectors, and (*c*) parallelling
connectors. When windings are required for large currents and low e.m.f.s,
the groups will not generally be connected in series round the whole of a
multipolar armature, but will be split up into sections each containing the
same number of similar coil-groups: the sections are then paralleled by suitable connectors.

Examples of coil and group connectors are given in Fig. 3.14, for part of
one phase of three typical 3-ph windings, viz. (*a*) single-layer, 60° groups;
(*b*) double-layer, 120° groups; and (*c*) double-layer, 60° groups. A winding
of type (*c*) in a sixteen-pole machine, for example, could have one, two, four,
eight, or even sixteen parallel circuits per phase by making provision for the
appropriate paralleling connectors. The number of circuits is equal to the
number of poles divided by the number of groups per circuit.

Nomenclature
A double pole-pitch covers 2π electrical radians. A machine with p pole-pairs
has $2\pi p$ electrical radians in its periphery. A distinction has therefore to be

made between mechanical (physical) and electrical angle. The following symbols are used in dealing with winding technology:

S total number of slots on a uniformly-slotted armature;

C number of coils;

e e.m.f. of a conductor;

e_t e.m.f. of a turn;

p number of pole-pairs;

$\gamma = 2\pi p/S$ slot pitch, electrical radians;

m number of phases;

$g = S/2p$ number of slots per pole;

$g' = S/2pm$ number of slots per pole and per phase;

σ = spread of a phase-group, electrical radians;

$\beta = 2\pi/m$ angle between e.m.f.s of successive phases, electrical radians.

S_{ab}, S_{ac} numbers of slots between given points AB, AC.

3.7   *Single-layer windings*

For more than one phase, a single-layer winding must be designed to allow the end-connections to be accommodated in separate tiers or planes. For a 3-ph winding, Fig. 3.15, the overhang may be arranged in two or in three planes.

Single-layer windings fall into two main classes: (i) *unbifurcated* windings in which the coils comprising a pair of phase groups in adjacent pole-pitches are concentric, and (ii) *bifurcated* windings in which each group is split into

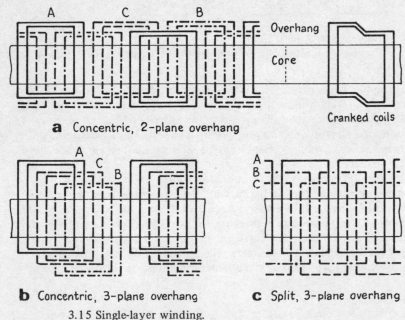

a Concentric, 2-plane overhang

b Concentric, 3-plane overhang   **c** Split, 3-plane overhang

3.15 Single-layer winding.
Two slots per pole and phase, narrow spread

two sets of concentric coils, each set sharing its return coil-sides with those of another group in the same phase. The windings (a) and (b) in Fig. 3.15 are unbifurcated (*concentric*), the former being a continuous and the latter a broken chain arrangement. This affects the number of overhang planes needed to accommodate the end-windings. The winding (c) is bifurcated (*split concentric*) type, requiring a three-plane overhang.

Single-layer coils may be of "hairpin" shape, Fig. 3.18(a), pushed through semi-closed slots, but the subsequent jointing is difficult and expensive. A single-layer constant-span coil arrangement is sometimes employed for small induction motors, using circular-section insulated wire for the conductors. This is a *mush* winding. Each coil is first wound on a former making one coil side shorter than the other. In winding, the wires are dropped individually into the slots, first the short side (the wires emerging beneath the long sides of other coils in adjacent slots), then the long sides. Round the armature, long and short sides occupy alternate slots, and the coils are of constant pitch with a whole number of slots per pole and per phase.

The number of coils in a single-layer winding is one-half the number of slots available, because each coil side completely occupies one slot: thus $C = \frac{1}{2}S$.

3.8 Double-layer windings

The armatures of nearly all synchronous generators and motors, and of most induction motors above a few kilowatts, are wound with double-layer windings. Windings with a whole number of slots per pole (g an integer) are called *integral-slot* windings. It is common, however, to find g fractional, and such arrangements (in which the number of slots S is not a multiple of the number of poles $2p$) are termed *fractional-slot* windings. These have a number of advantages. In all cases the number of coils is equal to the number of slots: i.e. $C = S$

Integral-slot windings

In Fig. 3.16(a, b) some basic arrangements are shown for integral-slot 3-ph windings. The letters indicate conductors to correspond with the phases ABC. Letters arranged vertically designate conductors in the same slot. The angular distance between successive slots is the angular slot-pitch γ in electrical radians. The coils are former-wound, and all have the same span.

Fig. 3.16(a) shows a winding with $g = 9$ slots per pole, $g' = 3$ slots per pole and phase, and full-pitch coils. The slot-pitch is $\gamma = 2\pi/18$ rad, or $20°$. Each phase-band spreads over 3 slot-pitches so that the phase-spread is $\sigma = 60°$. The conductors b are arranged $120°$ and conductors c $240°$ to the right of a, giving $\beta = 2\pi/3 = 120°$ for the angle between successive phase positions and for their corresponding e.m.f.s. The coils being of full pitch of π rad or $180°$, each slot contains coil-sides belonging to one phase only.

In order to reduce the end-connections and to reduce or suppress certain harmonics in the phase-e.m.f.s, the coils may be *chorded* or *short-pitched*. Such a winding is shown in Fig. 3.16(b). It has $g = 9$ and $g' = 3$, also $\beta = 120°$ as before, but the coil-span is 8 slots instead of 9 as in (a), giving a coil span (or coil-pitch) of 8/9 of a pole-pitch, i.e. $160°$. Certain slots now hold coil-

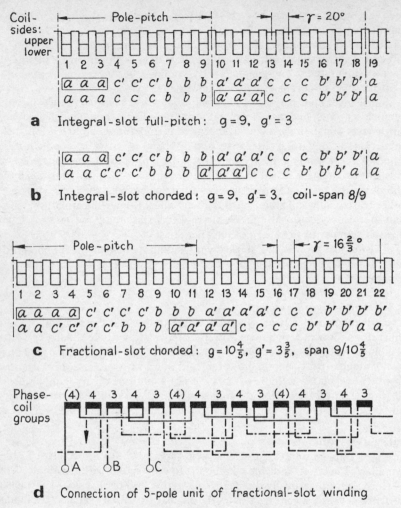

a   Integral-slot full-pitch: $g = 9$, $g' = 3$

b   Integral-slot chorded: $g = 9$, $g' = 3$, coil-span 8/9

c   Fractional-slot chorded: $g = 10\frac{4}{5}$, $g' = 3\frac{3}{5}$, span $9/10\frac{4}{5}$

d   Connection of 5-pole unit of fractional-slot winding

3.16 Double-layer winding.

sides of different phases. Chording reduces the phase-e.m.f., and the coil-span is rarely made less than $\frac{2}{3}$ pole-pitch because additional turns become necessary which offset the saving of overhang material.

Fractional-slot windings

These have non-integral slots per pole g and slots per pole and phase g': the value of g' is usually an improper fraction such as $\frac{7}{5}$, or in general $S'/2mp'$. In a unit of $2p'$ poles there are S' slots in all, and S'/m slots for each of the m phases, with S'/m integral. The fraction $\frac{7}{5}$ thus represents 7 slots per phase spread over 5 pole-pitches. In a 3-ph machine this means 21 slots for every unit of 5 poles, 42 for 10 poles and so on. All phase windings must be arranged for electrical and magnetic balance, with an exact $120°$ angle between phase e.m.f.s.

A fractional-slot winding is shown diagrammatically in Fig. 3.16(c) for one double pole-pitch. It will be seen that the coils cannot have a full span of $180°$, and that the phases are unbalanced. The balance is, of course, secured over the complete unit of poles ($2p' = 5$ in this case), after which the whole layout repeats for the next unit of poles, $2p'$, if any.

The coil-span that can be used has considerable freedom. That in Fig. 3.16(c) has a span of 9 slots in $g = 10\frac{4}{5}$ slots per pole; the equivalent $g' = 3\frac{3}{5}$ slots per pole and phase, although this number is not actually found, it representing only the average number in the 5 poles of the unit.

3.9 *Fractional-slot windings*

The apparent complication just described has theoretical advantages, reinforced by desirable simplifications in manufacture. These are: (a) the simple but effective reduction of slot harmonics, and (b) the use of a total number of armature slots that is not necessarily a multiple of the number of poles. Thus a particular number of slots for which notching gear exists may be used for a range of machines running at different speeds; this is of especial value where a wide range of numbers of poles may be called for, as with synchronous machines.

Fractional slotting is practicable only with double-layer windings. It limits the number of available parallel circuits because phase groups under several poles must be connected in series before a unit is formed and the winding repeats the pattern to give a second unit that can be paralleled with the first.

The simplest cases of fractional-slot windings are those for which g is an integer plus one-half, which employ alternate groups of coils differing by 1. Thus with $g = 7\frac{1}{2}$ slots per pole (or $g' = 2\frac{1}{2}$ slots per pole and phase) in a 3-ph winding with a $60°$ angular spread, alternate phase-groups will contain respectively 2 and 3 slots, and the two phase-groups, under two successive poles, connected in series will constitute the basic repeatable unit. The number of units is determined by the total number of poles $2p$.

As mentioned in Sect. 3.8, when g' is expressed as the fraction $S'/2mp'$ reduced to its lowest terms, S'/m represents the number of slots and coils per phase in the basic unit of $2p'$ poles. Thus for $g' = 2\frac{3}{4} = 11/4$, there are 11 coils per phase distributed among 4 successive poles, and the four groups may contain 3, 3, 3 and 2 coils in series. The basic unit of $2p'$ poles must naturally be a factor of the total number of poles $2p$, and if a is the number of parallel circuits required, the product $2ap'$ must also be a factor of $2p$.

EXAMPLE 3.2: (a) A 16-pole 3-ph machine with 108 slots and a $60°$ phase spread will have

$$g' = S/2pm = 108/16 \times 3 = 2\frac{1}{4} \text{ slots per pole and per phase.}$$

The phase groupings will be 3, 2, 2 and 2 coils in the basic unit of 4 poles. All 4-pole units may be series connected, or there may be 2 or 4 circuits in parallel.

(b) If a machine with 108 slots has 10 poles, then

$$g' = 108/10 \times 3 = 3\frac{3}{5} \text{ slots per pole and per phase}$$

the groupings being 4, 4, 4, 3 and 3 coils covering a basic 5-pole-pitches. Only one circuit is possible with both units in series, or one pair only of parallel circuits.

Symmetry

Fractional-slot windings must fulfil the requirements of e.m.f. balance. In each phase the group-sequence must be the same, and the groups must occupy slots so selected as to obtain the required phase-angle $\beta = 2\pi/m$.

Let $S_{ab}, S_{ac} \ldots$ be the number of slot-pitches displacement between a given slot in phase A and similar slots in phases B, C. . . . Then, with an angular slot-pitch $\gamma = 2\pi p/S$ electrical radians (i.e. the total electrical angle in the p pole-pairs divided by the total number of slots)

$$S_{ab}\gamma = 2\pi/m \text{ or } (2\pi/m) + x\,\pi$$

gives the position of the required slot in phase B, with x having any positive integral value. For the former expression the "polarity", i.e. the positive direction of the e.m.f., of the group in which the phase B slot lies will be the same as that in phase A. The same is true of the second expression if $x = 0, 2, 4, 6 \ldots$; but if $x = 1, 3, 5 \ldots$ the polarities will be opposite. This must be allowed for in making phase connections. In a similar way

$$S_{ac}\gamma = 4\pi/m \text{ or } (4\pi/m) + x\pi$$

and so on for the remainder of the m phases.

As already indicated, a fractional-slot winding may divide itself naturally into basic units of $2p'$ poles containing S' slots each, so that the values $S_{ab}, S_{ac} \ldots$ will be appropriate within each unit. Using for convenience the slots per pole and phase $g' = S/2pm = S'/2p'm = \pi/m\gamma$, the slot spacings are

$$S_{ab} = g'(mx + 2), \qquad S_{ac} = g'(mx + 4) \ldots$$

These always give whole numbers when an appropriate value of x is chosen except when $2p'$ is a multiple of m. Hence when $2p'$, in the improper fraction $g' = S'/2p'm$ reduced to its lowest terms, is a multiple of the number of phases, a balanced winding is not possible.

In 3-ph windings ($m = 3$) the grouping of phase B is S'/m slots from that of phase A when $2p' = 2, 5, 8, 11 \ldots$, and the phase C grouping is $2(S'/m)$ slots from phase A. When $2p'$ is 4, 7, 10, 13 . . . these displacements are interchanged.

When the starts (or, depending on the polarity, the finishes) of the various phases have been determined, it remains to fill in a succession of similar groups in each phase from these points. All subsequent slots in each phase will then be correctly located. The sequence of slot groups, e.g. 4, 4, 4, 3, 3 slots in Example 3.2(b), is determined in such a way as to give the greatest symmetry in the winding. One 5-pole unit of the winding might be tabulated as follows—

Phase A	(+4)		−4		+4		−3		+3	
Phase B		+3		(−4)		+4		−4		+3
Phase C	−4		+3		−3		(+4)		−4	
Pole-pitch	1		2		3		4		5	

The figures indicate the number of slots containing the *upper* coil-sides in each group with its appropriate polarity, those numbers in brackets being inserted first from the expressions given for S_{ab} and S_{ac}. A layout of the winding is drawn in Fig. 3.16(d). The diagram of two pole-pitches is shown in (c), the lower coil-sides in the left-hand pole-pitch corresponding to the arrangement of the upper coil-sides of pole-pitch 5 in the Table.

Coil span

When the grouping of the upper layer of coil-sides has been decided, it follows that the groupings of the corresponding lower coil-sides are determined also, for these have the same group sequence but displaced from the upper groups by the *winding pitch* or *coil-span*. The coil-span must be the same for all coils in the winding, and because g is fractional the span cannot be a full pitch. Thus with $g = 7\frac{1}{2}$ slots per pole the coils might span from slot 1 to slot 9 (8 slot pitches), or 1 to 8 (7 slot pitches): the former is greater and the latter is less than the pole-pitch of $7\frac{1}{2}$ slot-pitches. The span chosen is almost invariably short rather than long. With $g = 7\frac{1}{2}$, a span of 7 slots is 89% of full pitch, while 6 slots gives 80% of full pitch. Chording has technical advantages in the reduction of overhang copper, and of e.m.f. and m.m.f. harmonics.

3.10 *Types of double-layer winding*

A coil in a double-layer winding represents the entire set of conductors in one slot layer in association with the similar set in the other layer of another slot. The number of coils is therefore the same as the number of slots: $C = S$.

The end connections of a double-layer winding may have a variety of shapes, depending on the method employed to bring about the transition from the upper to the lower slot position. Basically there are only two types, viz. *lap*- and *wave-connected*. The coils may have one turn or several. In the former case single-turn coils require only one conductor in each layer, but this will generally be laminated to facilitate bending in the end-windings and to reduce eddy-current losses. These coils form *bar-windings,* and are commonly found in low-voltage machines with large phase currents. The choice of coil depends on the number of turns per phase required to generate the e.m.f., the current rating, and the number of parallel circuits obtainable.

Fig. 3.17 shows (a) a pair of single-turn (bar) lap coils, (b) two multi-turn

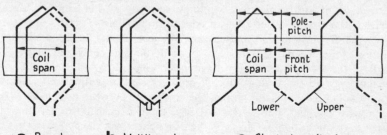

a Bar lap   **b** Multiturn lap   **c** Single-turn (bar) wave

3.17 Double-layer chorded lap and wave coils.

(bar) lap coils, and (c) two bar wave coils. Multi-turn wave coils are not used, because lap coils are more easily connected.

The connection of lap coils in groups is as indicated in Fig. 3.14(b) and (c). The wave-connection of bars is not so straightforward, since after one complete tour round the armature (i.e. after series connection of p coils) the winding would naturally close on to the start of the first coil. To avoid this, a connection is made to continue the winding into the second coil-side of the first group, and so on until all coil-sides of *alternate* groups in each layer are fully wound. Another independent wave winding is made to utilize the remaining half of the phase. The two parts can then be connected in series or parallel to form the complete phase winding.

When fractional-slot wave windings are used, the problem is complicated by the fact that isolated coils are left after the full number of complete turns round the armature has been made, necessitating special connectors.

3.11 *Choice of winding*
In the following items specifying a 3-ph winding, those features italicized are the most common:

   (a) Type of coil: concentric, *lap,* wave.
   (b) Overhang: *diamond,* multiplane, involute, mush.
   (c) Layers: single, *double.*
   (d) Slots: *open*, closed, semi-closed.
   (e) Connection: *star, mesh.*
   (f) Phase-spread: *60°*, 120°.
   (g) Slotting: integral, *fractional.*
   (h) Coil-span: full-pitch, *chorded.*
   (j) Circuits: series, or parallel.
   (k) Coils: single-turn, multiturn.

Multiturn coil windings. Machines with small numbers of poles and low values of flux per pole require a relatively large number of turns per phase: this is true also of high-voltage machines. Multiturn coils are needed, and the choice will lie between the double-layer lap winding and the single-layer winding. The former is dropped into open slots in all cases where good slot insulation is essential, although in small low-voltage machines the conductors may be fed through a narrowed slot-opening into a slot-liner which serves as the main insulation. Single-layer windings are rarely used in large machines except with partially-closed slots and push-through hairpin coils.

In general it may be said that modern practice favours the double-layer winding except where the slot-openings would be large compared with the length of the air-gap, as in high-voltage induction motors.

Bar windings. Where single-turn coils are necessary, as with turbo-generators or multipolar low-voltage machines, the choice lies between double-layer bar lap or wave windings. Both types may have the bars pushed through partially-closed slots and be bent to shape at the other end when the conductor section is moderate, but the heavy bars used in turbo-alternators must be completely formed before being inserted into open slots.

The chorded bar lap winding has the advantage of a shorter overhang at the connection end than the bar wave, and is more suited to cases where several parallel circuits are needed.

Coil making

Fig. 3.18(*a*) gives a view of a single-layer coil for a concentric winding, indicating its general shape before being pushed through a pair of slots. At (*b*) is shown stages in the production of a multiturn double-layer diamond coil of

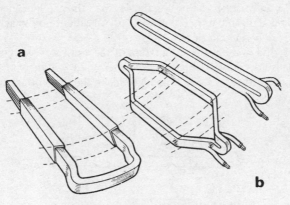

3.18 Single- and double-layer coils.

the *pulled* type. When such a coil is not too heavy it may be wound over two dowels as a flat loop, and the slot portion then gripped in a machine and pulled apart to the required coil span. More massive double-layer coils have to be built up on formers, and the overhang portion may be bent up to make an angle with the slot parts, to form an *involute* shape. When assembled on the stator the overhang presents a conical surface suitable for bracing.

Slotting

In very small machines, using round wires of the smaller gauges, the tooth shape is improved if tapered slots are used, Fig. 3.19. In all but these, parallel-sided slots are employed. The slot-openings serve to avoid excessive slot-leakage,

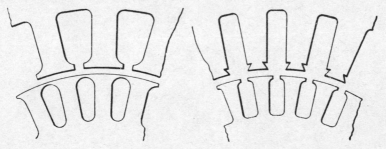

3.19 Types of slot.

but they cause an effective lengthening of the magnetic path across the air
gap, and tunnel slots are occasionally used for small induction motors. Open
slots (i.e. with the opening as wide as the slot) must be used where it is desired
to complete the coils outside the armature and drop them into the slots.

3.12 *E.M.F. of windings*

The e.m.f. generated in basic windings under idealized conditions was considered
in Sect. 3.3: these conditions were (i) a sinusoidal distribution of gap flux density, and (ii) a uniform spread of conductors over the gap periphery. The flux is
never exactly distributed as in (i), particularly in salient-pole machines, and the
windings (ii) are bedded in slots.

The e.m.f. of a phase is the summation of the e.m.f.s of slot conductors.
Such a phase is normally associated with other phase windings in star to give
a resultant line voltage, the effect being to make the line voltage more nearly

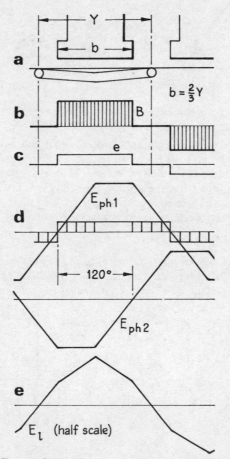

3.20 E.M.F. waveform due to rectangular gap-field distribution.

sinusoidal. The effects of e.m.f. summation and of phase interconnection are explained in Fig. 3.20, taking as an example a salient-pole machine with a pole-arc b two-thirds of the pole-pitch Y. The full-pitch coil in (a) is then associated with the rectangular gap-density distribution (b), and the e.m.f. in each coil-side has the waveform (c). A group of coils spread over $60°$ gives the phase e.m.f. E_{ph1}. (The stepping due to the location of coil-sides in separate slots is here ignored.) This phase e.m.f. subtracted from that of the associated phase, E_{ph2}, gives the line-to-line e.m.f. E_l in (e).

It was shown in Sect. 2.5 that a non-sinusoidal gap-density distribution could be expressed as a harmonic series

$$B = B_1 \sin \theta + B_3 \sin 3\theta + \ldots + B_n \sin n\theta$$

A machine can thus be considered to have $2p$ "fundamental" poles, together with families of $6p$, $10p \ldots 2np$ "harmonic" poles, all individually sinusoidal and all generating e.m.f.s in an associated winding. The magnitude of the fundamental e.m.f. is determined by (i) the fundamental flux, (ii) the phase spread and (iii) the coil-span. The harmonic e.m.f.s are similarly determined by (iv) the harmonic fluxes, (v) and (vi) the effective electrical phase spreads and coil-spans, and (vii) the method of interphase connection. There is a further effect, however: the slotting imposes local changes in the gap reluctance, so that besides the harmonic content of B, there is a superposed *slot-ripple*. We deal first with the response of a uniformly distributed winding to fundamental and harmonics in the wave of gap flux density, and then with the slot-ripple.

Uniformly distributed winding
A full-pitch coil of one turn, moving through a fixed fundamental sine-distributed flux Φ_1 at a speed corresponding to frequency f (as in Case B of Sect. 3.3) develops an e.m.f. of r.m.s. value $E_t = \sqrt{2}\pi f \Phi_1$. In a phase winding of N such turns spread uniformly over an electrical angle σ, the e.m.f. in each turn is E_t but its time-phase depends on its position in the winding. The total phase-winding e.m.f. E_{ph} of the N turns in series is not NE_t but is the *phasor* sum of the individual turn e.m.f.s, approximating the chord in Fig. 3.21(a). Then

$$\frac{E_{ph}}{NE_t} = \frac{\text{phasor sum}}{\text{arithmetic sum}} = \frac{\text{chord}}{\text{arc}} = \frac{\sin \frac{1}{2}\sigma}{\frac{1}{2}\sigma} = k_{d1}$$

where k_{d1} is the breadth coefficient or distribution factor for the fundamental of the gap flux.
Distribution factor. For space harmonics in the wave of flux density the 'harmonic poles' are narrower and the phase spread σ covers more of the wavelength: in fact, for the nth harmonic the effective phase spread is $n\sigma$. Hence in general for a harmonic flux of order n the distribution factor is

$$k_{dn} = \frac{\sin \frac{1}{2}(n\sigma)}{\frac{1}{2}(n\sigma)} \tag{3.15}$$

Compare (*a*) and (*b*) in Fig. 3.21 for the same phase spread: (*a*) is for the fundamental ($n = 1$) and (*b*) for the 5th harmonic ($n = 5$). The distribution factor for harmonics is generally different from that of the fundamental. If $\sigma = 2\pi/m$ where *m* is the number of phases in a multiphase winding, then although there may be a flux harmonic there will be no corresponding phase e.m.f. harmonic for the order $n = m$. The Table below gives values of fundamental and low-order harmonic distribution factors with the more usual phase spreads.

Uniformly distributed winding: values of k_{dn}

No. of phases *m*	Phase-spread σ	k_{d1}	k_{d3}	k_{d5}	k_{d7}	k_{d9}
3 or 6	60°	0.955	0.637	0.191	−0.136	−0.212
3 or 1	120°	0.827	0	−0.165	0.118	0
2	90°	0.900	0.300	−0.180	−0.129	0.100
1	180°	0.637	−0.212	0.127	−0.091	0.071

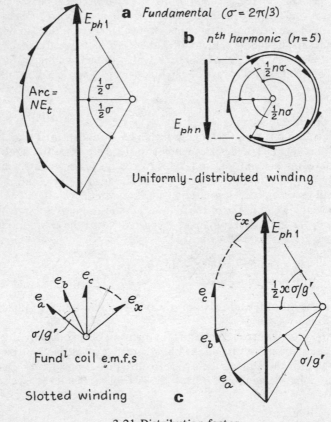

a *Fundamental* $(\sigma = 2\pi/3)$

b *n^{th} harmonic* $(n=5)$

Uniformly-distributed winding

Fund^l coil e.m.f.s

Slotted winding   **c**

3.21 Distribution factor

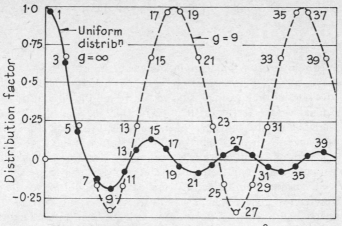

3.22 Distribution factor for $\sigma = 60°$

The distribution factors for the common phase spread $\sigma = 60°$ are plotted (full line and dots) in Fig. 3.22.

Coil-span or chording factor. If the span of a coil is not a full pole-pitch, but is chorded to cover $(\pi \pm \epsilon)$ electrical radians, the flux that it links is reduced by the factor $\cos \frac{1}{2}\epsilon$ for the fundamental, as shown by (*a*) and (*b*) in Fig. 3.23: its e.m.f. is reduced in proportion. In the case of harmonic flux components of order n, the effective chording angle is $n\epsilon$, so that the general coil-span or chording factor is

$$k_{en} = \cos \tfrac{1}{2}(n\epsilon) \tag{3.16}$$

Fig. 3.23(*c*) illustrates the effect on a 3rd harmonic of a chording angle of $60°$, i.e. a coil-span of $120°$. Such a coil covers two harmonic pole-pitches and links zero 3rd harmonic flux, so that 3rd harmonic e.m.f.s disappear from the phase e.m.f. In general, when a coil is pitched short or long by π/n, $3\pi/n \ldots$ no harmonic of order n survives in the coil or phase e.m.f.s.

Slotted winding

For an integral-slot winding there are g' slots/pole per phase, and individual coil e.m.f.s are represented by phasors displaced by the angle σ/g', as in Fig. 3.21(*c*) for the fundamental. A little trigonometry shows that the general distribution factor is given by

$$k_{dn} = \frac{\sin \tfrac{1}{2}(n\sigma)}{g' \sin \tfrac{1}{2}(n\sigma/g')} \tag{3.17}$$

For a fractional-slot winding g' is given by $S'/2mp'$, and a layout of a $2p'$-pole unit contains S'/m coils in S'/m slots per phase distributed evenly over a spread σ. The distribution factor is

$$k_{dn} = \frac{\sin \tfrac{1}{2}(n\sigma)}{(S'/m) \sin \tfrac{1}{2}(n\sigma)(m/S')} \tag{3.18}$$

For example, a 3-ph winding ($m = 3$) with a spread of $\sigma = 60°$ and $g' = 2\frac{1}{2}$ slots/pole per phase has

$$g' = S'/2mp' = 5/2$$

The unit comprises $2p' = 2$ pole-pitches in which there are $S'/m = 5$ coils or slots per phase and $S' = 15$ slots altogether. The effective angular displacement of the coils (and of their e.m.f.s) is $60°/5 = 12°$, and the distribution factor for the fundamental is $k_{d1} = \sin 30°/5 \sin 6° = 0.956$.

The variation of the permeance of the gap over a succession of slots and teeth may generate corresponding gap-flux harmonics in the wave of flux density, particularly if the number of slots per pole g is integral. Thus if there are 9 slots per pole ($g = 9$) and therefore 18 slots per pole-pair, there will be in the gap an 18th harmonic of flux density modulated in amplitude by the poles, giving the effect of constant-amplitude flux harmonics of orders 18 ± 1, that is a 17th and a 19th. Now it can be shown that the distribution factors for harmonics of orders n are the same as those of orders $6Ag' \pm n$, where A is any integer $0, 1, 2 \ldots$ Thus with $g = 9$ and $m = 3$, giving $g' = 3$, then for $A = 1$ we find that the distribution factors k_{d17} and k_{d19} are both equal to k_{d1}. Unless precautions are taken, such as skewing or the substitution of fractional for integral slotting, these harmonics will appear strongly in the phase e.m.f. The effect is shown by the circles (linked for clarity by the dotted line) in Fig. 3.22.

Coil-span factor. The appropriate factor is given by eq.(3.16). Fig. 3.24 gives a plot of k_{en} for harmonics up to the 15th. For example, for a coil-span of two-thirds of a pole-pitch, $\epsilon = \pi/3 = 60°$, the factor $k_{e1} = 0.866$. For the 3rd harmonic $k_{e3} = 0$; thus all 3rd (and *triplen*) harmonics are eliminated from the coil and phase e.m.f.s as we have seen already. The triplen harmonics in a 3-ph machine are normally eliminated by the phase connnection, and it is usual to select the coil-span to reduce the 5th and 7th harmonics instead. A pitch of $150°$ corresponding to a chording angle $\epsilon = 30°$ is useful in this respect, as it gives the following factors:

Harmonic order n	Fund.	3	5	7	9
Coil-span factor k_{en}	0.966	0.707	0.259	0.259	0.707

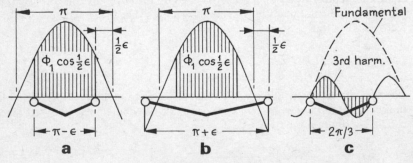

3.23 Chording.

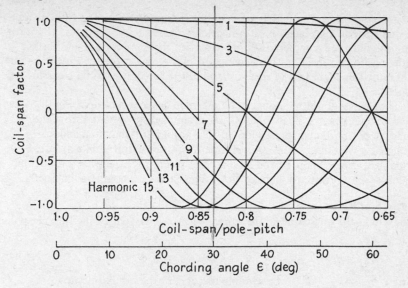

3.24 Chording (coil-span) factors.

Slotless Windings

Enclosure of active conductors in slots, a century-old practice from which the conventional modern machine has evolved, transfers interaction force from the relatively weak conductors to the teeth that flank them, and enables windings to be securely held in the slots by wedges. But there are drawbacks: the main airgap flux is concentrated into the teeth (which cover only about one-half of the periphery) leading to tooth saturation, higher magnetizing m.m.f., greater core loss, and greater leakage flux and reactance in the slot conductors, which have to be sectionalized to limit eddy-current loss.

A large modern turbogenerator has a long radial airgap — e.g. 150 mm — for reasons of stability. Using this space, it is possible to eliminate stator slotting altogether, and to fix the winding to the smooth stator bore while still leaving adequate rotor clearance. Two of the possible winding arrangements that have been proposed are shown in Fig.3.25. In (*a*) the whole stator winding, constructed and completed externally to the machine, is slid bodily into place and fixed suitably to the stator bore. In (*b*) the winding is

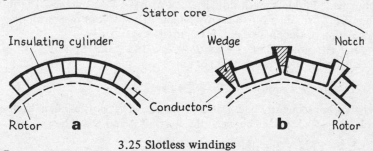

3.25 Slotless windings

constructed in sections which are held in position by wedges to form an arch-bound structure. In each case it is necessary subsequently to thread the rotor carefully into the completed stator. Although there are formidable design problems to be solved, both mechanical and thermal, savings are made in core and insulating materials, in core loss and in the avoidance of the core-slotting process.

The slotless stator winding can be regarded as a conventional winding free of slotting restraints. The winding factors are readily evaluated from the coil-side positions and spans adopted. The form in (*a*) may demand a concentric winding in order to facilitate its insertion; but unlike a slotted winding it can be chorded and so spaced as to improve the waveform of the induced e.m.f. Conventional and slotless machines are compared in Sect.12.7.

Phase E.M.F.
The e.m.f. of a phase due to the fundamental component of the flux per pole is

$$E_{ph1} = 4.44fk_{d1}k_{e1}N_{ph}\,\Phi_1 = 4.44fK_{w1}N_{ph}\,\Phi_1 \tag{3.19}$$

where $K_{w1} = k_{d1} \cdot k_{e1}$ is the winding factor. For the nth harmonic

$$E_{ph\,n} = 4.44nfK_{wn}\,\Phi_n \tag{3.20}$$

where $K_{wn} = k_{dn} \cdot k_{en}$. A more useful form is obtained from an analysis of the gap flux density distribution, Sect. 2.5; for $\Phi_1 = (2/\pi)B_1\,Yl$ and $\Phi_n = (2/\pi)B_n(Y/n)l = \Phi_1(B_n/nB_1)$. The nth harmonic and fundamental e.m.f. components are now related by

$$\frac{E_{ph\,n}}{E_{ph\,1}} = \frac{B_nK_{wn}}{B_1K_{w1}}$$

The r.m.s. phase e.m.f. is

$$E_{ph} = \sqrt{(E_{ph\,1}{}^2 + E_{ph\,3}{}^2 + \ldots + E_{phn}{}^2)}$$

In all ordinary cases this is indistinguishable from $E_{ph\,1}$. The line e.m.f. in star connection is $\sqrt{3}\,E_{ph}$ calculated with all triplen harmonics omitted.

EXAMPLE 3.3: A 3.75 MVA 10 kV 3-ph 50 Hz 10-pole low-speed hydro-generator has 144 slots containing a two-layer diamond winding with 5 conductors per coil-side in each slot. The coil-span is 12 slot-pitches. The flux per pole is 0.116 Wb, distributed as in Example 2.1. A layout of the winding and a calculation of the phase e.m.f. is required.

Winding. The slot angle is $\gamma = 2\pi p/S = 10\pi/144 = 0.218$ rad $= 12\frac{1}{2}°$. The fractional value of slots per pole per phase is

$$g' = S/2mp = 144/3 \cdot 10 = 24/5 = 4\tfrac{4}{5}$$

There will consequently be 24 slots per phase, or 72 slots in all, in each of two 5-pole units. The spacing between the starts of A and B will be

$$S_{ab} = g'(3x + 2) = 4\tfrac{4}{5}(3 \cdot 1 + 2) = 24 \text{ slots}$$

and for C

$$S_{ac} = g'(3x + 4) = 4\tfrac{4}{5}(3 . 2 + 4) = 48 \text{ slots}$$

To show the arrangement of coil groups, 5, 5, 5, 5 and 4 for each phase, the Table p. 102 can be drawn up. The slots are numbered 1, 2, 3 . . . 72 in order, with their angles $0°$, $12\tfrac{1}{2}°$, $25°$ . . ., values above $180°$ being reduced by $180°$ or integral multiples thereof. Coil-sides for phase A are then allocated to slots with angles from $0°$ up to (but not including) $60°$; phase B from $120°$; and phase C from $60°$. The letters a, b and c in square brackets are those located from the values of $S_{ab} = 24$ and $S_{ac} = 48$ already calculated; i.e. slots 1 for A, 25 for B and 49 for C.

The sequence of groups, from each starting point, is seen to be 5, 5, 5, 4, 5. The slots occupied by phase A cover the angle $0°-57\tfrac{1}{2}°$ in steps of $2\tfrac{1}{2}°$, so that the winding is uniformly spread with an angle of $2\tfrac{1}{2}°$ between coil e.m.f.s A winding Table, including polarities, is given below —

Phase A	[+5]		−5		+5		−4		+5	
Phase B		+5		[−5]		+5		−5		+4
Phase C	−5		+4		−5		[+5]		−5	
Pole-pitch	1		2		3		4		5	

The lower coil-sides will lie in slots 12 slot-pitches to the right of their corresponding upper coil-sides.

E.M.F.: Example 2.1 shows that the total flux per pole of 0.116 Wb can be accepted as Φ_1. The 3rd harmonic flux is ignored as it does not affect the line e.m.f. The coil-span is $12/(3 \times 4\tfrac{4}{5}) = 5/6$ pole-pitch or $150°$, so that the chording angle is $\epsilon = 30°$. From eqs.(3.18) and (3.16)

$$k_{dn} = \frac{\sin \tfrac{1}{2}(n60°)}{24 \sin \tfrac{1}{2}(n2\tfrac{1}{2}°)} \quad \text{and} \quad k_{en} = \cos \tfrac{1}{2}(n30°)$$

giving the winding factors $K_{wn} = k_{dn} \times k_{en}$:

$$K_{w1} = 0.956 \times 0.966 = 0.925 \qquad K_{w5} = 0.191 \times 0.259 = 0.049$$

and $\quad K_{w7} = (-0.136)(-0.259) = 0.035$

There is a material reduction in the harmonic factors. The e.m.f.s from eqs.(3.19) and (3.20) are

$$E_{ph1} = 4.44 \times 0.925 \times 50 \times 240 \times 0.116 = 5\ 750 \text{ V}$$
$$E_{ph5} = 5\ 750\ (0.049/0.925)(11.2/100.6) = \quad 34 \text{ V}$$
$$E_{ph7} = 5\ 750\ (0.035/0.925)(2.8/100.6) = \quad 6 \text{ V}$$

The harmonic e.m.f.s are quite negligible, so that the phase e.m.f. is $E_{ph} = 5.75$ kV, giving $E_l = 10$ kV with star connection.

Graded and interleaved windings
Excessive harmonics lead to vibration and noise, additional stator and rotor

Slot:	1	2	3	4	5	6	7	8	9	10	11	12	13	14	15
Angle:	0	12½	25	37½	50	62½	75	87½	100	112½	125	137½	150	162½	175
Phase:	[a]	a	a	a	a	c	c	c	c	c	b	b	b	b	b
Slot:	16	17	18	19	20	21	22	23	24	25	26	27	28	29	
Angle:	7½	20	32½	45	57½	70	82½	95	107½	120	132½	145	157½	170	
Phase:	a	a	a	a	a	c	c	c	c	[b]	b	b	b	b	
Slot:	30	31	32	33	34	35	36	37	38	39	40	41	42	43	44
Angle:	2½	15	27½	40	52½	65	77½	90	102½	115	127½	140	152½	165	177½
Phase:	a	a	a	a	a	c	c	c	c	c	b	b	b	b	b
Slot:	45	46	47	48	49	50	51	52	53	54	55	56	57	58	
Angle:	10	22½	35	47½	60	72½	85	97½	110	122½	135	147½	160	172½	
Phase:	a	a	a	a	[c]	c	c	c	c	b	b	b	b	b	
Slot:	59	60	61	62	63	64	65	66	67	68	69	70	71	72	
Angle:	5	17½	30	42½	55	67½	80	92½	105	117½	130	142½	155	167½	
Phase:	a	a	a	a	a	c	c	c	c	c	b	b	b	b	

loss, and (in induction motors) to parasitic torques. If a winding had a sinusoidal distribution of conductors per slot there would be no harmonics other than those caused by slotting. In special cases such a *graded* winding, with more conductors per coil in the phase centre than at the ends, can be used for space-harmonic reduction.

Another way, not so difficult to build, gets something of the same effect by *interleaving*. The end coils of each phase group are interchanged with the corresponding end coils of adjacent groups. A 3-ph winding with 5 slots/pole per phase in normal and in interleaved (or interspersed) form would appear as below, the letters indicating coil-sides belonging respectively to phases A, B and C:

Normal: $b'\ a\ a\ a\ a\ a\ c'\ c'\ c'\ c'\ c'\ b\ b\ b\ b\ b\ a'\ \ldots$
Interleaved: $a\ b'\ a\ a\ a\ c'\ a\ c'\ c'\ c'\ b\ c'\ b\ b\ b\ a'\ b\ \ldots$

The result is a distribution reducing the low-order harmonics, and for this example the distribution factors are

Harmonic order	Fund.	5	7	11	13	17
Normal	0.957	0.200	0.149	0.110	0.102	0.102
Interleaved	0.915	0.0	0.118	0.256	0.289	0.289

The fundamental is 5% reduced, but the 5th is zero. The 7th, somewhat reduced, could be chorded out. With a 2/3 pitch the overall design could actually be advantageous, especially for large machines in which stray load losses are important.

3.13 *Tooth harmonics*

In addition to the e.m.f.s generated in the windings by those space harmonics of flux which move at the same velocity as the fundamental flux relative to the conductors, certain harmonic voltages of a particularly undesirable order may be produced by the effect of the openings of the slots in which the conductors themselves are located. This effect is aggravated when the slot-openings are large compared with the tooth width and gap length, because these affect the air-gap permeance, to variations of which the harmonics are due. The ripples in the flux wave caused by variation of permeance from point to point in the air-gap are of the form indicated in the oscillogram of Fig. 3.26(*a*): but it is important to note that, since the ripples are due to slotting, they do not move with respect to the conductors but glide over the flux-distribution curve, always opposite the slots and teeth which cause them.

Pulsation of Total Gap Permeance. If the main flux wave were assumed to be rectangular (as in a salient-pole machine without fringing) and the ripple flux a pure sine-wave superposed on it, the particular case in which the pole-arc covers an integral-plus-one-half slot-pitches would give a total flux varying between the two extremes shown in Fig. 3.26(*b*) and (*c*), totalling ± the flux

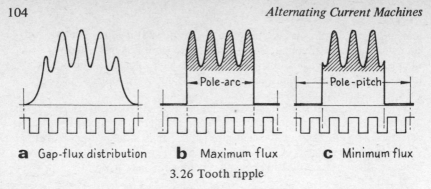

a Gap-flux distribution   **b** Maximum flux   **c** Minimum flux

3.26 Tooth ripple

contained in one loop of the ripple. The frequency of flux pulsation would correspond to the rate at which the slots cross the pole-face, i.e. $2gf$. Now this stationary pulsation may be regarded as two waves of fundamental space-distribution rotating at angular velocity $2g\omega$, one forwards and one backwards, where $\omega = 2\pi f$ corresponds to synchronous speed. The component fields will have velocities of $(2g \pm 1)\omega$ relative to the armature winding and will generate therein harmonic e.m.f.s of frequencies $(2g \pm 1)f$. But this is only a secondary cause of tooth-ripple.

Rotor Current Pulsation. The prime cause of tooth-ripple is the induction in rotor circuits of currents of frequency $2gf$. These flow in suitably-pitched damper bars in the poles of a salient-pole machine, in the brass or bronze wedges of a turbo rotor, and (to a minor extent) in the main rotor field windings.

The gap permeance at the stator surface has space undulations, and the variations from the average may be represented approximately by a sine fluctuation $\lambda_1 \sin 2g\theta$. The fundamental rotor m.m.f. acting at the stator bore rotates with respect to the stator slots and may be written, for a given point on the stator, as $F_1 \sin(\theta - \omega t)$. The flux density at the stator surface will have a fundamental corresponding to the average permeance, together with a superposed ripple due to the fluctuation given by

$$\lambda_1 F_1 \sin 2g\theta \ . \ \sin(\theta - \omega t)$$

Resolving this into its two oppositely-rotating components gives

$$\tfrac{1}{2}\lambda_1 F_1 \left\{ \cos[(2g - 1)\theta + \omega t] - \cos[(2g + 1)\theta - \omega t] \right\}$$

With respect to the stator these ripple fluxes vary with position in a slot-pitch, and with time at angular frequency ω: thus they cannot contribute any e.m.f. at tooth-ripple frequency.

The same two waves have, however, different velocities relative to the rotor. Any point designated by θ on the stator will require to be designated by $\theta + \omega t$ with reference to the rotor. Making this substitution gives

$$\tfrac{1}{2}\lambda_1 F_1 \left\{ \cos[(2g - 1)\theta + 2g\omega t] - \cos[(2g + 1)\theta + 2g\omega t] \right\} \ .$$

As these flux waves have a velocity $2g\omega$ relative to the rotor, they will generate currents of frequency $2gf$ in any closed rotor circuits of suitable pitch. Such currents will superpose an m.m.f. variation of $2gf$ on the resultant pole m.m.f. Resolving these again into forward- and backward-moving components of $2g\omega$ relative to the *rotor*, and therefore $2g\omega \pm \omega = (2g \pm 1)\omega$ relative to the *stator*, then stator e.m.f.s at frequencies $(2g + 1)f$ and $(2g - 1)f$ will be generated. These constitute the principal tooth ripples.

In the above only the fundamentals of the slot-permeance variation λ_1 and of the rotor m.m.f. F_1 have been considered. Further ripple effect may occur due to the harmonics of each. Quite objectionable ripple may be generated by the second harmonic of the slot permeance.

The suppression of slot ripples is accomplished by: (1) skewing the stator core; (2) displacing the centre-line of the damper bars in successive pole-shoes; (3) offsetting the pole-shoes in successive *pairs* of poles: these refer to the mitigation of ripple directly due to slot-permeance fluctuation. To counter the effects of harmonics in the field form, the methods are: (4) shaping the pole-shoes so that the maximum gap length is about twice the minimum, in salient-pole machines; and (5) the use of composite steel-bronze wedges for the slots of turbo rotors.

Tooth-ripple Loss

Pole-entry Loss. As a slot, moving with respect to a salient pole, enters the region of the pole-tip, rapid changes of flux occur which can give rise to parasitic and transient losses both in the slot conductors and in the tooth steel. The sudden change of slot-field pattern generates eddy currents and develops tangential forces on the teeth that may contribute an acoustic noise effect.

Pole-face Loss. The tooth-ripple flux causes surface pulsation loss in the pole-faces. An appreciable amount of this flux penetrates deeply into laminated pole shoes along the boundary surfaces of the (thick) pole-shoe punchings. Several analytical solutions to the problem of the surface loss produced have been suggested. One, by Greig and Freeman [19], when reduced and written in the symbolism of this book, gives the specific pole-face loss (power per unit surface area) as

$$p_e = \frac{B_t^2 y_s \omega_s}{8\pi\mu} \cdot k \tag{3.21}$$

in which B_t is the peak value of the fundamental of the ripple flux density of angular frequency ω_s, and y_s is the slot-pitch and μ the permeability of the pole-shoe material. The factor k depends on the kind of pole-shoe, and has the following values:

Solid pole-shoes: $k = \alpha(y_s/\pi)$

Laminated pole-shoes: $k = \frac{1}{3}(\alpha d)^2$ for $\quad 0 < (\alpha d) < 1.5$
                                  $k = \frac{1}{4}(\alpha d). f(\alpha d)$ for $\quad 1.5 < (\alpha d) < 5$
                                  $k = (\alpha d) \quad$ for $\quad 5 < (\alpha d) < 100$

where d is the thickness of a pole-shoe lamina, and α is from eq.(3.1) with

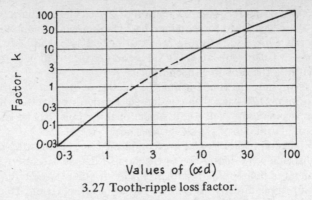

3.27 Tooth-ripple loss factor.

μ_0 replaced by μ and with $b/w_s = 1$. The function f(αd).is complicated, but can adequately be approximated by joining up the plots of k for the ranges 0–1.5 and 5–100, as indicated by the dotted line in Fig. 3.27. It is found that the second harmonic of the ripple flux density can add up to 25% loss: it can be evaluated from eq.(3.21) using appropriate values of ω_s and α.

Induction Machines. Large tooth pulsations can occur. Chalmers and Richardson [20] have given experimental results and compared them with formulae devised by Alger [21].

Skewed slots

Skew, as indicated in Fig. 3.28, is an angular 'twist' of a slot away from the axial direction by an electrical angle γ which is typically in the range 0.1–0.5 rad (5°–30°). Most cage induction motors have their rotor slots skewed by one slot-pitch to reduce those space harmonics in the airgap flux density that are introduced by the slotting. There are several consequences.

Winding Factors. A skewed rotor bar is, in effect, 'spread' over the angle γ and its induced e.m.f. is reduced in accordance with the winding factors

$$k_{d1} = (\sin \tfrac{1}{2}\gamma)/\tfrac{1}{2}\gamma \quad \text{and} \quad k_{dn} = (\sin \tfrac{1}{2}n\gamma)/\tfrac{1}{2}n\gamma$$

3.28 Skewing.

for the fundamental and the *n*th harmonic respectively of the gap-flux density distribution. Ideally, slot harmonics are eliminated because the skew gives an effective phase-spread of 2π rad at the slot-harmonic wavelength.

Reactance. Reduction of the bar e.m.f. is effectively a drop in the main-flux rotor linkage and a rise in the rotor leakage flux. Thus the rotor leakage reactance is increased.

Flux Distribution. Skew produces at each end of the axial length a displacement of $\frac{1}{2}\gamma$ between the stator and rotor m.m.f. patterns, which therefore "match" only in the mid region. The mismatch elsewhere causes a non-uniform axial distribution of gap-flux density by as much as 40%.

Axial Force. The skewed rotor-bar current has a circumferential component proportional to sin γ, developing an axially directed force which imposes an additional load on thrust bearings.

Losses. The non-uniform gap density increases core and rotor surface losses. Where (as is usual) the rotor bars are uninsulated, it is possible for currents to flow from bar to bar circumferentially, producing I^2R loss with no corresponding useful torque.

3.14 *Magnetomotive force of windings*

The airgap flux in a machine results from the combined m.m.f.s of all the windings, whether on the stator or on the rotor, that act on the gap region. We consider first the individual m.m.f.s of a number of typical windings.

Phase M.M.F.

The idealized m.m.f. of a uniformly distributed phase-type winding, considered as a uniform current sheet of spread σ, has the trapezoidal shape shown in Fig. 3.29(*a*). With N turns in the phase winding and a current i, the m.m.f. per pole is $\frac{1}{2}Ni$. A Fourier analysis of the trapezoidal waveform yields a series of harmonic sines of order $n = 1, 3, 5 \ldots$ in terms of the electrical angle x, each term having the form

$$F_n = \tfrac{1}{2}Ni \frac{4}{n\pi} \frac{\sin \tfrac{1}{2}(n\sigma)}{\tfrac{1}{2}(n\sigma)} = i \frac{2}{n\pi}(k_{dn}N)\sin nx$$

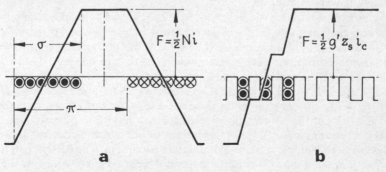

3.29 Phase m.m.f.

where k_{dn} is the distribution factor, eq.(3.15). If the winding is chorded the coil-span factor is introduced to give $k_{dn}k_{en}N = K_{wn}N$ for the effective turns of the phase winding.

With an alternating phase current $i = i_m \cos \omega t$, then each term of the Fourier series becomes

$$F_n = i_m \cos \omega t \frac{2}{n\pi}K_{wn}N \sin nx \qquad (3.22)$$

The m.m.f. wave rises, falls and reverses with the change of the current, but its axis remains fixed with respect to the winding.

The subdivision of practical windings into slots results in the 'stepped' wave of Fig. 3.29(b). The fundamental and low-order harmonics are scarcely affected, but slot harmonics become prominent in integral-slot windings. With g' slots/pole per phase each containing z_s conductors carrying a current i_c, the instantaneous amplitude of the m.m.f. wave is $\frac{1}{2}Ni = \frac{1}{2}g'z_si_c$.

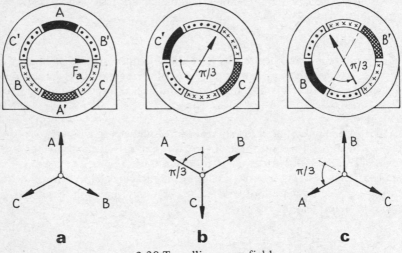

3.30 Travelling-wave field.

Travelling-wave M.M.F.

A symmetrical m-phase winding, with $m = 2, 3, 4 \ldots$ has its phase axes displaced successively around the airgap by an electrical angle $2\pi/m$, and is fed with a balanced set of m-phase currents. The resultant m.m.f. is nearly constant, and its axis moves over one double pole-pitch in the time of one period at almost constant speed. The very common 3-ph case ($m = 3$) is an example.

The stator in Fig. 3.30 carries a 3-ph narrow-spread 2-pole winding AA', BB', CC'. The current phasors for three successive instants are shown in (a), (b) and (c). Let the positive direction of current in ABC be outward (and in A'B'C' therefore inward). In (a) the current of phase A is a positive maximum, while in both B and C it is one-half negative maximum. The current pattern magnetizes along the axis of F_a. One-sixth of a period later, (b), phase C now has a maximum negative current, the whole current pattern shifting counterclockwise by 60°; a similar shift occurs after a further one-sixth of a period, as in (*c*). For all three instants the current pattern is identical in magnitude, and changes only in direction. The result is a travelling wave of m.m.f.

It would be incorrect to assume that the m.m.f. is constant between the 60° intervals, nor even that its angular rotation is at a constant speed; but it can be shown that the *fundamental* component is constant and moves at uniform speed.

Three-phase M.M.F.

Using eq.(3.22), we write the m.m.f.s of phases A, B and C with due regard (i) to the time-phase displacement of the currents, and (ii) to the space displacement of the phase-m.m.f. axes. The currents and angles are

Phase A: $i_m \cos \omega t$ nx
Phase B: $i_m \cos(\omega t - \frac{2}{3}\pi)$ $n(x - \frac{2}{3}\pi)$
Phase C: $i_m \cos(\omega t - \frac{4}{3}\pi)$ $n(x - \frac{4}{3}\pi)$

The resultant m.m.f. is the summation $F_a + F_b + F_c$ giving

$$F = \frac{3}{\pi} N i_m \left[K_{w1} \sin(x - \omega t) + \frac{1}{5} K_{w5} \sin(5x + \omega t) \right.$$
$$\left. + \frac{1}{7} K_{w7} \sin(7x + \omega t) + \frac{1}{11} K_{w11} \sin(11x + \omega t) \dots \right] \quad (3.23)$$

There is a constant fundamental, and harmonics of orders 5, 7, 11, 13, or $6a \pm 1$, where a is any positive integer. All multiples of 3 are absent. The fundamental travelling wave moves at a speed corresponding to ω, while the harmonics move with speeds proportional to the reciprocal of their order, in the same direction as the fundamental for $n = 6a + 1$, and opposite for $n = 6a - 1$.

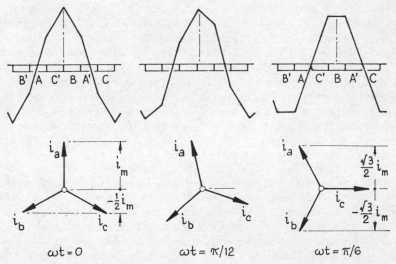

3.31 M.M.F. waveforms of travelling-wave field.

The m.m.f. wave naturally changes its shape from instant to instant, as exemplified by the three cases in Fig. 3.31 for a uniformly distributed 3-ph narrow-spread winding. For the extreme shapes we have

$$\omega t = 0: F = 2 \times \tfrac{1}{2}Ni_m \quad \text{and} \quad \omega t = \pi/6: F = \sqrt{3} \times \tfrac{1}{2}Ni_m$$

where $\tfrac{1}{2}Ni_m$ is the maximum m.m.f. of one phase.

Although the m.m.f. space harmonics are of practical importance it is sometimes permissible to consider the fundamental term alone. Its amplitude is $(6/\pi)\tfrac{1}{2}Ni_m$. With an r.m.s. current I_c in each phase conductor, and with a total of N_{ph} turns/ph all in series in a machine with p pole-pairs, the fundamental armature m.m.f. per pole is

$$F_a = (6/\pi)K_{w1}\tfrac{1}{2}g'z_s(\sqrt{2}I_c)$$
$$= 1.35\,K_{w1}N_{ph}I_c/p \tag{3.24}$$

Equivalent M.M.F.

The behaviour of a machine depends on the interaction of the m.m.f.s of its stator and rotor members. It is therefore necessary to be able to equate m.m.f.s that are contributed by different kinds of winding.

Two Three-phase Windings. For two 3-ph windings 1 and 2 it is only necessary to write eq.(3.24) for each, giving

$$K_{w1}N_{ph1}I_{c1} = K_{w2}N_{ph2}I_{c2}$$

where K_{w1} and K_{w2} are the *fundamental* winding factors for the respective windings.

Three-phase Winding and Distributed D.C. Winding. In a non-salient-pole synchronous machine (such as a turbo-generator) the field winding is distributed in slots, Fig. 3.11, giving the m.m.f. distribution (*b*) in Fig. 3.29. Apart from m.m.f. harmonics arising from the trapezoidal waveform, it is sufficient to equate the fundamentals of the field and armature m.m.f.s.

Three-phase Winding and Salient Pole System For salient-pole machines, Fig. 3.32, the method of equating m.m.f.s is not valid because the gap is not uniform over the pole-pitch. Equivalence has to be based on the fundamental fluxes that the 3-ph and salient-pole windings would separately produce in the airgap. Analysis of a salient-pole flux distribution is given in Sect. 2.5. The results so obtained, with the ratio b/Y of pole-arc to pole pitch between 0.65 and 0.75, give approximately

$$F_{ad} = 1.07\,K_{w1}N_{ph}I_c/p \tag{3.25}$$

for the pole excitation corresponding to the 3-ph winding current I_c which products m.m.f. in the *direct axis*, i.e. in the direction of the pole-centre. The component of 3-ph winding m.m.f. wave that is directed along the interpolar axis has no equivalent in the field winding. The m.m.f. component in this, the *quadrature axis,* is applied to the interpolar gap which presents a widely varying permeance, so that the quadrature-axis flux distribution has a large spatial third-harmonic. The usual star connection of the 3-ph winding prevents any

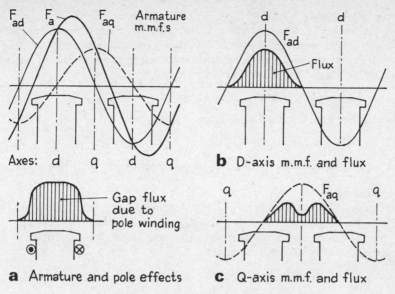

a Armature and pole effects   **c** Q-axis m.m.f. and flux

3.32 M.M.F. reactions in salient-pole machine.

corresponding 3rd-harmonic e.m.f.s from appearing at the line terminals. The fundamental of the q-axis flux does however generate an e.m.f. of fundamental frequency in the 3-ph winding, giving the q-axis reaction the nature of a leakage reactance. Evaluation of the q-axis flux fundamental from the profile of the airgap is based on flux-plotting or the application of gap-permeance coefficients.

Phase-sequence Components

The foregoing discussion relates to m.m.f. distributions produced by normal, positive-sequence currents. Under fault conditions, or unbalanced loading, or the deliberate application of asymmetrical voltages, currents of negative and (in certain cases) zero sequence will flow. Each will produce a corresponding m.m.f.

When the negative-sequence currents are present in the windings of a 3-ph machine, an m.m.f. is developed which rotates at synchronous speed in a direction opposite to that of the pole-system. Double-frequency e.m.f.s are consequently induced in the field winding by the negative-sequence flux. In a salient-pole machine the flux will not be constant, because of the strongly variable gap permeance presented to it as it passes in succession the direct and quadrature axes. This is especially true if the machine has laminated pole-faces and no damper windings, for the latter have a compensating effect tending to make the machine equivalent to a cylindrical-rotor construction. The effect is naturally accompanied by considerable pole-face and damper losses.

Zero-sequence currents, being co-phased in time in all three phases, produce normally no flux in the air-gap.

Two-phase Winding

A symmetrical 2-ph winding, AA' and BB', with N_{ph} turns in series per phase and with the winding axes separated by $\frac{1}{2}\pi$ elec. rad will develop a travelling wave of gap m.m.f. in a way closely similar to that in the 3-ph case, if its phase currents have also a 90° phase displacement, e.g.

$$i_a = i_m \cos \omega t \qquad \text{and} \qquad i_b = i_m \sin \omega t$$

The resultant m.m.f. has a fundamental of magnitude

$$F_a = 0.92 \, K_{w1} N_{ph} I_c / p \tag{3.26}$$

The harmonics are naturally rather more prominent than for a 3-ph winding.

Cage Winding

If the number of phases m in an m-phase winding is large, and if the phase currents are all of equal r.m.s. magnitude I_c with time-phase displacement $2\pi/m$ successively, the winding approximates to a sinusoidally distributed current sheet with the resultant m.m.f.

$$F_a = \sqrt{2}(m/\pi) K_{w1} N_{ph} I_c / p$$

A uniform cage winding with an adequate number of bars approximates closely to this condition. A cage in S slots (and therefore S/p slots per pole-pair) can be considered as a multi-phase winding with $m = S/2p$ phase windings each of a single turn. The winding factor $K_{w1} = k_{d1} k_{e1}$ is clearly unity. The fundamental of the m.m.f. becomes

$$F_a = \frac{1}{\sqrt{2}\pi p} S I_c \tag{3.27}$$

Harmonics of orders other than those close to $n = S/2p$ will normally be very small.

Winding Asymmetry

A symmetrical armature winding has all phases identical in layout, with their m.m.f. axes spaced at the proper electrical angle (120° for 3-ph, 90° for 2-ph). Such symmetry is important in large machines, but is not always readily attained. Asymmetry may arise in one of the winding re-connections for a 3-ph pole-change motor, and is inherent in 1-ph motors with auxiliary starting windings. In general, winding asymmetry results from (i) irregular space distribution of coils, and from differences in (ii) coil-span, (iii) slotting, (iv) numbers of coil turns, and (v) conductor sections. Some of the effects of winding asymmetry are discussed in Sect.12.3.

4
Loss Dissipation

4.1 *Dissipation of heat*

The processes of transfer in a transformer and interaction in a machine involve currents (and therefore I^2R loss) in conductors and fluxes (producing core loss) in ferromagnetic circuits. Stray leakage fluxes develop comparable losses in associated parts such as tanks, end-plates and covers, accounting for further 'load' loss. As all losses appear as heat, the temperature of each part so affected is raised above that of the ambient medium (normally the surrounding air). The temperature-rise above ambient is related (i) to the rate of heat production, (ii) the rate of cooling and (iii) the thermal capacity. The temperature-rise is paramount in determining the plant rating particularly through its effect on the life of the winding insulation, and is the subject of specified operating limits.

Heat is removed by a combination of *conduction* and *convection*, assisted by *radiation* from outer surfaces. Briefly, a cool fluid is passed through passages in a machine to conduct away the heat but a 'scrubbing' action of the coolant is relied upon for effective heat exchange. This is a kind of *forced convection*, whereas air-current dissipation from the outside surfaces is *natural convection* unless it is augmented by fans.

Convection by random air currents is mainly relied upon for a small machine, usually aided by rotor fans. For bigger machines the core volume is too large for purely surface cooling: radial and axial ducts must be provided to increase it, with means for directing the coolant through the labyrinth. The cooling system for the largest machines is complicated, but it is vital to the operating safety in view of the very large loss-rates.

4.2 *Ideal temperature-rise/time relation*

An ideal homogeneous body, internally heated and surface cooled, has a maximum surface temperature-rise proportional directly to the loss rate and inversely to that of the cooling. With simplifying assumptions and a constant rate of internal loss, the temperature-rise and the time are exponentially related. The symbols used are

p energy-loss rate, or loss power
G mass of body

113

c_p specific heat
S area of surface
λ emissivity
c cooling coefficient = $1/\lambda$
θ temperature-rise
τ time-constant

If the surface of the body is at some temperature θ above ambient, then in an element of time dt the heat energy $p \cdot dt$ is produced, a heat storage $Gc_p \cdot d\theta$ occurs due to an elemental rise dθ in temperature, and the heat $S\theta\lambda \cdot dt$ is dissipated. The energy balance is

$$Gc_p \cdot d\theta + S\theta\lambda \cdot dt = p \cdot dt$$

whence

$$\theta = \theta_m\,[1 - \exp(-t/\tau)] \tag{4.1}$$

where $\theta_m = p/S\lambda$ is the maximum temperature-rise, and $\tau = Gc_p/S\lambda$ is the *heating time-constant*. The basic assumption is that the dissipation is proportional to the temperature-rise above ambient.

EXAMPLE 4.1: A small totally-enclosed induction motor, of mass $G = 41$ kg, has on continuous full load a total loss $p = 110$ W. Its frame has a cooling surface $S = 0.10$ m^2 with an emissivity $\lambda = 29$ W/(m^{2}K). The specific heat of the machine averages $c_p = 420$ J/(kg K). Obtain the heating/time curve.

The final steady temperature-rise is $\theta_m = p/S\lambda = 110/0.10 \times 29 = 38\,^{\circ}$C and the time-constant is $\tau = Gc_p/S\lambda = 41 \times 420/0.10 \times 29 = 6\,000$ s (about 1.7 h). The temperature-rise/time relation is

$$\theta = 38\,[1 - \exp(-t/6\,000)]$$

The initial rate of rise is $\theta/\tau = 38/6\,000 = 0.0063\,^{\circ}$C/s, or about 23 $^{\circ}$C/h.

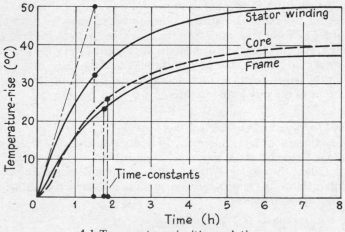

4.1 Temperature-rise/time relation.

The temperature-rise of the frame of the actual machine, taken by thermometer, is shown in Fig. 4.1, together with the core and winding temperatures. The frame temperature is seen to follow the expression above fairly closely, and comparable exponential relations apply to the core and windings, the temperature-rises of which are naturally greater.

It is an over-simplification to regard a machine or a transformer as a homogeneous body. There are many identifiable parts, each with a characteristic surface area, mass, heat capacity, thermal conductivity and loss. The temperature-rise of different parts (and of various points in the same part) may be very uneven, especially on intermittent load or with forced cooling, as discussed by Bates and Tustin [22]. Hot-spot temperatures are particularly important.

Rate of Temperature-rise

With the surface at θ above ambient, the rate of rise is

$$\frac{d\theta}{dt} = \frac{1}{Gc_p} (p - S\lambda\theta) = \frac{1}{\tau}(\theta_m - \theta) \qquad (4.2)$$

At the start of a heating cycle ($\theta = 0$) the rate of temperature-rise depends on the heat capacity and the loss only, and $d\theta/dt = \theta_m/\tau = p/Gc_p$. The same condition holds approximately when there is a sudden rise in the loss rate. A typical case occurs when a transformer is subjected to considerable overload or to a short-circuit lasting a time Δt. If the conductor material has a density δ, and a resistivity ρ_1 at the initial temperature θ_1, the specific loss (power per unit mass) for a current density J is $J^2\rho_1/\delta$: this may be augmented by the eddy-loss factor K_d as described in Sect. 3.2. Assuming that the heat is stored in the conductor material without dissipation, the temperature-rise is

$$\Delta\theta = \frac{K_d J^2 \rho_1}{\delta c_p} \Delta t \qquad (4.3)$$

If $\Delta\theta$ is large, then ρ_1 should be replaced by the resistivity at the mean temperature, $\theta_1 + \frac{1}{2}\Delta\theta$.

EXAMPLE 4.2: A 1 000 kVA transformer with a rated I^2R loss of 4 times the core loss operates in an ambient temperature of 25 °C. From cold it has a winding temperature-rise of 24 °C after 1 h and 38.5 °C after 2 h on rated load. Estimate (a) the final temperature-rise on rated load and the heating time-constant of the windings; (b) for a maximum winding temperature of 97 °C (i) the continuous overload that can be sustained, and (ii) the time for which 2 p.u. rated load can be taken after continuous operation on no load.

(a) From eq. (4.1), with time for convenience in hours:

$$24 = \theta_m [1 - \exp(-1/\tau)] = \theta_m [1 - k] \quad \text{and} \quad 38.5 = \theta_m [1 - k^2]$$

whence $38.5/24 = 1 + k = 1.6$; thus $k = 0.6$, $\exp(1/\tau) = 1.67$ and $\tau = 1.96 \approx 2$ h. The final steady temperature-rise becomes $\theta_m = 60$ °C, corresponding to a scale temperature of 85 °C.

(*b*) The allowable temperature-rise is now $97 - 25 = 72\,^{\circ}$C, which is 1.2 times that for normal continuous rating.

(i) The rated losses are proportional to $(1 + 4) = 5$; they can now be $5 \times 1.2 = 6 = (1 + 5)$. The load is therefore $\sqrt{(5/4)} = 1.12$ p.u. $= 1\,120$ kVA.

(ii) On no load with only core loss, i.e. 1/5 of rated loss, the steady temperature-rise is $60(1/5) = 12\,^{\circ}$C: this is the starting value. For 2 p.u. load the losses are proportional to $(1 + 4^2) = 17$, and a final steady temperature-rise of $60(17/5) = 204\,^{\circ}$C. But the rise is limited to a further $72 - 12 = 60\,^{\circ}$C: hence

$$60 = 204\,[1 - \exp(-t/2)]$$

giving $t = 0.7$ h $= 42$ min.

These estimates are valid only insofar as the transformer can be regarded as a homogeneous body. The greater overload $I^2 R$ loss causes changes in the circulation of the coolant and the conditions of heat-transfer, so that the transformer does not operate thermally in the same way as on rated load. The calculations do, however, show practical trends.

4.3 *Cooling of small units*

Small transformers and machines use the ambient air as a coolant, the units being small enough to incur little danger of the build-up of excessive hot-spot temperatures.

Transformers

Most small transformers are oil-immersed, and the surface of the tank provides heat-transfer to the ambient air. Plain sheet-steel tanks can transfer about 13 W/m^2 per $^{\circ}$C temperature-rise. Above about 25 kVA 3-ph rating the emitting surface must be increased, most commonly by setting vertical tubes of 40–50 mm diameter welded steel and 1.5 mm wall thicknesses into the sides of the tank. For ratings between 2 and 5 MVA, elliptical-section tubes are preferred as more can be accommodated.

Experiment shows that radiation accounts for about 6 W and convection for about 6.5 W per m^2 of plain tank wall, per $^{\circ}$C difference between tank and ambient air or surroundings. For a rise of 35°C a plain tank will consequently dissipate about 440 W/m^2. If this is insufficient, the total plain surface is increased x times by tubing. Then from 1 m^2 of plain tank there is a dissipation of $(6 + 6.5)$ W/$^{\circ}$C, and from the $(x - 1)$ corresponding area of tube wall a convection of $(x - 1)\,6.5 \times 1.35$ W/$^{\circ}$C, the factor of 1.35 accounting for the improved convection conditions obtained from tubes. Thus per unit of total area there is a dissipation

$$p = (8.8 + 3.7/x)\ \ \text{W/(m}^2\,{}^{\circ}\text{C)} \tag{4.4}$$

The tank top and bottom are usually ignored.

EXAMPLE 4.3: The tank of a 575 kVA 1-ph oil-immersed self-cooled transformer is 1.50 m in height and 0.65×1.05 m in plan. The rated loss is 6.5 kW. It is required to arrange 50 mm diameter tubes, spaced 75 mm between centres and averaging 1.275 m in length for a mean surface temperature-rise of 35 °C.

The vertical surface area of the plain tank is 5.1 m². With the area increased x times by the provision of tubes, the dissipation from eq. (4.4) is

$$6.5 \times 10^3/35 \times 5.1\, x = 36.5/x = (8.8 + 3.7/x) \quad W/(m^2\,°C)$$

from which $x = 3.75$. The total tube area is therefore $2.75 \times 5.1 = 14.0$ m². The wall area of each tube averages 0.2 m² so that $14.0/0.2 = 70$ tubes are required.

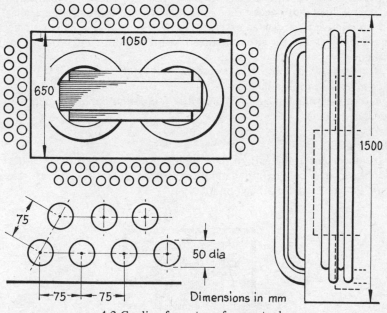

4.2 Cooling from transformer tank.

These are set out as shown in Fig. 4.2 on 75 mm centres in two rows, the arrangement being;

Long sides: 12 and 11 tubes, total 46 } 76 tubes
Short sides: 8 and 7 tubes, total 30 }

Rotating Machines
The scheme of air ventilation is closely related to the construction, in particular the type of enclosure:

Open-pedestal, in which the stator and rotor ends are in free contact with the outside ambient air, the rotor being carried on pedestal bearings mounted on the bedplate.

Open-end-bracket, where the bearings form part of the end-shields which are fixed to the stator housing, a common construction for small and medium-sized machines. The air is in comparatively free contact with the stator and rotor through the openings.

Protected, or end-cover types with guarded openings: the protection may be by *screen* or by *fine-mesh covers.* In the latter case, the machine is regarded as totally enclosed.

Drip-, Splash- or Hose-proof, a protected machine with the ventilation opening in the end-shields designed to exclude falling water or dirt, or jets of liquid.

Pipe- or *Duct-ventilated,* with end-covers closed except for flanged apertures for connection to pipes along which the cooling air is drawn.

Totally-enclosed, where the enclosed air has no connection with the ambient air: the machine is not necessarily airtight. Total enclosure may be associated with an internal rotor fan, an external fan, water cooling, or closed-air-circuit ventilation, in which the air is circulated to a refrigerator and returned to the machine.

Weatherproof, or *watertight,* as specified.

Flame-proof, for use in mines.

As a general rule, the more difficult the ventilation the lower must a given machine be rated.

Cooling-air Circuit

The arrangement of the path of the cooling air depends on the size of the machine (a larger core requiring more subdivision and more elaborate arrangements for gas distribution) and on its type.

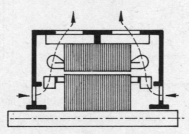

4.3 Radial ventilation.

Radial ventilation is commonly employed, because the rotor induces natural centrifugal movement of the air, augmented as required by rotor fans. Fig. 4.3 shows a typical method suitable for machines below 20 kW. The end-brackets are shaped to guide air over the overhang and the back of the core, and baffles are fitted to improve the fan efficiency. For larger machines, subdivison of the core is necessary and the air paths through the radial ducts are in parallel with those across the overhang convectors. A high rate of heat dissipation is possible in the gap between smooth-bore stators and rotors – unless this gap is very short – on account of high air speeds accompanied by turbulence.

Axial ventilation is suitable for machines of moderate output and high speed. The practically solid rotor construction and restricted spider, necessary to avoid undue centrifugal forces, make it difficult to provide adequate air-paths to radial ventilating ducts. Further, the tendency in design is to increase core lengths and to restrict core diameters in order to get a cheaper machine.

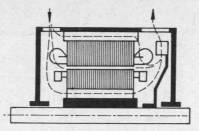

4.4 Axial ventilation.

Fig. 4.4 shows a method of applying axial ventilation to a small machine with plain cores. To increase the cooling surface, holes may be punched in the core plates to form through-ducts where considerable heat-dissipation occurs. This greatly improves the cooling, but requires a larger core diameter for the increased core depth necessary.

Combined radial and axial ventilation is employed for large machines. Fig. 4.5 shows the arrangement of an induction motor for mixed ventilation. The air is drawn in at one end, and encouraged to pass through the ducts by baffling the fan end of the rotor spider. The shaft-mounted fan ejects the air.

Multiple-inlet ventilation. With greater lengths of core, there is a tendency to starve the central radial ducts of air. With the multiple-inlet system it is possible to build machines with long cores and obtain effective centre cooling. The stator frame is divided into separate air circuits fed in parallel.

Totally-enclosed machines present a special problem in that the inside of the machine can have no air-connection with the outside. Two methods of fan cooling are shown in Fig. 4.6. At (*a*) a shaft-mounted fan external to the working parts of the machine blows air over the carcass through a space between the main housing and a thin cover plate. Internal air circulation is produced by an internal fan: this avoids the temperature gradient across the air gap. At (*b*) an internal fan circulates the heat to the carcass. Air is also blown over the outside of the carcass to improve the dissipation.

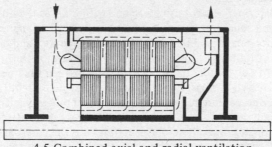

4.5 Combined axial and radial ventilation.

Cooling Coefficient

The maximum temperature-rise of a given part of a machine is given by $\theta_m = p/S\lambda = c(p/S)$, where c is the cooling coefficient. The Table gives typical

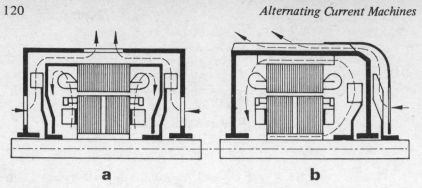

a   **b**

4.6 Ventilation of totally enclosed machine.

values of c in terms of a characteristic speed u (in m/s) and the dissipating surface area S (in m^2). The values can be applied only when justification is forthcoming from suitable tests.

Cooling Coefficient

Part	c	u	Notes
Cylindrical surfaces of stator and rotor	$\dfrac{0.03 \text{ to } 0.05}{1 + 0.1u}$	Relative peripheral speed	Lower figures for forced cooling
Back of stator core	0.025 to 0.04	0	
Rotating field coils	$\dfrac{0.08 \text{ to } 0.12}{1 + 0.1u}$	Peripheral speed	Based on total coil surface
	$\dfrac{0.06 \text{ to } 0.08}{1 + 0.1u}$		Based on exposed coil surface only
Ventilating ducts in cores	$\dfrac{0.08 \text{ to } 0.2}{u}$	Air velocity in ducts	u taken as 0.1 of peripheral speed of core

4.4   *Cooling of large units*

Consider a transformer or a rotating machine with k times the linear dimensions of another smaller but otherwise similar unit. Its core and conductor areas are k^2 times greater and its rating (with the same flux and current densities) increase k^4 times. The losses increase by the factor k^3 but the surface area is multiplied only by the factor k^2. Thus the loss per unit area to be dissipated is increased k times. Large units are therefore more difficult to cool than small ones, and require more elaborate methods. The cores are massive, the conductors are long, of large section and rated with high current densities. The coolant must therefore penetrate into the cores, and even into the conductors; for if not, interior heat must travel undue distance to the coolant, with the result that excessive temperature gradients occur and hot-spots develop. Forced-cooling systems are essential. Each part of the unit must be considered, both by itself and in relation to neighbouring parts, in respect of the conditions of thermal flow in the whole machine.

Transformers

The cooling of transformers differs from that of rotating machines in that no moving parts are available to assist the circulation of the coolant. If the natural thermal 'head' (resulting from the expansion and consequential upward flow of heated coolant) is not sufficient, forced circulation is necessary.

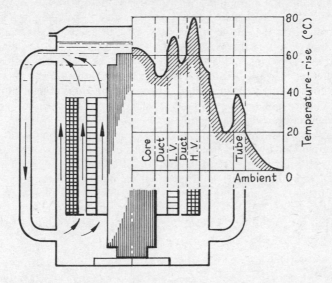

4.7 Oil-circulation and temperature distribution.

Natural Circulation. Fig. 4.7 is drawn to indicate on the left the thermal flow of oil in a transformer tank. The oil in the ducts, and at the surfaces of the coils and cores, takes up heat by conduction, and rises, cool oil from the bottom of the tank rising to take its place. A continuous circulation of oil is completed by the heated oil flowing to the tank sides (where cooling to the ambient air occurs) and falling again to the bottom of the tank. Oil has a large coefficient of volume expansion with increase of temperature, and a substantial circulation is readily obtained so long as the cooling ducts in the cores and coils are not unduly restricted. A typical temperature distribution is shown. On rated load the greatest temperature-rise will be in the coils. The maximum oil temperature may be about 10 °C less than the coil figure, and the mean oil temperature another 15 °C less.

Alternative to cooling tubes is a tank with narrow, deep corrugations, presenting a large surface area with a lower oil content and a more compact unit.

Forced Circulation. The choice of the method of cooling will depend largely upon the conditions obtaining at the site. Air-blast cooling can be used, a hollow-walled tank being provided for the transformer and oil, the cooling air being blown through the hollow space. The heat removed from the inner walls of the tank can be raised to five or six times that dissipated naturally, so that very large transformers can be cooled in this way, but it is more usual to employ radiator banks, of corrugated or elliptical tubes, separate from the transformer tank and cooled by fanned air circulation.

A cheap method of forced cooling where a natural head of water is obtainable is the use of a cooling coil, consisting of tubes through which cold water is circulated, inserted in the top of the tank. This method has, however, the disadvantage that it introduces into the tank a system containing water under a head greater than that of the oil. Any leakage will, therefore, be from the water to the oil, so that there is a risk of contaminating the oil and reducing its insulating value. Fins are placed on the copper cooling tubes to assist in the conduction of heat from the oil, since heat passes three times as rapidly from the copper to the cooling water as from the oil to the copper tubing. The inlet and outlet pipes are lagged to avoid water from the ambient air condensing on them and getting into the oil.

For large installations the best cooling system appears to be that in which the oil is circulated by pump from the top of the transformer tank to a cooling plant, returning when cold to the bottom of the tank. When the cooling medium is water, this has the advantage that the oil can be arranged to work at a higher static head than the water, so that any leakage will be in the direction of oil to water. The system is suitable for application to banks of transformers, but for reliability not more than, say, three tanks should be connected in one cooling pump circuit.

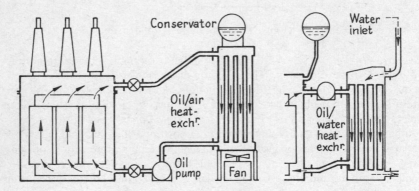

4.8 Transformer cooling methods.

Fig. 4.8 shows two methods of cooling with separate heat-exchanger banks. It is common to bring in the fan and oil-circulating pump at predetermined load levels or when the oil temperature in the transformer tank has reached a set limit; this improves the overall efficiency at small loads.

Cores and Coils. In small transformers the dissipation of core loss is simple, as the core surface/volume ratio is high. Very large cores, on the other hand have a relatively small surface/volume ratio, so that the cooling surface must in some way be augmented by additional ducts. There are two ways of arranging ducts – either parallel or perpendicular to the direction of the laminations. The first is easy, the second requires special punching. Unfortunately the first method does not present to the oil any additional plate *edges*. Heat flows twenty times more readily along the laminations to the edges than from plate to plate across the intervening insulation, which has a low thermal conductivity.

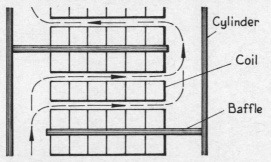

4.9 Directed oil-flow in transformer winding.

It is important that oil circulation is such that there is a flow over the main coil cooling surfaces. This is relatively simple when these surfaces are vertical, as with helix coils. But with disc coils in large transformers the major surfaces are horizontal. Oil-flow deflectors, Fig. 4.9, are necessary to reduce the conductor/oil temperature gradient.

Rotating Machines

Forced Air Cooling. For large machines, which may require many tonnes of cooling air per hour, forced ventilation permits the cleaning of the air by suitable filters, avoiding the clogging of the ducts. In an arrangement for a large induction motor, the end covers are airtight, and air from an external fan is fed in from a duct. On high-speed machines rotor-mounted fans may be sufficient. The air is filtered or washed with a water spray, then baffled against flooded scrubbing surfaces to precipitate the dirt. It is then dried by passing over a series of dry scrubbing plates. A more complete means of securing clean cooling air is the *closed-circuit* system, as employed for the smaller ratings of turbo-generators. The hot air from the machine is passed through a water-cooled heat-exchanger and then returned to the machine by a centrifugal fan. The arrangement is shown in Fig. 4.10.

Hydrogen Cooling. Turbo-generators have forced ventilation and total enclosure to deal with the large losses and the high rating per unit volume. Certain advantages are gained by the use of hydrogen in place of air as a coolant: hydrogen has $\frac{1}{14}$ of the density, reducing windage loss and noise; 14 times the specific heat; $1\frac{1}{2}$ times the heat-transfer, so more readily taking up and giving up heat; 7 times the thermal conductivity, reducing temperature gradients; reduces insulation corona; and will not support combustion so long as the hydrogen/air mixture exceeds 3/1. As a result, hydrogen cooling at 1, 2 and 3 atm can raise the rating of a machine by 15, 30 and 40% respectively.

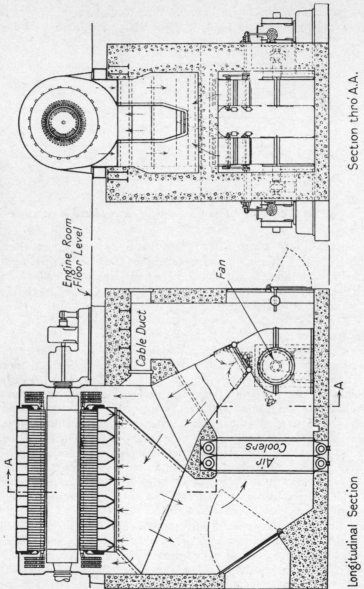

Engine Room
Floor Level

Cable Duct

Fan

Air Coolers

Longitudinal Section

Section thro' A.A.

4.10 Forced air ventilation.

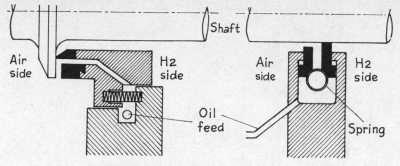

4.11 Hydrogen shaft seals.

The stator frame must be gas-tight and explosion-proof. Oil-film gas-seals at the rotor shaft ends are necessary. Two forms are shown in Fig. 4.11: each must accommodate axial expansion of the rotor shaft and stator frame. Oil is fed to the shaft and the flow is split, part towards the interior (gas) side and part to the air side. The latter mingles with the bearing oil, while the former is collected and degassed.

Fans mounted on the rotor circulate hydrogen through the ventilating ducts and internally-mounted gas-coolers. The gas pressure is maintained above atmospheric by an automatic regulating and reducing valve controlling the supply from normal gas cylinders. When filling or emptying the casing of the machine, an explosive hydrogen-air mixture must be avoided, so air is first displaced by carbon dioxide gas before hydrogen is admitted: the process is reversed for emptying. It is usual to provide a drier to take up water vapour entering through seals. The hydrogen purity is monitored by measurement of its thermal conductivity.

Turbo-generators operating at hydrogen pressures just above atmospheric (so that leaks will be outwards) require about 0.03 m^3 per MW of rating per day. This rises to about 0.1m^3 for hydrogen pressure of 2 atm. The gas consumption of synchronous capacitors, which do not need shaft seals, is very much less.

Hydrogen cooling results in substantial increase in rating for a given temperature-rise, and the reduction in windage may add 0·5—1·0% to the efficiency of a 100 MW machine. Fig. 4.12 gives a diagram of the auxiliary equipment required for a hydrogen-cooled machine.

Direct Gas Cooling. Large stator cores for turbo-generators are provided with both axial and radial ducts, Fig. 4.13. Large rotors may have comparable ducting as indicated in Fig. 4.14(*a*), but for machines of 100 MW or more the temperature gradient over the conductor insulation is high enough to call for direct contact between the coolant and the material of the conductors themselves. In Fig. 4.14(*b*) the rotor conductors comprise rectangular tubes, ventilated by a cooling circuit separate from that of the stator, the hydrogen gas being admitted to the tubes through insulating flexible connections at the ends from a centrifugal impeller mounted on the outboard end of the rotor shaft. The slot tubes are electrically connected at the overhang by suitably

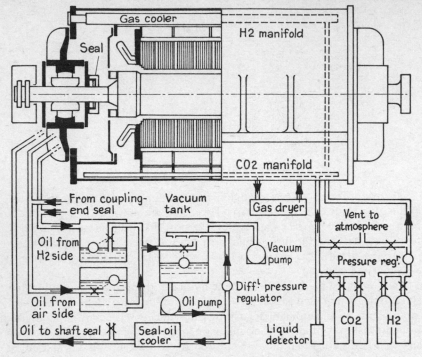

4.12 Hydrogen-cooling system.

shaped copper bars forming inlet or outlet ports. The hollow conductors are of hard-drawn silver-bearing copper, with synthetic-resin-bonded glass-cloth laminate insulation.

The direct gas cooling of stator and particularly of rotor windings permits of much higher electric loadings. With ratings of 1 000 MW even this is not enough to cope with the very large gas flow-rate and pressure, which might be 15 m³/s of hydrogen at 1.2 atm to absorb an excitation loss of 5 MW.

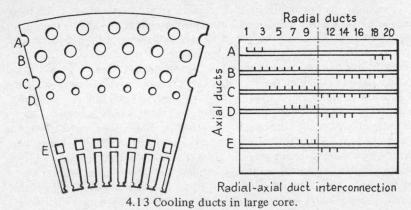

4.13 Cooling ducts in large core.

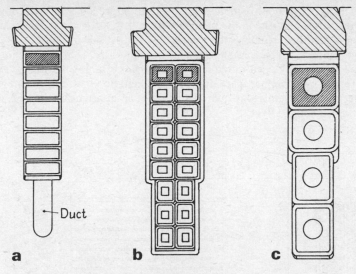

4.14 Turbo-rotor conductor cooling.
(*a*) Conventional; (*b*) Direct gas; (*c*) Direct water.

Direct Water Cooling. Turbo-generators of the highest ratings so far contemplated are likely to have hydrogen-cooled stator cores, and direct water-cooled stator and rotor windings. The thermal capacity of water is 750 times that of hydrogen at 500 kN/m^2, but there are limits on the permissible speed of water flow in channels (e.g. 2.5 m/s) to avoid erosion and cavitation. Nevertheless, the conductor water-duct area may be small, allowing more space for copper.

In the direct water cooling of *stator* windings, one problem is to devise flexible water-tube connections with insulation against the high winding voltages and to preserve a low water conductivity (less than 0.5 mS/m). Water cooling of the *rotor* winding, even more desirable because of its high electric loading, offers more mechanical difficulty. Fig. 4.14(*c*) shows a rotor slot-conductor arrangement, the slot-width being reduced near the root to allow for greater tooth width to sustain the centrifugal force. To enable water to enter the bore of the shaft, an inlet seal is attached to the outboard (non-coupling) end. The coolant travels axially to a point in the plane of the end-winding, where its flow becomes radial into manifolds extending·axially under the end-windings (which are reasonably accessible before the end-bells are fitted). The shaft passages are lined with stainless steel to prevent corrosion and water contamination. Flexible insulating connectors transfer the water from the stainless steel manifolds to the winding, entering at the coil ends and leaving at the mid point of each coil-side. On emerging from the conductors, the water flows radially inward to the rotor bore channel, then axially towards the coupling end. The water outlet is a simple catcher box surrounding the shaft. The external water system comprises an air-release tank, heat-exchangers, filters and circulating pumps. Reservoir tanks maintain the necessary water levels. The purity is monitored, and can be adjusted by a demineralization unit connected across a bypass in the main circuit.

Water-cooled field windings have been used for the salient poles of large hydro-generators, a typical arrangement being shown in Fig. 4.15. The interpolar space, normally open to the cooling air in an orthodox machine, can be closed to give more copper space and a lower excitation loss. The mass of cooling water required is only one-quarter of that of air at atmospheric pressure for the same cooling effect.

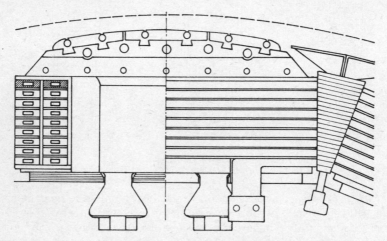

4.15 Direct water-cooling of salient-pole winding.

Superconducting Rotor Windings. Feasibility studies have been made on machines based on a substantially conventional water-cooled stator winding and a superconducting rotor winding. In one proposal by Lorch [106] for a 660 MW turbo-generator, the rotor has no magnetic steel but is a hollow cylinder of glass reinforced epoxy resin with steel ends. The superconducting windings are in slots and wrapped with glass epoxy resin. The windings are cooled to a temperature of about 5 K by a flow of helium entering and leaving the rotor on the axis at the outboard end. A copper cylinder shell rotates with the rotor and screens it from harmful a.c. magnetic fields; this shell is thermally insulated from the rotor by a sealed stainless steel vacuum enclosure and is maintained at a temperature between 80 and 135 K. Liquid nitrogen is passed through a stationary system of cylinders within the stator bore to form a cold screen at about 77 K. The gap between the fixed and rotating members is filled with hydrogen at 0.1 mmHg to reduce 'windage' but not unduly lower the thermal conductivity. In this way the complexity of a liquid nitrogen supply to the rotor is avoided. The exciter is conventional, but has to supply 2.5 MW only for a few seconds on initial excitation. The field is lowered by reversing the exciter or by means of a discharge resistor.

4.5 Heat transfer
Heat transfer is a complex phenomenon presenting formidable analytical dif-

ficulties. The loss heat is transferred by a combination of conduction, radiation and convection processes to the ambient air and surroundings which are considered in the ultimate to constitute a heat sink having the character of an 'infinite thermal busbar'. The symbols used in this section, and typical values of relevant physical quantities, are set out below:

a	area	δ	density
b	gas pressure ratio	ϵ	radiation factor
c_p	specific heat at constant pressure	θ	temperature-rise
		θ	Celsius (scale) temperature
d	tube diameter		
k	thermal conductivity		*Subscripts*:
k	coefficient	a	absorption
p	power	c	conduction
S	surface	i	iron
T	Kelvin (abs.) temperature	r	radiation
w	width	s	space
x	dimension	t	tank
		u	tube
		v	convection

Material	Specific heat c_p J/(kg K)	Density δ kg/m^3	Thermal conductivity k W/(m K)	Thermal resistivity $1/k$ (m K)/W	Expansion coeff. x 10^{-6}/K	
Aluminium	920	2 700	200	0.0050	25.5	†
Copper	390	8 900	380	0.0026	16.7	†
Steel	500	7 800	150	0.0067	11.0	†
Air*	1 000	1.3	0.025	40	3 660	‡
Hydrogen*	3 400	0.09	0.18	5.5	3 660	‡
Oil	1 900	850	0.16	6.3	1 000	‡
Water	4 200	1 000	0.63	1.6	–	
Mica	–	–	0.33	3	–	
Micanite	–	–	0.12	8	–	
Paper	–	800	0.12	8	–	

* At s.t.p.   † Linear expansion   ‡ Volume expansion

Conduction

Internal flow of heat, from the region in which it originates to the surface at which it is transferred to the coolant, is important in determining hot-spot temperatures and the thermal conditions to which associated insulants are subjected. A 50 °C hot-spot temperature might be made up of a rise of 11 in the coolant, 25 at the heat-exchange surface, and 14 °C internally from the surface to the hot-spot.

Heat conduction is a function of the temperature gradient $d\theta/dx$, the thermal capacity of the path and the thermal conductivity of the material, the latter being the power transferred across a path of unit length and unit area per kelvin of temperature difference. For one-directional flow the basic equation is

$$\frac{\partial \theta}{\partial t} = \frac{k}{c_p \delta} \cdot \frac{\partial^2 \theta}{\partial x^2}$$

with corresponding forms for 2- and 3-dimensional flow. These govern transient heating. Under steady-state conditions $\partial\theta/\partial x$ is constant and there is no further thermal storage; the rate of energy flow through a path of length x between thermal equipotential planes of area a and having a temperature-difference $\Delta\theta$ is then

$$p_c = \Delta\theta \cdot k(a/x) \tag{4.5}$$

EXAMPLE 4.4: A transformer core, of plate width $w = 0.50$ m and with a stacking (space) factor $k_s = 0.94$, has a uniformly distributed core loss of $p_i = 3.0$ W/kg. The thermal conductivity of the steel is $k = 150$ W/(m K) and the surface temperature is 40 °C. Estimate the maximum core temperature if the heat flow is (i) all to one end of the core, (ii) one-half to the surface at each end.

The heat flow is assumed to be along the laminae. The specific loss p_i corresponds to $p = p_i k_s \delta = 3.0 \times 0.94 \times 7\,800 = 22\,000$ W/m^3.

(i) At a point distant x from one cooling surface, the power-flow density is $p(w-x)$ from the region of the core beyond x. The temperature gradient is $(d\theta/dx) = p(w-x)/k$, whence

$$\theta = px(w - \tfrac{1}{2}x)/k + 40$$

For $x = w$ the core temperature is

$$\theta_m = (pw^2/2k) + 40 = 18 + 40 = 58 \,^\circ\text{C}$$

(ii) The power density at x is now $p(\tfrac{1}{2}w - x)$, and for $x = \tfrac{1}{2}w$

$$\theta_m = (pw^2/8k) + 40 = 5 + 40 = 45 \,^\circ\text{C}$$

Perfect contact between a hot body and the coolant at the interface is rare. The temperature drop $\Delta\theta$ is strongly dependent on the fluid-flow conditions (stream-line or turbulent) and upon the condition of the surface. There may be interfering oxide films, gas bubbles in liquids, and a stagnant surface layer of fluid. In any case the thermal conductivity of coolants is much lower than that of metals. If the surface temperature is θ_s and the bulk temperature of the coolant is θ_f, the thermal conduction across the interface can be written

$$p_c = k_c (\theta_s - \theta_f) \; \text{W/m}^2 \tag{4.6}$$

where the coefficient k_c is 50–1 500 for oil, and 10–50 for air.

Quantity of Coolant. Where heat is conducted away from a surface by a coolant, a flow of 1 kg/s can absorb a power θc_p for a temperature-rise of θ. For a power p the required rate of flow in mass or in volume per unit time is

$$q = \frac{p}{c_p \theta} \text{ kg/s} \quad \text{or} \quad q = \frac{p}{c_p \delta \theta} \text{ m}^3/\text{s} \tag{4.7}$$

For *liquid* coolants these evaluate to

Water: $\frac{p}{\theta} 24 \times 10^{-3}$ kg/s $\qquad \frac{p}{\theta} 24 \times 10^{-8}$ m^3/s

Oil: $\frac{p}{\theta} 53 \times 10^{-3}$ kg/s $\qquad \frac{p}{\theta} 62 \times 10^{-8}$ m^3/s

For *gases* the volumes depend on absolute temperature and pressure so that

$$q = \frac{p}{c_p \theta} \text{ kg/s} \quad \text{or} \quad q = \frac{p}{c_p \delta \theta} \cdot \frac{T}{273} \cdot \frac{1}{b} \text{ m}^3/\text{s}$$

For both *air* and *hydrogen* the volume of coolant becomes

$$q = \frac{p}{\theta} \cdot \frac{T}{273} \cdot \frac{1}{b} 8 \times 10^{-4} \text{ m}^3/\text{s} \tag{4.8}$$

but hydrogen has important advantages in thermal conductivity.

Radiation

The radiant energy interchange between two surfaces involves the 'view' that they have of each other, their surface emissivity factors and their respective absolute temperatures. According to the Stefan-Boltzmann law, a body with a surface emissivity factor ϵ_1 and absolute temperature T_1 within a full radiator ('black body') of temperature T_2 radiates

$$p_r = 5.7 \times 10^{-8} \epsilon_1 (T_1{}^4 - T_2{}^4) \text{ W/m}^2 \tag{4.9}$$

For metal surfaces ϵ is low and dependent on the surface roughness; for non-metals (e.g. paints) ϵ approaches that of a full radiator, and generally exceeds 0.8. The net heat exchange between two plane surfaces, of temperatures T_1 and T_2 and emissivity factors ϵ_1 and ϵ_2 respectively, is

$$p_r = 5.7 \times 10^{-8} \frac{\epsilon_1 \epsilon_2}{\epsilon_1 + \epsilon_2 + \epsilon_1 \epsilon_2} (T_1{}^4 - T_2{}^4) \text{ W/m}^2$$

Radiation does not normally occur by itself in machines and transformers, being accompanied in almost every case by convection. It is therefore often sufficient to use the expression

$$p_r = 2.9 \, \epsilon \theta^{1.17} \text{ W/m}^2 \tag{4.10}$$

for a body with the temperature θ above ambient in the range 0–100 °C.

Emissivity and Absorption Factors. The emissivity factor ϵ depends on the surface and its treatment. It is higher for lead paints than for metallic paints, and varies with the colour. The Table gives typical values for ambient conditions between 0 and 100°C.

Emissivity and absorption factors

Factor	Al	Cu	Steel rough	Steel sheet	Metal paint Al	Lead paints white	Lead paints grey
Emissivity ϵ	0.10	0.15	0.24	0.55	0.55	0.90	0.95
Absorption ϵ_a	0.15	–	–	–	0.55	0.25	0.75

Some plant, particularly transformers, may have to operate in the open, and may absorb radiant heat from the sun by *insolation.* The earth's outer atmosphere receives about 1.3 kW/m², and up to about two-thirds of this may reach its surface when climatic and air conditions are favourable. The effect depends on the absorption factor ϵ_a: if it approaches unity the insolation may raise the surface temperature by several degrees, but if it is low some of the sun's radiation is re-emitted. The factors ϵ and ϵ_a will normally differ because they refer to different heat wavelengths.

Ambient Temperature. For radiation, the ambient temperature is that of the sky and the ground to which the body radiates. For other forms of heating to the atmosphere, the ambient temperature is that of the bulk of the air at a distance remote enough to be unaffected by the thermal field of the body (a few metres). Strictly, the use of the air temperature as the ambient is justified only when convection is the major cooling process.

Convection

Convection is a transfer of heat from a surface to a coolant fluid, causing the fluid to expand and to create a thermal head and a consequent circulation. It is a complicated phenomenon involving many variables: the power density; the temperature difference between the surface and the bulk of coolant; the height, orientation, configuration and condition of the emitting surface; the thermal conductivity, density, specific heat, viscosity and coefficient of volume expansion of the fluid; and the value of the gravitational constant. It is the custom to deal with convection problems in terms of dimensionless parameters (Nusselt, Grashof and Prandtl), some of which are kept constant while others are varied, and formulae drawn up on an empirical basis for calculation. Even then the formulae have to be used with great caution.

For the temperature range and structure of machines and transformers, the following simplified formulae have been developed:

For vertical planes of height not less than 1 m and with a temperature difference θ between the surface and the ambient air of pressure b atm:

$$p_v = 2.0b^{1/2}\theta^{5/4} \text{ W/m}^2 \tag{4.11}$$

For vertical tubes of diameter d:

$$p_v = 1.3d^{-1/4}\theta^{5/4} \ \text{W/m}^2 \tag{4.12}$$

EXAMPLE 4.5: A 7.5 MVA transformer has a rated core $+I^2R$ loss of $14 + 46 = 60$ kW. The tank is 1.4 m x 2.5 m in plan, and 3.0 m in height, fitted with a total of 925 m of 0.05 m diameter tubes. Find the surface temperature-rise θ above 25 °C (298 K) ambient.

Eqs. (4.9), (4.11) and (4.12) are summed and solved by iteration for θ, giving $\theta = 45$ °C rise (corresponding to a surface temperature of 343 K). For this value the calculations check as follows:

Radiation. The radiating area is the tank walls and top, giving $23.4 + 3.5 = 29.6 \ \text{m}^2$ (the tubes making no additional contribution). With $\epsilon = 0.9$

$$p_r = 5.7 \times 10^{-8} \times 0.9(343^4 - 298^4) = 303 \ \text{W/m}^2$$

$$P_r = p_r S_t = 303 \times 29.6 = 9\,000 \ \text{W} = 9.0 \ \text{kW}$$

Convection: The tank surface (walls only) is $23.4 \ \text{m}^2$. Assuming normal atmospheric pressure, $b = 1$,

$$p_{vt} = 2.0 \times 45^{5/4} = 2.0 \times 117 = 234 \ \text{W/m}^2$$

$$P_{vt} = p_{vt} S_t = 234 \times 23.4 = 5\,500 \ \text{W} = 5.5 \ \text{kW}$$

The tube area is $S_u = \pi 0.05 \times 925 = 145 \ \text{m}^2$, so that

$$p_{vu} = 1.3 \times (0.05)^{-1/4} \times 45^{5/4} = 319 \ \text{W/m}^2$$

$$P_{vu} = p_{vu} S_u = 319 \times 145 = 46\,200 = 46.2 \ \text{kW}$$

The total dissipation is $9.0 + 5.5 + 46.2 = 60.7$ kW, as required.

4.6 *Flow of coolant in transformers*

In the classical theory of isothermal steady flow of a fluid in a long straight tube, the flow can be (i) *streamline* (or laminar) with a parabolic velocity profile having a mean equal to 0.5 of the maximum, or (ii) *turbulent* in which the mean velocity is about 0.8 of the maximum and there remains a thin annular streamline flow at the tube-wall boundary. Actual conditions are rarely as straightforward, but turbulence has to be encouraged if effective heat-transfer is required.

Circulation of the coolant fluid — air or hydrogen — is relatively simple with rotating machines, which can be fitted with rotor-mounted fans. Although fan efficiencies are rather low (e.g. 0.2−0.4), rapid and directed air streams can be produced in the cooling ducts, the nature of which stimulates turbulence. In large machines, especially hydrogen-cooled turbo-generators, gas pressures up to 2 atm may be used to obtain the high-speed flow of large quantities of coolant.

In *natural* cooling by thermal head, the method employed in most oil-immersed transformers, the flow of the coolant is complex. The fluid 'drags' at the walls of the channel, constituting a partial heat barrier, and the heat dissipation produces complicated velocity and temperature profiles. The temperature falls sharply across the walls of the cooling surface; and although there is a small bulk velocity in one direction, the 'core' of the column of oil may actually move in the opposite direction. While the details are thermally and hydrodynamically very intricate, the system as a whole may be looked upon as operating on a combination of thermal and gravitational 'heads'.

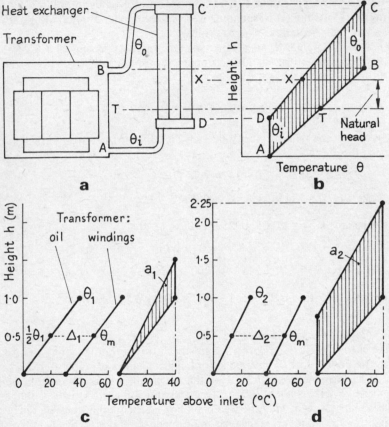

4.16 Temperatures in transformer and heat-exchanger.

Consider the transformer/exchanger system in Fig. 4.16(a). With the oil having the same temperature throughout the system the gravitational forces balance and there is no circulation. With the transformer on load, oil in the region AB absorbs the core and load losses; the temperature in AB rises and the oil density falls, unbalancing the columns and initiating circulation in the direction ABCD. At the base A the tank inlet oil temperature is θ_i. At B and C (ignoring dissipation from the pipework BC) the outlet temperature is θ_o. Assuming for simplicity that the height/temperature relation is linear, the h/θ diagram in Fig. 4.16(b) can be drawn. The shaded area is a measure of the effective head causing oil circulation, for if multiplied by $g(d\delta/d\theta)$, where g is the acceleration due to gravity and δ is the oil density, the result is the pressure-difference or 'head'. For a given circuit the oil flow-rate can be taken as proportional to the head. Thus the circulation is determined (i) by the fall of oil density resulting from the temperature-rise $(\theta_o - \theta_i)$, and (ii) by some function of the height of the exchanger with respect to the tank.

Clearly there can be no circulation without temperature-rise, and equally there can be no 'density head' unless the 'centre of cooling' (X) of the heat-exchanger is above that (T) of the transformer. In practice the designer is limited by specification, Sect. 4.6, in his choice of the temperature-rise $(\theta_o - \theta_i)$, and he must therefore arrange that X is adequately above T. Truncating the heat-exchanger at its lower end may actually result in some improvement in oil circulation if it raises point X. Even in transformers carrying tubes on the tank walls (for which case the circulating 'head' is somewhat marginal) a small gain may be had by raising the tube entry level at the lower end.

EXAMPLE 4.6: In a 20 MVA transformer on steady full load, oil at the inlet enters the tank from the exchanger at 15 °C above ambient, flows upward by a vertical distance of 1.0 m through the oil ducts and reaches a temperature of 55 °C above ambient. It then flows to the header of a heat-exchanger bank 1.5 m high. The mean temperature-rise of the windings (estimated from a resistance/time curve taken after shut-down) is 65 °C. Estimate the additional load obtainable if the heat-exchanger is raised bodily by 0.75 m for the same mean winding temperature as before, assuming that the oil inlet temperature remains at 15 °C above ambient.

The original temperature condition is shown in Fig. 4.16(c). It is assumed that the temperature difference Δ_1 between the oil and the windings is constant. Taking the inlet temperature as reference, then the transformer top-oil temperature is θ_1, the mean oil temperature is $\frac{1}{2}\theta_1$, the mean winding temperature is θ_m and the temperature-difference is $\Delta_1 = \theta_m - \frac{1}{2}\theta_1$. The rate of loss transfer from the transformer to the oil is proportional to Δ_1. The area a_1 of the height/temperature diagram is a measure of the oil flow-rate. The numerical values are

$$\theta_1 = 40; \tfrac{1}{2}\theta_1 = 20; \theta_m = 50; \Delta_1 = 50 - 20 = 30 \text{ °C};$$

$$a_1 = \tfrac{1}{2} \times 0.5 \times 40 = 10 \text{ °C-m}.$$

With the heat-exchanger raised by 0.75 m the shape of the height/temperature diagram changes to that in (d). The new values are

$$\theta_2, \text{ unknown}; \theta_m = 50; \Delta_2 = 50 - \tfrac{1}{2}\theta_2;$$
$$a_2 = \theta_2 \times \tfrac{1}{2}(0.75) + 1.25) = \theta_2.$$

The mean winding temperature is

$$\theta_m = 50 = \tfrac{1}{2}\theta_2 + \Delta_2 = \tfrac{1}{2}\theta_2 + \Delta_1(a_2/a_1)(\theta_2/\theta_1)$$

$$= \tfrac{1}{2}\theta_2 + 30\frac{\theta_2}{10}\frac{\theta_2}{40}$$

whence $\theta_2 = 23\,°C$ and $\Delta_2 = 50 - 11\frac{1}{2} = 38\frac{1}{2}\,°C$. The loss rate is increased by the factor $\Delta_2/\Delta_1 = 1.3$. The implication in terms of load rests on several factors, such as the ratio of core to I^2R loss on rated load, the relative effectiveness of heat transfer to the oil from the core and the windings respectively, the reduction of the oil fluidity as a result of the lowering of the oil temperature, and the reduction of heat dissipation from the tank itself. Nevertheless, within the validity of the assumptions, the advantage of raising the centre of cooling of the heat-exchanger is evident.

It is also clear that a further increase in loss dissipation could be obtained (i) by forced oil circulation to reduce Δ, and (ii) by placing fans below the heat-exchanger to increase thermal emission to the ambient air.

Multiflow Principle
The thermal rating of power transformers is based on the average temperature-rise of the windings as determined from the increase in resistance (Sect. 4.7), because this is the best practical measurement. The insulation life, however, depends on the temperature of the hottest region of the windings, which exceeds the average by some value dependent on the individual design characteristics; and it is not measurable directly.

Even for natural cooling the assumptions made in Fig. 4.16(*a*) are an oversimplification. With normal forced-oil cooling, the greater part of the flow will be outside the windings, between their perimeter and the inner wall of the tank. Only a small part of the oil will actually pass through the windings because of the greater hydraulic resistance. The top-oil temperatures of the streams will differ before they mix and leave the top of the transformer. On this *multiflow principle* it would be possible to identify, for example, 43 °C rise in oil temperature at the top of the h.v. windings, 52 °C for the l.v. winding, and 39 °C for the oil flow in the tank external to the windings, with the combination of the three oil streams to give a 42 °C rise of the oil outlet to the cooler. Carruthers and Norris [25] have shown that the hottest-spot temperature-rises can be more closely estimated on the multiflow principle from measurements made in a single heat run, and the estimate freed from the basis of temperature-rise of windings by resistance. As a result they show that it is feasible in forced-oil cooled transformers to allow an increase from 65 to 75 °C in the average winding temperature-rise without any increase in either the hottest-spot or hottest-oil temperature. The multiflow principle has not created this increase; rather it has shown the possibility by a more sophisticated examination of the cooling conditions.

4.7   *Limits of temperature-rise*
The maximum working temperature of a transformer or a machine depends upon the thermal quality of its insulation. It is commonly remarked that a rise of 6°C in the operating temperature halves the life of insulation, and although the figures may be arguable, the basic truth is amply supported by experience. The classes into which insulating materials are divided are set out in Sect. 3.1 together with their maximum working temperatures. The temperature-rise

limits are such as to keep within these levels from arbitrary ambient or inlet coolant temperatures [25, 26].

Measurement of temperature
The temperature of parts of a transformer or electrical machine may be measured by

T: thermometer, applied to the surface of the part. This gives the temperature of the *surface* at one point only;
E: embedded temperature detector (thermo-couple or resistance coil), which gives the temperature at one *internal point*;
R: resistance, involving the measurement of the resistance both cold and hot, and estimating the *mean* rise by use of the resistance temperature coefficient: it is available for windings only.

The designations T, E and R are used in the following Tables to indicate the methods by which the limits of temperature-rise specified are to be determined.

These methods give results that do not refer to the same thing and furnish different bases for estimating hot-spot temperatures. If an embedded temperature detector can be placed, during construction, at a point at which experience shows the temperature to be highest, it will give an indication of the hot-spot temperature. In practice, however, the problem of insulation prevents the detectors from being placed too near the conductors.

Although methods are available for measuring the resistance of a winding in operation, it is more usual to make resistance (or thermometric) measurements immediately after shut-down. Extrapolation is then necessary to cover the period between actual shut-down and the start of temperature measurement. The temperature/time relation during cooling is plotted and extrapolated backward to the instant of shut-down by means of a simple application of the exponential cooling curve, Sect. 4.2.

Transformers
The temperature-rises given below are for transformers (other than miniature or other special types) operating at altitudes not exceeding 1 000 m above sea-level. For water cooling, the inlet water temperature must not be greater than 25 °C; for air cooling the limits of ambient temperature are −25 and +40 °C, with the proviso that the air temperature shall not exceed a mean of 30 °C in any one day, nor 20 °C over a year. Adjustments are made for conditions outwith those stated. The cooling nomenclature (i.e. the designations AN, AF, etc.) is detailed in Sect. 5.17.

Industrial rotating machines
The Table on page 139 refers to machines other than those intended for ship, aircraft or traction application, turbo-generators, and fractional-kilowatt machines. Temperature-rises are based on an inlet coolant temperature of 40 °C, and continuous rating at a voltage not exceeding 11 kV. Temperature-rises are reduced by 1 °C per kV for the excess of the working voltage above

Transformer temperature-rise, °C

Part	Cooling method	Oil circn.	Insulation class					
			A	E	B	F	H	C
DRY								
Windings	R	AN, AF	–	60	75	80	100 125 150	
Cores near windings	T	any	–	60	75	80	100 125 150	
not near windings	T	any	–	non-injurious value				
OIL-IMMERSED								
Windings	R	NA, FA W(int)	N	65				
	R	FA W(ext)	F	65				
Top oil	T	–	–	60 sealed or with conservator				
	T	–	–	65 unsealed and without conservator				
Cores	T	–	–	non-injurious value				

11 kV, and may be increased by agreement for short-time ratings. Adjustments are necessary for abnormal coolant temperatures and for altitudes above 1 000 m.

4.8   *Thermal rating of transformers*

The nominal rating of a transformer is a conventional value of apparent power conforming to the specification and test requirements, and establishing the 'rated' primary and secondary currents. But the performance and useful life of a transformer are functions of the temperature of its insulation, so that loads in excess of the nominal rating can be sustained for limited periods provided that the temperature limitations are not exceeded. This is a consequence of the thermal lag provided by the heat capacity of the transformer as expressed by eq.(4.1). However, the observations of numerous investigators have shown that, over the temperature range 80–140 °C, the rate of reduction of life-expectancy of transformer insulation is doubled for every increase of 6 °C in its hot-spot temperature. Overload conditions are therefore classified as (i) those that do not impair the 'normal' life-expectancy of some tens of years, and (ii) those in which a more rapid deterioration is acceptable. Estimation of the crucial hot-spot temperature θ_h is based on the simple thermal diagram of Fig. 4.16(c), using the top-oil temperature θ_1 and the temperature-difference Δ in the expression

$$\theta_h = \theta_1 + 1.1\,\Delta$$

For (i) the hot-spot temperature must not exceed 98 °C. For a 20 °C ambient this corresponds to a hot-spot temperature-rise of 78 °C, viz. 13 °C above the mean winding temperature-rise of 65 °C. For (ii) the hot-spot temperature must not exceed 140 °C, and the current must not be greater than 1.5 p.u. of rated value (except in emergency) in order to avoid difficulties that would otherwise arise from the limited thermal capacity of associated bushings, tap-changers, cable-ends and circuit-breakers.

Loading tables [107] appropriate to various forms of natural and forced cooling methods are available, from which acceptable over-loading conditions can be determined. They are based on a simplified daily load cycle of constant load K1 of duration $(24 - t)$ hours and a load K2 for t hours. For parti-

Industrial machine temperature-rise, °C

| Part | | Insulation class | | | | |
---	---	A	E	B	F	H
A.C. WINDINGS						
Air-cooled:						
Output 5 MW or more, or core-length						
1 m or more	RE	60	70	80	100	125
Smaller than above	R	60	75	80	100	125
Indirect hydrogen-cooled:						
At absolute pressure (bar) 1–1.5	E	–	–	80	100	
2	E	–	–	75	95	
3	E	–	–	70	90	
4	E	–	–	65	85	
Direct-cooled:						
By hydrogen or liquid	E	–	–	80	100	
COMMUTATOR WINDINGS	T	50	65	70	85	105
	R	60	75	80	100	125
D.C. FIELD WINDINGS						
Air- or hydrogen-cooled:						
High-resistance, insulated conductors	T	50	65	70	85	105
	R	60	75	80	100	125
Low-resistance single-layer with						
insulated conductors	TR	60	75	80	100	125
Exposed bare conductors	TR	65	80	90	110	135
OTHER PARTS						
Short-circuited windings with insulated conductors; cores in contact with insulated windings	T	60	75	80	100	125
Commutators and slip-rings	T	60	70	80	90	100
Short-circuited windings with uninsulated conductors; cores not in contact with insulated windings	T	non-injurious value				

cular load cycles not tabulated it is possible to estimate the conditions by interpolation.

4.9   *Thermal rating of machines*

The Table of temperature-rise limits θ_m for industrial machines is interpreted in accordance with the duty that the machine is called upon to perform. Although this is rarely predictable in precise terms, it may be comparable with one of a number of internationally agreed standard duty cycles (BS 2613), designated as follows:

S1	Continuous running at rated load
S2	Short-time operation
S3	Intermittent periodic operation
S4	As S3 but with starting
S5	As S3 but with electric braking
S6	Continuous cyclic operation
S7	As S6 but with electric braking
S8	As S6 but with related load/speed characteristic

The operating conditions are illustrated with associated temperature-rise/time curves in Fig. 4.17. The conditions are denoted

N	operation under rated conditions of load
R	machine at rest and de-energized
D	starting duty
F	braking duty
V	operation at no load, but rotating

Cyclic-duration Factor: For the more elaborate duty cycles a cyclic-duration factor is defined: it influences the class of rating. The definitions, referred to the diagrams in Fig. 4.17, are

| S3: | $N/(N + R)$ | S4: | $(D + N)/(D + N + R)$ |
| S5: | $(D + N + F)/(D + N + F + R)$ | S6: | $N/(N + V)$ |

Rating Classes

These are brief indications of the rating on which the design of the machine is based.

Maximum Continuous (MCR): continuous operation as in S1.

Short-time (STR): the load durations in S2 are usually 10, 30, 60 and 90 min, and the limits of temperature-rise may, by agreement, be increased by 10 °C.

Equivalent Continuous (ECR): this is for test purposes in the case of duties S1 and S2.

Duty Type (DTR): corresponding to S3–S8. The time for a cycle should be 10 min, and the cyclic duration factors should be 0.15, 0.25, 0.40 or 0.60. The information required for determining the d.t.r. includes the stored kinetic energy of the motor and of the load, the duration of the cycle and the cycle-duration factor.

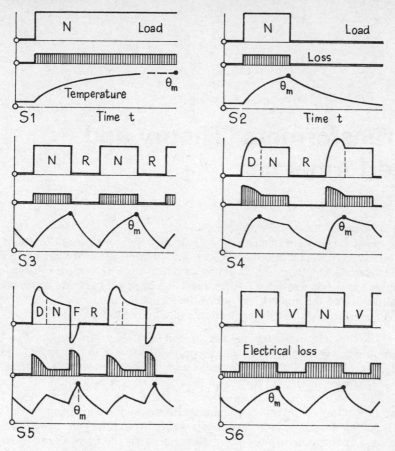

4.17 Standard duty-cycles for rotating machines.

Under the stated conditions the maximum temperature-rise must not exceed the appropriate value in the Table for the part of the machine considered; except that for cage machines operating on a d.t.r., the cage temperature-rise in the *middle* of the final operating cycle following the attainment of equilibrium thermal conditions shall not exceed the value given in the Table.

5

Transformers: Theory and Performance

5.1 *Theory of the power transformer*

The essential behaviour of an *ideal* power transformer under conditions of sinusoidal excitation is discussed in Sect. 1.9. In brief, the behaviour is expressed in terms of the characteristic ratios set out in eq.(1.8), using the conventions in Fig. 1.6 in which the positive direction of power (whether active or reactive) is into the primary and out of the secondary windings.

An actual transformer differs from the ideal in the following respects:

(i) The core is not infinitely permeable, and to establish the working flux it requires a magnetizing m.m.f. F_0. The m.m.f. balance is no longer $F_1 + F_2 = 0$, but now becomes $F_1 + F_2 = F_0$. As the core has a non-linear B/H relation, the magnetizing m.m.f. depends upon the degree of core saturation.

(ii) The core is the seat of loss arising from alternating magnetization, so that the exciting (primary) winding must also supply the core-loss power.

(iii) The windings have individual resistances between terminals, of values determined by the resistivity of the conductor material and augmented by eddy-current loss.

(iv) The windings are not perfectly coupled, so that not all the flux developed by the primary links the secondary, and the latter produces further flux not linking the primary. The effect is taken into account by assigning to each winding the property of leakage inductance. As the leakage fluxes are established in paths partly external to the core, the leakage inductances can be taken as constant. At a given operating frequency they can be represented in an equivalent circuit by leakage reactances.

The actual transformer can now be taken as an ideal transformer with resistance r_1 and leakage reactance x_1 in series with the primary winding, and correspondingly r_2 and x_2 in series with the secondary, as in Fig. 5.1 (a), to take account of the defects (iii) and (iv). The m.m.f. F_0 is introduced by means of a magnetizing reactance x_m shunted across the primary terminals of the ideal transformer to take a magnetizing current I_{0r} such that $N_1 I_{0r} = F_0$. The magnetizing current is in time phase with the core flux, and is taken for simplicity to be sinusoidal, an assumption untrue for practical levels of core saturation but acceptable because I_{0r} is usually a very small fraction of the rated current.

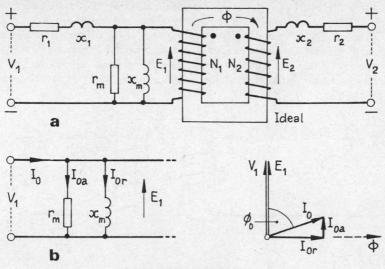

5.1 Transformer on no load.

To represent core loss, a resistance r_m is connected in parallel with x_m; again this is an approximation, for the loss is not strictly proportional to the square of the flux. The shunt resistance r_m takes an active current I_{0a} such that $r_m = P_0/I_{0a}^2$ where P_0 is the core loss.

We have in this way taken account of the defects (i)–(iv) by representing them in an equivalent circuit, Fig. 5.1 (*a*), containing an ideal transformer with r_1 and x_1, r_2 and x_2, r_m and x_m external to it. The combinations $z_1 = (r_1 + jx_1)$ and $z_2 = (r_2 + jx_2)$ are referred to as the *leakage impedances.*

Let the ideal-transformer part of the system carry a sinusoidally alternating mutual (or working) magnetic flux of peak value Φ_m and angular frequency $\omega = 2\pi f$. By definition this flux links every turn, inducing therein the e.m.f. $e_t = d\Phi/dt$ having the peak value $\omega\Phi_m$ and the r.m.s. value

$$E_t = \omega\Phi_m/\sqrt{2} = \sqrt{2}\pi f\Phi_m = 4.44 f\Phi_m \tag{5.1}$$

The total primary e.m.f. in N_1 turns in series is $E_1 = N_1 E_t$, and for the secondary correspondingly $E_2 = N_2 E_t$. We shall first consider the secondary terminals to be open-circuited: this is the *no-load* condition.

No-load Condition
The actual core requires the magnetizing m.m.f. F_0, which in equivalent-circuit terms is the current in x_m, namely I_{0r}. Further, the core loss is provided in the equivalent circuit by the current I_{0a} in r_m. The e.m.f. E_1 balances the primary applied voltage V_1 (neglecting the very small volt drops due to the magnetizing and loss currents flowing through the primary leakage impedance z_1). Thus $E_1 = V_1$ and the no-load primary input current is

$$I_0 = \sqrt{(I_{0r}^2 + I_{0a}^2)}$$

at a phase angle to V_1 (or E_1) of ϕ_0. Typically, I_0 is 0.05 p.u. of normal rated current. The phasor diagram corresponding to no-load conditions in Fig. 5.1(*b*) shows the active and reactive components of the primary no-load current, with the volt drops due to I_0 in r_1 and x_1 neglected. These drops could be typically 0.0005 and 0.003 p.u. respectively of normal primary voltage. The no-load power is therefore almost identical with $V_1 I_{0a} = P_0$, which is the core-loss power.

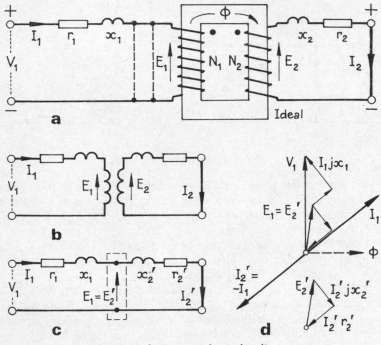

5.2 Transformer on short circuit.

Short-circuit Condition

Suppose the secondary terminals to be short-circuited, as in Fig. 5.2(*a*). The secondary e.m.f. E_2 is absorbed entirely in circulating I_2 through the leakage impedance $z_2 = r_2 + jx_2$, and the current is likely to be very large. To limit I_2 to the level of normal rated current it will be necessary to limit the primary voltage V_1 to about 0.1 or less of normal. The primary current of the ideal transformer is $I_1 = I_2(N_2/N_1)$ and its e.m.f. $E_1 = E_2(N_1/N_2)$. The primary voltage is employed in providing E_1 and the volt drop in the primary leakage impedance z_1. As the primary and secondary per-unit leakage impedances are roughly equal, it follows that E_1 is about one-half of V_1. With V_1 equal to 0.1 p.u. then E_1 is about 0.05 p.u., which means that the mutual flux Φ_m is of the same order. The core loss, which is proportional approximately to the square of the flux density, is consequently about 0.0025 of what it was on no load; further, the magnetizing current is greatly reduced, partly by reason of the reduced flux and partly because of the low saturation level. In the equivalent circuit it is legitimate to ignore completely the shunt magnetization branches r_m and x_m, to give the simplified equivalent circuit of Fig. 5.2(b) and the corresponding phasor diagrams for primary and secondary. The turns-ratio can have any designed value: if it is large, $E_1 \gg E_2$ and $I_2 \gg I_1$, making the phasor diagrams somewhat awkward. We therefore adopt the unity turns-ratio equivalent.

Unity Turns-Ratio: Suppose that $N_1 = N_2$ and that consequently $E_1 = E_2$ in the ideal-transformer part of the equivalent circuit. Energy is now transferable from primary to secondary without any change of voltage or current, so that there is no need to have the ideal transformer at all. To retain such simplicity when $N_1 \neq N_2$ it is necessary to imagine the actual secondary winding of N_2 turns to be replaced by an equivalent winding of N_1 turns with the proviso that there is no change in the I^2R loss nor in the effective leakage reactance. For this the equivalent secondary must have a resistance r_2' and leakage reactance x_2' such that $I_1{}^2 r_2' = I_2{}^2 r_2$ (as the secondary current is now $I_2' = I_1$), and $I_1{}^2 x_2' = I_2{}^2 x_2$: hence the resistance and leakage reactance of the equivalent N_1-turn secondary are

$$r_2' = k_n{}^2 r_2 \quad \text{and} \quad x_2' = k_n{}^2 x_2 \tag{5.2}$$

where $k_n = (N_1/N_2) = E_1/E_2 = I_2/I_1$. This result could be inferred directly from the impedance transformation of an ideal transformer, discussed in Sect. 1.9.

The equivalent circuit for a power transformer with secondary short-circuited can now be reduced to that in Fig. 5.2(c) and the phasor diagrams (d) for primary and secondary combined. Obviously, if the magnetizing branches are ignored, the condition becomes that of a primary voltage V_1 applied to a simple series circuit of total resistance R_1 and leakage reactance X_1 given by

$$R_1 = r_1 + r_2' \quad \text{and} \quad X_1 = x_1 + x_2' \tag{5.3}$$

As the terms 'primary' and 'secondary' describe merely which winding accepts electrical energy and which delivers it, the argument above can equally well be applied to a transformer in which the actual primary turns N_1 are assumed to be changed to N_2 and the primary current from I_1 to I_1'. It is immaterial which winding is used as the 'base'.

EXAMPLE 5.1: A 400 kVA 11/0.415 kV 3-ph delta/star-connected transformer has on rated load an h.v. I^2R loss of 2.46 kW and l.v. loss of 1.95 kW. The total leakage reactance is 0.055 p.u. Find the ohmic values of resistance and leakage reactance.

The phase voltages are 11 kV and 240 V. The phase currents are 12.1 A and 555 A, and the I^2R losses per phase are 0.82 kW for the h.v. side and 0.65 kW for the l.v. side. Hence $r_1 = 820/12.1^2 = 5.6\ \Omega/\text{ph}$ and $r_2 = 650/555^2 = 2.1\ \text{m}\Omega/\text{ph}$. Referred to the h.v. winding, the total resistance is

$$R_1 = 5.6 + 0.0021(11/0.240)^2 = 5.6 + 4.4 = 10.0\ \Omega/\text{ph}$$

giving $I_1{}^2 R_1 = 12.1^2 \times 10.0 = 1.47$ kW/ph, which is of course the sum $(0.82 + 0.65)$ kW/ph. Referred to the l.v. side

$$R_2 = 5.60(0.240/11)^2 + 0.0021 = 0.0027 + 0.0021 = 4.8 \text{ m}\Omega/\text{ph}$$

The reactance referred to the h.v. side is such that $I_1{}^2 X_1/V_1 I_1 = I_1 X_1/V_1 = 0.055$, whence

$$X_1 = 0.055 \times 11\ 000/12.1 = 50\ \Omega/\text{ph}$$

and referred to the l.v. side

$$X_2 = 50(240/11\ 000)^2 = 24.0 \text{ m}\Omega/\text{ph}$$

or obtained directly, $X_2 = 0.055(240/555) = 24.0$ mΩ/ph. The division of X_1 or X_2 between the h.v. and l.v. windings is entirely arbitrary: a common way is to put one-half in each. Then

$$x_1 = 25\ \Omega, x_2{}' = 25\ \Omega \quad \text{or} \quad x_2 = 12\ \text{m}\Omega, x_1{}' = 12\ \text{m}\Omega.$$

5.2   Power transformer on load

If a load impedance Z is connected across the secondary terminals, the e.m.f. E_2 will drive an output current I_2 through it, and the secondary winding will impress the m.m.f. $F_2 = N_2 I_2$ on the magnetic circuit. To maintain the working flux Φ_m and the primary e.m.f. E_1 (which must always be in near balance with the applied voltage V_1) the primary must take an input current I_1 to develop an m.m.f. $F_1 = N_1 I_1$ to balance F_2 and at the same time to provide for the magnetizing m.m.f. F_0. Thus a secondary current causes a counterpart to flow in the primary, and the combined primary and secondary m.m.f.s must result in a net magnetizing m.m.f. F_0.

If we adopt the equal-turns-ratio condition with $N_2 = N_1$, primary and secondary currents can be used to represent also the winding m.m.f.s, making the phasor diagram more easily drawn. It is shown for a partially inductive secondary load in Fig. 5.3. The equivalent secondary current is $I_2{}'$. The primary current I_1 is the phasor sum of $-I_2{}'$ and I_0, the latter corresponding very nearly to the no-load current as the mutual flux is only slightly less than on secondary open-circuit. For load conditions the phasor diagram incorporates the basic features of both the no-load and short-circuit conditions.

The applied primary voltage V_1 balances E_1 and provides for the primary resistance and leakage reactance components $I_1 r_1$ and $I_1 j x_1$. On the secondary side the e.m.f. $E_2{}'$ circulates $I_2{}'$ through the equivalent load impedance and provides for the secondary leakage impedance volt drops $I_2{}' r_2{}'$ and $I_2{}' j x_2{}'$. The secondary phase angle is ϕ_2, the secondary operating in a 'generator' mode. The primary phase angle ϕ_1 differs from ϕ_2 by reason of the necessary provision of reactive (magnetizing) power in addition to that required by the secondary load. The current I_0 and the leakage impedance volt drops in the phasor diagram correspond to those that might be expected in a small transformer: for a large unit they are very much smaller.

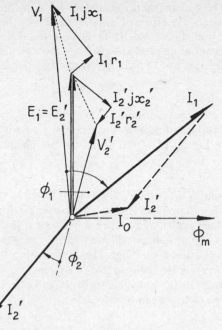

5.3 Transformer on load: phasor diagram.

5.3 *Losses and efficiency*

The transformer has no moving parts, and its efficiency has far less significance than its losses, which are:

Core Loss resulting from the alternating core magnetization.
Dielectric Loss in insulating materials, particularly in the oil and solid insulation of high-voltage units.
I^2R *Loss* in the windings.
Load (Stray) Loss, largely the result of leakage magnetic fields inducing eddy currents in the tank walls and conductors.

A close approximation to the losses can be obtained from open- and short-circuit tests, without operating the transformer on load.

No-Load Loss

With open-circuited secondary, the primary current is I_0. The I^2R loss due to it is in most cases quite negligible (e.g. 1/400 of the rated normal I^2R loss). The no-load power input is consequently concerned with core and dielectric loss, the latter having significance only for transformers operating at very high voltages. The e.m.f. E_1 is substantially identical with V_1, the conditions being represented by the equivalent circuit of Fig. 5.1 in respect of the core magnetization. If the dielectric effect is negligible the no-load power input $P_0 = V_1 I_{0a}$ is the loss P_i in the iron of the core. In large h.v. transformers, however, it will also include the loss P_d in the dielectric material. In the latter case, too, the core magnetizing current I_{om} is accompanied by a leading capacitive current I_{od}, the effect of which is to reduce the lagging reactive component I_{or} of the no-load current. Thus the core and dielectric losses are measurable by means of an *open-circuit test* at normal voltage and frequency. They are generally assumed to be *constant* and independent of the load.

Short-Circuit Loss

If rated secondary current is circulated on short circuit by means of a low value of primary voltage of normal frequency, the circuit in primary terms is effectively that in Fig. 5.2(*d*), from which the complexor relation is

$$V_1 = I_1 [(r_1 + r_2') + j(x_1 + x_2')] = I_1 (R_1 + jX_1) = I_1 Z_1$$

and the power input $I_1^2 R_1$, which is the total I^2R loss in the primary and secondary windings. In practice the active power input includes (stray) load loss, for with normal primary and secondary currents the leakage fluxes are normal (or nearly so), producing eddy-current loss in metallic parts of the transformer adjacent to the windings. Thus a *short-circuit test* at rated current and frequency can establish the load loss to a close approximation. It also gives information about the leakage impedance and effective resistance, from which the leakage reactance can be calculated.

Power Efficiency

The peak flux normally varies so little between no-load and full-load conditions that the core loss can be considered as constant at the no-load value P_i. If the short-circuit loss at rated current be P_c, the loss at any fraction k of the rated load S is $k^2 P_c$. The total loss at load kS of power factor $\cos \phi$ is $P_i + k^2 P_c$, and the output/input power ratio, i.e. the power efficiency, is

$$\eta = \frac{kS \cdot \cos \phi}{kS \cdot \cos \phi + P_i + k^2 P_c} = 1 - \frac{P_i + k^2 P_c}{kS \cdot \cos \phi + P_i + k^2 P_c} \tag{5.4}$$

Solving $d\eta/dk = 0$ gives $P_i = k^2 P_c$ for the maximum efficiency condition, i.e. a load of the fraction $k = \sqrt{(P_i/P_c)}$ of full-load rating for which the variable I^2R loss is equal to the constant core loss.

Eq.(5.4) shows that the efficiency is dependent upon the load power factor. Greatest efficiency is naturally obtained with a power factor of unity; for a purely reactive load the efficiency is zero, but the losses are the same. It is for this reason that the losses are more significant than the efficiency.

EXAMPLE 5.2: A 300 kVA transformer has a core loss of 1.5 kW and a full-load I^2R loss of 4.5 kW. Calculate its efficiency for 25, 58 (maximum efficiency load), 75, 100 and 125% of rated load at power factors respectively of unity, 0.8 and 0.6.

The core loss is 1.5/300 = 0.005 p.u. and the full-load I^2R loss is 0.015 p.u. Calculations in per-unit values based on eq.(5.4) with the full-load rating $S = 1.0$ p.u. and the full-load active power $P = 1.0 \cos \phi$ p.u. are given in the Table:

Load, p.u.		0.25	0.58	0.75	1.0	1.25
Output p.f.	1.0	0.25	0.58	0.75	1.00	1.25
power:	0.8	0.20	0.46	0.60	0.80	1.00
	0.6	0.15	0.35	0.45	0.60	0.75
Core loss		0.005	0.005	0.005	0.005	0.005
I^2R loss		0.0009	0.005	0.0084	0.015	0.023
Total loss		0.0059	0.010	0.0134	0.020	0.028
Effici- p.f.	1.0	0.968	0.983	0.982	0.980	0.978
ency:	0.8	0.971	0.979	0.978	0.976	0.972
	0.6	0.962	0.972	0.971	0.968	0.963

Energy Efficiency

A transformer in circuit continuously has a steady core-loss rate. Its I^2R energy loss depends on the square of the load. The annual cost of losses therefore depends on the hour-to-hour variation in the transformer loading. Fig. 5.4 shows two hypothetical daily load curves each of the same *peak* and the same *average* daily value, the load factor (mean/maximum) being 0.5 for both. But they produce different I^2R losses, averaging 0.5 of the maximum for (*a*) but only 0.3 for (*b*). For minimum daily or yearly energy loss, the I^2R/core loss-ratio should be chosen in accordance with the type of load variation. This is by no means easy to predict; a rough relation between the load-factor of the load and the corresponding load-factor of the I^2R loss in typical supply network loadings is:

Load-factor of transformer load: 0.30 0.40 0.50 0.60 0.70 0.80
Load-factor of I^2R loss: 0.16 0.25 0.35 0.47 0.60 0.72

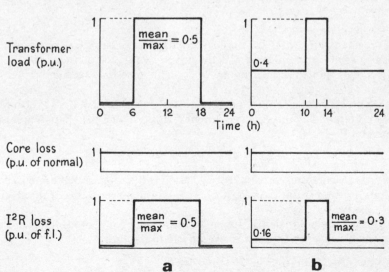

a   b

5.4 Load- and loss-energy relation.

The total loss in a transformer working on a load with a load-factor of 0.5 may be expected to consist of the normal constant core loss together with 0.35 p.u. of the rated I^2R loss. On the basis of energy-loss equality this suggests that a transformer with a full-load I^2R equal to $1.0/0.35 \simeq 3$ times the core loss is likely to be the most economical on the assumption that the cost of wasted energy is the same per unit for each loss. This is not necessarily so, for the I^2R loss predominates when the system load is high and energy most valuable. Evidently the choice of a transformer on economic grounds is a complex matter. Briefly, it is necessary to consider capital cost and the charges thereon, life, depreciation, load factor, loss factor, loss ratio and energy cost: a formidable list. A computer solution is obtainable if all these can be analytically related.

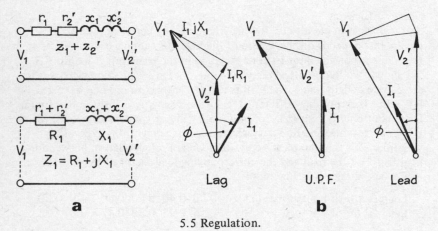

5.5 Regulation.

5.4 Regulation

The regulation is the fall (or rise) of secondary terminal voltage between no-load and full-load conditions with the primary voltage at constant rated value. It is usually quoted as a percentage or per-unit value for a specified power factor.

On no load the secondary terminal voltage is very nearly $V_2' = V_1$. On full load corresponding to a primary current I_1 it is given by the phasor expression based on Fig. 5.5(a):

$$V_2' = V_1 - I_1(z_1 + z_2') = V_1 - I_1 Z_1$$

ignoring magnetizing current. The regulation is the *scalar* difference between V_1 and V_2', and as shown in (b) it depends on the power factor of the secondary load.

It is convenient to define the *per-unit resistance* (or per-unit full-load I^2R loss) and the *per-unit reactance* respectively in the form

$$\epsilon_r = I_1{}^2 R_1/V_1 I_1 = I_2{}^2 R_2/V_2 I_2 = P_c/S$$

$$\epsilon_x = I_1{}^2 X_1/V_1 I_1 = I_1 X_1/V_1 = I_2 X_2/V_2 \tag{5.5}$$

These can be combined to give the *per-unit leakage impedance*

$$\epsilon_z = \sqrt{(\epsilon_r{}^2 + \epsilon_x{}^2)} \tag{5.6}$$

Multiplied by 100 these give the percentage resistance, leakage reactance and leakage impedance respectively. The voltage phasor diagram in Fig. 5.6 uses

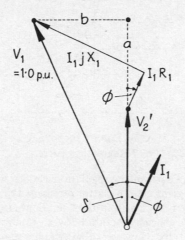

5.6 Assessment of regulation.

per-unit values to obtain the regulation. With $V_1 = 1.0$ p.u., then $a = \epsilon_r \cos \phi + \epsilon_x \sin \phi$ and $b = \epsilon_x \cos \phi - \epsilon_r \sin \phi$, where ϕ is the phase angle of the secondary load. The angle δ between V_1 and $V_2{}'$ is defined by $\delta \cong b/V_1 = b$, and $\cos \delta = (V_2{}' + a)/V_1 = V_2{}' + a$. Expanding $\cos \delta = \cos b$ as a series, $(V_2{}' + a) = 1 - \frac{1}{2}b^2 + \ldots$, whence the per-unit regulation is given by the *scalar* equation

$$\epsilon = V_1 - V_2{}' = 1 - V_2{}'$$

$$= \epsilon_r \cos \phi + \epsilon_x \sin \phi + \tfrac{1}{2}(\epsilon_x \cos \phi - \epsilon_r \sin \phi)^2 - \ldots \tag{5.7}$$

If ϵ_z is less than 0.1 (or 10%) the squared term can be ignored to give

$$\epsilon = \epsilon_r \cos \phi + \epsilon_x \sin \phi \tag{5.8}$$

The full-load regulation is a maximum when $\phi = \arctan(\epsilon_x/\epsilon_r) = \arctan(X_1/R_1)$; this is at a moderately low lagging power factor. It is zero for $\phi = -\arctan(\epsilon_r/\epsilon_x)$ corresponding to a high leading power factor, and for lower leading power factors the secondary voltage *rises* between no-ioad and full-load condition. A typical case for $\epsilon_r = 0.01$ p.u. $= 1\%$ and $\epsilon_x = 0.04 = 4\%$ is shown in Fig. 5.7, the diagrams in Fig. 5.5(*b*) providing an explanation in phasor terms.

Regulation is a numeric, not a complex quantity, so that the regulation produced by two impedances connected in series can be combined algebraically. Thus we could write

$$\epsilon = \epsilon_{12} = \epsilon_1 + \epsilon_2$$

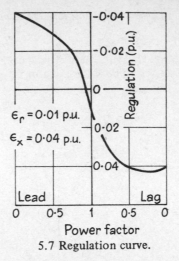

5.7 Regulation curve.

where ϵ_1 and ϵ_2 are the parts of the regulation associated with the individual impedances. In the present case the latter represent the primary and equivalent secondary leakage impedances in the equivalent circuit of a two-winding transformer. There is a clear distinction between r_1 and r_2', but the division in $X_1 = x_1 + x_2'$ is arbitrary. However, if the windings are regarded as having distinct leakage reactances, each can be credited with individual leakage impedances and, consequently, per-unit regulations ϵ_1 and ϵ_2. This takes care of all mutual effects and is extremely convenient where the transformer has more than two windings.

5.5   Operational equivalent circuits
The transformer is a device in which electromagnetic phenomena occur with electric and magnetic fields of considerable structural complexity interacting to transfer energy from primary to secondary circuit; in which intricate heat-exchanges take place between hot metals and fluid coolants; and in which the mechanical structure is subject to forces of thermal, magnetostrictive and magnetic origin, and responds by deflection, deformation and vibration.

Within normal limits of load and working conditions, the transformer behaves electromagnetically as a simple low-loss voltage-changer, but during faults or transient system conditions its reaction can be complicated. In consequence, a transformer may have to be considered from a viewpoint appropriate to particular conditions of operation. Much labour can be saved by including only those characteristics that are essential and omitting those whose influence is small. This is a matter of selecting a 'model' to suit the conditions of immediate interest. Typical models are listed below.

Mechanical. The laminated core, not a very rigid structure, has cyclic dimensional changes due to magnetostriction and magnetic forces of repulsion between laminae, in both cases at twice excitation frequency. Mechanical forces are developed on and between coils. Coils may deform, and cores vibrate in complex modes.

Thermal. The thermal transfer and the flow of coolant are vital in determining maximum working temperatures and the consequent life of the insulation. Generally the heat dissipation system can be considered as a problem almost on its own.

Electromagnetic. The behaviour of a transformer subjected to steep-fronted voltage surges is determined initially by the consequent distribution of the electric field, subsequently by the interchange of electric and magnetic field energies. The model can thus be represented in terms of a capacitor-inductor network with damping.

Magnetic. Switching-in phenomena are markedly affected by magnetic saturation, making it necessary to include some form of *B/H* relation in the appropriate model. Even in normal operation the flux and e.m.f. harmonics resulting from saturation may have to be taken into particular account, and not all harmonic fluxes have a path wholly within the core.

Normal Operation. This model, developed in Fig. 5.1, is the network equivalent using leakage impedance concepts in terms of 'lumped' parameters. As has already been shown, it may itself be simplified to the forms of Fig. 5.1 (*b*) for no-load conditions (core loss and magnetizing current) and Fig. 5.2 for short-circuit conditions (I^2R loss and regulation).

System Network. In the analysis of supply-system operation, a transformer may be represented as a two-port, or even as a simple series impedance. In more precise studies account may be taken of the fact that the turns-ratio and the no-load voltage ratio are not quite the same, the divergence being greater with smaller ratings. Morris [23] has shown that *any* transformer can be accurately represented by the equivalent circuit in (*a*) of Fig. 5.8, constructed from direct measurements. The chain-dotted line represents a loss-free transformer which introduces the complex no-load voltage-ratio. The parameters are Z_m, a complex impedance found from the primary input voltage and current with the secondary open-circuited; Z_s, an impedance measured between the secondary terminals with the primary short-circuited; and t_v, the *complex* primary/secondary open-circuit voltage-ratio. The circuit (*a*) then follows from the Helmholtz-Thevenin theorem. An equivalent approach, shown in (*b*), is to invoke the theorem of passive quadripoles in which

$$V_1 = AV_2 + BI_2 \text{ and } I_1 = CV_2 + DI_2$$

where the quadripole parameters *ABCD* can all be found from open-circuit and short-circuit tests.

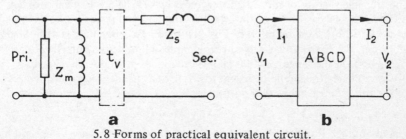

a   **b**

5.8 Forms of practical equivalent circuit.

5.6 *Vibration and noise*

The rapid growth of domestic electrification and the ubiquity of large power transformers for local distribution, make noise produced by such apparatus a problem of social consequence. Under no-load conditions, the 'hum' developed by energized power transformers originates in the core, where the laminations tend to vibrate by magnetic forces. The noise is transmitted through the oil to the tank sides and thence to the surroundings. The essential factors in noise production are consequently: (*a*) magnetostriction (e.g., an extension of 0.000 12% for a flux density of 1 T); (*b*) mechanical vibration developed by the laminations, depending upon the tightness of clamping, size, gauge, associated structural parts, etc.; (*c*) mechanical vibration of the tank walls; and (*d*) the damping.

This problem has been studied experimentally. It is established that magnetostriction is the first cause of transformer hum, but much of the noise depends on the natural frequency of vibration of the mechanical parts. Any constructional change which removes this frequency outside the audible range will be *ipso facto* an improvement. Transformer radiators and plain unreinforced tank walls are not usually troublesome as their natural frequencies lie as a rule below 100 Hz. Stiffening and clamping may do more harm than good if the effect is to raise the natural vibratory period into the audible range.

In general, the total noise emission may be reduced by (*a*) preventing coreplate vibration, which necessitates the use of a lower flux-density and attention to constructional features such as clamping bolts, proportions, and dimensions of the 'steps' in plate width, tightness of clamping and uniformity of plates; (*b*) sound-insulating the transformer from the tank by cushions, padding, or oil-barriers; (c) preventing vibration of the tank walls by suitable design of tank and stiffeners; and (*d*) sound-insulating the tank from the ground or surrounding air. There is no complete solution to the problem.

Core Vibration

If a core consisted of homogeneous steel, magnetostriction would cause vibration only in a plane parallel to that of the core. But as the laminations are not homogeneous, differential movements take place within the core, exciting vibrations in a direction perpendicular to the plane of the core. A core structure, having distributed mass and stiffness, possesses an infinite number of natural frequencies each associated with a particular mode. A few of these may lie in the range of magnetostrictive or magnetic-force harmonics, but the questionable rigidity of mitred corner joints makes the mechanical problem difficult to analyse. Some of the results computed and measured by Henshell and others [24] are illustrated in Fig. 5.9: (*a*) shows three modes of in-plane vibration, and (*b*) a typical out-of-plane mode, all modes having vibration frequencies in the range 40–300 Hz. Those modes of frequency near to a drive frequency (magnetic or magnetostrictive) can contribute considerably to the total transformer noise output.

5.7 *Harmonics*

The use of high flux densities in the cores of power transformers, imposed by

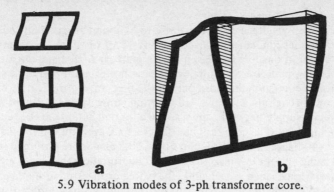

5.9 Vibration modes of 3-ph transformer core.

design requirements and the reduction of size, means that the saturation level is high. The typical magnetizing current waveform for a sinusoidally varying flux density, Fig. 2.14, clearly contains strong harmonics, chiefly those of order 3rd and 5th. Some of the implications can be seen by considering only the fundamental and the 3rd harmonic, as in Fig. 5.10(a), in which a flux density consisting only of the fundamental, of peak B_{1m}, induces the sinusoidal e.m.f. e but requires an exciting current i_{or} with a 3rd harmonic. If, as in (b), a purely sine magnetizing current were supplied, the flux density B would be flattened by saturation, introducing a 3rd harmonic flux density B_3. Neglecting hysteretic phase-shift, the flux density can be taken as the sum of a fundamental $B_{1m} \sin \omega t$ and a 3rd harmonic $B_{3m} \sin 3\omega t$, whence the e.m.f. per turn embracing a net core area A_i is

$$e = \omega A_i [B_{1m} \cos \omega t + 3B_{3m} \cos 3\omega t]$$

a peaked waveform as indicated in Fig. 5.10(b). We examine the consequences of this.

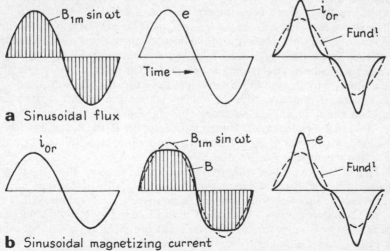

a Sinusoidal flux

b Sinusoidal magnetizing current

5.10 Flux, e.m.f. and magnetizing current harmonics.

Single-phase Transformer

A sinusoidal supply voltage cannot give rise to harmonic magnetizing currents directly; it supplies fundamental reactive power, part of which is converted to harmonic power in the core by reason of its nonlinear B/H characteristic. The flow of a 3rd harmonic current implies the presence of a 3rd harmonic e.m.f. and therefore a corresponding harmonic in the flux waveform. The harmonic currents flow through the effective 3rd harmonic impedances of the primary winding and its supply network, and also (to a limited extent) in the secondary winding and its load. The core flux has therefore a waveform intermediate between the two shapes shown shaded in Fig. 5.10, being more flattened (and the magnetizing current less peaked) if the 3rd harmonic impedances are high. If the *network* impedance is negligible, the whole of the 3rd harmonic e.m.f. is expended in driving the harmonic current through the appropriate triple-frequency impedance of the transformer and no harmonic voltage appears across the terminals of the primary; but if not, the supply voltage will show a 3rd harmonic distortion. The higher harmonic effects due to core saturation are similar to those described for the 3rd harmonic.

Three-phase Bank of Single-phase Transformers

Each of the three separate cores must carry the flux demanded by its operating conditions, but the effects are now modified by the phase interlinkage. In the following both supply and load are balanced and star-connected. The connection symbolism of the transformer bank is explained in Sect. 5.9.

(a) *Dd Connection.* In a symmetrical 3-ph voltage system, the fundamental and the harmonics of order $(6n \pm 1)$, where $n = 1, 2, 3 \ldots$, are all representable by phasors with a $120°$ displacement. But those of order $3n$ (for brevity termed the *triplen* harmonics) are voltages co-phasal in all three phase windings and peak simultaneously. Delta connection provides a closed path to triplen e.m.f.s, which circulate triplen currents in the closed mesh and are absorbed by the triplen leakage impedance drop: there is no resultant triplen voltage at the line terminals. In the Dd case, therefore, the action is as follows. The supply voltage provides only a fundamental magnetizing current, excites a flattened flux waveform with a triplen content and therefore induced triplen e.m.f.s. The latter circulate triplen current in the delta, to add to the sinusoidal supply current as in Fig. 5.10(a) and to restore the flux waveform nearly to a pure sinewave. Thus the primary voltage and current are sinusoids, the flux is nearly so, and the necessary triplen magnetizing current is circulated around the closed mesh. This condition applies to the 3rd, 9th, 15th . . . harmonics. Harmonics of other than triplen order (i.e. 5th, 7th . . .) behave like sets of balanced 3-ph quantities.

(b) *Yd and Dy Connection without Neutral.* Either connection operates as in (a), except that with a closed mesh on one side only, the triplen impedance is greater and the flux waveform diverges rather more from the sinusoidal.

(c) *Yy Connection without Neutrals.* The triplen e.m.f.s are all directed simultaneously away from or towards the star point and therefore cancel between any pair of lines. Consequently no triplen currents can flow, and (apart from the 5th, 7th . . . harmonic components) the input magnetizing currents are

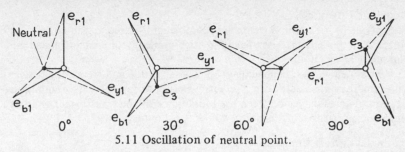

5.11 Oscillation of neutral point.

sinusoidal. The flux waveform is flattened, its triplen harmonics providing
the triplen e.m.f.s. Balance between lines of the supply voltage and the e.m.f.
is maintained, but the effect on the star-point voltage is to make it exhibit
the *oscillating neutral* phenomenon, Fig. 5.11. The fundamental e.m.f.s,
e_{r1}, e_{y1} and e_{b1}, are mutually displaced by $2\pi/3$ rad. The superposed 3rd
harmonic e_3 is co-phasal in all three legs, and the resultant phase voltages
are shown by the dotted lines. The diagram shows four successive instants displaced by 1/12 period, it being noted that the phasor e_3 rotates thrice as fast
as the fundamental system. All the line-to-star-point voltages fluctuate,
causing 'neutral oscillation'. The non-triplen harmonics form balanced systems
with $2\pi/3$ displacement and appear across the lines.

(*d*) *Yy Connection with Neutrals.* The presence of the neutral connection
effectively separates the three 1-ph transformers, the neutrals carrying the co-
phasal triplen compensating currents.

(*e*) *Other Yd Connections.* All variants of the possible connections will fall
under one or other of the cases (*a*) to (*d*) above. A delta connection always
provides a closed path for triplen currents; a neutral connection to the supply
allows triplen currents to flow into the supply network.

(*f*) *Stabilizing D (Tertiary) Winding.* This is an auxiliary winding on each
transformer, connected in delta and applied to Yy or Yz connected 1-ph
banks to provide a closed path in which triplen currents can circulate. In effect,
it decreases the zero-phase-sequence impedance.

Three-phase Unit Transformer

It is essential to distinguish between the cases where the phases are magnetically separate, and those where they are magnetically (as well as electrically)
interlinked. In the 3-ph shell-type transformer the magnetic circuits are separate and do not interact, and the discussion above for the bank of 1-ph transformers will apply. In the more common 3-ph 3-limb core-type unit, however,
this is not so, Any triplen flux harmonics are all upward or all downward at a
given instant in the limbs, and their return path must be external (i.e. in the
air or oil or the walls of the tank). These paths have a high reluctance and
there is a strong tendency to suppress triplen harmonic core fluxes. The resi-
duals may cause eddy-current loss in the tank, an effect mitigated if a copper
band is fitted around the inside perimeter of the tank.

In five-limbed cores, the end limbs provide a magnetic path for triplen
fluxes.

Harmonic Effects

The effects of harmonic *currents* are: additional I^2R loss due to circulating currents; increased core loss; magnetic interference with protective gear and communication circuits.

Harmonic *voltages,* depending on their magnitude and frequency, may cause: increased dielectric stress; electric-field interference with communication circuits; resonance between the inductance of the transformer windings and the capacitance of a feeder to which it is connected.

Harmonic Current Compensation

For core densities exceeding 1.5 T, the harmonic current content markedly rises. It may in some cases be reduced by special compensating arrangements.

A saturated Yy isolated-star-point core-type transformer may have a magnetizing-current waveform like (*a*) in Fig. 5.12. The 3rd harmonic is precluded by the connection, the 5th is predominant; further, the high reluctance

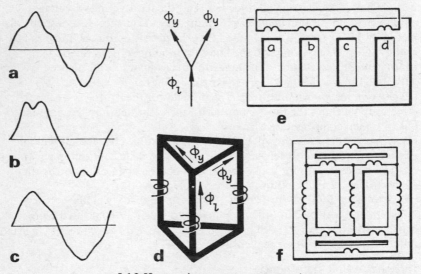

5.12 Harmonic current compensation.

to 3rd harmonic fluxes forces the flux towards the sinusoidal. If the transformer has a 5-limb core (*e*), however, a 3rd harmonic flux path is provided by the end limbs and the flux waveform flattens; the magnetizing current waveform (*b*) shows that the 5th harmonic is reversed and the fundamental slightly reduced. Comparison of (*a*) and (*b*) suggests that if the end limbs had a suitable reluctance, there would be zero 5th harmonic in the magnetizing current, as in (*c*). The method applies only to isolated star/star and to star/zigzag connections.

The parallel connection of a star/star with a delta/delta transformer results in suppression of 5th and 7th harmonic currents since these are equal and opposite in phase, cancelling in the combined input. The same effect in a single transformer may be obtained by providing in one magnetic circuit a 'star' and a 'delta' path. If a symmetrical 3-phase transformer, Fig. 5.12(d), is considered, it is seen that the limb flux Φ_l splits into two yoke fluxes Φ_y differing in phase each by 30° with respect to the limb flux, and requiring 5th and 7th harmonic currents displaced respectively by $\pm 5 \times 30° = 150°$ and $\pm 7 \times 30° = 210°$ with respect to the current components in the limb. The two yokes together, by symmetry, take 5th and 7th harmonic excitation in phase opposition to that of the limb to which they are magnetically connected, and the combination may be devoid of harmonic magnetizing currents if the reluctances are suitably chosen unless high saturation densities are used, in which case harmonic yoke excitation can be provided by a yoke winding. This magnetic star-delta arrangement is inherent in the 5-limb core (e), and could also be obtained in a 3-limb core (f) by slitting the yokes. A more practical method would be to wind round the yokes a delta-connected winding to force the fluxes into the required relationship in spite of any difference in reluctance. In (f) there is in addition an auxiliary star-connected winding interlinked with the delta winding on the yokes to aid production between limbs and yokes of the 5th and 7th harmonic currents.

5.8 Transients

Apart from thermal transients, short-duration switching currents and line-surge voltages may occur in transformer operation, of such magnitude as to impose high mechanical stresses or insulation damage.

Switching

When the primary winding of an unloaded transformer is switched on to normal voltage supply, it acts as a nonlinear inductor. Neglecting resistance, the applied voltage must be balanced at every instant by the e.m.f. induced by the magnetizing current. The behaviour depends on the magnitude, polarity and rate of change of the applied voltage at the instant of switching. Assume first that the core is initially unmagnetized and that the switch is closed at a *voltage peak*.The balancing e.m.f. demands a flux having maximum rate of change, so that the flux starts to grow from zero in the required direction, as also does the magnetizing current. But such conditions are those of normal no-load operation, Fig. 5.13(a), and there is no transient. If, however, the switch closes at a *voltage zero* as in (b), the voltage is positive during the first half-period, throughout which the flux must therefore increase, with a total *change* corresponding to that of a 'normal' half-period. It therefore reaches a maximum of twice the normal peak, i.e. $2\Phi_m$, a phenomenon called the *doubling effect*. In subsequent half-periods the effect of losses rapidly reduces the flux waveform to symmetry about the time-axis.

Flux doubling implies a peak density of twice the normal steady-state value. With the high saturation levels employed for modern transformers, the magnetizing current required for twice normal flux density is very great. For a peak density of 1.4 T, for example, the excitation is about 3 kA-t/m; for twice this density it might be increased 100-fold. A normal magnetizing current of 0.05 p.u. is for such conditions raised to 5 p.u., producing electro-magnetic forces 25 p.u. of normal. The doubled flux density will not in fact be reached, for with such overcurrents the transformer resistance drop is significant, and further limitation is imposed by the source impedance of the supply network. The transformer switching transient is referred to as the *inrush current*. It causes difficulties in the application of current-balance protection.

Remanent flux in the core can aggravate the condition. If the primary voltage is applied at a zero instant and in such a direction that the rising field augments the existing remanent flux Φ_r, as in Fig. 5.13(c), the sinusoid of flux variation has still the amplitude $2\Phi_m$, but is entirely offset from zero and rises to a peak value $2\Phi_m + \Phi_r$.

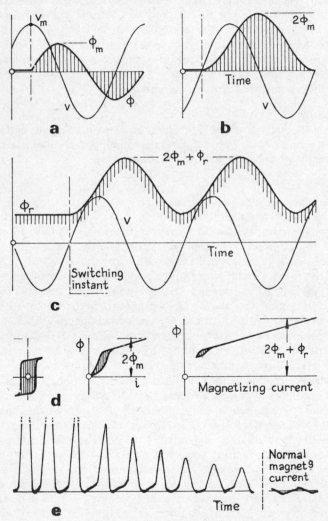

5.13 Inrush current transients.

Hysteresis loops in terms of flux and current are shown in (*d*) for cases (*a*), (*b*) and (*c*). The first is normal and symmetrical, the second is unidirectional and the loop area is actually reduced. The third, for case (*c*), shows super-saturation and an almost negligible hysteresis loss.

An analytical method of predicting inrush currents using an exponential series to represent the magnetic nonlinearity (excluding hysteresis and eddy-current effects but including interphase coupling) has been devised by Macfadyen and co-workers [108]. Analysis is complicated by the drastic change in flux path. At low densities the flux is substantially confined to the core, but at high saturation the permeability of the core steel is so low that the core shares much of the flux with what are normally leakage paths. The peak magnetizing current approximates to that determined by the inductance L of the primary winding considered as an air-cored coil. The current peaks thereafter fall rapidly because the rate of decay is a function of the ratio R/L, where R is a large loss resistance and L is subnormal. As the flux density reduces towards its normal excursions the loss resistance reduces but the inductance increases, slowing the rate of decay and causing current distortion to persist for several seconds, as shown by the typical oscillogram in Fig. 5.13(*e*).

A condition akin to remanent flux can occur in a 3-limb transformer if one primary terminal is energized later than the other two, and in the neutral connection large inrush currents characterized by an exponentially decaying d.c. component may occur.

Surges

Surge overvoltages, initiated on overhead transmission lines by switching, faults or lightning discharges, may be imposed on the system transformers. The surges may have a very steep wavefront with a rate of rise 1 000 times as great as the peak rate of normal voltage at operating frequency, imposing intense and rapidly changing electric stresses within the transformer. For such a condition the appropriate model represents the transformer in terms of an equivalent capacitance network, the capacitance values being too small to affect normal-frequency behaviour. The problem has often been analysed, e.g. by Lewis [28], by Dent et al. [29] and by Adamson and Mansour [30].

Model. The network adopted is that of Fig. 5.14(*a*), in which *c* is the total distributed series capacitance through the winding of axial length *l*. If the winding has *n* evenly spaced turns (or coils), then *c* can be regarded as the series sum of *n* turn-to-turn (or coil-to-coil) capacitances each of *nc*. The total distributed shunt capacitance of the winding to the tank, core and l.v. winding (i.e. in effect to earth) is *C*. The winding has a distributed series inductance *L* and resistance *r*.

Response to Unit-function Surge Voltage. The winding inductance offers such a great opposition to rapid current changes that its presence can under these conditions be ignored. The model can then be simplified initially to the distributed series-parallel capacitance network (*b*). Consider an elemental length d*x* at a distance *x* from the far end: it consists of a series capacitance *cl*/d*x* and a shunt capacitance $C \cdot$ d*x*/*l*. Then for the charge on the element of the winding

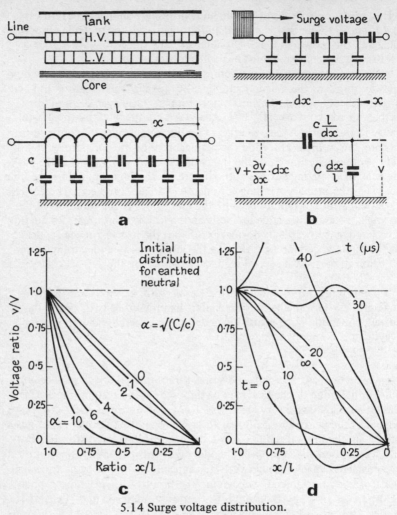

5.14 Surge voltage distribution.

and its voltage to earth we can write

$$\frac{\partial q}{\partial x} = \frac{C \cdot dx}{l} \cdot \frac{\partial v}{\partial t} \quad \text{and} \quad \frac{\partial v}{\partial x} = \frac{dx}{cl} \cdot \frac{\partial q}{\partial t}$$

Differentiating the former with respect to t and the latter with respect to x gives

$$\frac{\partial^2 v}{\partial x^2} = \frac{1}{l^2} \frac{C}{c} \frac{\partial^2 v}{\partial t^2} = \frac{\alpha^2}{l^2} \cdot \frac{\partial^2 v}{\partial t^2}$$

where $\alpha = \sqrt{(C/c)}$. The solution takes the form

$$v = A \exp\left[\alpha(x/l)\right] + B \exp\left[-\alpha(x/l)\right]$$

If the remote end of the winding is earthed, then $v = 0$ at $x = 0$ gives $A = -B$. At the line end $x = l$ and $v = V$, the magnitude of the unit-function surge voltage. With these boundary conditions the voltage at a point x in the winding is

$$v = V \frac{\sinh \alpha(x/l)}{\sinh \alpha} \tag{5.9}$$

A similar treatment for the case in which the remote end of the winding is isolated gives

$$v = V \frac{\cosh \beta(x/l)}{\cosh \beta}$$

where $\beta = \sqrt{(2C/c)}$. This case is less common.

Initial Voltage Distribution. Fig. 5.14(c) shows the *initial* distribution of a surge voltage over a uniform winding with the far end earthed (at $x = 0$). For $\alpha = 0$, which means negligible capacitance to earth, the distribution is uniform, but in the presence of large earth capacitance, e.g. $\alpha = 10$, most of the voltage is dropped across a small fraction of the line end of the winding, in which the voltage between turns becomes excessive. Because the surge voltage V may itself be several times the working voltage, the first few turns (or coils) may be subjected to an electric stress of several hundred times normal.

Final Voltage Distribution. After the transients have died away, the tail of the surge is the equivalent of a direct sustained voltage, and the voltage distribution obviously becomes

$$v = V(x/l) \tag{5.10}$$

which is identical with that in (c) for $\alpha = 0$.

Intermediate Transient. Between the initial and final distributions, eqs.(5.9) and (5.10), an interchange of stored energy between capacitors through coil inductances generates complex oscillations at a variety of natural frequencies. As a result the voltage distribution changes characteristically with time. As an example, a transformer with $\alpha = 6$ might exhibit the distributions shown in Fig. 5.14(d) at intervals of 10 μs following the application of a unit-function voltage at $t = 0$, at which instant the initial distribution is that in (c) for this value of α. Strongly fluctuating electric stresses are thrown upon various parts of the winding insulation, until the oscillations die out by loss attenuation to give the uniform final distribution (marked $t = \infty$).

Travelling Waves. Another approach to the problem is to analyze the surge into a Fourier series of component sines impressed on one end of a ladder network, Fig. 5.15(a). At a critical frequency corresponding to the rejector resonance of the elements L' and c', the ladder offers infinite impedance (neglecting loss). At *super-critical* frequencies the component sines are presented with the equivalent circuit (b), resulting in the initial surge-voltage distribution already discussed. At sub-critical frequencies the component sines encounter the equivalent circuit (c) and travel through the h.v. winding as through a low-pass filter. Both L' and c' are frequency dependent, and affect the velocity and attenuation of the travelling-wave components differently: this modifies

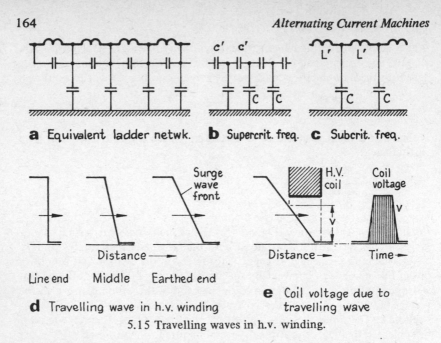

a Equivalent ladder netwk.   **b** Supercrit. freq.   **c** Subcrit. freq.

Line end   Middle   Earthed end

d Travelling wave in h.v. winding

e Coil voltage due to travelling wave

5.15 Travelling waves in h.v. winding.

the wavefront as it passes through the winding, as in (*d*). At the end of the winding reflection takes place in accordance with the termination (e.g. earth or open circuit). The voltage to earth at any point and at a given instant is then the resultant of the initial and reflected voltage waves.

The voltage across an individual section or coil of the winding is seen from Fig. 5.15(*e*) to have a pulse form due to the passage of the travelling wave.

As the surge voltage distribution is affected by the several capacitances, and consequently by the insulation design, it is essential to co-ordinate the insulation with the effect that it will have on the voltage distribution. *Surge Protection.* Before the effect of surge voltages was understood it was the practice to re-inforce the insulation of the line end of a winding to withstand the impulsive voltage gradients; but by reducing the series capacitance this actually intensified the trouble. External *surge-absorbers* may be connected between the transmission-line and transformer terminals to reduce the steepness of the wavefront and to dissipate some of the surge energy, but it is preferred to design the transformer in such a way that an approximation to the $\alpha = 0$ condition is achieved.

The provision of *static shields* aims to neutralize the coil-to-earth capacitance C by introducing coil-to-line capacitance, on the same principle as in grading a transmission-line insulator string. Suitable coil-to-line capacitance C_1 provides the coil-to-earth currents through C, leaving the inter-coil currents in c all equal and giving uniform voltage distribution over the winding. This may be done by mounting metal shields around the winding and connecting them to the line terminal, as in Fig. 5.16(*a*). The effect on the electric field is shown in (*b*), the figures indicating percentage equipotentials.

An alternative (or additional) method is the use of *interleaved coils*, (*c*).

Two normal disc coils are shown: if the order of turn-connection is altered to increase the normal voltage between turns, the effect is to raise the inter-turn series capacitance c, and so reduce the ratio $\alpha = \sqrt{(C/c)}$. Windings of interleaved double-disc type are better able to withstand impulse voltages than non-interleaved double-disc coils; partially-interleaved windings combine the two, with the interleaved section normally at the h.v. end, but they introduce surge-voltage reflections. The problem of optimizing such windings is discussed by Schleich [31].

EXAMPLE 5.3: Compare the effective series capacitance of the interleaved disc-coil pair in Fig. 5.16(c) with that of the normal pair.

Let v be the voltage per coil-pair, and n the number of turns (in this case $n = 10$). The effective capacitance between the coil terminals (turns 1 and 10) can be found only by a field plot of the rather complicated electric-field pattern; but as the turns are much closer together than are the discs, we may as an approximation consider only the turn-to-turn capacitance C_t.

Normal Coil: The voltage between turns is $v_t = v/10$. The electric-field energy between successive and adjacent turns is

$$w_t = \tfrac{1}{2}C_t \, (v/10)^2$$

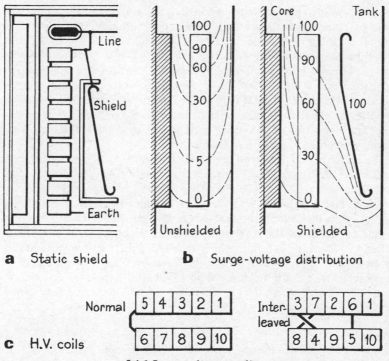

a Static shield

b Surge-voltage distribution

c H.V. coils

5.16 Surge voltage grading.

and the total for the complete pair of discs (8 interturn gaps) is

$$w_n = 8 \times \tfrac{1}{2} C_t (v/10)^2 = \tfrac{1}{2} C_n v^2$$

where C_n is the total effective series capacitance between turns 1 and 10, whence

$$C_n = C_t(8/100) = 0.08\, C_t$$

Interleaved Coil: The geometry is unchanged, and so therefore is C_t; but the effect of interleaving is to increase the voltage between adjacent (but no longer successive) turns to $5v_t$ (e.g. turn 1 to turn 6, 2 to 7 ...) or to $4v_t$ (e.g. 6 to 2, 7 to 3 ...). The total field energy is consequently

$$w_i = \tfrac{1}{2} C_t [4(5v/10)^2 + 4(4v/10)^2]$$
$$= 164 \times \tfrac{1}{2} C_t (v/10)^2 = \tfrac{1}{2} C_i v^2$$

and the effective series capacitance is

$$C_i = C_t (164/100) = 1.64\, C_t$$

This markedly reduces the coefficient α and improves the surge-voltage distribution.

5.9   Three-phase connections

With windings that can be connected in star, delta, zigzag or vee, and with primary, secondary, tertiary and auto-connected windings, the variety of possibilities is considerable.

A choice is possible between a 3-ph unit and a bank of three 1-ph transformers. The unit may cost about 15% less than the bank and will occupy much less space. There is little difference in reliability, but it is cheaper to carry spare 1-ph than 3-ph transformers if only one installation is concerned. Single-phase units may be preferred in mines to ease underground transport. The possibility of operating two 1-ph units in open delta at 58% of nominal full load is a practice more favoured in the U.S.A. than in Europe.

The choice between star and mesh connection merits separate consideration in each case. Star connection is cheaper, since mesh connection needs more turns and more insulation. The difference is small, however, at voltages below 11 kV. With very high voltages a saving of 10% may be effected, mainly on account of the insulation. An advantage of the star-connected winding with earthed neutral is that the maximum voltage to the core (frame or earth) is limited to 58% of the line voltage, whereas with a delta-connected winding the earthing of one line (due to fault) increases the maximum voltage between windings and core to the full line voltage. Technically the mesh-connected primary is essential where the l.v. secondary is a star-connected four-wire supply to mixed 3-ph and 1-ph loads.

Nomenclature

Transformer terminals conforming to BS 171 are brought out in rows, h.v. on one side and l.v. on the other, and are lettered from left to right facing the h.v. side. The h.v. terminals have capital letters (e.g. ABC); the l.v. terminals have small letters (e.g. *abc*); tertiary windings are lettered in capitals with the figure 3. Neutral terminals precede line terminals. Each winding has two ends designated with the subscripts 1, 2; or if there are intermediate tappings these are numbered in order of their separation from end 1. Thus an h.v. winding with four tappings is labelled $A_1, A_2 \ldots A_6$, with A_1 and A_6 forming the ends of the phase winding. If the e.m.f. of an h.v. phase winding has, at a given instant, the direction $A_1 A_2$, then at the same instant the e.m.f. direction in

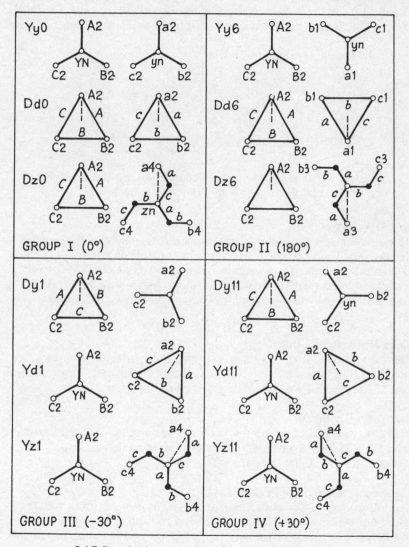

5.17 Standard connections for 3-ph transformers.

the associated l.v winding is $a_1 a_2$. The diagrams in Fig. 5.17 show the connections and e.m.f. phasors for a number of common arrangements.

Polyphase transformers are all allotted symbols giving the type of phase connection and the angle of *advance* in passing from the phasor representing the e.m.f. of the h.v. winding to that of the l.v. phasor at the corresponding terminal. The angle is indicated in terms of a clock face, the h.v. phasor being at 12 o'clock (zero) and the corresponding l.v. phasor at the hour-hand number. Thus 'YzD3 11' represents a (h.v. star/l.v. zigzag/tertiary delta)-connected 3-ph transformer with the l.v. (secondary) e.m.f. in a given phase combination at 11 o'clock, i.e. 30° in advance of the 12 o'clock position of the h.v. phasor. The groups into which 3-ph transformers are classified are as follows:

Group	Phase displacement	Two-winding connections		
1	Zero	Yy0	Dd0	Dz0
2	180°	Yy6	Dd6	Dz6
3	30° lag	Dy1	Yd1	Yz1
4	30° lead	Dy11	Yd11	Yz11

Connections
The main features of the more common connections are noted below. References to harmonic behaviour are dealt with in Sect. 5.7.

Star/Star (Yy0 or Yy6)
This is economical for small h.v. transformers as it minimizes the turns/ph and the winding insulation. With star points available on both sides it is possible to provide a neutral connection. Triplen voltages are absent from the lines, and (unless there is a neutral connection) no triplen currents flow. The neutral voltage may oscillate, and triplen voltages may be high in shell-type units.

Delta/Delta (Dd0 or Dd6)
This suits large l.v. transformers as it requires more turns/ph of smaller section. Large load unbalance can be tolerated, and triplen voltages are damped out. The absence of a star point may be a disadvantage.

Star/Delta (Dy or Yd)
This is very common for supply networks. It has a star point for a neutral to serve mixed 1-ph and 3-ph loads, and a delta winding that can carry triplen currents and so stabilize the starpoint voltage.

Interconnected-star/Star (Yz1 or Yz11)
The interconnected (or 'zigzag') star arrangement reduces triplen voltages and is not sensitive to conditions of unbalanced loading. But the zigzag is restricted to comparatively low-voltage windings, and as the phase voltages are composed of half-voltages with a 60° displacement, 15% more turns

are required for a given phase terminal voltage compared with a normal star. The connection in sometimes employed in rectifier supply.

5.10 *Three/two- and three/one-phase connections*
Phase conversion is needed in special cases, such as the supply to 2-ph electric furnaces.

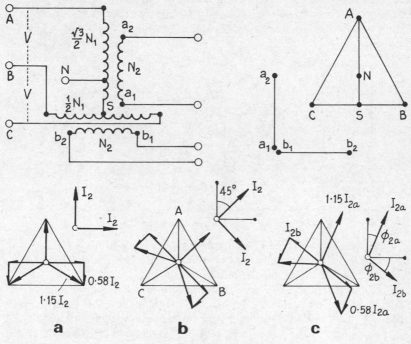

5.18 Scott connection.

Three/two-phase Connection
Scott Connection. This requires two transformers with different tappings and ratings, although they may be constructed for interchangeability. Consider in Fig. 5.18 the N_1-turn primary of a 1-ph ('main') transformer connected between terminals B and C of a 3-ph supply of line voltage V. The voltage between line A and the mid point S of BC is $(\sqrt{3}/2)\,V = 0.87\,V$. If now the primary of a second ('teaser') transformer with $0.87\,N_1$ turns is connected across AS, the voltage per turn is the same for both main and teaser primaries. Then two identical secondaries with N_2 turns have terminal voltages of equal magnitude but phase quadrature. The 3-ph neutral point N is located on the teaser primary at one-third of the winding distance from S, where the voltage is $V/\sqrt{3} = 0.58\,V$ in AN and $0.29\,V$ in SN.

Let equal currents I_2 at unity p.f. be taken from the two secondary phases, Fig. 5.18(*a*), and let magnetizing current be neglected. There must, in each transformer, be an m.m.f. balance. In the main transformer the primary balancing current is $2I_2N_2/\sqrt{3}N_1 = 1.15\,I_2(N_2/N_1)$. The teaser primary current is in phase with the star voltage AN. The total current in the main primary is the resultant of two components: the first is $I_2(N_2/N_1)$ to balance the secondary current, the second is one-half of the teaser primary current in either direction from S (and having a zero net m.m.f. in consequence). The currents in lines B and C are shown in Fig. 5.18(a), taking $N_1 = N_2$ for clarity; their phasors, combining the rectangular components I_2 and $0.58I_2$, are co-phasal respectively with the star voltages NB and NC, and are equal to the teaser primary current. Thus for a balanced 2-ph load of unity p.f. the 3-ph side is balanced.

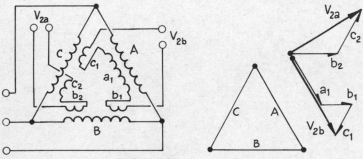

5.19 Le Blanc connection.

Case (b) is for a balanced 2-ph load of p.f. 0.71. Again the 3-ph load is balanced, and the main transformer loading is 15% greater than that of the teaser. Case (c) is for an unbalanced 2-ph load.

Le Blanc Connection. The 3-ph side consists of three windings in star or delta, wound on an ordinary 3-limbed core. If the 2-ph side is to have a phase voltage V, the limbs ABC in Fig. 5.19 are provided respectively with windings a_1 of voltage $\frac{2}{3}\,V$; b_1 of $\frac{1}{3}\,V$ with b_2 of $V/\sqrt{3}$; and c_1 of $\frac{1}{3}\,V$ with c_2 of $V/\sqrt{3}$. The windings when grouped give phase I of the 2-ph supply with a_1, b_1 and c_1 in series; and phase II with b_2 and c_2 in series. The phasor sums show that the 2-ph side has equal voltages in phase quadrature. The use of b_1 and c_1 in phase I is to balance the m.m.f.s around the three interlinked magnetic circuits and to give 3-ph input balance for a balanced 2-ph load. The connection has the advantage of using a standard 3-ph core.

Three/one-phase connection

Single-phase power pulsates twice per period, but a balanced 3-ph load takes a power that, over the three phases together, is constant. It is therefore not possible to obtain 3-ph balance when supplying from it a 1-ph load unless there is some energy-storing device (such as rotating machine or a capacitor-inductor system) to give a 'flywheel' effect. Fig. 5.20 shows some simple direct methods not employing storage. The actual division of the 1-ph current can be made to assume any proportions between 1:1:0 and $1:\frac{1}{2}:\frac{1}{2}$. Connection ($a$) is the simplest; ($b$) is one phase of a Scott connection; (c) is an open delta; and (d) uses both units of the 2-ph Scott arrangement.

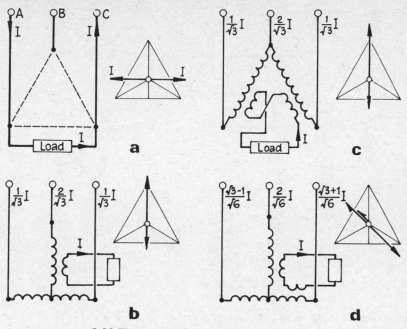

5.20 Three-phase/single-phase connection.

The 3-ph supply, because of its single-phase loading, carries equal positive-and negative-sequence currents. The p.p.s. currents supply the 1-ph load, while the n.p.s. currents, ineffective with the p.p.s. voltage, simply increase the line loss.

5.11 *Auto connection*

An auto transformer has windings common to primary and secondary, the input and output circuits being electrically connected as one continuous winding per phase, with input and output currents superposed in the common section. The principal application is to cases of voltage ratio not less than about 2:1 where electrical isolation of the primary from the secondary circuit is not essential; such applications include starters and the interconnection of h.v. transmission lines. The advantage gained is the saving in transformer size, rating and loss.

Neglecting magnetizing current, let the 1-ph auto-transformer in Fig. 5.21 have N_1 primary turns tapped at N_2 for a lower-voltage secondary. For an output of I_2 at voltage V_2, the input is $I_1 = I_2(N_2/N_1)$ at voltage $V_1 = V_2(N_1/N_2)$ if the losses can be ignored. The current in the common part of the winding is $I_2 - I_1$ so that here the conductor section may be smaller. With the same conductor current density throughout, the ratio G_a/G_2 of conductor material in the auto and two-winding cases is

$$\frac{G_a}{G_2} = \frac{I_1(N_1 - N_2) + (I_2 - I_1)N_2}{I_1N_1 + I_2N_2} = 1 - \frac{V_2}{V_1}$$

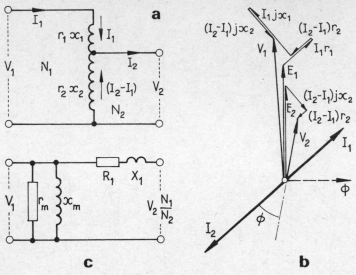

5.21 Auto transformer.

The saving in material and cost is less than this, as the core is nearly as large as in the two-winding transformer. For a voltage ratio of 2:1, about 50% saving in winding material might be obtained, and the overall cost of the unit might be 60–70% of that of a normal transformer of the same input rating.

Equivalent Circuit

Fig. 5.21 shows at (*a*) the equivalent circuit ignoring magnetizing and core-loss current, and at (*b*) the corresponding phasor diagram. For a load phase angle ϕ the primary and secondary voltages are approximately

$$V_1 = E_1 + I_1(r_1 \cos \phi + x_1 \sin \phi) + (I_1 - I_2)(r_2 \cos \phi + x_2 \sin \phi)$$

$$V_2 = E_2 - (I_2 - I_1)(r_2 \cos \phi + x_2 \sin \phi)$$

Combining these and writing $E_2 = (N_2/N_1)E_1$ and $I_2 = (N_1/N_2)I_1$ yields the relation

$$V_1 = (N_1/N_2)V_2 + I_1(r_1 + r_2{}') \cos \phi + I_1(x_1 + x_2{}') \sin \phi$$

$$= (N_1/N_2)V_2 + I_1 R_1 \cos \phi + I_1 X_1 \sin \phi$$

where

$$r_2{}' = \left(\frac{N_1}{N_2} - 1\right)^2 r_2 \quad \text{and} \quad x_2{}' = \left(\frac{N_1}{N_2} - 1\right)^2 x_2$$

and R_1 and X_1 are the equivalent resistance and leakage reactance referred to the primary. A test with the secondary short-circuited gives $Z_1 = R_1 + jX_1$ as the ratio of the applied primary voltage and input current, just as for a two-winding transformer. The magnetizing branch included in Fig. 5.21 (*c*) is obtained from an open-circuit test.

Three-phase Auto Connection. This is readily obtained, as in Fig. 5.22. To avoid undue fault voltages it is usual to employ a star connection. The Scott connection is also possible, but it is not permissible to interconnect the 2-ph side as a 3-wire 2-ph system.

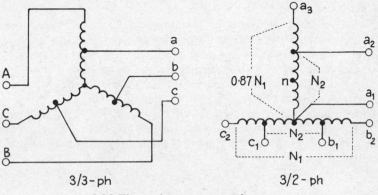

3/3-ph   3/2-ph

5.22 Three-phase auto transformers.

An auto-transformer starter for cage induction motors consist of a 3-ph star-connected unit with one or two (occasionally more) tappings per phase; it is designed for high current and flux densities for economy as the duty is required normally for only a few seconds at starting.

The interconnection of high-voltage supply networks (e.g. 275 and 132 kV) by auto-connected transformers results in a substantial saving of bulk and cost compared with two-winding transformers.

Comparison

Auto-transformers are not generally employed when the voltage-ratio exceeds 3:1, except for motor-starting duty, as the disadvantages preponderate. The latter are due to the direct electrical connection between the two sides, by reason of which disturbances on one system are liable to affect the other more seriously than when (as in the normal two-winding arrangement) there is only a magnetic link. The reactance is inherently low on account of (*a*) the 'coalescence' of the primary and secondary windings and (*b*) the smaller currents, so that the conditions on short circuit are more severe.

5.12   *Three-winding connection*

Transformers may be built with *tertiary* windings (i.e. additional to the primary and secondary) for any of the following reasons:

(*a*) To supply a small additional load at a different voltage.
(*b*) To supply phase-compensating devices (e.g. capacitors) requiring a different voltage or connection.
(*c*) In star/star or star/zigzag transformers, a delta-connected tertiary winding

reduces the zero-phase-sequence impedance and allows adequate earth-fault current to flow for the operation of protective gear, and it limits voltage unbalance when the main load is asymmetric.

(*d*) To indicate voltage in a h.v. testing transformer.

(*e*) To load split-winding generators.

A tertiary winding for (*a*) is called *auxiliary*; for (*c*) it is termed a *stabilizing* winding.

Tertiary windings are frequently delta-connected: consequently, when faults and short-circuits occur on the primary or secondary sides (particularly between lines and earth), considerable unbalance of phase voltage may be produced, compensated by large tertiary circulating currents. The reactance of the winding must be such as to limit the circulating current to that which can be carried without overheating.

With given secondary and tertiary load currents I_2 and I_3 and with windings of N_1, N_2 and N_3 turns/ph, the primary current I_1 is readily found by phasor summation of the secondary, tertiary and magnetizing m.m.f.s or equivalent currents.

Equivalent Circuit

Properly interpreted, a three-winding transformer can be represented by the equivalent 1-ph circuit of Fig. 5.23, with each winding assumed to have an equivalent resistance and separate leakage reactance. All values are reduced to a *common-voltage and common-rating* basis, and because the secondary output is not in general substantially the same as the primary input, the simple secondary equivalents used in the two-winding transformer are no longer valid: the load division between secondary and tertiary is, in fact, arbitrary.

Consider the primary and secondary operating as a two-winding transformer. The total per-unit resistance voltage Ir_{12} is the sum of the drops in the two windings separately. A similar statement is true for any pair of windings, and can be stated also in appropriate terms for the reactance drop. Thus we have

$$Ir_{12} = Ir_1 + Ir_2, \qquad Ix_{12} = Ix_1 + Ix_2,$$
$$Ir_{23} = Ir_2 + Ir_3, \qquad Ix_{23} = Ix_2 + Ix_3,$$
$$Ir_{13} = Ir_1 + Ir_3, \qquad Ix_{13} = Ix_1 + Ix_3,$$

for the possible combinations of pairs of circuits taken together as two-circuit transformers. Using these relations,

$$r_1 = \tfrac{1}{2}(r_{12} + r_{13} - r_{23}) = \tfrac{1}{2}\Sigma r - r_{23}.$$

The other corresponding relations are similar, whence

$$r_1 = \tfrac{1}{2}\Sigma r - r_{23}, \qquad x_1 = \tfrac{1}{2}\Sigma x - x_{23},$$
$$r_2 = \tfrac{1}{2}\Sigma r - r_{13}, \qquad x_2 = \tfrac{1}{2}\Sigma x - x_{13},$$
$$r_3 = \tfrac{1}{2}\Sigma r - r_{12}, \qquad x_3 = \tfrac{1}{2}\Sigma x - x_{12},$$

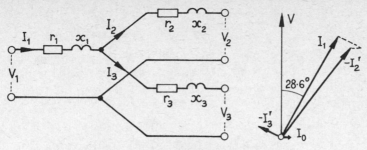

5.23 Three-winding transformer.

Thus from a series of normal two-winding short-circuit tests, these expressions give, for the assigned voltage and rating, the ohmic values r_1, r_2, r_3 and x_1, x_2, x_3 in Fig. 5.23 taking completely into account all mutual effects.

Further, suppose the secondary to be short-circuited, the tertiary being left on open circuit. The total voltage applied to the primary will be the sum of the primary and secondary impedance voltages. A voltmeter across the tertiary output terminals is connected in effect across the junction of primary and secondary: it therefore indicates the secondary impedance voltage (after adjustment to the common voltage basis). In general, the voltage across one circuit when a second supplies short-circuit apparent power to the third, is the impedance voltage of the short-circuited winding.

Regulation. This is best considered from the individual contribution of each winding, the total for two windings in combination being the algebraic (not phasor) sum. For the primary alone, using eq.(5.8) adjusted for the ratio $k = $ (load/base ratings),

$$\epsilon_1 = k(\epsilon_{r1} \cos \phi_1 + \epsilon_{x1} \sin \phi_1)$$

Summation of the part-regulations of two circuits together is an addition if power flows from one to the other, a subtraction if not. For a transformer with primary fed and both secondary and tertiary loaded, the regulations become

$$\epsilon_{12} = \epsilon_1 + \epsilon_2 \qquad \epsilon_{13} = \epsilon_1 + \epsilon_3$$
$$\epsilon_{23} = \epsilon_2 - \epsilon_3 \qquad \epsilon_{32} = \epsilon_3 - \epsilon_2$$

It is quite possible for these to be zero or negative.

EXAMPLE 5.4: A 3-ph 6 600/400/110 V star/star/delta transformer with a magnetizing current of 5.5 A has balanced 3-ph loads of 1 000 kVA at p.f. 0.8 lagging on the secondary, and 200 kVA at p.f. 0.5 leading on the tertiary winding. The per-unit resistance and reactance coefficients on a base of 6 600 V and 1 000 kVA are

$$\epsilon_{r1} = 0.0053 \qquad \epsilon_{r2} = 0.0063 \qquad \epsilon_{r3} = 0.0083$$
$$\epsilon_{x1} = 0.0265 \qquad \epsilon_{x2} = 0.0250 \qquad \epsilon_{x3} = 0.0332$$

Find the per-unit regulations.

Working in primary terms, we have

$$I_0 = -j\,5.5 \text{ A}, \quad -I_2' = 70 - j\,52.5 \text{ A}, \quad -I_3' = 8.8 + j\,15.1 \text{ A}$$

which sum to $I_1 = 73.8 - j42.9 = 89.7 \angle -28.6°$ A. The primary load (neglecting losses) is therefore $S_1 = 1025$ kVA at p.f. cos 28.6 ° = 0.88 lagging. The phasor relations for the currents are shown in Fig. 5.23.

The primary regulation component with $k = 1025/1000 = 1.025$ is

$$\epsilon_1 = k(\epsilon_{r1} \cos \phi_1 + \epsilon_{x1} \sin \phi_1)$$

$$= 1.025(0.0053 \times 0.88 + 0.0265 \times 0.48) = 0.018 \text{ p.u.}$$

For the secondary, $k = 1\,000/1\,000 = 1.0$, whence

$$\epsilon_2 = 1.0(0.0063 \times 0.80 + 0.0250 \times 0.60) = 0.020 \text{ p.u.}$$

For the tertiary, $k = 200/1\,000 = 0.2$, and thus

$$\epsilon_3 = 0.2(0.0083 \times 0.50 - 0.0332 \times 0.87) = -0.005 \text{ p.u.}$$

The primary-secondary and primary-tertiary regulations are therefore

$$\epsilon_{12} = \epsilon_1 + \epsilon_2 = 0.038 \text{ p.u.} \qquad \epsilon_{13} = \epsilon_1 + \epsilon_3 = 0.013 \text{ p.u.}$$

The secondary-tertiary and tertiary-secondary regulations are

$$\epsilon_{23} = \epsilon_2 - \epsilon_3 = 0.025 \text{ p.u.} \qquad \epsilon_{32} = -\epsilon_{23} = -0.025 \text{ p.u.}$$

Stabilization by Tertiary Windings

Star/star transformers comprising 1-ph units, or 3-ph *shell-* or *five-limb-core-* type units, have the disadvantages (*a*) that they cannot readily supply unbalanced loads between line and neutral, and (*b*) that their phase voltages may be distorted by triplen harmonic e.m.f.s. These arise because their phase magnetic circuits are not interlinked, and out-of-balance (zero-phase-sequence) currents have an iron path for the production of z.p.s. flux. The z.p.s. impedance is therefore high. The flow of earth-fault current is severely restricted, and may be insufficient for the satisfactory operation of protective gear. The zero-sequence impedance to out-of-balance or earth-fault currents must include the very high impedance (0.5–5 p.u.) between the secondary winding of the loaded or faulted phase and the primary windings of the other two phases. Abnormal voltages therefore occur when z.p.s. current flows. The output load or fault current concerns one phase only, whereas the corresponding input load current has, in the absence of a primary neutral, to be conducted through both of the other phases, which act as inductors. The voltage of the loaded phase is reduced, that of the others being raised. By use of a delta-connected tertiary winding, induced currents are caused to circulate in it, apportioning the load more evenly over the three phases. The action is, in brief, to interlink electrically the separate magnetic circuits.

The delta-connected tertiary provides a path for triplen magnetizing current. The disadvantages mentioned above are mitigated in 3-limb core-type transformers because, with z.p.s. flux forced out of the limbs, the z.p.s. impedance is lower. Nevertheless it is usual to provide a delta-connected stabilizing tertiary winding, rated basically by its heat capacity under fault conditions.

5.13 Zero-phase-sequence impedance

The impedance of a transformer to positive- and to negative-sequence voltages is the same, as the phase order makes no difference to the operation. In determining the impedance to zero-phase-sequence currents, account must be taken of winding connections, earthing and, in some cases, the constructional form.

Fig. 5.24 sets out the connections and zero-sequence conditions, in A for two- and in B for three-winding transformers, neglecting resistance. In A the reactance x_{12} refers to $x_1 + x_2'$ or $x_1' + x_2$; in B the symbols x_{12}, x_{23} and x_{13} have the significance given to them in Sect. 5.12.

Two-winding Connections

(a) *Yy (both star points isolated)*. There being no earth circuit, an infinite impedance is offered to z.p.s. currents.

(b) *Yy (secondary neutral earthed)*. An earth fault on the primary side will not result in z.p.s. current. The same is nearly true for an earth fault on the secondary side if the transformer has no magnetic phase interlinkage (shell-type or three 1-ph units), but in a core-type unit the fluxes produced by z.p.s. currents find a high-reluctance path outside the core.

(c) *Yy (both neutrals earthed)*. Provided that the primary circuit is complete for z.p.s. currents, a secondary earth fault will give rise to z.p.s. components whose counterparts flow in the primary circuit; and similarly for an earth fault on the primary. The phase reactance to z.p.s. currents is three times the reactance of the 3-ph windings in parallel, i.e. the reactance of one phase. It is therefore identical to the p.p.s. and n.p.s. values. With 3-ph core-type transformers the effect described in (b) may reduce the z.p.s. reactance value.

(d) *Yd (star point isolated)*. No z.p.s. currents can flow to or from the primary or secondary, but they can circulate within the delta.

(e) *Yd (neutral earthed)*. For a secondary earth fault the z.p.s. reactance is infinite. For a primary earth fault, compensating currents flow in the secondary delta. The z.p.s. reactance is as for (c).

(f) *Dd*. No z.p.s. current to or from either side or in the delta.

(g) *Z earthing transformer*. To p.p.s. and n.p.s. currents the reactance is infinite; to z.p.s. currents the reactance is due only to leakage as the two m.m.f.s on each limb are self-cancelling.

Three-winding Connections

(a) *Yy (both star points isolated)*. The z.p.s. reactance is infinite.

(b) *Yy (secondary neutral earthed)*. As for A(b). Compensating current flows in the tertiary delta, the reactance being x_{23}.

(c) *Yy (both neutrals earthed)*. With a fault on either primary or secondary, z.p.s. currents can flow because of tertiary compensating currents. Currents can flow also in an unfaulted primary or secondary if the z.p.s. circuit is complete.

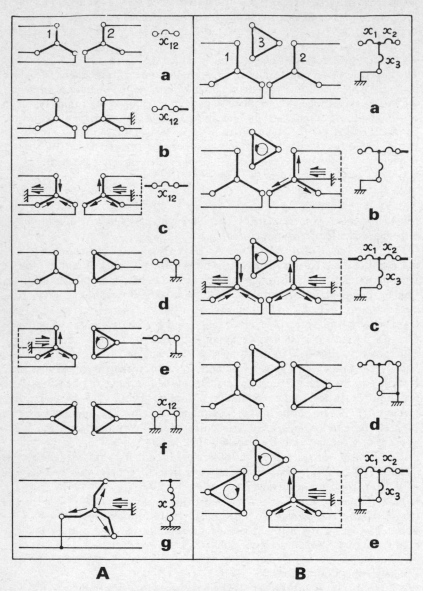

5.24 Zero-phase-sequence connections.

(*d*) *Yd (star point isolated).* There is no path for z.p.s. currents.

(*e*) *Dy (neutral earthed).* A secondary earth fault permits z.p.s. currents to flow, with compensating currents in primary and tertiary; the z.p.s. reactance is $x_2 + (x_1 x_3 / x_{13})$. For primary faults the z.p.s. reactance is infinite.

5.14 *Tap-changing*

Voltage control in a supply network is required for: (i) adjustment of consumers' terminal voltage within the statutory limits; (ii) control of active and reactive power flow in the network; (iii) seasonal (5–10%), daily (3–5%) and short-period (1–2%) adjustments in accordance with the corresponding variations of load. Much of the adjustment is done by altering the effective turnsratio of the system transformers by tappings on the windings. Occasional adjustments can be made by *off-circuit* tap-changing, the common range being ± 5% in $2\frac{1}{2}$% steps. Daily and short-time adjustment is generally by means of *on-load* tap-changing gear.

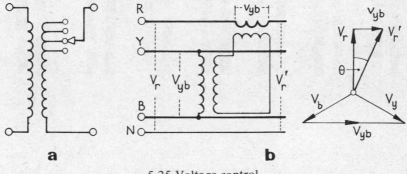

a   b

5.25 Voltage control.

The essentials are indicated for one phase in Fig. 5.25. Simple up or down voltage regulation by tappings is shown in (*a*), while (*b*) is for a control of phase angle in which a voltage v_{yb} derived from the line voltage V_{yb} gives an output voltage V_r' differing from V_r by the angle θ. It can be used for the control of reactive power flow on interconnector feeders.

Tappings

The *principal* tapping is that to which the rating of the winding is related. A *positive* tapping includes more, and a *negative* tapping less, turns than those of the principal tap. Tap-changing may be arranged in accordance with one of three conditions, namely (i) voltage variation with constant flux and constant turn-e.m.f.; (ii) with varying flux; and (iii) a mixture of (i) and (ii). In (i) the percentage tapping range is identical with that of voltage variation.

Location of the tapped part of a winding is partly a constructional question. With tappings near the line ends the number of bushing insulators is reduced; with tappings near the neutral ends the insulation conditions between phases are eased. Where a large voltage variation is required, tappings should be near the centres of the phase windings to reduce magnetic asymmetry, but this arrangement cannot be used on l.v. windings placed next to the core.

It is not possible to tap other than an integral number of turns, and this may cause difficulty with l.v. windings. A 260 V phase winding with 15 V/ turn cannot be tapped closer than 5%. It is consequently necessary to tap the h.v. windings, a further advantage in a step-down transformer being that, on light load with lowest secondary voltage, the maximum number of turns is included on the h.v. side, reducing the e.m.f. per turn, the flux density and the core loss.

An effect of importance is that of unbalanced forces on short circuit when tappings have produced magnetic asymmetry. Some methods of locating tappings are shown in Fig. 5.26. Taps at one end (*a*) are permissible in small

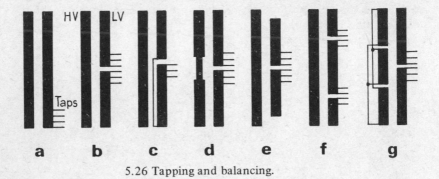

5.26 Tapping and balancing.

transformers. In larger units they are placed centrally, both for (*b*) delta and (*c*) star-connected windings. Axial m.m.f. unbalance is mitigated by 'thinning' the l.v. winding (*d*), or by balancing parts of the winding more symmetrically (*e*) and (*f*). The untapped winding may be split into several parts (*g*) corresponding to the tapped side, and connected in parallel. For very large tapping ranges a special tapping coil may be used.

Tap-changing involves changes in leakage reactance, I^2R loss and core loss, and possibly some difficulty in operating dissimilar transformers in parallel. The response of the h.v. winding to a steep-fronted surge may also be impaired.

Off-circuit Tap-changing

This adjustment is obtained by tapping the respective windings as required and taking connections therefrom to some position near the top of the transformer. Change of tapping is effected by means of hand holes in the cover. Alternatively the reconnection can be made by carrying the tapping leads through the cover for changing by hand or by manually operated switch. Two common types of switch are:

Vertical Tapping Switch. This comprises an insulated bar, one for each phase, carrying knife contacts, to be raised or lowered by handwheel and gearing to make connection to the tapping contacts.

Faceplate Switch. The spindles of movable contact arms are geared to a shaft with a handwheel external to the tank. The faceplate switches may be mounted on the upper yoke, or located nearer to the tapped positions on the windings.

On-load Tap-changing

The essential feature is the maintenance of circuit continuity throughout the tap-change operation. Momentary connection must be made simultaneously to two adjacent taps during the transition, and the short-circuit current between them must be limited by some form of impedance. Inductors have been used for this purpose, but in modern designs the current limiting is almost invariably obtained by means of a pair of resistors. Fig. 5.27 shows the method, which has a straightforward switching sequence. Back-up main contactors are provided which short-circuit the resistors for normal operation.

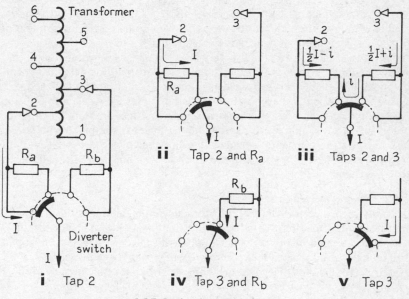

5.27 On-load tap changing.

Resistor transition requires one winding tap for each operating position. The resistors are short-time rated and it is essential to minimize their time of duty. Some form of energy storage must be incorporated in the driving mechanism to ensure that a tap change, once initiated, shall be completed irrespective of failure of the auxiliary control supply. All modern devices use springs, which operate to reduce the time that a resistor is in circuit to a few periods. Such a tap-changer is compact, and the high-speed break limits contact wear. The current break is eased by the fact that the short-circuit resistor current has unity p.f.

Irrespective of the form of transition, on-load tap-changers have one or other of the following switching arrangements:

1. A separate contactor for each winding tap, and camshaft operation to ensure the proper sequence.
2. Winding taps connected to a selector switch with an associated pair of contacts. Make and break occurs at the selector switch contacts, and some

degree of carbonization and wear is unavoidable. This 'single-compartment' tap-changer is common for transformers rated up to 20 MVA.

3. For very large transformers the tap selector switches do not move when carrying current: make and break is carried out by two separate diverter switches, usually in a separate compartment to minimize main-oil pollution by carbon. The diverters are mechanically interlocked with the selectors, which move (only when not current loaded) to provide the correct sequence of tap connection. Vacuum switches may be used as diverters, and as they do not contaminate oil they can be mounted in the same compartment as the selectors.

On-load tap-changer control gear can vary from simple push-button initiation to complex automatic control of as many as four transformers operating in parallel. The object of control is to maintain a set voltage level within a specified tolerance, or to raise it with load to compensate for line drop. The main components are an automatic voltage regulator, a time-delay relay, and compounding elements. The time-delay prevents initiation of a tap-change by a minor transient voltage fluctuation: it can be set for a delay up to 60 s.

Line-drop Compensation. This is essentially an artificial transmission line consisting of adjustable resistor and inductor elements combined with current- and voltage transformers to provide a signal to the voltage relay coil. Variation of the load current then gives compensation for the line drop through the on-load tap-changing equipment.

5.15 *Parallel operation*

The satisfactory performance of transformers connected on both sides in parallel requires that they have the *essential* features of the same polarity, the same phase-sequence, and zero relative phase-displacement argle; a *near identity* of voltage ratio; and a *limited disparity* in per-unit impedance.

Polarity. This can be either right or wrong. If wrong it results in a dead short circuit.

Phase-sequence and relative phase-displacement. These questions arise with polyphase transformers. Only a few of the possible connections (Sect. 5.9) can be worked in parallel without excessive circulating current on small loads; for example, the secondary voltages of star/star and star/delta transformers have a phase difference of $30°$, making parallel connection inadmissible. The various connections produce various magnitudes and phase displacements: magnitudes can be adjusted by tappings, but phase divergence cannot be compensated.

The phase sequence must be identical for two parallelled transformers. If three secondary terminals $a_1 b_1 c_1$ of transformer 1 are to be parallelled with $a_2 b_2 c_2$ of transformer 2, the polarity and ratio being correct, then a_1 may be connected to a_2. If the result is a potential difference across $b_1 b_2$ or $c_1 c_2$, then the two phase-sequences differ.

The following are typical of the connections for which, from the viewpoint of sequence and phase divergence, transformers can be connected in parallel:

Transformer 1:   Yy   Yd   Yd
Transformer 2:   Dd   Dy   Yz

Thus all transformers in the same group can be paralleled. Further, transformers with a $+30°$ and a $-30°$ angle can be paralleled by reversing the primary and secondary phase-sequence of one.

Voltage Ratio

Equal voltage-ratio (not necessarily precisely the same as equal turns-ratio) is necessary to avoid no-load circulating current when transformers are in parallel on both primary and secondary sides. The leakage impedance being low, a small voltage difference may cause considerable no-load circulating current and additional I^2R loss. On load, the circulation is masked, but may cause over-current on one transformer when the parallelled group is loaded to the full combined rating.

Leakage Impedance

The leakage impedances of transformers required to operate in parallel may differ in magnitude and in quality (i.e. in reactance/resistance ratio). It is necessary also to distinguish between per-unit and ohmic impedance: the currents carried by two transformers are proportional to their ratings if their ohmic impedances are inversely proportional, and the per-unit impedances directly proportional, to those ratings.

A difference in the quality of the per-unit impedance results in a divergence of phase angle of the two currents, so that one transformer will be working with a higher, and the other with a lower, power factor than that of the combined output.

Load-sharing

Equal Voltage Ratios. If the voltage ratios are equal and the voltages coincident in phase, both sides of a pair of transformers can be connected in parallel and no current will circulate between them on no load. Neglecting magnetizing admittance, the case can be represented by the equivalent circuit of Fig. 5.28(a), in which the leakage impedances (with respect to either primary or secondary side) are parallel connected. Let them be Z_1 and Z_2, and let the two currents be I_1 and I_2 (at a common voltage V) which together provide the total load current I. The volt drop is

$$v = I_1 Z_1 = I_2 Z_2 = I Z_{12}$$

where $Z_{12} = Z_1 Z_2 / (Z_1 + Z_2)$ is the parallel equivalent of Z_1 and Z_2. Then $I_1 = I(Z_{12}/Z_1) = I \cdot Z_2/(Z_1 + Z_2)$; and similarly for I_2. Multiplication by V gives the transformer loadings $S_1 = VI_1$ and $S_2 = VI_2$, and the combined load $S = VI$. Then

$$S_1 = S\frac{Z_2}{Z_1 + Z_2} \quad \text{and} \quad S_2 = S\frac{Z_1}{Z_1 + Z_2} \tag{5.11}$$

All these are complexor or phasor expressions, giving the loadings in magnitude and phase angle. Eq.(5.11) holds for per-unit loads and leakage impedances provided that all are expressed with reference to a common base.

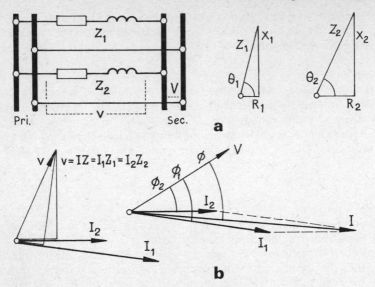

5.28 Parallel connection.

Fig. 5.28(a) shows the leakage impedances $Z_1 = (R_1 + jX_1)$ with the angle $\theta_1 = \arctan(X_1/R_1)$, and Z_2 in similar terms. The currents (b) must be such as to make I_1Z_1 equal in magnitude and phase to I_2Z_2, and sum to the total load current I. The actual output phase angles ϕ_1 and ϕ_2 differ by $(\theta_1 - \theta_2)$.
Unequal voltage ratios. Using the symbolism of the foregoing with the addition of Z, the load impedance across the secondary terminals, and E_1 and E_2, the no-load secondary e.m.f.s, then

$$v = IZ_{12} = (I_1 + I_2)Z_{12}$$

as before. The transformer e.m.f.s are equal to the total voltages in their respective circuits, i.e.

$$E_1 = I_1Z_1 + IZ \quad \text{and} \quad E_2 = I_2Z_2 + IZ$$

whence $E_1 - E_2 = I_1Z_1 - I_2Z_2$. On no load with $I = 0$ this gives between the two transformers the circulating current

$$I_1 = -I_2 = \frac{E_1 - E_2}{Z_1 + Z_2}$$

On load we have

$$I_1 = \frac{(E_1 - E_2) + I_2Z_2}{Z_1}$$

and by substitution the two currents become

$$I_1 = \frac{E_1Z_2 + (E_1 - E_2)Z}{Z_1Z_2 + (Z_1 + Z_2)Z}, \quad I_2 = \frac{E_2Z_1 - (E_1 - E_2)Z}{Z_1Z_2 + (Z_1 + Z_2)Z} \tag{5.12}$$

Normally E_1 and E_2 are co-phasal (or very nearly so); but eq.(5.12) could be used to show the severe results of incorrect parallel connection of 3-ph transformers with a phase-angle divergence.

The load-sharing of any number of transformers in parallel can more readily be found by applying the Millman (or parallel-generator) theorem: the common terminal voltage V of a number of transformers or generators, connected in parallel across a load of impedance Z, and having respective internal impedances $Z_1, Z_2, Z_3 \ldots$ is $V = I_{sc} Z_0$, where I_{sc} is the sum of the generator short-circuit currents and Z_0 is the parallel sum of $Z, Z_1, Z_2, Z_3 \ldots$ In symbols

$$V\left[\frac{1}{Z} + \frac{1}{Z_1} + \frac{1}{Z_2} + \ldots\right] = \frac{E_1}{Z_1} + \frac{E_2}{Z_2} + \frac{E_3}{Z_3} + \ldots \tag{5.13}$$

The square bracket is $1/Z_0$ and the right-hand side is the sum of the individual short-circuit currents.

Reactive Power Absorption. Operating paralleled transformers on different ('staggered') tappings is a way of absorbing lagging reactive power at points in a supply network. In the case of two transformers each of leakage impendance $Z = R + jX$, with tappings staggered to give a voltage difference ΔV, the circulating apparent power is $S_c = (\Delta V)^2/2Z$. For example, two 125 MW units each with $R = 0.004$ and $X = 0.125$ p.u., operated on load with 5% voltage difference ($2\frac{1}{2}$% up and down), circulate $0.05^2/0.25 = 0.01$ p.u. apparent power, i.e. 1.25 Mvar. The additional I^2R loss is 0.04 MW, or 8% of the full-load I^2R.

EXAMPLE 5.5: A 500 kVA 1-ph transformer with 0.010 p.u. resistance and 0.05 p.u. leakage reactance is to share a load of 750 kVA at p.f. 0.80 lagging with a 250 kVA transformer with per-unit resistance and reactance of 0.015 and 0.04. Find the load on each transformer (i) when both secondary voltages are 400 V, and (ii) when the open-circuit secondary voltages are respectively 405 V and 415 V.

(i) The per-unit impedances expressed on a common base of 500 kVA are

$$Z_1 = 0.010 + j0.05 = 0.051 \angle 79°$$

$$Z_2 = 2(0.015 + j0.04) = 0.085 \angle 69°$$

$$Z_1 + Z_2 = 0.04 + j0.13 = 0.136 \angle 73°$$

The load is $S = 750(0.8 - j0.6) = 750 \angle -37°$ kVA. Applying eq.(5.11),

$$S_1 = 750 \angle -37° \frac{0.085 \angle 69°}{0.136 \angle 73°} = 471 \angle -40° = 359 - j305 \text{ kVA}$$

$$S_2 = 281 \angle -31° = 241 - j305 \text{ kVA}$$

The total active power is $359 + 241 = 600$ kW ($= 750 \times 0.8$), and the total reactive power is 450 kvar ($= 750 \times 0.6$). The 250 kVA transformer operates with a $12\frac{1}{2}$% overload because of its smaller per-unit leakage impedance.

(ii) With the Millman theorem it is necessary to work in ohmic values of impedance

$$Z_1 = 0.0032 + j0.0160 = 0.0163 \angle 79° \; \Omega$$
$$Z_2 = 0.0096 + j0.0256 = 0.0275 \angle 69° \; \Omega$$

It is also necessary to estimate the load impedance Z: assuming an output voltage on load of 395 V, then $Z = 0.208 \angle 37° \; \Omega$. From eq.(5.13)

$$\frac{E_1}{Z_1} + \frac{E_2}{Z_2} = \frac{405 \angle 0°}{0.0163 \angle 79°} + \frac{415 \angle 0°}{0.0275 \angle 69°} = 39\,700 \angle -75.2° \; A$$

$$\frac{1}{Z_0} = \frac{1}{0.208 \angle 37°} + \frac{1}{0.0163 \angle 79°} + \frac{1}{0.0275 \angle 69°}$$

$$= \frac{1}{0.0099 \angle 73.5°}$$

The secondary terminal voltage is

$$V = (39\,700 \angle -75.2°)(0.0099 \angle 73.5°) = 393 \angle -1.7° = 393 - j12 \; V$$

The internal volt drop in the first transformer is $E_1 - V = 405 - (393 - j12) = 17 \angle 45° \; V$, and in the second is $22 - j12 = 25 \angle 29° \; V$, whence

$$I_1 = \frac{17 \angle 45°}{0.0163 \angle 79°} = 1\,040 \angle -34° \; A \quad \text{and} \quad I_2 = 910 \angle -40° \; A$$

The loads are $S_1 = VI_1{}^*$ and $S_2 = VI_2{}^*$, giving

$$S_1 = 340 - j220 = 410 \; kVA \text{ and } S_2 = 270 - j220 = 350 \; kVA$$

The combined load is $S = 610 - j445 \; kVA \; (\simeq 600 - j450)$ and the 250 kVA transformer is overloaded by 40%. The secondary circulating current on no load is

$$(E_1 - E_2)/(Z_1 + Z_2) = 230 \; A$$

corresponding to about 95 kVA and a considerable waste in I^2R loss.

5.16   Cooling methods

The considerable variety of possible methods of heat dissipation make necessary a concise standard designation. The letter symbols employed are:
Medium: air A, gas G, synthetic oil L, mineral oil O, solid insulation S, water W.
Circulation: natural N, forced F.
Order: up to four letter symbols are used for each system for which the transformer is assigned a rating. The order is (1) the medium and (2) the circulation of the coolant in contact with the windings; and (3) the medium and (4) the circulation of the coolant in an external heat-exchanger system. The common forms are listed below. The limits of temperature-rise in each case can be obtained from the Table in Sect. 4.7.

Air Cooling

AN: This has the ambient air as coolant, with natural circulation by convection. The development of high-temperature insulating materials (glass and silicone resins) makes the method suitable for ratings up to 1.5 MVA, and for special conditions such as those in mines.

AF: With forced air circulation and improved heat dissipation, the specific loadings can be raised.

Oil-immersed, Oil Cooling

ONAN: This method of natural oil circulation and natural air flow over the tank (and tubes) is very common, and can be applied to transformers rated up to 5 MVA.

ONAF: Air is blown on to the tank surface, so that less surface is required for a given rating.

OFAN: This is uncommon, but can be useful where the coolers have to be well removed from the transformer. The oil is pumped round the system, and high current densities can be used in the transformer windings.

OFAF: The forced oil- and air-circulation method is the usual one for transformers of 30 MVA and upward. Mixed cooling may be used, with an ONAN condition up to 0.5 p.u. rating, temperature-sensing elements initiating oil pumps and air fans for the forced cooling at upper conditions of load.

Oil-immersed, Water Cooling

ONWF: Copper cooling coils are mounted in the tank above the level of the transformer core, but below the oil surface.

OFWF: Oil/water heat-exchangers are external to the transformer. The advantages over the ONWF system are that the transformer is smaller and the tank does not have to contain cooling coils; there are no condensation troubles; leakage of water into the oil is improbable if the oil pressure is greater than that of the water. Where the cooling water has considerable 'head', it is usual to employ cascaded heat-exchangers (oil/water and water/water), with the intermediate water circuit at low pressure. The use of water as a coolant is common in generating stations, particularly hydro stations, where an ample supply is available.

5.17 *Protection*

Apart from fire, the hazards to which a power transformer may be subjected are listed in the table following. The types of fault (other than those arising in the associated power networks) are:

Internal: Earth; phase and interturn short- and open-circuits in h.v., m.v., and l.v. windings; core-insulation breakdown; low oil level; internal tap-changing gear.

External: Cooler; connections; auxiliaries; external tap-changing gear.

Buchholz Protection

This is applicable to oil-immersed transformers. Internal breakdowns violently generate gas or oil vapour. Short circuits increase the tem-

Transformer Protection

OVERCURRENTS: affecting insulation life, temperature-rise and mechanical stress.

Overloads exceeding rated value:
1. Winding temperature indication and alarm, allows full use of transformer overload capacity.
2. Oil temperature indication and alarm, winding temperature may be excessive.
3. Time-delay overcurrent relays, may isolate too rapidly on large and too slowly on small overcurrent.

External Faults in connected systems:
1. System protection to isolate faulty network, transformer impedance to limit throughput.

Internal Faults:
1. Buchholtz relay to give alarm or to initiate isolation.
2. Relays or fuses, high setting necessary for discrimination; restricted earth-fault relays give full discrimination but respond only to earth faults.
3. Balance methods, fully discriminative.

OVERVOLTAGES: affecting insulation stressing.

External, due to atmospheric causes and switching operations:
1. Insulation co-ordination and grading.
2. Protective gaps to lower surge crest voltage, disturb service continuity.
3. Cable connections to overhead lines, lower steepness of surge wave-front.

Internal, due to harmonics and/or resonance with system capacitance:
1. Delta-connected tertiary stabilizing winding.
2. System insulation co-ordination.

---

perature of the windings, particularly the inner layers, and pocketed oil is there vaporized. Discharges due to insulation weakness, e.g. by deterioration of the oil, will also cause oil dissociation accompanied by the generation of gas. Core faults, such as short circuits due to faulty core-clamp insulation, produce local heating and generate gas.

This generation of oil vapour or gas is utilized to actuate a relay. The relay is a hydraulic device, arranged in the pipe-line between the transformer tank and the separate oil conservator, Fig. 5.29(a): the relay is shown in greater detail at (b). The vessel is normally full of oil. It contains two floats, b_1 and b_2, which are hinged so as to be pressed by their buoyancy against two stops. If gas bubbles are generated in the transformer due to a fault, they will rise and traverse the pipe-line towards the conservator, and will be trapped in the upper part of the relay chamber, thereby displacing the oil and lowering the float b_1. This sinks and eventually closes an external contact which operates an alarm.

A small window in the wall of the vessel shows the amount of gas trapped and its colour. From the rate of increase of gas an estimate can be made of the severity and continuance of the fault, while from the colour a diagnosis of the type of fault is possible.

If the rate of generation of gas is small, the lower float b_2 is unaffected. When the fault becomes dangerous and the gas production violent, the sudden displacement of oil along the pipe-line tips the float b_2 and causes a second contact to be closed, making the trip-coil circuit and operating the main switches on both h.v. and l.v. sides.

Gas is not produced until the local temperature exceeds about 150 °C. Thus momentary overloads do not affect the relay unless the transformer is already hot. The normal to-and-fro movement of the oil produced by the cycles of heating and cooling in service is insufficient to cause relay operation.

5.18 *Testing*

A transformer may be subjected to a range of tests for a variety of purposes, including—

(*a*)   Routine tests after manufacture;
(*b*)   Acceptance tests, heat runs, etc.;
(*c*)   Specialized investigations on particular details of design, performance or operation.

Phasing and Polarity

In manufacture, where all connections are traceable, the phases on primary and secondary sides may readily be grouped. If not, all phases are short-circuited

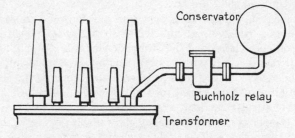

a

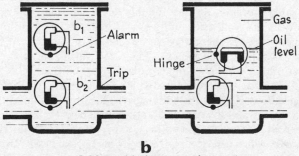

b

5.29 Buchholz protection.

except a primary and a supposedly corresponding secondary. A small direct current is circulated in the primary and a voltmeter is connected across the secondary. A momentary deflection of the voltmeter when the primary current is made and broken confirms that the two windings concerned belong to the same phase.

For this test all phase-ends should be separate. Difficulty may be experienced with internally connected zigzag (interconnected star) arrangements.

Polarity has to be in accordance with B.S. 171: see Fig. 5.17.

Voltage Ratio

Most manufacturers employ bridge methods, for checking coils before assembly.

Assembled transformers may be checked on a *ratiometer*, which is essentially a potential divider excited from the same supply as the transformer under test, and subdivided so as to read the l.v. voltage in terms of the h.v. Balance is obtained by connecting the ratiometer tapping to the l.v. winding through an ammeter and adjusting the former until the current is zero.

D.C. Resistance

Any usual method may be used. For large low-voltage transformers a low resistance bridge method is necessary. The d.c. resistance must be made at known temperature, and may be used to check the design or to estimate the eddy-current loss ratio.

Magnetizing Current and Core Loss

This test is made on a transformer complete with tank and oil, as it is performed at normal rated voltage and electric stress. One winding (usually the h.v.) is left on open circuit and the other connected to a supply of normal frequency. Readings are taken of current and power at rated voltage. The power includes the core loss, the I^2R loss (often negligible) and the dielectric loss: the latter may be significant for transformers of extra-high voltage. The conditions are discussed in Sect. 5.1.

If a test is made at normal frequency and variable voltage V, the active and reactive components of the no-load current appear as in Fig. 5.30. As the peak flux density B_m is proportional to V and the core loss roughly to $B_m{}^2$, i.e. to V^2, the power/voltage curve is to a first approximation parabolic, while the active current I_{oa} is very roughly linear. The magnetizing component I_{or} rises initially at low densities until the core attains its condition of maximum permeability, and thereafter rises steeply with the onset of saturation around normal voltage. The values in Fig. 5.30, which apply to transformers in the range 100–1 000 kVA, are expressed in per-unit terms.

Core-loss Voltmeter. Hysteresis loss (which accounts for 70–80% of the core loss) depends upon B_m, but the eddy-current loss depends upon the waveform of the flux density and therefore upon the r.m.s. density. The core loss should be measured with a sine applied voltage, but to avoid errors due to changes in the peak value, which can occur with negligible change in the r.m.s. voltage, the voltage is measured as a *mean* value, to which the peak flux density is pro-

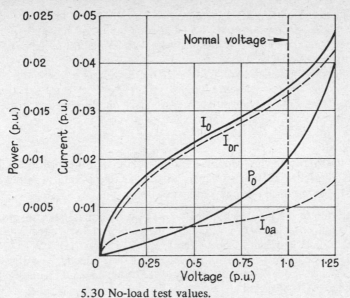

5.30 No-load test values.

portional. The mean voltage is adjusted to $2\sqrt{2}/\pi = 0.90$ times the rated r.m.s. voltage. A rectifier type of voltmeter is suitable.

Magnetizing Current Unbalance. The magnetic path length associated with the central phase of a 3-ph 3-limbed core-type transformer is significantly shorter than that of either of the outer phases. The configuration [ⵊ] shows that the outer path lengths include two half-yokes additional to the limb. As a consequence, the magnetizing current and core loss values are asymmetric, to an extent depending on the path-length ratio. If the centre path length is one-half that of either outer, then its magnetizing current is likely to be about 30% less, and this is independent of the peak flux density level. The measurement of the total core loss by 2-wattmeter method is made difficult by the fact that it is the small difference between two large readings on account of the very low power factor: the measurement also depends upon which phase currents pass through the wattmeter current-circuits, and even on the phase sequence if the wattmeter volt-circuits have inductance errors. The most satisfactory way of measuring the core loss is to use one wattmeter for each phase.

Leakage Impedance and Load Loss

One winding is short-circuited, and the other (usually the h.v. to reduce the current and increase the voltage to be measured) is connected through voltmeters, ammeters and wattmeters to a supply of normal frequency but of a fraction (5–10%) of normal voltage. With rated full-load current circulated, the load loss is measured and adjusted to standard reference temperature, i.e. to 75 °C for insulation classes AEB, and to 115 °C for FHC. The power measurement includes a very small core loss (which is generally neglected) and the eddy and stray losses. The terminal short-circuit connections must be carefully arranged, as l.v. winding resistances may be very low. In adjusting the measured loss to reference temperature it may be necessary to take account of the fact that while the ohmic I^2R loss increases with temperature, the eddy-loss ratio, eq.(3.1), actually falls.

The leakage impedance voltage is given by the voltage V_{sc} required to circulate rated current at reference temperature. The ratio V_{sc}/I_{sc} gives the leakage impedance Z_1 (if the measurements are made on the primary), from which the leakage reactance can be calculated.

Total Loss

This is the sum of the no-load loss at rated voltage, and the load loss at rated current adjusted to reference temperature. In the case of multiwinding transformers it is stated for one pair of windings.

EXAMPLE 5.6: An 11/0.44 kV 50 Hz 300 kVA 3-ph mesh/star transformer gave the following test results at rated voltage and current: No-load: 440 V 21.1 A, 1.30 kW (l.v. side). Short-circuit: 630 V, 15.7 A, 3.08 kW (h.v. side, winding temperature 30 °C). Evaluate the rated loss, impedance voltage, per-unit resistance and leakage reactance, and the efficiency and regulation on full load at p.f. 0.8 lagging.

The no-load loss is $P_0 = 1.30$ kW (0.043 p.u.); the no-load current is $I_0 = 21.1$ A (0.053 p.u.)

The s.c. test impedance voltage is $630/11\,000 = 0.057$ p.u. and the resistance is $3.08/300 = 0.010$ p.u., giving the leakage impedance $\epsilon_x = \sqrt{(0.057^2 - 0.010^2)} = 0.057$ p.u. The adjusted load loss is $3.08\,(235 + 75)/(235 + 30) = 3.60$ kW, whence $\epsilon_r = 0.012$ p.u. and $\epsilon_z = 0.057$ p.u.

The total adjusted loss is $1.30 + 3.60 = 4.90$ kW. At full load and p.f. 0.8 lagging the efficiency and regulation are

$$\eta = 1 - \frac{4.9}{300 \times 0.8 + 4.9} = 0.980 \text{ p.u.}$$

$$\epsilon = 0.012 \times 0.8 + 0.057 \times 0.6 = 0.043 \text{ p.u.}$$

With large transformers it is necessary to treat the ohmic and eddy-loss I^2R effects separately in determining the load loss.

Temperature-rise

This is measured during the performance of a heat-run intended to reproduce as far as is practicable the conditions of rated continuous load. The methods include (*a*) Back-to-back connection; (*b*) Delta/delta connection; (*c*) Equivalent open- or short-circuit run.

Back-to-back Connection. Two identical transformers A and B, Fig. 5.31 (*a*), are connected in parallel on one side to a supply of normal voltage and frequency. Disregarding the effect of the auxiliary transformer C, wattmeter W_1 reads the core loss of the two transformers together. The other sides are now properly paralleled, and a circulating voltage introduced by C to circulate full-load current in the primaries, and therefore also in the secondaries.

The circulating power is measured by wattmeter W_2, which records the load loss. If the two transformers have suitable tappings it may be possible to dispense with C by deliberate unbalance. The arrangement for 3-ph transformers is shown in (*b*). Because of the phasing of the magnetizing and circulating currents, there is a slight difference in the load loss of the pair; this may be averaged out if the auxiliary supply frequency is lower than normal.

Delta-delta Connection. The primary side is excited normally: the secondaries, in open delta, are joined to the auxiliary supply to circulate full-load current, achieving the same result as the back-to-back connection.

Equivalent Run. Where the I^2R loss is several times the core loss, a heat-run may be made by a short-circuit test at a current sufficiently greater than full-load value to cover the (absent) core loss. With high-frequency transformers it may be possible to approximate to normal loss by an open-circuit test at overvoltage. A *normal* heat-run at low power factor may require a large synchronous compensator, possibly backed with a static capacitor bank. A 500 MVA transformer with 0.2 p.u. reactance would require a test equipment capable of 100 Mvar for the temperature-rise test.

Insulation

The integrity of the insulation to earth and between turns, coils, tappings, windings and phases is tested (i) by power-frequency induced voltage, (ii) by power-frequency applied voltage, and (iii) by impulse applied voltage. *Induced voltages* are generated by the transformer itself by open-circuit test

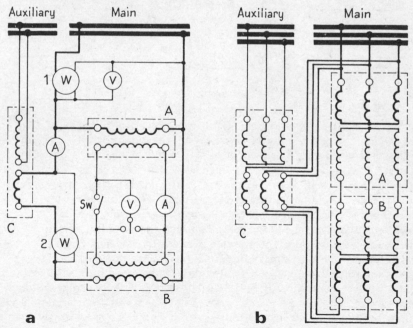

5.31 Back-to-back tests.

at a frequency up to twice normal in order to limit the level of core saturation. The peak induced voltage v_m of a given winding is measured, and $v_m/\sqrt{2}$ is taken as the r.m.s. voltage that must match the specified B.S. value for the system rated voltage level. Although the test voltage is greater than normal, its distribution is the same as in service.

Applied voltages are derived from test transformers. The whole of a tested winding is raised to terminal level, unlike the effect of induced voltage.

Impulse voltages are applied to the h.v. windings of a transformer by means of a surge or impulse generator. The test voltage wave takes the forms shown

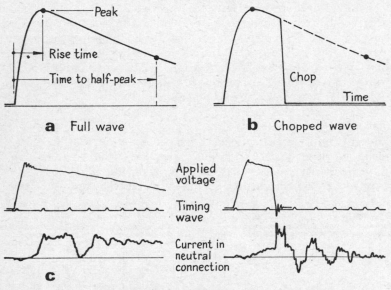

5.32 Impulse tests.

in Fig. 5.32: (*a*) is a full wave, and (*b*) a chopped wave obtained by parallel-ing the impulse generator by a rod spark-gap, which has a delayed breakdown. The standard full wave is designated 1.2/50 μs, indicating that the initial rise time is 1.2 μs, and that it decays to one-half peak value in 50 μs. A test sequence might be as follows: one full wave at reduced voltage; one full wave at test level; two chopped waves in succession; one final full wave at test level. For each test simultaneous oscillographic records are made of the applied impulse voltage and of the neutral current, the latter from the drop across a non-reactive resistor connected between the tank and earth. If an initial test is made at a voltage substantially below the expected impulse strength of the transformer, any failure that occurs during the normal test routine will show an anomaly in the current record. With modern techniques and considerable experience, failures can be detected with near certainty even when there are no other signs (e.g. noise or smoke). Fig. 5.32(*c*) shows typical oscillographic records.

Test Voltages

These are based on the highest r.m.s. system voltage V_s. Test voltage levels V_t (in r.m.s. for power-frequency tests and in peak for impulse tests) are given in BS 171. They approximate to the following:

Test Voltages, kV

A: Air-insulated windings, and oil-immersed windings not subject to impulse testing.
O: Oil-insulated windings designed for impulse testing.

System voltage V_s	Power-frequency test A	O	Impulse test O
Less than 1.1	2.5	–	–
3–36	2.1 V_s	–	–
17.5–72.5	–	7 + 1.8 V_s	20 + 4.2 V_s
100–245 (Std. 1)*	–	1.9 V_s	4.5 V_s
100–420 (Std. 2)*	–	1.5 V_s	3.5 V_s

* Standard 1 for non-effectively-earthed windings; Standard 2 for effectively-earthed windings.

Overcurrent

System short-circuit currents have immediate mechanical effects, and rapidly developing thermal effects, on transformers. The most severe case is that of a terminal short circuit, for which the result for a transformer of leakage impedance z per-unit is a current comprising the components

 (i) a symmetrical alternating current of r.m.s. value $1/z$ per-unit, and
 (ii) an exponentially decaying direct current with an initial value that may reach almost the peak of the a.c. component, as a result of the 'doubling effect'.

The overcurrent limits in BS 171 specify the symmetrical r.m.s. current I_{sc} in (i) for a range of ratings S:

S MVA :	0.63	1.25	3.15	6.3	12.5	25	100
I_{sc} p.u. :	25	20	16	14	12	10	8
z p.u. :	0.04	0.05	0.0625	0.0715	0.0835	0.10	0.125

A transformer will in general be designed to have leakage impedance corresponding to z.

Thermal Effects. If I_{sc} is not greater than 20 p.u. the transformer must with stand the heating effect for 3 sec without exceeding a specified temperature limit. For I_{sc} greater than 20 p.u. the time is reduced to 2 sec. Taking the initial winding temperature θ_0 to be the maximum coolant temperature (40 °C) plus the permissible temperature-rise given in Sect. 4.7, then the maximum winding temperature θ_m reached must not exceed the following values (in °C) corresponding to the insulation class and conductor material indicated:

Type and insulation class:	Dry	A	E	B	FHC	Oil A
Windings: copper		180	250	350	350	250
aluminium		180	200	200	–	200

The calculation is made on the basis of eq.(4.3) in the form

$$\theta_m = \theta_0 + aJ^2 t \times 10^{-3}$$

for a short-circuit current of duration t (in sec) producing in the windings the current density J (in A/mm^2). The factor a takes account of the resistivity, temperature coefficient, density and thermal capacity of the conductor material and is expressed as a function of the average temperature $\frac{1}{2}(\theta_m + \theta_0)$:

$\frac{1}{2}(\theta_m + \theta_0)$, °C	140	160	180	200	220	240
Factor a, copper	7.41	7.80	8.20	8.59	8.99	9.38
aluminium	16.5	17.4	18.3	–	–	–

The conditions imposed settle the maximum short-circuit current densities, which range from 40 A/mm^2 for a 3-sec short circuit in a dry transformer with aluminium windings up to 100 A/mm^2 for an oil-immersed water-cooled transformer with copper windings and a 2-sec duration.

Mechanical Forces. The transformer must withstand the mechanical forces on short-circuit. The value of $F = Ni$ in eqs.(2.19) and (2.20) is found with $i = 1.8(\sqrt{2} I_{sc})$, the asymmetrical initial peak current assuming a small decrement.

Partial Discharges

In manufacture, the insulation structure of a transformer is first dried and then vacuum-impregnated. Imperfect impregnation leaves residual gas-filled cavities or 'voids'. As the permittivity within a void is lower than that of the surrounding insulant, the electric stress in it is higher, so that *partial discharges* may take place with consequent deterioration of the adjacent insulating material. Partial discharges may also occur in the larger oil spaces, if gas or air bubbles develop. Further, fibre and moisture contaminants may bridge oil channels and lead to the formation of voids, and even the oil/solid-insulant interfaces may influence discharge. The limiting electric stresses within a h.v. transformer are reached when disruptive processes develop voids in which partial discharges take place. The breakdown of a void is akin to the short-circuit of a small capacitor, and it imposes small change in the test voltage or current, detectable by oscillograph, radio-frequency pick-up and several other means. Jones [116] gives a survey of these. But the difficult problem is the *location* of discharge sites within the insulation structure, to which the terminal observation of pulses gives no clue. Acoustic pick-up devices, lowered into the oil ducts, have been suggested.

5.19 Power network transformers

Transformers for particular applications present specific problems of their own. Here we consider units for power-supply networks. Special features of a number of industrial applications are discussed in Sect.12.2.

System Transformers

Units of 1000 MVA (3-ph) and 3000 MVA (1-ph) can be built for high-voltage transmission networks. Outline diagrams showing the overall dimensions of a 600 MVA, 50 Hz, 750/420/33 kV transformer, constructed (*a*) as a 3-ph unit and (*b*) as a bank of 1-ph units, are given in Fig.5.33. For (*a*) the total mass in service (including oil) is 380 tonne, of which the core alone accounts for 160 tonne. In manufacture, the shifting of such loads requires substantial crane capability, or the 'floating' of the unit on a flat floor by means of compressed air. For transport to site, most manufacturers have developed schemes for shipping very large transformers by subdivision into three main parts: (1) the tank base with the bottom yoke and the three (or five) core limbs, (2) a complete winding assembly within the tank body, and (3) the upper yoke and tank top with bushings removed. The scheme must allow for ready reassembly and erection on a concrete platform in the open air under any likely weather conditions, and must simplify on-site testing. Weight reduction may demand the use of aluminium tanks; these in the presence of polluted atmospheres or salt-spray contamination must have a paint or surface processing treatment.

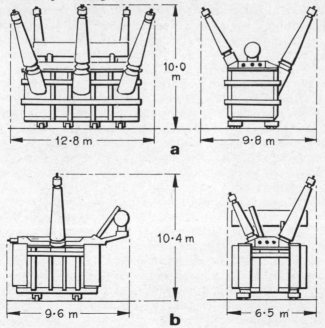

5.33 Dimensions of 600 MVA 50 Hz 750/420/33 kV transformer.

Many h.v. system transformers for linking transmission networks of differing voltage are auto-connected. The problem of voltage regulation is concerned with the choice between the h.v. and the m.v. windings for tap-changing, the tapping range required, the type of tap changer, the variation with tapping of the leakage reactance, the arrangement of windings on the core, and the cost. Refined computer-aided methods are essential for the prediction of the leakage reactance and the local distribution of the temperature-rise. Mechanical forces developed on short circuit are dynamic, and the consequent movement of windings is affected by their inertia and frictional damping.

Specifications. For large transformers these include guarantees (with tolerances) on $I^2 R$ and core losses, leakage reactance, temperature-rise and noise. Design is also concerned with the magnetizing current and its harmonic content for normal and maximum system operating voltages. Typical values (in p.u. of normal for voltage and current) are:

System voltage	(p.u.) :	0.95	1.00	1.05	1.10	1.15
Peak flux density	(T) :	1.52	1.60	1.68	1.76	1.84
Magnetizing current	(p.u.) :	0.80	1.00	1.30	3.00	5.50

Load (Stray) Loss. A large transformer has a high level of winding m.m.f.; inherent self-protection demands a large per-unit reactance and leakage flux. The latter gives rise to eddy-current power loss in clamps, tank walls and cores that, unrestricted, could exceed the winding I^2R. The eddy-loss mechanism, idealized in Fig.3.6, is exemplified in Fig.5.34 for the lower end of a typical transformer structure: it shows the leakage flux penetrating the core, yoke-clamps and tank wall, and the distribution of the resulting eddy currents. Where the flux enters the core-plate edges, the eddy current is limited by the laminar surface presented; but where it enters the broad core-plate surfaces the eddy current is much less restricted, and hot-spots are likely to occur. In large units it becomes necessary not only to include core ducts but also to slit the core along the magnetic axis or partially to slit each plate. Hot-spots can damage insulation, producing oil vapour and operating the gas-alarm relay. Eddy losses can be limited in five ways: reducing the *leakage flux density* (i) by use of low-permeability material, (ii) by shunting it through a parallel path of low loss, (iii) by diverting it into a highly-conductive screen; and reducing the *eddy-current density* by use of (iv) laminated or (v) high-resistivity material. For method (i), ductile non-magnetic cast irons of high resistivity have been employed for yoke clamps. Aluminium tanks may not require screening, but steel walls may be protected by method (ii) using shields of 0.5 mm low-silicon iron plates bonded together and bolted to the tank wall, or method (iii) with a copper or aluminium plate. The tank-loss problem has been treated analytically by Valkovic [117]. Other authors [18, 118, 119] have dealt with core heating and the shielding of tank walls.

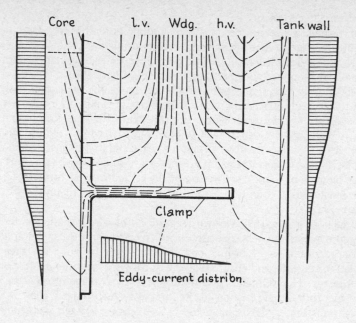

5.34 Leakage field and eddy-current distribution.

Insulation. For a maximum system voltage V r.m.s., the peak voltage to earth with a h.v. star connection is 0.81 V, the basic insulation design level is about 3.5 V and the overvoltage test is about 2.5 V. The insulation system is concerned with three main areas — conductor, coil (inter-turn and inter-coil) and major insulation — and is directed towards an arrangement of insulating layers everywhere normal to the direction of the electric field, fitting tightly at points of high stress to eliminate voids. In disc coils with interleaving (Fig.5.16), turns geometrically adjacent may be several equivalent turns apart in voltage to aid surge-voltage stress control, but for normal conditions the interleaving results in a distortion of the local electric field. The *major insulation* scheme includes wrappings, cylinders, washers, flanges, and barriers between windings and to earth. It is necessary to scarf, overlap or interleave all joints in washers, channels and angle-pieces, and to provide static shields where sharp corners could lead to overstressing. The design must take account of the different electric strengths and permittivities of the various layers 'in series', together with their thermal properties and mutual compatibility. The electric field distribution is complex in large h.v. transformers, which may have up to six coils per limb (outer and inner h.v., m.v., l.v., tapping and tertiary windings). Barriers divide the oil path and prevent line-up of chains of contaminants. Electric stresses have to be determined separately for insultation assemblies and sub-assemblies (coil sections and oil ducts) for normal-frequency overvoltage and surge-voltage conditions.

Generator Transformers

A transformer in a generator/transformer combination has to match the generator voltage to that of the connected transmission system. In some the transformer is designed with a higher protective leakage reactance and is provided with on-load tap-changing; in others, either off-load tap-changers are fitted, or no taps at all. A cost study will determine the choice between a fixed transformer turns-ratio with a wide range of generator excitation, and a constant-voltage generator with a transformer of the necessary tapping range. Usually, generator terminal voltage excursions are limited to about ± 5% with a suitable transformer tapping range and its leakage reactance must be limited to about 0.125 p.u. The hazard of a phase/phase short circuit can be reduced by careful separation of the generator/transformer interconnections, a condition favouring the adoption of a 1-ph transformer bank.

Some large generators have two separate stator windings to ease design and improve performance. The associated transformer must so combine the two outputs as to secure proper load division, minimize out-of-balance circulating currents, and ensure that the stator windings operate at the same power factor. This demands careful balance of the leakage reactances.

When a turbogenerator runs excited but not synchronized, there is a risk that its transformer will be subjected to an excessive voltage/frequency ratio, i.e. to an abnormally high core-flux density. As the normal rated flux density is typically 1.6–1.7 T, higher densities result in greater core loss, high peaks of magnetizing current, and rapid heating in associated metal parts.

Distribution Transformers

Units immersed in mineral oil are the most common, and for outdoor siting have a virtual monopoly. For location indoors they can be tolerated provided that fire-fighting equipment and effective oil drainage are installed. An alternative is to substitute an askarel (polychlor) or a silicone fluid. Askarels can produce toxic pollution. Silicone liquid is chemically stable, compatible with most solid insulants, more viscous and more costly than a hydrocarbon oil, but it can be worked at a higher temperature. A sealed construction is normally preferred.

Almost all transformers of 5MVA upwards are fitted with on-load tap-changing gear using resistor transition. The 'flick' mechanism is so well developed that transductors and vacuum switches for this service have lost much of their former attraction.

Mild-steel tanks are designed for maximum stresses of the order of 160 MN/m^2 under pressure or vacuum test, giving a factor of safety of about 2.5 and a stress below the yield point. An important problem in tank design is the form and dimensions of the many welded joints. Tanks installed in vaults underground are subject to corrosion: either mild-steel with high-quality paint or cathodic protection, or a corrosion-resistant steel, may be used.

For supply networks, transformer ratings range from 5 kVA 1-ph to 10 MVA 3-ph at primary voltages between 3.3 and 33 kV. The main types are (i) pole-mounted (up to 100 kVA), (ii) generating-station auxiliaries, and

(iii) network and industrial substation. All 3-ph transformers are connected delta/star. Up to 11 kV, most units are free-breathing, but conservators are fitted to generating-station auxiliaries.

Inductors

Inductors ('reactors'), connected in series for limiting fault levels and in shunt for reactive-power compensation in power networks, may be built with laminated cores. Three forms, all normally immersed in oil, are shown in Fig.5.35.

Gapped-core (i). This follows transformer practice but has the wound limb sectionalized to stabilize the inductance. Fringing fluxes in the gaps have components that enter the core sections at right-angles causing eddy loss and vibration. The effect is mitigated by use of a greater number of shorter gaps (making clamping more difficult) or by 'fanning-out' the core plates so that fringing flux enters mainly the plate edges.

Coreless, Magnetic Shield (ii). If the shield has an adequate cross-section, almost perfect inductance linearity can be achieved.

Coreless, Conducting Shield (iii). The screen provides a path for induced currents that suppress the external flux that would otherwise penetrate the tank walls, but at the same time reduces the winding inductance by typically 10%. The design is a compromise between loss of inductance and the tank size and oil content. Under fault conditions the force between coil and screen may be considerable.

A large inductor for a 275 kV network has typically a rating of 600 kvar at 50 Hz, a normal current of 1.5 kA (overcurrent 2.0 kA for 20 min), a reactance of 0.12 p.u (16.6. Ω) and a rated loss of 36 kW.

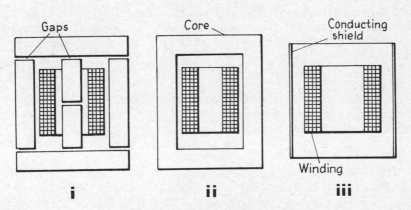

5.35 Inductive reactors.

6
Transformers: Construction and Design

6.1 *Constructional features*
The main constructional elements are: cores, comprising limbs, yokes and
clamping devices; primary, secondary and sometimes also tertiary phase windings, coil formers, spacers and conductor insulation; interwinding and winding/
earth insulation and bracing; tanks, coolers, dryers, conservators and other
auxiliaries; terminals and bushings, connections and tapping switches.
Dry Transformers. For indoor and well-protected sites, and for ratings up to
about 1.5 MVA at 11 kV, 'dry' type transformers may be advantageous. They
have class C insulation, and hot-spot temperatures up to 150 °C. They may
have simple natural air cooling, or be sealed into cabinets with nitrogen filling.
Mica-glass tape is employed for conductor insulation, while parts closely
associated with the windings are of micalex or porcelain. These materials are
stable at working temperatures. Dry units can be free-standing or be set within gated enclosures.
Oil-immersed Transformers. The great majority of industrial power transformers are immersed in oil for cooling, insulation and mechanical protection.
Certain of the details below, in particular for tanks and oil, refer specifically
to this type.

6.2 *Cores*
European practice is to standardize on core-type constructions. Some 1-ph
shell types are built, but the 3-ph shell type is rare. General details of transformer magnetic circuit construction are given in Sect. 2.2 and 2.4. Core
plates are slit and sheared from c.r.o.s. material and given a subsequent stress
relief.

   The limbs and lower yoke are assembled with the top yokes removable for
admitting the windings. As many as 20 widths of steel strip may be used in a
large core: Fig. 6.1 shows the dimensions of three typical cores. The advantage
of multi-stepped cores in respect of active core cross-section has to be balanced
against the intricacy of the building program. The insulated-lamina form of
the core means a loss of 5—10% of the gross area, and to avoid further loss of
active area requires clamping pressure to minimize gaps due to waviness of the
core plates. With varnish as a plate insulation, a compressed core has little

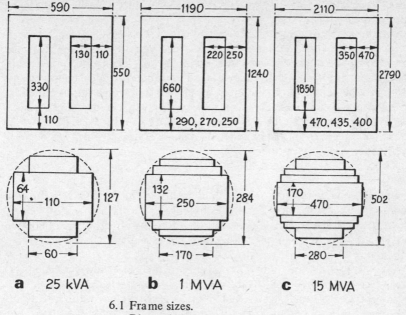

a 25 kVA   **b** 1 MVA   **c** 15 MVA

6.1 Frame sizes.
Dimensions in mm; not to scale.

'spring' and remains substantially at its compressed dimensions. Core clamping
may be effected by outside stiffeners and insulated non-magnetic through-
bolts, but to retain the advantage of c.r.o.s. even very large cores may be bound
under pressure by synthetic-resin-bonded glass or Terelene tape which is subsequently heated to set the resin. Rigidity is aided by using varnish insulation
for the outer packets of the core. Mechanical stresses imposed during assembly
and in subsequent service are taken by a cradle comprising top and bottom
yoke clamps, main tie-bolts, and substantial feet beneath the bottom yoke.
Large bound cores have a high electrical resistance through the stack and must
be provided with earth strips to prevent the static discharges that would otherwise occur during an impulse test.

Small cores wound with continuous varnished c.r.o.s. strip as in Fig. 2.7 do
not need binding or clamping.

6.3   Tanks and oils
Small tanks are welded from steel plate, larger ones assembled from boiler-
plate or cast-aluminium parts, usually mounted on a shallow fabricated steel
base. A tank must withstand the stresses imposed by jacking and lifting, and
should be no larger than is necessary to accommodate the core, windings and
internal connections with appropriate electrical clearance. If the tank can
support a partial vacuum, drying out on site (if required) is facilitated. The
tanks of large transformers may be shaped to the outline plan of the core and
windings to give a robust form not needing stiffener bars and minimizing the
oil content. This 'form-fitting' shape is not possible for a 3-ph unit having the

tap-changer sited at one end because of the multiplicity of horizontal connectors needed.

Conservators. Transformer oil expands by about 7% over the temperature range from 0 to 100 °C. Conservator tanks are required to take up the cyclic expansion and contraction without allowing the oil to come in contact with the ambient air, from which it may absorb moisture. A typical conservator is an airtight cylindrical drum mounted on or near to the tank lid, or a flexible corrugated-disc container. The main tank is filled when cold, and expansion of the oil is taken up by the conservator. Displacement of the conservator air by oil takes place through a breather containing silica gel, which extracts moisture from ingoing air.

Oils

Transformer oil, which has to be both an insulator and a coolant, has to fulfil certain specifications, including the following.

Viscosity. This affects the oil-flow and therefore the effectiveness of cooling. A high viscosity causes sluggish flow through the many small orifices in the windings.

Electric Strength. Clean hydrocarbon oil normally has a fully adequate breakdown stress. More important is the severe reduction in electric strength due to the presence of water, which must be rigorously excluded. Further, if dust and small fibres are present in the oil, paths of low resistivity (and therefore of weakness) may be formed.

Flash Point. This is the temperature at which oil vapour ignites spontaneously. A flash point higher than 160 °C may be required.

Fire Point. The temperature at which an oil will ignite and burn is about 200 °C, and about 25% above the flash point.

Purity. The oil must not contain sulphur or its compounds, which cause corrosion of metal parts and accelerate the production of sludge.

Sludge. Sludging is the slow formation of semi-solid hydrocarbons as deposits on windings and tank walls; it is due to heat and oxidation, and is self-aggravating because the sludge deposits increase the thermal gradients and block the oil-circulation paths. Sludge seems to form more quickly in the presence of bright copper surfaces, and its acidic nature makes it corrosive.

Acidity. Among the products of oxidation are carbon dioxide, volatile water-soluble organic acids, and water. These in combination can attack and corrode steel and other metals. The rate of acidification is reduced by the use of breathers and conservators.

The deterioration of transformer oil during its working life can be retarded by anti-oxidants, usually of phenolic or amino type, which convert chain-forming molecules in the oil into inactive molecules, being gradually consumed in the process. Such inhibitors prolong the active life of *inhibited oil* by delaying the onset of deterioration by acid and sludge formation.

Fire- and Explosion-proofing. When it is essential to reduce the risks of fire and explosion, a synthetic oil (*askarel*) may be used. It is chemically stable, non-oxidizing, volatile, heavier than water and with greater dielectric strength than the normal petroleum-based mineral oil. Its relative permittivity is 4.5, nearly twice that of normal oil, so that the distribution of electric stress between the liquid and solid insulants is affected. An askarel is a solvent of varnishes, binders and paints, but is resistant to the migration of moisture. When decomposed by an arc the chief product is hydrogen-chloride gas, which can form deleterious hydrochloric acid if water is present. However, with askarels a transformer may be economic (or the non-toxic silicone oils) because fire-extinguishing equipment and fireproof masonry are not needed.

6.4   *Windings and insulation*

Types of coils and windings are described in Sect. 3.4. Their direct connection with electric power networks makes windings the most vulnerable parts of the transformer. The main design and constructional features are:

(i) The provision of adequate *insulation* strength to withstand power-frequency applied or induced test voltages, both to earth and between coils, turns and phases; and full-wave and chopped impulse test voltages to prove capability against switching surges and lightning.

(ii) The limitation of *load loss* ($I^2 R$ and stray) to the guaranteed performance figure.

(iii) The provision of *cooling* to meet guaranteed limits of winding temperature-rise.

(iv) The provision of the required *leakage inductance*.

(v) The achievement of sufficient *mechanical strength* to withstand short-circuit forces.

For transformers of small and medium size, the inner (low-voltage) winding is insulated from the core by pressboard or by a synthetic-resin bonded paper cylinder, with axial bars arranged around it to form cooling ducts for the inside cylindrical surface of the winding. With disc or disc-helix coils, the bars have a wedge-shaped section on which intercoil or interturn dovetail-slotted spacers can be threaded. Similar bars are placed around the outer surface of the l.v. winding and (if necessary) between the layers of a helix winding. The main h.v./l.v. insulation is provided by another pressboard or s.r.b.p. cylinder, and the bars and spacers repeated for the h.v. winding. Insulation at the ends of the windings comprises blocks keyed to the axial bars in line with the spacers to form a series of columns by which the winding can be clamped.

For operating voltages demanding thick insulation it is customary to use concentric thin-walled cylinders spaced by axial bars, the material being carried round the ends of the outer (h.v.) winding by flanged collars as in Fig. 6.2. The 'solid' form (*a*), for 110 kV upward, has the collars as extensions of the cylindrical pressboard or paper winding insulation, the h.v. winding being assembled direct over the outer layer, with cooling ducts transferred from the inside turns at suitable points. In (*b*) the collars are separate. For multilayer h.v. helix windings the interlayer insulation may be a combination of paper or pressboard wraps, with bars arranged axially to form cooling ducts. These wraps may be formed in tapered halves when internal interlayer connections are required, providing maximum insulation thickness at the appropriate places.

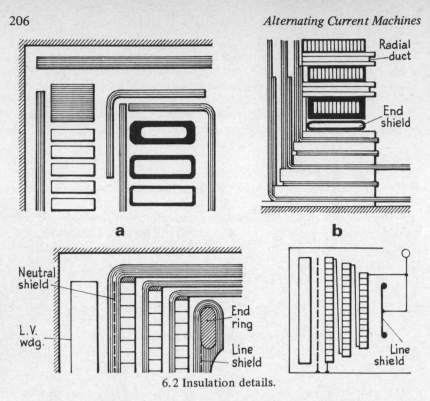

6.2 Insulation details.

Conductor insulation is a tape, of paper for oil-immersed and glass for dry transformers. The tape is machine wrapped and lashed with cotton strands to keep it in place.

Surge-voltage Distribution. Methods of improving the distribution are given in Sect. 5.8. A multilayer winding has a higher series capacitance than a disc winding, but stress shielding may still be necessary at the ends of the layers. The choice between disc-coil and multilayer windings depends on the specified impulse level and the rating.

Load Loss. The conductor I^2R loss is calculable. The load (stray) loss arises from eddy currents in connectors, conductors (if not adequately transposed), tank walls and clamping structures. Where for a high reactance the leakage flux is large, the tank must be shielded from it.

Cooling. Most transformers are cooled by thermal circulation of oil over core and coil surfaces. The arrangements for heat dissipation are discussed in Sect. 4.2, 4.4 and 4.6.

Leakage Inductance. A given arrangement of core and windings has a 'natural' value of leakage inductance, Sect. 2.7. By dimensional adjustments this can be varied within limits without undue influence on cost and performance, but considerable changes involve expense and some reaction on load loss, surge-voltage distribution and other parameters subject to specification.

Short-circuit Mechanical Strength. The radial and axial forces to which transformer windings are subjected under short-circuit conditions are discussed in

Sect. 2.7. Thermal stresses can also be developed when the windings carry sustained overcurrents, and the temperature-rise in such cases is calculable from eq.(4.3), Sect. 4.2.

6.5 Terminals and fittings
Connections to the windings are copper straps, insulated wholly or in part, with the terminal connections taken to bushings on the tank top. The size and shape of the connectors may have to take account of electric stress, and sharp edges and corners avoided.

Bushings
Up to 33 kV, simple porcelain insulators can be used. Above this voltage, oil-filled or capacitor bushings are required. The *oil-filled* bushing consists of a hollow porcelain cylinder with a conductor, usually a cylinder, through its axis. The space between the conductor and the inner surface of the porcelain is filled with oil, and at the top there is a small expansion chamber to accommodate the variation of oil bulk due to changes in operating temperature. A capacitor bushing is constructed of thick layers of s.r.b.p. alternating with the graded layers of tinfoil to form between the outer layer and the inner conductor a series of capacitors of substantially equal capacitance, so giving an approximation to a uniform electric stress distribution over the radius.

For use outdoors, a bushing is covered by a porcelain rain-shed, corrugated circumferentially to provide a longer and partially protected leakage path.

Fittings
Among the devices normally provided for oil-immersed transformers are the following:
Cable Boxes for direct connection of paper- or plastic-insulated supply cables.
Oil Conservator, Oil Gauge and Oil-temperature Indicator.
Breather for transformers and conservators not fully sealed, provided with a chemical or refrigerant to remove moisture from the intake air.
Buchholz Relay. See Sect. 5.17.
Winding-temperature Indicator, essentially a thermometer with the bulb associated with a heater coil carrying a current proportional to the load current. The thermometer is immersed in the oil, and its indication, raised above the ambient oil by the heater, gives an analogue indication of the winding temperature.
Tapping Switch, either on- or off-load, Sect. 5.14.
Explosion Vent, relieving dangerous rises of internal pressure by a spring-loaded device or by a thin non-metallic diaphragm which breaks under excess pressure.
Base; Skids or Wheels.

6.6 Distribution transformers
The preferred ratings for steady loads are related approximately by the factor $2^{1/3} = 1.26$. They are

5	6.3	8	10	12.5	16	20	25	31.5	40
50	63	80	100	125	160	200	250	315	400
500	630	800	1000	...					

A range of distribution transformers for batch manufacture might comprise basic sizes of 63, 100, 160, 250, 400, 630 and 1 000 kVA rating. Laminations for the limbs and yokes are produced in the press shop from coils of grain-oriented sheet in one operation, heat treated in a stress-reducing tunnel oven, and then transported to the assembly shop. The h.v. and l.v. windings are produced on special-purpose machines, either as complete coils or as assemblies. Terminals and connections are prepared in advance, and the coils are aged and ready for installation.

The tank walls, bases and frames are welded together in the tank assembly shop, and pressure tested with hot oil for leaks. In the final production stage the tanks pass through steam cleaning and de-greasing plant, and are then painted.

The core and coil assembly line involves the following operations: placing the cores on trolleys; dismounting the top yoke; fit coils; re-fit yokes; assemble cover with bushings, switches and other equipment; fit cover on coil-core unit and secure electrical connections. The coils and cores are heated for some hours in an evacuated vessel, checked for tightness, fitted into the tank and bolted in position. Finally the tank is filled with dried and degassed oil.

Standard tests are carried out in the following sequence: resistance, voltage ratio, polarity; no-load tests; short-circuit test. The measurements are fed to a computer which processes them and prints out a test report.

The following Example traces the evaluation of the main features of a medium-rated distribution transformer. Normally all the calculations would be computed from suitable programs with the necessary specific data fed in.

EXAMPLE 6.1: Three-phase 3-limbed core-type transformer to BS 171; rating 400 kVA, 6.6/0.41−0.45 kV in $2\frac{1}{2}\%$ steps, delta/star, oil-immersed, natural cooled. See Fig. 6.3.

Main Dimensions. The frame size (Sect. 1.14, Fig. 1.13(b)) has a stepped core of net area $A_i = 0.027$ m^2 contained in an overall limb diameter $d = 0.21$ m, with the limb centres spaced $D = 0.37$ m. The yoke area is the same as that of a limb. With a peak flux density $B_m = 1.45$ T, the voltage per turn from eq.(1.13) is $V_t = 8.7$ V. The primary phase voltage is $V_1 = 6\ 600$ V; the secondary phase voltage at maximum flux density is $V_2 = 450/\sqrt{3} = 260$ V. The turns per phase are therefore

$$N_2 = 260/8.7 = 30 \quad \text{and} \quad N_1 = 30 \times 6\ 600/260 = 760$$

N_1 is increased by 76 turns to a total of 836 for the lower limit of secondary voltage. The rated primary current per phase is $I_1 = 133/6.6 = 20.0$ A. With a current density $J = 3.0$ MA/m$^2 = 3.0$ A/mm^2, and a window space factor $k_w = 0.29$, the required window area from eq.(1.15) is $A_w = 0.071$ m^2. The depth of the window is $l = A_w/(D-d) = 0.55$ m. The main dimensions are thus

$$d = 210 \text{ mm}; \quad D = 340 \text{ mm}; \quad l = 550 \text{ mm};$$
$$W = 2D + 0.9d = 870 \text{ mm}; \quad H = 930 \text{ mm}.$$

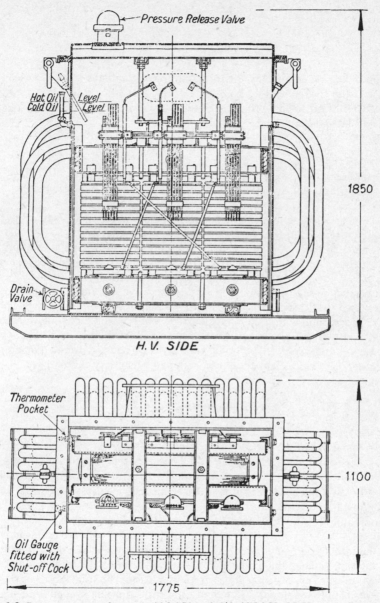

6.3 Core-type transformer: 400 kVA, 6.6/0.415 kV, 50 Hz, delta/star.

Magnetic Circuit. The material is c.r.o.s., and the yoke area is the same as the limb area. The total core volume is 0.093 m^3 and its mass is 700 kg. The specific loss (Fig. 2.5) is 1.50 W/kg, so that the total core loss (with a building factor of 1.3) is

$$P_i = 700 \times 1.50 \times 1.3 = 1.37 \text{ kW}$$

From the curves in Fig. 2.3, the total m.m.f. for the limbs is $3 \times 0.55 \times 150 = 250$ A-t, and for the yokes is $2 \times 0.87 \times 150 = 260$ A-t, a total of 510 A-t, raised to 600 A-t to allow for the greater reluctance of the joints. The magnetizing current per primary phase is

$$I_{om} = 600/3 \times \sqrt{2} \times 760 = 0.19 \text{ A}$$

corresponding to a magnetizing reactive power per phase of $q_0 = 1\,260$ var, or a total of 3 800 var. An estimate based on Fig. 2.14 gives 4.6 var/kg, i.e. 3 200 var.

Low-voltage Winding. With a current density $J_2 = 2.85$ A/mm^2, a little lower than the mean to improve the inner cooling, the secondary phase current $I_2 = 133/0.26 = 512$ A requires a conductor area $a_2 = 512/2.85 = 180$ mm^2. Taking an available winding depth of 465 mm, space is required to accommodate $30 + 1$ turns, giving 15 mm depth per turn for the helix winding. The conductor must be subdivided to reduce eddy currents. A suitable conductor comprises four parallel strips, each 13.5 mm x 3.5 mm with corners rounded, each strip being wound with 0.25 mm paper tape and the whole conductor assembly taped over with 0.5 mm paper. Wound on an s.r.b.p. tube of inside diameter 215 mm and thickness 2.5 mm, the winding is 254 mm in overall diameter. The mean turn has a diameter of 237 mm and a length of 0.75 m. The phase resistance at 75 °C is

$$r_2 = 0.021 \times 30 \times 0.75/180 = 0.0026 \ \Omega$$

and the I^2R loss is $512^2 \times 0.0026 = 0.69$ kW/ph

High-voltage Winding. The conductor section $a_1 = I_1/J_1 = 20/3.1 = 6.45$ mm^2 is suitable for circular-section wire of diameter 2.85 mm, insulated by paper wrap to 3.15 mm diameter. The h.v. winding is sectionalized into nine coils of 72 turns, two end coils of 55 turns with reinforced insulation, a central 76-turn coil tapped at each end and at three intermediate points, and two gunmetal support rings to form the outmost turn at each end; the total is $(9 \times 72) + (2 \times 55) + 76 + 2 = 760 + 76 = 836$ turns. The normal coils are arranged with 9 conductors axially and 8 radially. The axial length of each coil-section is 31.5 mm including taping. The coils are spaced 6 mm apart by pressboard U-pieces and the support rings are each 12 mm thick. The total depth of the h.v. winding is 480 mm, leaving 35 mm at each end for insulation and bracing.

The outside diameter of the l.v. winding is 254 mm. With an oil duct 12 mm wide to facilitate the cooling of the l.v. winding, the inner diameter of the s.r.b.p. cylinder carrying the h.v. winding is 278 mm. The h.v. winding is fitted over 5 mm axial spacer bars of pressboard, its diameters being 294 mm (inner),

348 mm (outer) and 320 mm (mean). The resistance of the h.v. winding at 75 °C per phase of 760 turns of mean length 1.00 m is

$$r_1 = 0.021 \times 760 \times 1.0/6.45 = 2.48 \ \Omega$$

and the I^2R loss is $20.0^2 \times 2.48 = 0.99$ kW/ph.

Cooling. The core loss is $P_i = 1.37$ kW, the I^2R loss is $P_c = 3(0.69 + 0.99) = 5.04$ kW and the total full-load loss is 6.41, say 6.5 kW. Allowing 10 mm between adjacent h.v. windings and adequate clearances to earth for connections and tappings, the transformer can be accommodated in a tank 1.15 m x 0.55 m in plan. The core height is 0.90 m; allowing 50 mm for the base and 250 mm of oil above, the oil depth is 1.2 m, and with a further 250 mm for connections to the tank top, the total tank-wall height is 1.45 m. The surface area of the tank walls is 5.0 m^2, making it necessary to fit cooling tubes. An array of 80 tubes each 50 mm diameter and mean length 1.05 m gives a total tube surface of 13.2 m^2. Using the method of Sect. 4.3, eq.(4.4), the ratio x is 3.6 and the specific dissipation rate is $p = (8.8 + 3.7/3.6) = 9.83$ W/(m^2 °C), whence the tank temperature-rise is

$$\theta_t = 6\ 500/[9.83 \times (5.0 + 13.2)] = 36\ °C$$

Impedance. From eq.(5.5), the per-unit resistance is $\epsilon_r = 6.5/400 = 0.016$ p.u. The terms for the per-unit leakage reactance in eq.(2.16) are:

$$
\begin{array}{lllll}
f & = 50 \text{ Hz} & F & = N_2 I_2 = 30 \times 512 = 15\ 360 \text{ A-t} \\
E_t & = 8.7 \text{ V} & L_{mt} & = \tfrac{1}{2}(0.75 + 1.00) = 0.88 \text{ m} \\
L_c & = 0.47 \text{ m} & a & = 0.020 \text{ m} \quad b_1 = 0.026 \text{ m} \quad b_2 = 0.019 \text{ m}
\end{array}
$$

and these give $\epsilon_x = 0.046$ p.u. The total leakage impedance is 0.0485 p.u.

Mass. The mass of active ferromagnetic material is $325 + 392 = 717$ kg. The copper in the windings has the mass $108 + 144 = 252$ kg. The ratio $G_i/G_c = 2.85$.

6.7 Large transformers

Rise in service voltage and increase in rating impose more stringent technical demands, making necessary more refined methods of assessing both electromagnetic and mechanical problems. In transformers of 500 MVA rating and upward, the specific design problems are concerned with very large currents, powerful stray fields, the removal of heat in large quantities, and the short-circuit mechanical strength. Solutions must recognize the restraints imposed by transport facilities and the voltage levels.

The mass of a 1 000 MVA transformer core alone may exceed 200 t. Dimensional limits generally make a 5-limbed core essential, for loading-gauge reasons as well as those of the magnetic circuit. The 5-limbed core can be used to advantage for ratings above 100 MVA if the technical specification permits of optimum geometric proportions. The 5-limbed core is less massive than a 3-limbed core, but its mean specific core loss is higher because the yoke fluxes are no longer sinusoidal; however, the stray loss of a 5-limbed core is greatly reduced because of the magnetic path provided.

Winding design may demand double- or multiple-concentric windings to carry very large currents (particularly, of course, on the l.v. side). The main insulation can, however, follow existing techniques based on paper, board and oil.

The stray magnetic field is responsible for the generation of eddy-current loss and of disruptive mechanical forces on short circuit. To determine and ameliorate the effects it is necessary to know the stray-field distribution pattern. Considerable advance has been made in field-pattern calculation by digital computer in the difficult cases of complex boundaries and of non-linearities. Screening is sometimes necessary to limit stray loss in tank walls by transferring the eddy-currents to aluminium or copper shields. Even more important than the absolute loss level is its concentration locally: critical sites in this respect are the ends of the coils, the stress rings, the tank sides, and the clamping beams and bolts.

The high power density and the necessarily compact construction of the windings necessitate intensive internal cooling, achieved only by forced oil-flow through the windings to reduce hot-spots and lower the temperature gradient between windings and oil. The basic construction is such that all con-centrically arranged windings are hydraulically in parallel, and the flow in each path is determined by its hydrodynamic characteristics: these can vary widely, and may demand some baffle or valve control over the oil flow so that the individual paths may be assigned oil-flow rates proper to the heat-transfer de-manded by each.

Avoidance of mechanical breakdown under short-circuit electromagnetic forces requires accurate calculation of the axial forces by digital methods, and mechanically rigid construction of each component winding with the end supports, spacers, cylinders and caps fixed to prevent them from working loose. Pressure elements pre-stress the windings to accommodate the effects of the drying-out and impregnation processes, and to ensure the maintenance of proper axial pressure in service. The windings must have carefully assembled supports to prevent the conductors from buckling due to inward radial forces. These means call for a high degree of dimensional accuracy, a demand not easy to satisfy because the thickness of the spacers varies during construction with the humidity in the winding shop. The use of forced oil circulation, necessary for proper cooling, enables coils to be sited more closely, but results in bigger radial forces: it is therefore necessary to make the inner coils more stable against inward force, and the outer coils resistant to expansion.

6.8   *Design*

Because it does not rotate, the transformer presents certain mechanical simpli-fications compared with rotating machinery, although it has a number of special features of its own. Considerable advances have been made in devising digital-computer design programs. Synthesis programs have been in regular use for several years. The analysis of surge-voltage distribution in windings, and the optimization of h.v. insulation were developed at an early stage, and later served as components of a full design program, in particular the basic dimen-sions directly from the customer's specification. Such programs can be run by computer staff having no detailed design knowledge, but minor adjustments by designers are almost always made to improve details before the design is passed to production

The digital computation of particular features (such as leakage reactance) is normal. The more sophisticated the initial analysis in its use of design parameters, the more useful and accurate is the program.

Principles

The design clearly depends on the voltage and rating, less obviously on the type, service conditions, and the relative costs of active materials, labour, manufacturing equipment and works organization. The optimum design achieves the minimum total cost in relation to the capital cost, energy loss, thermal and mechanical depreciation (resulting from fault-generated stresses), the last few considerations not being readily expressible in money terms. Constraints are imposed by depreciation rates, transport to site, facilities for repair and labour expertise. The design process is an experienced manipulation of the many interrelated variables, and can be exceedingly complex. There are several possible starting points.

From eq.(5.4), a transformer has a *maximum power efficiency* for the fraction k of its rating S when the I^2R loss P_c and the core loss P_i are related by $k = \sqrt{(P_i/P_c)}$. Let G_i and G_c be the total active masses of core and conductor material, and p_i and p_c the specific losses (in power per unit mass); then for maximum efficiency the fractional load is such that $G_i p_i = k^2 G_c p_c$. It is possible to relate the specific losses explicitly to the main magnetic and electric loadings by writing

$$p_i = a + bB^n \quad \text{and} \quad p_c = J^2 \rho/\delta_c$$

where the exponent n is based on tests on core strips adjusted by the building factor (e.g. 1.07), and ρ and δ_c are respectively the resistivity and density of the conductor material. The ratio G_i/G_c can then be expressed in terms of k and the loadings.

Quite different criteria could be used, based on the *aggregate cost* of material (weighted to include processing costs) to secure the cheapest transformer. Writing γ as the cost per unit mass of the processed and assembled parts, the total cost is the sum $G_i\gamma_i + G_c\gamma_c$. Although G_i and G_c are individually variable, it turns out that the cheapest transformer is one in which the ratio G_i/G_c is approximately the same as γ_c/γ_i.

If both maximum-efficiency and lowest-cost criteria are to be simultaneously satisfied, the requirement is

$$k^2 = \frac{\gamma_c}{\gamma_i} \cdot \frac{\delta_c(a + bB^n)}{J^2\rho}$$

With k known, and no other unaccounted constraints can exert an influence, the optimum values of J and B can be computed; but if further conditions have to be imposed (such as transport dimensions) they must somehow be related to the specific magnetic and electric loadings, a process that may be rather complicated. For the largest transformers, the imposition of dimensional and leakage-reactance constraints forces the designer to choose the maximum flux density and to let the current density emerge from the equation.

If k is unknown (the more common case), then B can be settled on saturation, harmonic or noise-limit conditions, and J on the cooling system necessary; but now k has little significance.

The specific loadings are continuously variable, but many data are variable only in steps (choice of coil type, conductor size, cooling method, core stepping, etc.). The current rating, impulse voltage test level and transport conditions dictate the type of winding, and if tappings are necessary a decision must be made as to where they are to be placed and the consequent effect on leakage, asymmetry and short-circuit forces. Insulation and shielding are related in a complicated way to impulse strength. Flow of coolant affects temperature-rise. From such magnetic, dielectric, thermal and conduction considerations an array of significant parameters is selected: the type, form, section and construction of the core; the self and mutual arrangement of the windings; the dimensional allowances to meet the electrical and thermal boundary conditions.

Provided that analytical relationships can be worked out, the calculations can be made by means of a digital computer program. Many such programs have been devised and are in use in design offices. They are constantly refined in accordance with experience and become more comprehensive, and they are of great value in providing rapid appraisal of the consequences of changes in dimensions and in the cost and properties of new materials.

7
Polyphase Rotating Machines

7.1 *Synchronous and induction machines*

The basic concepts relating to rotating machines in general are discussed in Sect. 1.10–1.13, in which it is assumed that the distribution in the airgap of the flux produced by sine-distributed current sheets is itself sinusoidal. It is shown in Sect 3.14 that balanced currents in a 2- or 3-ph winding approximate to sine-distributed current sheets, and have in addition the further and important property of developing an airgap flux in the form of a travelling wave that moves one double pole-pitch in one period of the a.c. supply.

The torque set up in such an arrangement is a function of the stator and rotor m.m.f.s F_1 and F_2, and of the torque angle λ between them. The condition that the torque shall be maintained unidirectionally under steady-state conditions is that for given stator and rotor m.m.f.s the torque angle shall remain constant. Thus both F_1 and F_2 must be waves travelling at the same speed. With the stator carrying a polyphase winding connected to a supply of angular frequency ω_1 then F_1 travels at the angular speed ω_1 in a 2-pole winding, or at the synchronous speed $\omega_s = \omega_1/p$ in a winding with p pole-pairs. The rotor m.m.f. F_2 must therefore travel at ω_s with respect to the stator. The speed of rotor rotation, ω_r, then depends upon the movement at speed ω_2 of F_2 with respect to its own rotor windings. If F_2 is 'fixed' by supplying d.c. rotor excitation, then the rotor must rotate at $\omega_r = \omega_s$ and can produce steady-state torque at no other speed: the arrangement gives a *synchronous* machine. If, however, the rotor m.m.f. is produced by a 2- or 3-ph winding carrying currents of frequency ω_2, then the rotor rotation must be such as to give $\omega_r = \omega_1 \pm \omega_2$ in a 2-pole machine, or $\omega_r = (\omega_1 \pm \omega_2)/p$ in a $2p$ - pole machine. By far the commonest arrangement to give this effect is the *induction* machine.

In this Chapter the polyphase synchronous and induction machines are considered in general terms, with particular reference to ways in which they can be modelled for analysis. Such an analysis must deal not only with *steady-state* conditions but must also be able to take account of *transience*. In the former the machine can be taken as having a purely electromagnetic nature, but in the latter the model must be elaborated to include the mechanical effects of speed and load changes and the response of control and regulating systems. These result from the nature of the machine as a link between electrical and

215

mechanical networks, the responses of which vitally effect the behaviour of the machine.

Representative phase for steady-state conditions

Each stator phase of a balanced m-phase machine deals with $1/m$ of the total active and reactive power input to (or output from) the machine; and all have the same magnitudes respectively of voltage, current and phase angle, differing only in the time-phase displacement of $2\pi/m$ rad between the voltage or current of one phase and the next. It is useful to refer the machine as a whole to the measurable voltage V_1 and current I_1 of one *representative* phase winding. In Fig. 7.1(a) let only the stator be excited by connection to a polyphase

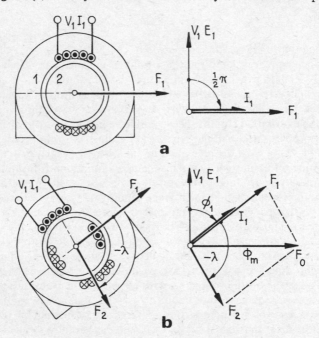

7.1 Polyphase machine: representative phase.

supply. At any instant that phase carrying *peak* current, $i_{1m} = \sqrt{2}I_1$, fixes the axis of the m.m.f. F_1 of the complete polyphase stator winding. In a time phasor diagram for this representative phase winding, its current I_1 can be drawn in the direction of the total-m.m.f. axis F_1 as an arbitrary, but pictorially advantageous, assumption. Then the stator applied voltage V_1, and the e.m.f. E_1 generated in the phase winding, must be drawn in leading phase-quadrature with I_1 because the representative phase (as, indeed, the stator-winding as a whole) is simply an inductor taking only reactive magnetizing power (if loss can be ignored). Thus we can combine the phasor and m.m.f. diagrams into one that conventionally associates space conditions in the gap with time phasors of one of the contributing phase windings.

Now let both members be excited. The rotor is assumed to have a polyphase winding identical with that of the stator (i.e. it is 'referred' to the stator), and each rotor phase winding carries a current I_2 developing an m.m.f. F_2. With F_2 aligned on the axis of that rotor winding carrying peak current at the instant considered, let F_2 make the torque angle $-\lambda$ to F_1, corresponding to a motor mode. It is now the *resultant* m.m.f. F_0 that determines the gap flux Φ_m. The conditions are shown in Fig. 7.1(b), the stator having been tilted in the diagram in order to make the axis of F_0 and Φ_m horizontal. As Φ_m induces E_1 in the representative stator phase, both E_1 and V_1 lead Φ_m by $\frac{1}{2}\pi$ rad. But the phase angle ϕ_1 between V_1 and I_1 is now less than $\frac{1}{2}\pi$ and the representative stator phase is accepting a complex power $S_1 = V_1 I_1 (\cos \phi_1 + j \sin \phi_1) = P_1 + j Q_1$, where P_1 is $1/m$ of the total active input power, and Q_1 is $1/m$ of the total magnetizing reactive power of the stator.

Fig. 7.1 applies only to steady-state balanced conditions. Any transient or fault condition that disturbs the phase balance must be dealt with by considering the machine as an entity, with the individual phases making individual contributions to the operation in accordance with the appropriate constraints imposed by the circumstances that obtain.

Operating conditions
An electromagnetic machine, connected to an electrical supply system at its terminals and to a mechanical load or prime-mover at its shaft, is a device of great complexity. It is therefore customary to adopt conditions of operation which are simplified as far as possible while still retaining the essentials for a representation adequate to bring out the characteristics sought. The electrical supply system can, for some purposes, be taken as an *infinite busbar* system. If speeds can be considered as constant, then the mechanical system is reducible to a 'dead' load torque. If the behaviour can adequately be considered as consisting of small cyclic oscillations about a steady mean speed, the electrical side of the machine can be dealt with in terms of constant speed with appropriate electrical perturbations, and the mechanical side by cyclic fluctuations of stored kinetic energy together with any changes of load torque arising from the fluctuation of speed. But if the electrical machine is of large rating, and if substantial speed changes occur, the analysis must be comprehensive enough to cope with many interacting variables, electrical and mechanical, involving the supply network and the mechanical controls (such as governing) as well as the electromagnetic machine itself. In these circumstances an analytic solution is well-nigh impossible, and recourse must be had to the step-by-step solution of each specific problem by computer.
Infinite busbar. For steady-state conditions, and particularly for small machines that do not greatly affect the supply system, the machine can be assumed to run from infinite busbar providing a constant-voltage constant-frequency supply capable of supplying or absorbing electrical energy without limit and at any rate demanded of it.

7.2 *Rotational effects*
A rotating electromagnetic machine must be a 'machine' in the mechanical as

well as in the electrical sense, and is subject to mechanical effects that may greatly influence its design and behaviour.

Centrifugal force

The centrifugal force on a body of mass m revolving at angular speed ω_r is given by $m\omega_r^2 R$, where R is the radius of gyration.

Consider a cylindrical steel rotor, such as that of a turbo-generator, of diameter 1.0 m and rotating at 50 r/s: the centrifugal force on each 1 kg of mass at the surface is $(2\pi 50)^2 \times 0.5 = 50$ kN. Such a rotor is slotted to take copper windings, and the teeth have to withstand the centrifugal force of the windings as well as that arising from their own mass. Let the rotor be uniformly slotted with parallel-sided slots 35 mm wide and 200 mm deep, cut at an angular pitch of 1/30 of the circumference. At its root, each tooth is consequently 28 mm wide, and has to contain the tooth and slot content above it against the centrifugal force. Taking the density of the slot content (copper and conductor insulation) as averaging the same as that of the steel (7 800 kg/m^3), the centrifugal force on an element of radius r and thickness dr for a 1 m axial length, is $\omega_r^2 r \cdot$ dm where dm = 7 800 $(2\pi r/30)$dr. Over the tooth depth (i.e. from $r = 0.3$ m to $r = 0.5$ m) this integrates to 5.3 MN, a centrifugal force spread over a tooth-root area of 0.028 m^2 to give a tensile stress of 190 MN/m^2. If, as is usual, the rotor is designed to withstand a 25% overspeed, the stress is raised by the factor $(1.25)^2$ to 300 MN/m^2. As a result there are mechanical limitations imposed on the diameters of large high-speed machines.

Hoop stress

Turbo-generator rotor end-windings are retained by end-bells, and the rotor rims of hydro-generators carry the rotor poles. Rotating about their axis these develop hoop stress due to centrifugal force, both self-developed and imposed by the inner or outer masses that have to be retained.

An annular ring of radius R and unit cross-section develops a self centrifugal force of $\delta \omega_r^2 R$ per unit length of circumference which imposes a tensile hoop stress amounting to $2\delta \omega_r^2 R^2$ in the ring, where δ is the density of the material. For example, let the rotor windings of a turbo-generator to be retained at each end by rings 0.35 m in axial length and 45 mm in radial thickness, and of mean radius 0.49 m; and the windings have a mass of 300 kg at a mean radius of 0.35 m. The self-produced hoop stress is 7 800 $(2\pi 50)^2 \times 0.49^2 = 190$ MN/m^2 and the copper adds a further 100 MN/m^2 to give a total of 290 MN/m^2, of which nearly two-thirds is self-produced. The rotor itself has a diameter of 1 m and runs at 50 rev/s; with a greater diameter or a higher speed, the limits of the retaining rings would be approached and they would be unable to retain any useful mass — perhaps not even their own.

Critical speed

As a structure with elastic mass, a rotor has a natural frequency of vibration. Consider a shaft running in two bearings without restraint, and carrying a disc of mass m having its mass-centre displaced ϵ_0 from the axis of rotation. At a speed ω_r the centrifugal unbalanced force on the mass is $m\omega_r^2\epsilon_0$. This will produce a shaft deflection ϵ, increasing the unbalanced force to $m\omega_r^2(\epsilon_0 + \epsilon)$. The deflection is resisted by the elastic shaft, which develops a counter force equal, say, to $k\epsilon$. For equilibrium the two forces are equal, giving

$$\epsilon = m\omega_r^2\epsilon_0/(k - m\omega_r^2)$$

At some critical angular speed ω_c the denominator becomes zero and the deflection theoretically infinite. In practice ϵ is limited by friction and flexure loss, but at this *critical speed* will tend to be large. If there is no eccentricity ϵ_0 the expression becomes indeterminate, and a machine may indeed run through its critical speeds without vibration; but chance disturbance of centrality is probable and it is hazardous to operate a machine near to one of its critical speeds. The elementary case considered does not apply to practical systems. There may be more than one shaft span, bearing constraint, the development of nodes, and the complication of a rotor not equally stiff across all diameters (as in a slotted turbo-rotor). In small machines the critical speed is well above normal running speed, but the shaft must be stiff enough to withstand unbalanced magnetic pull arising from eccentricity.

Balance

Lack of static and dynamic balance may produce disastrous vibration. Large rotors are usually adjusted on a balancing machine at moderate speeds, at which they behave substantially as rigid bodies. In operation, however, they show a compound deflection curve between the first and second critical speeds through which they may have to pass. For this reason subsequent balancing on site at rated speed may be indispensable. The electrical sensing and compensating circuits of the balancing machine must then be imitated by some suitable mechanical artifice.

Bearings

Roller and ball bearings are common in small machines, and give satisfactory service with little attention. Conditions are very much more severe in large generators. In turbo-generators the journal bearing rubbing speeds may exceed 70 m/s, the pressures 2 MN/m^2 of the projected area, and the friction loss of the order of several hundred kilowatts. Advanced bearing design is necessary, as the oil-feed pressures are insignificant in comparisons with the bearing pressure, and serve only to keep the bearing flooded and the oil in circulation through a cooler. In hydro-generators the thrust bearing is of major importance, particularly in vertical-shaft machines where the bearing supports the weight of the rotor and turbine runner as well as a downward hydraulic thrust. A thrust bearing 2.5 m in diameter may carry a load of 1 000 Mg with a rubbing speed of 20 m/s normally or twice this value on overspeed. If the machine is shut down for some hours or days the oil film is squeezed out, in which condition a start would seriously damage the bearing. To restore the oil film, the rotor and and runner may be initially jacked up by about 5 mm, or raised by means of a pressurized hydrostatic oil system by about 0.05 mm, using pressures up to 10 MN/m² which are reduced when the machine has been started and has reached about one-tenth of normal running speed.

Kinetic energy

Inertia, and the stored kinetic energy associated with it, are factors of importance in machine operation when starting and stopping, and in oscillatory and transient conditions. The necessary kinetic energy must be supplied (electrically and/or mechanically) to a machine to bring it from rest to running speed; and unless the machine is braked regeneratively the energy is eventually lost in friction. Speed fluctuations impose the release and absorption of amounts of kinetic energy, and in some cases strong oscillations can be sustained unless there is some way in which the oscillation energy can be dissipated by damping.

The inertia concerned is, of course, that of all the coupled rotating parts. In large modern turbo-generators the inertia of the steam-turbine rotor is dominant. In hydro-generators the inertia of the water-turbine runner is comparatively small, and additional inertia may have to be built into the electrical machine or incorporated in a flywheel, for a minimum inertia is necessary because of the sluggishness of hydraulic governing. If a dropped electric load is not to result in excessive rise in speed, endangering the structure by centrifugal forces, the inertia may have to be very great (especially if the normal speed is low).

Inertial Coefficients. A mass of inertia J subjected to a torque M accelerates at rate $\alpha = M/J$. At an angular speed ω_r it has a momentum $m_0 = J\omega_r$ and a stored kinetic energy $w_s = \frac{1}{2} J\omega_r^2$. The 'flywheel effect' in a given machine is expressed as the ratio of w_s [J] at rated speed ω_r [rad/s] to the apparent-power rating S [VA], giving the *stored-energy coefficient* (or *inertia constant*) $H = w_s/S$ [J/VA, or sec]. For generators the ranges of H are typically: turbogenerator, $7 - 3$ (2-pole), $10 - 5$ (4-pole); hydrogenerator, $4 - 2$ (20-pole), $2\frac{1}{2} - 1\frac{1}{2}$ (60-pole); compensator, $1\frac{1}{2} - 1$.

Related to H are two other quantities: T, the time for the machine to accelerate from rest to rated speed ω_r with rated torque $M = S/\omega_r$; and K, relating per-unit torque $M_{pu} = M/(S/\omega_r)$ to angular acceleration. The three quantities are thus

$$H = w_s/S = \frac{1}{2}J\omega_w^2/S = \frac{1}{2}m_0\,\omega_r/S \qquad [\text{sec}]$$
$$T = \omega_r/\alpha = \omega_r/(M/J) = J\omega_r^2/S = 2H \qquad [\text{sec}] \qquad (7.1)$$
$$K = M_{pu}/\alpha = J\omega_r/S = (2/\omega_r)H \qquad [\text{sec}^2/\text{rad}]$$

The stored-energy coefficient (or inertia constant) H is the most common. For example, a generator suddenly shedding full load will, neglecting losses and damping, accelerate at rate $\alpha = \omega_r/2H$; and H represents one-half of the time T.

7.3 Drive dynamics

The torque balance of a machine connected between an electric system and a mechanical load or prime-mover is given at every instant by the equation

$$M_e = M_a + M_s + M_c + M_l \qquad (7.2)$$

which relates the electrical input torque to the mechanical accelerating, elastic, damping and dead-load torques. The rate of change of angular speed of the rotating system of inertia J is

$$\alpha = d\omega_r/dt = M_a/J \qquad (7.3)$$

whence the time for a small speed change $d\omega_r$ is

$$dt = (J/M_a) \cdot d\omega_r \qquad (7.4)$$

If M_a is constant, the system accelerates from speed ω_p to a speed ω_q in a time $t = (\omega_q - \omega_p)/\alpha$.

Electrical torque, M_e. This is a function of time, voltage, current, speed, frequency, slip, load angle and the relevant circuit parameters of inductance and resistance. Sometimes, if the mechanical rotary system has a large inertia, it can be assumed that the electrical transients imposed by some sudden change of condition have vanished before the mechanical response begins to take effect; but this assumption, which greatly simplifies the analysis, is by no means always valid.

Elastic torque, M_s. This is associated with the elastic deformation of shaft and stator frame, and comes into play when an impulsive torque is suddenly applied to one end of a shaft which has an inertia load at the other, or with flexible mechanical couplings, or with the winding of mine cages at the end of long suspension cables. Except in the last-named case, the extent of the deformation is small and its effect on speed insignificant, but a substantial amount of strain energy may be imparted to shafts and structures by it, with such consequences as failure by shearing and the impact on frames and foundations.

Loss Torque, M_c. Losses in friction, windage, eddy-currents in damping windings or in solid parts of the magnetic circuit, develop heat energy and impose a corresponding torque on the system.

Load torque, M_l. Mechanical input or output torques, considered as distinct from inertia (accelerating) torque, are determined by the characteristics of the mechanical system. These are normally speed-dependent.

EXAMPLE 7.1: A synchronous-motor drive with a moment of inertia $J = 420$ kg-m^2 and a friction torque $M_f = 135$ N-m (assumed to be independent of speed) runs on no load at 500 r/min. The drive is dynamically braked to standstill by connecting the stator windings to a bank of fixed resistors, the field current being held at a value that gives an initial electrical braking torque $M_e = 5.4$ kN-m, the torque thereafter being proportional to speed. Estimate the time to brake the machine to rest, and the number of revolutions of the rotor in that time.

The synchronous speed is $\omega_s = 52$ rad/s, so that the electric-braking torque can be expressed for any speed ω_r as $K\omega_r = 5.4 \times 10^3 (\omega_r/52) = 104\omega_r$. The total braking torque is $M_b = (M_f + K\omega_r)$, the acceleration is $d\omega_r/dt = -M_b/J$, and therefore $t = -(J/K)\ln(M_f + K\omega_r) + \text{const}$. As $\omega_r = \omega_s$ at $t = 0$, then

$$t = \frac{J}{K}\ln\left[\frac{M_f + K\omega_s}{M_f + K\omega_r}\right]$$

The time t_1 to reach zero speed, $\omega_r = 0$, is

$$t_1 = \frac{420}{104}\ln\left[\frac{135 + 5\,400}{135 + 0}\right] = 15 \text{ s.}$$

The total angular rotation θ is the time-integral of ω_r. The expression above for t gives

$$(M_f + K\omega_r) = (M_f + K\omega_s)\exp(-K/J)t,$$

and with $\theta = 0$ for $t = 0$, the total angular rotation θ_1 in time t_1 is

$$\theta_1 = \frac{1}{K}\left[\frac{J}{K}(M_f + K\omega_s)\left\{1 - \exp\left(\frac{K}{J}\right)t_1\right\} - M_f t_1\right]$$

$$= 192 \text{ rad} = 31 \text{ revolutions}$$

Below $\omega_r = 1.3$ rad/s the electrical braking torque is less than the friction torque, and the machine is finally stopped by friction alone.

Electrical torque

Expressions for the electrical torque of a prototype machine in two-axis terms are given in Sect. 7.5. They are based on axis currents and inductances, but the currents are affected by a change of speed and also in practice the inductances themselves are current dependent as a result of saturation. All the terms in eq.(7.2) are in fact speed-dependent, and solutions of the equations of behaviour have to be obtained by an iterative process, a formidable task even with computer aid. Particular cases can, however, often be simplified by making appropriate assumptions or adopting special *ad-hoc* techniques.

In purely electrical problems it is common to select a chosen rating (not necessarily that of the machine under consideration) as the per-unit base of power. The technique is equally applicable to electromechanical cases. The inertia constant H of eq.(7.1) is expressed as a ratio. The torque corresponding to a rating S at a normal (synchronous) speed ω_s is $M_1 = S/\omega_s$, and per-unit torques can be stated as the ratio of the actual torque to M_1. Similarly the per-unit speed is ω_r/ω_s for a given actual speed ω_r, and the per-unit acceleration is $\alpha_1 = \frac{1}{2}M_1/H$ (in per-unit speed-change per second). As this acceleration is normalized, conclusions from it are also normalized, making it possible to obtain normalized response curves for typical values of inertia and of torque/speed characteristics in per-unit terms. Such curves enable generalized performance to be derived, for application to specific problems after reduction to per-unit values.

Energy-rate balance

An electromechanical machine system in terms of instantaneous input powers is described by eq.(1.6). With an inertia J, an elastic torque $(1/K)\,\theta$ arising from a displacement angle θ and a compliance K (which may be electrical as well as mechanical in origin), and specific forms of loss, some of the terms on the right-hand side of eq.(1.6) can be written

Rate of change of stored energy:
$$\frac{dw_s}{dt} = \left[J\frac{d^2\theta}{dt^2} + \frac{1}{K}\theta \right] \omega_r$$

Rate of energy loss:
$$p = C_1\omega_r + C_2 + \Sigma\, I^2 R$$

The loss power comprises a term proportional to speed $\omega_r = d\theta/dt$, and typical of both mechanical viscous-friction and electrical eddy-current loss; a constant; and I^2R loss in the machine windings. Eq.(1.6) includes the term dw_f/dt for the rate of change of magnetic-field energy. The energy-rate balance equation for the system can then be written

$$p_e + p_m = \frac{dw_f}{dt} + \left[J\frac{d^2\theta}{dt^2} + \frac{1}{K}\theta \right] \omega_r + C_1\omega_r + C_2 + p_c \qquad (7.5)$$

where for convenience the I^2R loss is designated p_c. In the case of large machines it may be permissible to ignore p_c as being insignificant in comparison with other terms, and sometimes the change of magnetic energy can also be ignored.

Oscillations

The combination of an inertial and an elastic restoring torque makes possible the generation of oscillations by reason of the interchange of energy between the two. The basic idealized situation in which only the J and K terms are operative is one in which the system has a natural undamped oscillation angular frequency $\omega_n = 1/\sqrt{(JK)}$. Idealized damping proportional to the instantaneous angular velocity of the oscillation, and having the form $C(d\theta/dt)$, modifies the oscillation in accordance with its sign: if positive it reduces the amplitude with time according to an exponential law, and if it has the critical value $C_0 = 2\sqrt{(J/K)}$ its effect is just to suppress oscillation altogether; if the damping term is negative it increases the oscillation amplitude with time, again exponentially. Thus a torque M applied to a rotary inertial-elastic system yields the torque equation

$$M = Jp^2\theta + (1/K)\theta + Cp\theta = (p^2 + \omega_n{}^2)\theta + 2c\omega_n p\theta \qquad (7.6)$$

where $c = C/C_0$ is the damping ratio.

The natural frequency of oscillation of an electromagnetic machine is usually low (one or two oscillations per second) and does not vary greatly with the size of the machine unless there is a shaft load of abnormal inertia. The compliance, however, may change with the excitation on the electrical side of the electromechanical system.

Oscillatory conditions may arise in several ways, especially where there are forced oscillations at a frequency near to that of the natural frequency. Oscillation amplitudes may be substantially increased by mechanical torques pulsating at low frequency, as in a generator driven by a diesel engine, or a motor driving a reciprocating air-compressor. In ship propulsion with direct electric drive, fluctuating load torque may be impressed by the screws, and oscillation is possible in mill shafting connected to rolls some considerable distance apart. Rope elasticity in the vertical or inclined haulage of mine winders may resonate with the load mass at a frequency near to the natural frequency of the machine or of its closed-loop control devices.

Induction and synchronous machines respond in different ways to a change of load. The *induction* machine responds by a change of slip (and therefore of speed), and oscillations are of purely mechanical origin. The *synchronous* machine, when operating in parallel with other synchronous machines or on supply busbars, responds by adjusting its torque angle without change in *mean* speed: any divergence between the actual and equilibrium angles calls into play a synchronizing torque proportional to a function of the angle of divergence and so adding to the system an elastic torque of electrical origin. A case of interest is that for which the load change is cyclic.

Cyclic disturbing torque. Cyclic irregularity in the driving torque of a generator (as when it is driven by a diesel engine), or in the load torque of a motor (as for an air-compressor drive), or in a control system (e.g. governor 'hunting'), excite small fluctuations of speed. With induction motors the speed variations are constrained by the system inertia and the inherent damping by I^2R loss is large. With synchronous machines the synchronizing torque adds electrical elasticity. and provides a combination with a natural oscillation frequency. If this is close to the frequency of the cyclic disturbance, and if the damping is small, the conditions may be such as to result in considerable phase-swing. Suppose that the cyclic disturbing torque can be expressed as a simple sinusoidal variation $M_d \sin \omega_d t$ of angular frequency ω_d: the resulting acceleration α is proportional to the torque and in phase with it; the speed change ω is the time-integral of α; and the angular divergence δ is the time integral of ω. Hence we have

disturbing torque:	$+ M_d \sin \omega_d t$
acceleration:	$+ \alpha_d \sin \omega_d t$
speed-change:	$- \omega_d \cos \omega_d t$
angular divergence:	$- \delta_d \sin \omega_d t$
synchronizing torque:	$+ (\delta_d/K) \sin \omega_d t$

the last expression being obtained from a synchronizing torque proportional to the angle of divergence and acting so as to reduce the displacement to zero. It is seen that the synchronizing torque is in phase with the disturbing torque, and reinforces it. Writing $p = j\omega_d$ and $\theta = \delta_d$ in eq.(7.6) for the sinusoidal conditions here assumed, the displacement due to the disturbance is

$$\delta = \frac{KM_d}{1 - \omega_d{}^2(JK) + j\omega_d(KC)} = \frac{KM_d}{(1 - k^2) + j\,2ck} \tag{7.7}$$

where $k = \omega_d/\omega_n$ is the ratio of the disturbing frequency to the natural undamped frequency of the system. Clearly, if the damping is small and ω_d approaches ω_n the divergence may grow large enough to swing the machine out of step. It must also be noted that cyclic disturbing torques may be rich in harmonics, one of which may resonate. Low-frequency harmonics are as a rule more likely to be troublesome; in engine-driven generator sets, for example, there can be a torque harmonic at one-half engine speed, corresponding to the speed of the camshaft.

Graphical methods

Motor starting. The speed/torque characteristics of motors and loads can be complicated, but they can be graphed and integrated in steps to give the run-up speed/time relation, it generally being necessary to ignore the short-lived electrical switching transient torques and to employ 'steady-state' motor characteristics. The construction shown in Fig. 7.2 shows at (*a*) the relation for a given

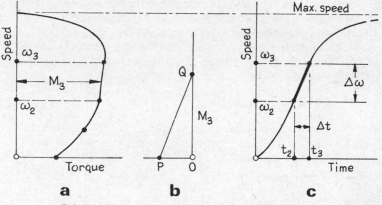

7.2 Graphical integration for speed/time relation.

case of the speed ω_r to the net accelerating torque M_a, derived as in eq.(7.2) by deducting $(M_s + M_c + M_l)$ from M_e. Then eq.(7.4) is solved by finite differences by finding $\Delta\omega_r/\Delta t = M_a/J$ for the slope of the speed/time curve (*c*) over selected speed steps ω_0 to ω_1, ω_1 to ω_2 . . ., using the auxiliary slope PQ in (*b*). Pole P is placed so that OP = J to the same scale that OQ = M_a. The run-up time, theoretically infinite, can be taken as that for which the speed reaches 99% of the final steady-state value.

Swing Curves. In a polyphase synchronous machine the electrical torque M_e (or the electrical power $P = M_e\omega_s$) is roughly a sine function of the load angle δ, as shown in Fig.7.3(*a*) for the motor mode. Any change in torque demand implies a corresponding change in the load angle. With the machine operating at a torque $M_e = M_a$ and load angle $\delta = \delta_a$, let the mechanical load torque demand be suddenly increased to M_b, for which the steady-state load angle is δ_b: the torque unbalance ΔM then does work on the rotor system (of inertia J) to retard the rotor and increase the load angle, the rotating parts releasing kinetic energy in the process. The speed drop carries the load angle beyond δ_b to δ_c. During this overshoot ΔM reverses, the spread begins to rise with the rotating parts regaining kinetic energy,

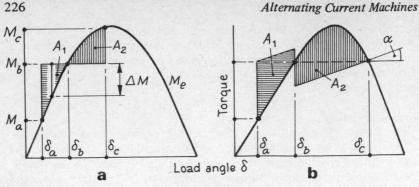

7.3 Equal-area criterion.

and the load angle decreases. Oscillation about the equilibrium angle δ_b continues until damping effects dissipate the cyclic component of the kinetic energy. If damping can be taken as developing a pure asynchronous ('induction') counter-torque proportional to the rate $(d\delta/dt)$ at which δ varies (i.e., to the oscillation velocity about the mean synchronous speed ω_s), then from eq.(7.5)

$$\Delta M = J(d^2\delta/dt^2) + C(d\delta/dt)$$

The *equal-area criterion* can give a simple approximate solution to the stability problem, that of deducing whether or not the rotor can swing through the angular excursion from δ_a to δ_c without falling out of step. Let the damping be negligible so that the oscillating system is represented by $\Delta M = J(d^2\delta(dt^2)$. Then if the two hatched areas A_1 and A_2 are equal, this condition defines the excursion of the load angle for stable operation. Each area is an integral of $\Delta M.d\delta$ and is proportional to the swing energy, so that the kinetic energy lost in the change from δ_a to δ_b is fully regained in the change from δ_b to δ_c.

The swing curve in Fig.7.4(a) shows such a *stable* condition, expressed for convenience in terms of the power *P*. In (b) the over-swing energy represented by A_2 can just match A_1, a condition of *critical* stability. In (c) there is insufficient torque during overshoot and the load angle passes into a region in which the rotor is actively retarded to give an *unstable* loss of synchronism.

The effect of induction damping torque, $C(d\delta/dt)$, can be included in the equal-area criterion by the modification in Fig.7.3(b)'. The areas A_1 and A_2 are augmented by triangles bounded by lines inclined at angle $\alpha = \arctan$ $[C^2(p/J)]$, where *p* is the number of pole-pairs. So long as the total area A_2 is greater than A_1 the machine will remain in step, the oscillation being initially between δ_a and an angle between δ_b and δ_c. Clearly the damping torque has a stabilizing effect.

The equal-area criterion, which does not depend on the torque/angle characteristic being a sinusoid, can be applied to a generator. However, stability is then likely to be critical under conditions of fault when the sytem voltage falls (reducing the peak of the torque/angle curve) and when both governor and voltage-regulator actions effect the transient behaviour.

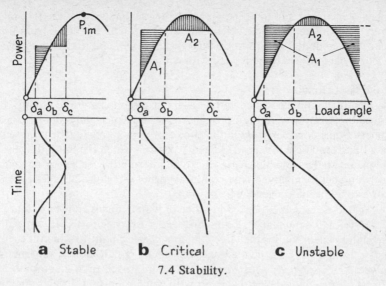

a Stable **b** Critical **c** Unstable

7.4 Stability.

7.4 Models

In the present context a model is a formulation of part or of the whole of a machine system in terms of appropriate parameters and on the basis of some acceptable theory of behaviour. The formulation may be (i) *mathematical*, as a set of integro-differential simultaneous equations or of transfer functions, or (ii) *analogue*, comprising hardware arranged to represent the prototype to some scale of size and time, and including such devices as analogue computers. In either case the aim is to observe the response of the model to a given stimulus, so that its behaviour can then be translated into the comparable performance of the prototype.

The model must have the appropriate *parameters*, and enough of them to give useful results. The choice of parameters is dictated by the stimulus to be applied and the response expected; and there must be some theory of behaviour into which the parameters will fit.

Some considerations in the choice of a model for an electrical machine are briefly discussed in Sect. 1.4. An ever-present difficulty is magnetic saturation, which produces physical effects that are not readily made analytic in the mathematical model, but can sometimes be incorporated into hardware by simulation.

A model adequately representing a *complete* system of electrical supply, electromechanical machine and mechanical attachments is extremely complex. A usual method of simplification lies in the approximate representation of those parts of the system that are not concerned with a specific investigation, but modelling the parts of interest in more detail. Full-scale tests on a supply system under load, fault and transient conditions can sometimes be carried out (e.g. during holiday week-ends). Organization of such tests is elaborate and expensive, takes months of preparation and involves very comprehensive programming and instrumentation. When evaluated the results can be used to check model parameters and to refine the modelling technique. But the fact remains: only the characteristics of a machine or system can be evaluated from the model by computation, not the actual system itself, and the answers given by a model have always to be treated with caution.

Equivalent circuits

An electric circuit associated with a magnetic circuit can, up to a point, be described in purely electrical terms by means of resistance and inductance parameters. Some considerations in the choice of parameters for the representation of a transformer are discussed in Sect. 5.5, from which the limitations of a simple resistance-inductance model can be inferred. Nevertheless, so long as such a model is employed only within its region of validity, it does provide a powerful concept in machine behaviour.

An electric circuit expressed in terms of parameters is an idealized model of a physically comparable circuit. An *electric equivalent circuit* in machine technology incorporates the effects of a magnetic circuit to which it is coupled. As Laithwaite [73] has pointed out, a magnetic circuit models a magnetic prototype, and a *magnetic equivalent circuit* could be set up to incorporate the effects of the coupled electric circuit with equal validity. Consider the coupled circuits in Fig. 7.5: the current i in the electric circuit is the source of the

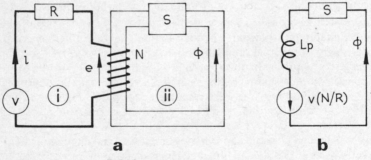

a   b

7.5 Magnetic equivalent circuit.

m.m.f. Ni which develops the flux Φ in the magnetic circuit of reluctance S. The basic behaviour is expressed for the respective circuits (i) and (ii) in (*a*) by

(i) $e = N \, \mathrm{p}\Phi$ and   (ii) $Ni = S\Phi$

The 'Kirchhoff' law in each case is

(i) $v = Ri + N\,\mathrm{p}\Phi = (R + L\,\mathrm{p})i$ and   (ii) $i = (S/N)\,\Phi$

where $L = N^2/S$ is the inductance. Eliminating i,

$$v(N/R) = S\Phi + (N^2/R)\mathrm{p}\Phi = (S + \mathrm{Lp})\Phi$$

and $\mathrm{L} = N^2/R$ is the *transference,* or magnetic equivalent of inductance. The result expresses a magnetic equivalent circuit (*b*) in which the coupled effects of the associated electric circuit are included. The viewpoint has advantages in that a machine has only one main flux but several associated windings. Fiennes [97] works out a generalized machine theory using magnetic equivalent circuits. At present, however, analysis in electric-circuit terms is well established and more familiar.

Mathematical models for electrical machines

Comparatively simple models can be set up for specific parts of a machine, more elaborate ones for comprehensive analysis.

Electric circuit model. The representation of the electric circuits of a machine in two-axis terms is discussed in Sect. 7.5.

Electromechanical models. When both electrical and mechanical interactions are to be studied, the problem of modelling becomes more complicated. Particular examples concern generator stability, control-system design, the starting of synchronous motors, the behaviour of boiler-turbine-generator sets, the effect on supply systems of large induction machines during faults, and the optimization of generating-station performance.

The steam turbo-generator system has been analysed several times in the literature. In a classic paper [74] Shackshaft established a 'general-purpose' model for the system shown diagrammatically in Fig. 7.6(*a*), representing a real system which was subjected to full-scale field tests in order to verify the selected parameters or to derive them by such techniques as curve-fitting. The generator itself was modelled by the methods of Sect. 7.5. The voltage regulation and excitation network, (*b*), required the setting up of equations and transfer functions for the voltage sensor with its amplifiers and stabilizers, and for the exciter response; the governing system (*c*) required equations for the action of the centrifugal governor, relays and valves, and for the turbine power and torque; the transmission network was represented by lumped resistance and inductance. Humpage and Saha [75] gave a model suitable for digital computation of the same system as that field-tested by Shackshaft. Optimization of the running of a boiler-turbine-generator set by mathematical modelling, and for a multi-machine supply system, have been investigated by Nicholson [76]. These several references demonstrate the formidable task of modelling such multivariable systems, the justification being the great economic importance attached to an understanding of the performance of very large machines under abnormal conditions, as well as of their efficient operation on a system load cycle.

Micromachines

A micro-machine is specially designed to represent, on the basis of per-unit parameters, a large machine. A 2 kW low-voltage machine can be made (though with some difficulty and with auxiliary equipment to compensate for IR drop) to represent, say, a 1 000 MW synchronous generator. With alternative rotors the micro-machine can simulate many alternative large-machine designs. A driving motor can be controlled to represent a mechanical load, or a prime-mover with appropriate turbine and governor characteristics, and by the addition to the shaft of an appropriate flywheel can give the model the required in-

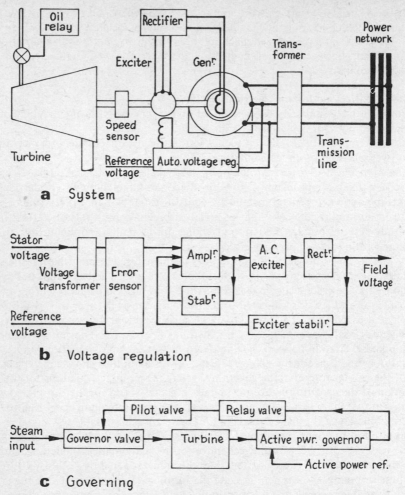

a System

b Voltage regulation

c Governing

7.6 Turbo-generator control system model.

ertia. The field resistance of a micro-machine is inevitably of higher per-unit resistance than that of a large machine, and for proper simulation must be provided with an adjustable negative resistance externally, operated by electronic feedback devices. Voltage regulators, either actual full-scale equipments or electronic simulators, can be connected to the field circuit. In such ways it is possible to represent very large machines by miniscule machines, greatly facilitating the prediction of large-machine behaviour under abnormal conditions with very detailed measurements. The investigations made possible are more comprehensive than can be achieved by computation based on machine parameters, for the direct effects of e.g. saturation and eddy-current induction can be included. The following are typical of the practical problems that can be investigated by micro-machine installations.

Generator short circuit. Unbalanced short-circuit currents, the analytical computation of which is difficult, can be investigated directly. The measurement of transient torques can readily be carried out if accelerometers and torque-meters are available.

Asynchronous operation. It is now recognized that a generator which pulls out of synchronism need not necessarily be disconnected provided that appropriate measures are taken to control the conditions and to restore synchronous running. Such an investigation with a real large generator is hazardous, but the small scale of the micro-machine arranged to simulate the generator conduces to exhaustive measurement of the transient currents and torques, and also to a study of the optimum control characteristics.

Solid rotors. The normal concepts of transient and subtransient reactance are based on the assumption that the damper system is equivalent to two discrete axis dampers. With a solid rotor the damping is by eddy-currents and the paths are complicated. A comparable problem arises in a salient-pole synchronous motor when the starting torque depends upon eddy-current effects in the solid pole-shoes, with the additional complexity of saliency and a full range of rotor frequency from standstill to synchronous speed. Both these cases are susceptible to direct measurement on micro-machines.

Induction-motor transients. The effect of large cage induction motors on power-system operation can be of importance in the prediction of system stability under conditions of fault. Analysis is particularly difficult by reason of the changing load angles of the synchronous plant and the rapid fall in speed of induction machines. A typical system can, although somewhat limited in extent, be set up with micromachines to represent, say, a synchronous generator, a transmission line and an infinite busbar. A fault on the line can be imposed, the phase-swing of the synchronous machine and the rise in slip of the induction motor directly observed, and the stability of the system resulting from the interaction of these effects deduced, with comprehensive instrumentation and under the advantageous conditions of a laboratory.

A description of the design of a micro-generator using negative-resistance devices, and dealing with its dimensional scaling, is given by Hammons and Parsons [79].

Power-system modelling

A quarter-century ago it was sufficient to model a power system electrically by 'network analysers', with transmission links represented by resistors, inductors and capacitors, and synchronous machines by voltage sources (variable in magnitude and phase) connected in series with 'internal' resistance and inductance. Later a more complex two-axis representation was introduced for the generators. Modern micromachines provide the most satisfactory solutions for qualitative problems, but it is not easy to obtain close adjustment of machine parameters and the number of machines that can be simulated is rather restricted.

Transient-stability studies at first required only more detailed versions of the two-axis equations. The modern digital computer can be programmed to provide numerical simulations of several hundred machines and several thousand busbars. The calculations cannot (as yet) be performed in real time, but the digital machine is ideally suited to routine studies. The development of high-level languages, more comprehensive computers, multi-access systems and graphical output facilities will, however, lead almost certainly to optimization and the complete on-line control of the extensive power-supply system.

7.5   *Dynamic circuit theory*
It has for some time been recognized that rotating electrical machines differ not in their fundamental action, but in the arrangement and excitation of their windings. Early attempts to co-ordinate the piecemeal treatments of separate machine types led to what is called the *generalized* approach through dynamic circuit theory. The concept was founded by Park in a classic paper [50] on synchronous machines in 1929. Gabriel Kron developed Park's ideas to deal comprehensively with all rotating electrical machines by means of tensor analysis [75]. In the modern version of Park's method, all active circuits are described by resistance and inductance parameters. By writing the circuit equations, the required behaviour can be found in terms of voltages, currents and torques. As any given circuit in a machine may be magnetically coupled with any other, the circuit equations are complicated. Substantial simplification is achieved if each circuit is so transformed as to represent separately its effect on the direct and quadrature axes of the machine, for it can then be assumed that no equivalent circuit whose m.m.f. acts along one axis can have any magnetic coupling with any circuit acting along the other axis. Further, the two-axis concept can realize in a simple manner the normal travelling-wave field of a polyphase machine, and is therefore equally applicable to steady-state conditions.

The process, in brief, is to *transform* a winding and its voltage and current into an equivalent coil on the d- or the q-axis (or on both) with the polyphase quantities thus converted into their axis equivalents; to set up the circuit equation for each such circuit; to impose the operating conditions; to solve the equations; and finally to convert the solutions back into terms of the currents and voltages of the actual machine windings.

There are some sacrifices. Manipulation of the equations is tractable only on the assumption of constant inductance and resistance parameters; saturation effects must usually be ignored; speed changes may render the equations non-linear in the mathematical sense; and secondary phenomena — such as brush-contact resistance or flux and m.m.f. harmonics — must be neglected. In spite of these drawbacks, the generalized two-axis dynamic circuit theory is very powerful, as can be appreciated from the work of Adkins [65] and Jones [66]. A simple introduction is given by Say [1].

Basic two-axis three-winding machine
This is a 2-pole unit with not more than one salient element, either stator or rotor. The d-axis is chosen as the axis of saliency. For this purpose the polarized field system on the cylindrical rotor of a turbo-type synchronous machine is treated as equivalent to a saliency. In a polyphase induction machine there is a less restricted choice of d-axis.

For convenience of reference framing, the d-axis is taken as horizontal, the q-axis as vertical and the rotation as counter-clockwise. As only relative motion is significant, the salient element is stationary, the moving element (or armature) rotating around or within it. The windings on each element are grouped or resolved so as to magnetize along one or other axis. Positive current in any axis winding then magnetizes in the positive axis-direction. With d- and q-axis coil groups thus in space quadrature there is no magnetic coupling between coils on different axes; but rotational e.m.f.s can be developed in coils on one axis that can be *considered to move* in the flux developed by coils on the other axis.

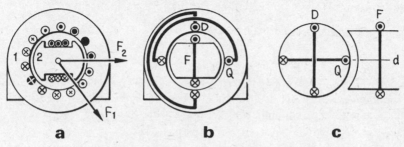

a   **b**   **c**

7.7 Basic two-axis three-winding machine.

The two-axis concept was briefly described in Sect. 1.11 and 12. The machine of Fig. 1.11 is shown schematically in Fig. 7.7(*a*), in which the m.m.f. axes of stator and rotor are emphasized. The m.m.f. of the polyphase stator current is resolved into two separate m.m.f.s in coils D and Q, as in (*b*), with the d-axis rotor m.m.f. produced by winding F. An equally valid (and more convenient) representation is that in (*c*). Magnetically, coils D, Q and F can be taken as single turns, but the fluxes due to them are nevertheless assumed to be distributed sinusoidally.

Quasi-stationary coils. The armature coils D and Q are fixed: but they represent the m.m.f. axes of a *real winding* that *rotates*. It is necessary to consider them to be the seat of rotational e.m.f.s in spite of the condition that they are stationary with respect to the d- and q-axes. The coils are termed *quasi-* (or *pseudo-*) *stationary.* Rotational e.m.f.s appear in Q due to d-axis flux, and in D due to q-axis flux.

Armature E.M.F.s. Consider coil D in Fig. 7.7(*c*). It is coaxial with the 'field' coil F, and any change of its linkage ψ_d will produce the pulsation e.m.f.

$$e_{pd} = \mathrm{d}\psi_d/\mathrm{d}t = p\psi_d$$

writing p for d/d*t*. Because of its orientation there can be no rotational e.m.f. in D due to ψ_d; but if there were a q-axis linkage ψ_q a rotational e.m.f.

$$e_{rd} = \omega_r \psi_q$$

would result. In the case of coil Q, with its axis at right-angles to the d-axis of coil F, no pulsational e.m.f. can be induced by any change in ψ_d, but it will have an e.m.f. of rotation in ψ_d when the winding it represents rotates at angular speed ω_r, of value

$$e_{rq} = -\omega_r \psi_q$$

The negative sign results from the fact that a q-axis coil with reference to a d-axis flux bears the opposite relation to that of a d-axis coil in a q-axis flux. *Armature applied voltages.* Remembering that terminal voltages are considered as applied to the windings and that positive currents flow in a corresponding direction, then such voltages must account for the e.m.f.s resulting from axis linkages, and for volt drops in resistance and leakage inductance. Hence for armature coils D and Q the equations for the voltages applied are

$$v_d = r_d i_d + L_d p i_d + p\psi_d + \omega_r \psi_q$$

$$\text{(7.8)}$$

$$v_q = r_q i_q + L_q p i_q + p\psi_q - \omega_r \psi_d$$

where r represents a winding resistance and L its leakage inductance.
Field applied voltage. The stationary coil F is not the seat of any rotational e.m.f., and it has no coupling with the q-axis: thus its applied voltage is given by

$$v_f = r_f i_f + L_f p i_f + p\psi_d \qquad\qquad \text{(7.9)}$$

General multi-winding machines

While the D and Q coils are adequate to represent the armature, the field winding F is, for generality, augmented by at least one further coil KD on the d-axis as indicated in Fig. 7.8(a). KD represents, for example, a closed-circuit field collar or a damping winding in a synchronous machine, or the equivalent of the eddy-current path in a solid pole-shoe. For a like reason it is necessary to include a q-axis coil KQ, as in (b). Under transient conditions, currents induced in pole-shoes and dampers have an important effect on the flux of the machine, and

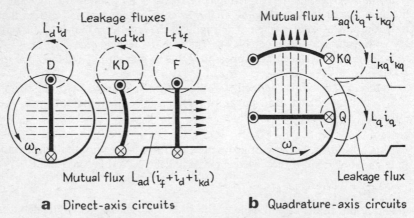

a Direct-axis circuits   **b** Quadrature-axis circuits

7.8 Axis coils: mutual and leakage fluxes.

the problem is to find the axis fluxes resulting from currents in all coils of both axes.

D-axis fluxes. There are three mutual inductances concerned, i.e. D with KD, KD with F and F with D. A useful simplification is to take these as all equal to a mutual inductance L_{ad}. For this purpose it is necessary to work with single-turn equivalent coils. Each coil has a leakage inductance of its own, and the total coil inductances will consequently be

$$\text{D: } (L_{ad} + L_d) \quad \text{KD: } (L_{ad} + L_{kd}) \quad \text{F: } (L_{ad} + L_f)$$

The self flux of F due to an instantaneous current i_f will thus be $(L_{ad} + L_f)i_f$, and similarly for the other d-axis coils. The mutual linkage with armature coil D when there are currents in all the axis coils is

$$\psi_d = L_{ad}(i_f + i_{kd} + i_d) \tag{7.10}$$

Q-axis fluxes. Considering the group of q-axis coils in Fig. 7.8, and calling the common mutual inductance coefficient L_{aq}, the mutual linkage in coil Q is

$$\psi_q = L_{aq}(i_{kq} + i_q) \tag{7.11}$$

If the leakage inductances are respectively L_q and L_{kq}, the total inductances of Q and KQ will be $(L_{aq} + L_q)$ and $(L_{aq} + L_{kq})$ respectively.

Circuit equations. It is now possible to write down the voltage equations for all five circuits in the machine shown diagrammatically in Fig. 7.9. If for some

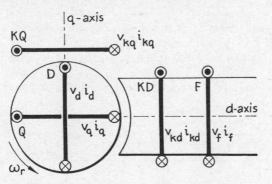

7.9 Basic two-axis five-winding machine.

purpose any further circuits are to be included, they are dealt with in precisely the same manner. Each circuit has an applied voltage v and an input current i.

F: This coil has a resistance r_f, a leakage inductance L_f and a total inductance $(L_{ad} + L_f)$. The current in any other d-axis coil will affect the linkage of F, but no q-axis coil can affect it, nor do any rotational e.m.f.s appear. Hence the applied voltage v_f has the components

$$v_f = r_f i_f + (L_{ad} + L_f)p i_f + L_{ad} p i_{kd} + L_{ad} p i_d \tag{7.12}$$

KD: This differs in no essential from F, so that

$$v_{kd} = r_{kd}i_{kd} + (L_{ad} + L_{kd})pi_{kd} + L_{ad}pi_f + L_{ad}pi_d \qquad (7.13)$$

KQ: This has a resistance r_{kq} and a total inductance $(L_{aq} + L_{kq})$. It has no rotational e.m.f., but is affected magnetically by any current i_q in armature coil Q: thus any applied voltage must satisfy the condition

$$v_{kq} = r_{kq}i_{kq} + (L_{aq} + L_{kq})pi_{kq} + L_{aq}pi_q \qquad (7.14)$$

In practice, coils KD and KQ usually represent short-circuited paths with an applied voltage of zero.

D and Q: The armature coils D and Q have the additional property of generating rotational e.m.f.s in the flux of the other axis. Their applied voltages are given in eq.(7.8), and using the expressions for ψ_d and ψ_q in eqs. (7.10) and (7.11)

$$v_d = r_d i_d + (L_{ad} + L_d)pi_d + L_{ad}pi_f + L_{ad}pi_{kd}$$
$$+ L_{aq}\omega_r i_{kq} + (L_{aq} + L_q)\omega_r i_q \qquad (7.15)$$

$$v_q = r_q i_q + (L_{aq} + L_q)pi_q + L_{aq}pi_{kq}$$
$$- L_{ad}\omega_r i_f - L_{ad}\omega_r i_{kd} - (L_{ad} + L_d)\omega_r i_d \qquad (7.16)$$

We have now obtained a set of circuit equations for the machine represented by three d-axis and two q-axis coils. More coils can be added in a similar way.

Matrix form
The circuit equations can be written down accurately and rapidly in the matrix form [V] = [Z] · [I] by aid of a few simple rules. For the five coils F, KD, KQ, D and Q, a 5-row 5-column impedance matrix is set out as in Fig. 7.10.

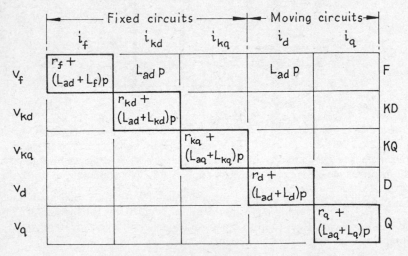

7.10 Construction of impedance matrix.

The rows are related to the voltages and the columns to the corresponding currents. Compartments in the main diagonal are emphasized: these contain the self-impedances. Thus r_f is entered only in row F and column F because only current i_f can produce a voltage drop in r_f, no resistance carrying more than one current. For the same reason the total self-inductances are also entered in the compartments of the main diagonal.

The remaining compartments contain the mutual impedances where any exist. Thus in row F, the second compartment contains $L_{ad}p$, the third zero, the fourth $L_{ad}p$ and the fifth again zero. There are no entries in row-F compartments lying in q-axis columns. Now, by inspection of the F-row, v_f is read as the sum of the products of the compartment impedances with their appropriate currents. This gives eq.(7.12).

The same procedure is employed to fill and read each row of the matrix. But in the case of the fourth and fifth rows (and only in these rows), rotational-e.m.f. terms will appear. Every coefficient of p in row Q appears again in row D in the same column, but with p replaced by $-\omega_r$. Similarly, every coefficient of p in row Q appears again in row D in the same column, but with p replaced by $+\omega_r$. The matrix is now complete, and if read off as already described it gives the circuit equations, eq.(7.12) to (7.16).

	F	KD	KQ	D	Q
F	$r_f+(L_{ad}+L_f)p$	$L_{ad}p$		$L_{ad}p$	
KD	$L_{ad}p$	$r_{kd}+(L_{ad}+L_{kd})p$		$L_{ad}p$	
KQ			$r_{kq}+(L_{aq}+L_{kq})p$		$L_{aq}p$
D	$L_{ad}p$	$L_{ad}p$	$L_{aq}\omega_r$	$r_d+(L_{ad}+L_d)p$	$(L_{aq}+L_q)\omega_r$
Q	$-L_{ad}\omega_r$	$-L_{ad}\omega_r$	$L_{aq}p$	$-(L_{ad}+L_d)\omega_r$	$r_q+(L_{aq}+L_q)p$

7.11 Impedance matrix of five-winding machine.

Fig. 7.11 gives the complete impedance matrix. For further coils, it is only necessary to add further rows and columns to correspond.

Torque

Torque is developed only in windings which, with the convention here adopted, move with respect to the fixed d- and q-axes. The torque produced by any winding is such that $M\omega_r = e_ri$, so that M is obtained from the rotational voltage multiplied by the current of the winding concerned, and divided by ω_r.

Consider row D of the matrix in Fig. 7.11. The rotational terms are $L_{aq}\omega_r$ associated with i_{kq}, and $(L_{aq} + L_q)\omega_r$ associated with i_q. The torque developed by armature coil D is therefore

$$M_d = [L_{aq}i_{kq} + (L_{aq} + L_q)i_q]i_d.$$

In the same way, for the armature coil Q,

$$M_q = [-L_{ad}i_f - L_{ad}i_{kd} - (L_{ad} + L_d)i_d]i_q.$$

The total armature torque is the sum of these two components giving

$$M = L_{aq}(i_{kq} + i_q)i_d - L_{ad}(i_f + i_{kd} + i_d)i_q - (L_d - L_q)i_d i_q \qquad (7.17)$$

The last term is due to the saliency, which makes $L_d > L_q$.

Mechanical torque is defined as that *applied* to the shaft, so that the torque developed by an electrical input becomes a mechanical output and is therefore reckoned as negative. The total torque of a machine will, of course, include both electrically-developed and mechanically-applied torques. These normally balance (i.e. they sum to zero), but if they do not, an acceleration torque $Jp\omega_r$ is released to alter the speed of the rotating parts of inertia J. In this case the circuit equations become non-linear and a solution offers some difficulty.

Synchronous machine

A 3-phase winding rotates with reference to the d- and q-axes, and it is necessary to show (i) that it can be replaced by fixed D and Q armature coils, and (ii) the relation between the actual phase currents and their axis counterparts. *Armature phase inductance.* Consider phase A of a 3-phase winding. Its inductance is a combination of its own self inductance L_{aa} with its mutual inductance L_{ab} and L_{ac} with phase windings B and C. All three inductances vary with the relative position of stator and rotor because of saliency. When the axis of phase

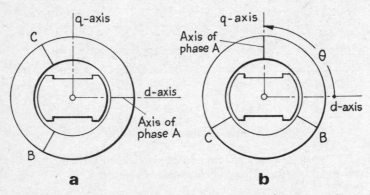

a   **b**

7.12 Effect of position on phase inductance of 3-ph machine.

A coincides with the d-axis, Fig. 7.12(*a*), its self inductance is a maximum; and when coincident with the q-axis, (*b*), it is a minimum. Thus L_{aa} fluctuates twice per double pole-pitch. Fig. 7.12 also shows that the mutual inductance of phases B and C is a maximum at (*a*) and a minimum at (*b*). Both self and mutual inductances, being functions of the angle θ, can be expressed as Fourier series to take into account any 'shape' of inductance/angle relation. The simplest variations are

$$L_{aa} = L_0 + L_2 \cos 2\theta$$

$$L_{bc} = L_{cb} = -L_{m0} + L_{m2} \cos 2\theta \qquad (7.18)$$

From the latter we can, by symmetry, write

$$L_{ab} = L_{ba} = -L_{m0} + L_{m2} \cos(2\theta - 2\pi/3)$$

$$L_{ac} = L_{ca} = -L_{m0} + L_{m2} \cos(2\theta + 2\pi/3)$$

(7.19)

We thus have expressions for all three components of the total inductance of phase A, and can write similar expressions for phases B and C.

Axis inductances. With balanced 3-phase currents and normal-speed rotation corresponding to the frequency, the phase currents can be written in terms of angles θ and γ. The position $\theta = 0$ corresponds to the coincidence of the axis of phase A with the rotor pole-centre, i.e. the d-axis; and γ is the phase angle of lag of the phase A current with respect to this instant:

$$i_a = i_m \sin(\theta - \gamma)$$
$$i_b = i_m \sin(\theta - \gamma - 2\pi/3)$$
$$i_c = i_m \sin(\theta - \gamma + 2\pi/3)$$

The total flux linkage of phase A is

$$\psi_a = L_{aa}i_a + L_{ab}i_b + L_{ac}i_c$$

Inserting the currents, together with the inductance coefficients from eqs.(7.18) and (7.19), gives the linkage ψ_a. If i_m is put equal to unity, the linkage per unit current is the effective inductance of phase A:

$$(L_0 + L_{m0}) \sin(\theta - \gamma) - (\tfrac{1}{2}L_2 + L_{m2}) \sin(\theta + \gamma)$$

A third-harmonic space variation has been ignored in this result.

Suppose that $\gamma = 0$; then phase A has peak current at the instant when $\theta = \tfrac{1}{2}\pi$, i.e. when its axis coincides with the q-axis. Its effective inductance is

$$L_0 + L_{m0} - \tfrac{1}{2}L_2 - L_{m2} = L_{aq}$$

If $\gamma = \tfrac{1}{2}\pi$, then phase A has peak current when on the d-axis with $\theta = 0$; for this the effective inductance is

$$L_0 + L_{m0} + \tfrac{1}{2}L_2 + L_{m2} = L_{ad}$$

Thus the flux per unit armature current (the 'armature reaction') is affected by saliency, and depends upon how the stator (or 'armature') m.m.f. is directed with respect to the d- and q-axes. The larger flux per unit phase current L_{ad} applies when the m.m.f. axis and the d-axis coincide; the smaller, L_{aq}, when the m.m.f. of phase A is along the q-axis. These differ by

$$L_{ad} - L_{aq} = L_2 + 2L_{m2}$$

that is, by the second-harmonic space-distribution coefficients. If there is no saliency, then $L_{ad} = L_{aq}$. It is to be noted that even with cylindrical-rotor machines there may be a magnetic saliency by reason of the distribution of the windings in slots and also, in practice, because of saturation effects.

Three-phase/two-axis transformation. Under *steady-state* conditions the m.m.f. set up by the currents in phases A, B and C is stationary with respect to the d-

and q-axes of the rotor. Consequently the fictitious coils D and Q, the quasi-stationary coils that reproduce the same m.m.f., must carry the *direct* currents i_d and i_q. The instantaneous axis currents are related to the instantaneous 3-phase stator currents and the instantaneous position θ of the axis of phase A by the expressions

$$i_d = \tfrac{2}{3}[i_a \cos\theta + i_b \cos(\theta - 2\pi/3) + i_c \cos(\theta - 4\pi/3)]$$

$$i_q = \tfrac{2}{3}[i_a \sin\theta + i_b \sin(\theta - 2\pi/3) + i_c \sin(\theta - 4\pi/3)]$$

$$(7.20)$$

The axis voltages v_d and v_q are related to the stator phase voltages v_a, v_b and v_c by expressions of identical form. The reverse transformations are

$$
\begin{aligned}
i_a &= i_d \cos\theta &&+ i_q \sin\theta \\
i_b &= i_d \cos(\theta - \tfrac{2}{3}\pi) &&+ i_q \sin(\theta - \tfrac{2}{3}\pi) \\
i_c &= i_d \cos(\theta - \tfrac{4}{3}\pi) &&+ i_q \sin(\theta - \tfrac{4}{3}\pi)
\end{aligned}
\qquad (7.21)
$$

and again the expressions for the voltage have an identical form. If there are zero-sequence components, then the further equation $i_0 = \tfrac{1}{3}(i_a + i_b + i_c)$ and a corresponding expression for the voltage are added to eq.(7.21); the reverse transformations have simply the addition of i_o (or v_o) to the right-hand side.
Power. Taking unit power as that for the machine as a whole, then with per-unit voltages and currents in a 3-phase machine the per-unit input power is $p_1 = \tfrac{1}{3}(v_a i_a + v_b i_b + v_c i_c) = \tfrac{1}{2}(v_d i_d + v_q i_q)$.
Summary. The analysis in circuit form of a synchronous machine for any condition of operation requires first the reduction of all windings to d- and q-axis groups. Fig. 7.13 shows the basic 5-circuit machine. Coil F is always a d-axis

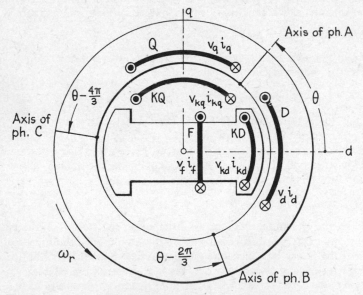

7.13 Five-winding synchronous machine.

winding. The armature (normally the stator) winding is transformed into the axis coils D and Q. The windings KD and KQ, associated with the presumed fixed field system, represent damping windings, eddy-current paths, single-turn equivalents of field collars and pole-shoes; they are normally closed and have no applied source voltage. If there are several distinguishable eddy-current paths it may be necessary to include additional coils, but here it is sufficient to assume one in each axis. The behaviour of the equivalent machine is determined from the conditions of voltage imposed, and the voltages and currents in the actual machine circuits derived therefrom, using transformations where phase windings are concerned.

Induction machine

With both stator and rotor cylindrical and symmetrical, the d-axis can be chosen arbitrarily as the axis of stator phase A. The 3-phase winding is transformed to a 2-phase 2-axis winding, so that the axis of the second equivalent phase becomes the q-axis. The stator winding then becomes two fixed-axis coils, 1D and 1Q.

The rotor carries a slip-ring winding or a short-circuited cage. In the former case it is possible to convert the winding to fixed d- and q-axis coils, just as for the armature of the synchronous machine.

A cage rotor is rather more complex; but insofar as it can, and does, produce an armature reaction m.m.f. substantially identical (apart from space harmonics) with that of a slip-ring winding, it can also be represented by d- and q-axis coils. These latter are called 2D and 2Q.

Disregarding zero-sequence voltages and currents, the conversion of a 3-phase stator winding to a 2-phase axis winding is by

$$v_{1d} = \tfrac{2}{3}[v_a - \tfrac{1}{2}v_b - \tfrac{1}{2}v_c]$$
$$\text{and} \quad v_{1q} = \tfrac{2}{3}[-(\sqrt{3}/2)v_b + (\sqrt{3}/2)v_c]. \tag{7.22}$$

Identical coefficients relate the currents. For any zero-sequence voltage, $v_{10} = \tfrac{1}{3}(v_a + v_b + v_c)$. The reverse transformation is

$$v_a = v_{1d}; \qquad v_b = -\tfrac{1}{2}v_{1d} - (\sqrt{3}/2)v_{1q};$$
$$v_c = -\tfrac{1}{2}v_{1d} + (\sqrt{3}/2)v_{1q}; \tag{7.23}$$

and similarly for the currents. Any zero sequence voltage v_0 is added to each right-hand side.

For the rotor, the appropriate transformations are those given in eqs. (7.20) and (7.21).

The impedance matrix is set up as for the synchronous machine, appearing as in Fig. 7.14. The stator and rotor resistances and inductances are r_1 and r_2, L_1 and L_2 for each axis. Coils 1D and 2D have the mutual inductance L_m, and similarly for 1Q and 2Q. Rotational e.m.f.s appear only in the rotor coils. Thus 2D has a quasi-rotation in the mutual flux of 1Q and also in the total flux of 2Q. The voltage equations are read directly from the impedance matrix in Fig. 7.14.

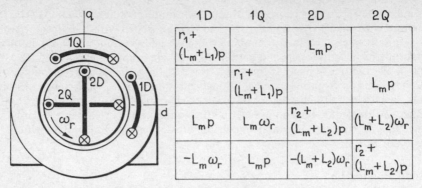

7.14 Four-winding induction machine.

Torque. This is obtained from the rotational-voltage terms, to give

$$M = L_m(i_{1q}i_{2d} - i_{1d}i_{2q}).$$ (7.24)

Speed changes

If a change of rotor speed is introduced, the matrix equations become mathematically nonlinear, making a step-by-step solution necessary. Where speed change is small, however, the calculus of variables can be applied. It is assumed that steady-state voltages v, currents i, electrical torque M and speed ω_r are changed by Δv, Δi, ΔM and $\Delta \omega_r$ respectively about constant mean values; and if these perturbations are small enough it is possible to enlarge the purely electrical matrix of Fig. 7.11 by one further row and column M, to represent torque and accleration acting on the rotor mechanical axis (i.e. that of the shaft). The dynamic electromechanical matrix becomes that in Fig. 7.15, which permits of a solution to the behaviour of the machine when subjected to small impressed voltages or torques. Gibbs [77] discusses the application of the method to several cases.

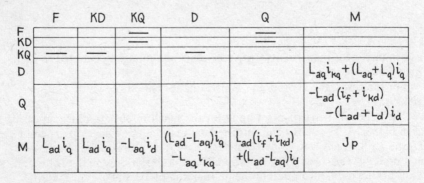

7.15 Motional impedance matrix.

The blank compartments contain the corresponding terms in Fig.7.11.

Applications

Dynamic circuit theory based on 2-axis transformations has a number of restrictions. It is not applicable to machines with saliency on both sides of the airgap because in such structures it is not possible to identify axes that present substantially constant reluctances to the axis m.m.f.s. It is a linear method based on sinusoidal distributions and is not capable without some elaboration of dealing directly with harmonics or with mechanical asymmetry. In particular, it cannot cope (except over restricted operating ranges) with magnetic saturation.

The problem of saturation needs special consideration. The generalized theory assumes linear relationships, not because saturation is unimportant but because there is no possible alternative. Any direct analytical treatment of nonlinear conditions, except in very simple cases, becomes impossibly complicated. It is an accepted method, used over the whole of engineering science, to develop a basic linear theory, which is modified where needed to allow for nonlinear effects. In the field of electrical machines the method gives good results if used with discretion. A full physical appreciation of all the ways in which the idealized machine differs from the real one is, however, essential if the method is to be applied successfully; moreover, repeated experimental verification of any deductions made must be available.

Exclusion from the theory of certain practical and important effects, such as mechanical soundness or the detailed distribution of flux and losses, means that design of a machine calls for skill and experience. However, provided that undesirable side effects are avoided by practical means, and that the most appropriate parameters can be selected, the generalized theory provides a reliable means of predicting machine performance. Naturally the design should not be subservient to a need for an amenable theoretical analysis; but in practice it turns out that the best design is often that one which conforms most nearly to the pattern of the idealized machine.

The method is very successful in analysing such problems as the short-circuit behaviour of large generators, the transient starting torques in induction motors and the run-up torques of synchronous motors. It forms an effective tool for the elucidation in particular of conditions of transience — conditions that are quite beyond the capability of simple steady-state theories to tackle.

Phase co-ordinate analysis

The two-reaction method described has the advantages that it makes evident that a machine has a single working flux produced by the action of all winding currents in combination, can be used for machines with any number of balanced phases, and is applicable directly (i.e. without transformations) to machines such as those with commutators in which the action depends on there being two distinct and recognizable physical d- and q-axes. But it has some disadvantages for a.c. machines, in that certain assumptions have to be made in regard to the distribution of gap-flux density resulting from the contributing windings. A salient-pole field winding or a damping winding, for example, is assumed to give a sinusoidal flux distribution, an assumption that is clearly untrue. The variations of self and mutual inductance expressed in eqs. (7.18) and (7.19) are simplified by the neglect of all space-distribution harmonics except that of second order, whereas in fact there are higher-order harmonics actually present, some of which are of significant importance. Finally, it is necessary to transform stator axis-coil currents back into their phase equivalents after the basic solution because the only measureable currents are phase currents.

It was to avoid the use of position-dependent inductance coefficients that Park developed the two-axis dynamic circuit theory. With the advent of powerful digital computers, however, it is no longer essential to use phase-to-axis transformations, and by adequate iteration it is possible to calculate directly with eqs. (7.18) and (7.19) and to include, if necessary, higher space harmonics to express more exactly the variation of the inductance coefficients. The solutions for all three phase currents are then obtained. For practical purposes, apart from its heavy demands upon computer time, the 'phase co-ordinate' method has clear advantages. In effect, the matrix of Fig. 7.11 is changed by replacing the stator coils D and Q by phases A, B and C, and incorporating the appropriate angle-dependent inductance coefficients in place of the axis values. The method is particularly useful for cases of unbalance, whether of steady state or fault, conditions for which the two-axis method requires rather more computation.

7.6   *Industrial machines*

In specifying an electric motor for an industrial drive, many factors must be considered besides the voltage and frequency. The duty, starting performance, type of protection and ambient conditions may all have a direct influence on the machine to be selected. Sometimes a difficult decision is to settle which of the many alternatives is best: induction or synchronous, cage or slip-ring, pole-change or thyristor-controlled. By far the commonest industrial machine is the cage induction motor, which gives a cheap, robust and simple single-speed drive; but in its most elementary form its starting current is high and its starting torque low. With relatively minor rotor modifications using deep-bar, T-bar or double-cage designs, machines can offer significantly better starting performance, and two-speed operation is possible by suitable elaboration of the stator winding.

Where the drive demands an adjustable speed rather than one or two 'fixed' speeds, the choice of motor type widens considerably. Slip-ring induction motors with rotor resistance control are often adopted for intermittent use (e.g. for cranes and hoists) where the additional rotor-circuit losses at low speeds are acceptable. If they are not, and the load requirements justify the choice, use may be made of static frequency-converters to feed induction motors.

Small conventional synchronous motors are, compared with the simpler induction machines, expensive; but if motors of large output are required, the synchronous machine may be cheaper, is certainly more efficient, and is capable of operation at leading power factor. Modern synchronous-machine designs are generally of the 'brushless' type, devoid of slip-rings, with the d.c. exciter replaced by diodes carried on the rotor and fed from an a.c. excitation system. Brushless synchronous motors are somewhat more complicated than brushless generators in that during run-up the diodes need protection from slip-frequency e.m.f.s induced into the field system, employing additional rotor resistors and solid-state switches.

Ratings and dimensions

Modern motors of small and medium rating are built to the standards of IEC72, which lists a coherent range of main structural dimensions with centre heights between 56 and 1 000 mm. The outputs from standard motor frames are periodically reviewed to take account of technological advances. BS 3939 gives the standard ratings for the U.K. Some data are tabulated below.

Recommended Ratings and Dimensions of Rotating Machines
IEC Publications 72 and 72A

Preferred Ratings (kW):

0.06	0.55	5.5	30	110	220	355	500	710
0.09	0.75	7.5	37	132	250	375	530	750
0.12	1.1	11	45	150	280	400	560	800
0.18	1.5	15	55	160	300	425	600	850
0.25	2.2	18.5	75	185	315	450	630	900
0.37	3.7	22	90	200	335	475	670	950
								1000

Preferred Dimensions (mm):

FOOT-MOUNTED MACHINES, distance from shaft centre to mounting surface:

56	63	71	80	90	100	112	132	160	180	200	225	250
280	315	355	400	450	500	630	710	800	900	1000		

FLANGE-MOUNTED MACHINES, various flange pitch-circle diameters over the range 55 to 1080.

Cooling and enclosure

The simple construction of electric motors, more particularly of the cage induction machine, is readily cooled by straightforward circulation of the ambient air. Total enclosure has the merit that the windings are to a considerable degree protected from unfavourable environments; however, the cooling is restricted, and while there is little difference between the dimensions of open and totally-enclosed motors in ratings up to 15 kW, it becomes increasingly difficult to obtain adequate cooling-surface area for t.e. machines above 300 kW, and separate heat-exchangers may be necessary.

Modern systems of winding insulation are tolerant of dampness; nevertheless some minor refinements to the enclosure are generally worthwhile to suit specific operating conditions. Drip-proof, splashproof and hoseproof enclosures serve for the environmental conditions implied. Weatherproofing to the extent of preventing the ingress of moisture by spigots, flanges and bearing seals can be applied for motors to work in the open air without further protection. The nomenclature for machine enclosures is embodied in the international IP and IC codes, indicating respectively the degree of protection afforded and the type of cooling employed. The basic IP code comprises two digits, the first indicating the method of protection against accidental contact with live or moving parts, the second that against the ingress of water or of foreign bodies. The IC code is similar, but there is a greater variety and complexity in cooling systems.

Some machines have to work in a corrosive atmosphere. Simple cage motors are relatively resistant to such an ambient medium, but machines with carbon brushes and sliding contacts rely on the maintenance of an oxide film on the sliding surfaces for satisfactory current collection. Total enclosure of the machine or brush-gear, or cooling by pressurized clean air, may be necessary. Operation in hazardous atmospheres of a flammable type may call for non-sparking cage motors with modifications to the enclosure aimed at minimizing risk. For mines, certificated flashproof motors are essential, the basis of acceptance being that if flammable gases should ignite within the machine, the explosion is contained in such a way that gases external to the machine are not ignited.

Noise

Much of the acoustic noise emitted by a machine is generated by the cooling air as it is passed through or over the machine by its fan. Some noise reduction may be possible by substituting unidirectional trailing-bladed fans for the normal radial-bladed form. Further reduction may require intake and outflow silencers or the adoption of closed-air-circuit ventilation.

The varying forces and pressures in the cooling medium responsible for the acoustic noise cannot be eliminated: they are inherent in the processes of circulation and heat-exchange. In very large machines the noise level may exceed 110 dB and be substantially unavoidable. The frequency spectrum of the noise has usually a broad band with a few superposed pure tones (which may be harmonics of the supply frequency or of the speed or rotation). The noise-power field around a machine is affected by the shape and dimensions of the outer surface of the carcase, and is subject to such phenomena as reverberation or reflection from hard surfaces.

A further cause of noise is magnetic in origin, and is dependent on the stator and rotor slotting. It often occurs in machines with fractional numbers of open slots per pole, producing during rotation a relatively large variation of airgap flux density. This may produce a penetrating note at approximately slot frequency (e.g. 0.5 − 1.5 kHz). The amplitude of the stator permeance variation appears to be the causative factor, but the amplitude also depends on high-order rotor m.m.f. space harmonics. An empirical rule, supported by experiment, suggests that the frequency of the noise is twice the product of the supply frequency with that integer nearest to the number of slots per pole.

Voltage levels

The user of a large machine may have a choice of more than one system voltage.

Motors can be wound for a wide range of supply voltages, but generally the higher the voltage the more costly is the motor, although the price differential may be swamped by the cost of switchgear and cabling in such a way that these exert a determining influence on the choice of voltage. It is naturally uneconomic to specify a high working voltage for a small machine, and the lower limits of rating might be taken typically as 150 kW for 3.3 kV, 300 kW for 6.6 kV and 1 MW for 11 kV.

Insulation

The temperature classification for machine insulation is given in Sect. 4.7. In order to maximize the rating for a given frame, most small- and medium-sized machines use class E insulation. Large machines almost always employ class B. The increasing availability of class F materials has led to a change from bitumastic bonded to synthetic-resin bonded insulants based on polyester or epoxy compounds for 6.6 and 11 kV machines.

Protection

The primary function of the commonly used current-operated protective devices is the recognition of abnormally large currents. The matching of these devices to the thermal characteristics of the machine may be difficult. As an alternative, *thermistors* (temperature-sensitive resistors) may be used to activate a protective switch when the winding temperature exceeds a predetermined value. The present internationally accepted method (where the machine is important enough to justify it) is to embed one thermistor in each phase section of the stator winding, i.e. three thermistors in a 3-phase machine for each speed range. The devices are normally connected in series into a bridge network which is balanced at temperatures not exceeding the assigned limit. For excessive temperatures the bridge unbalances and operates a control contactor. It is still advisable to use overcurrent relays to protect the connections and to act if necessary as a back-up.

All modern overcurrent relays are thermally operated devices that use the heating effect of phase currents to deflect bimetal elements. Tripping is caused by large deflection for overcurrent, and by differential deflection between phase elements to protect against single-phasing.

The protection afforded by thermistors is very effective when the rate of temperature-rise is slow, but for high rates (as when a motor is stalled) there may be too much thermal lag. In small machines, in which the stator temperature rises more rapidly than that of the rotor (a 'stator-critical' condition), thermistors in the stator winding may provide adequate protection. Large machines are more likely to be 'rotor-critical', and stator thermistors may not give proper protection.

Generators

Many of the points so far discussed apply also to generators, except that these machines are usually large, and in most cases are of 'one-off' design. (A turbo-generator of high rating takes some years to design, build, instal and commission).

Almost all a.c. generators, regardless of scale, are 3-ph star-connected synchronous machines, and are provided with elaborate equipments for control, synchronizing, voltage regulation, protection, thermal dissipation, excitation, governing, etc.

7.7 Machines and power systems

In the following Chapters the behaviour of machines is related in the first instance to operation on infinite busbars. This, though useful, is an oversimplification. Modern supply networks are very extensive and have enormous generator power; but there has been an increase in the size of electric motor loads. Attention must therefore be given to the effects on a supply network of the operating characteristics of motors connected to it and of the reaction of the network on the machines. To have practical validity, the machine performance must take account of starting, control, protection, load characteristics, fault consequences and system impedance. Particular aspects of performance are:

Normal: efficiency, temperature-rise, starting, control, noise.
Special: cyclic loading, pulsating torques, limitations on current level imposed by local system impedance.
Abnormal: operation and reaction during system faults, voltage depressions, excessive ambient temperatures, unexpected physical conditions (e.g. flooding).

The basic factors listed by Stephen [72] as concerned in the interaction of a large machine and its network are: (i) transient magnetizing-current inrush pulses caused by energizing an unmagnetized machine; (ii) transient currents produced, as in synchronizing, by voltage divergence between the machine and the supply; (iii) voltage surges produced by the interruption of machine currents; (iv) current pulsations produced by a synchronous machine when running asynchronously; (v) mechanical oscillations resulting from sudden voltage changes or transient load fluctuations; (vi) system voltage distortions due to large rectifying plant.

Voltage depressions

Engine-driven generators. These are increasingly installed for peak-lopping service, the supply to construction sites, and the maintenance of essential services in the event of dislocation of the public supply by industrial conflict. Such installations must usually feed induction motors, some of which may be of rating large enough to impose a considerable load when starting. The engine-driven generator invariably has some form of automatic voltage regulator to hold the steady-state voltage variation within about 2%; but on the application of a large load increment the terminal voltage dips considerably, and the time-constants of the generator and regulator delay the recovery. The effect on induction-motor loads may be considerable.

Supply network. The Supply Authority is required by statute to limit voltage variation at the point of connection of a consumer's load to ± 6% of the declared value. However, the starting of large motors can cause temporary voltage dips of 10% or more, and greater but less frequent dips result from system faults.

Effects of voltage dips. A prolonged depression is significant more on account of its duration than of its magnitude. The effect on motors is governed initially by the transient characteristics of the electromechanical system: in some cases it is trivial, but the tripping of one machine in an interlocked continuous industrial process may trip the whole plant with serious effect.

No general induction motor characteristics can be specified to counter voltage dips, as the load and the supply system are also concerned, and the economic value of continuity of drive is an individual matter. The salient-pole synchronous motor is at some disadvantage if it pulls out of step on voltage depression, for apart from severe power fluctuations (reflected into the supply system at a time when they are particularly undesirable), the reaction on 'brushless' excitation circuits can be very damaging unless suitable protection is provided.

The frequency variation that accompanies voltage dip is usually small, but may on occasion be as much as 3%, adversely affecting both synchronous and induction machines.

Power-factor improvement

The power factor of an industrial load can be raised by the use of overexcited synchronous drives for certain loads, or of synchronous motor-generator sets which can compensate for standard induction motors driving pumps, fans, compressors and process auxiliaries. Apart from voltage regulation with load, the matter is basically economic, and p.f. improvement is justified only if there is a related tariff concession.

To be chosen in preference to a static capacitor, a synchronous motor must have several factors in its favour. Although the efficiency is less affected by starting and accelerating torque requirements than the cage induction motor (because the starting cage can be designed almost independently of the main-field running winding), the ideal condition is one of roughly constant load for long periods and not a high-inertia load requiring frequent starts. A disadvantage is that the synchronous motor will contribute to the current flowing into a short-circuit fault on the system and may call for additional expenditure in switchgear.

Variable-speed Drives

Apart from considerations of supply (availability) and environment (temperature, humidity, explosive atmosphere) the main factors that affect the choice of a variable-speed drive are: *cost* (including control gear); *load characteristics* (particularly inertia); *speed range* (with power and torque); *operational requirements* (braking, reversal, regeneration, speed accuracy). Sophisticated drives are usually based on logic devices controlling power-electronic equipments. These react on the supply system by introducing harmonic distortion and an increased demand for reactive power.

8
Induction Machines: Theory and Performance

8.1 *Development*

Nikola Tesla exhibited an elementary form of polyphase induction motor in 1891. Two years later, Dobrowolsky described a machine with a cage rotor and a distributed stator winding. The slip-ring machine was developed at the turn of the century. Since then, many detailed improvements to the 'plain' induction machine have been made. At the same time, attempts to circumvent the inherent speed and power-factor limitations have engaged the attention of inventors, but most of the ingenious devices developed have passed into history. The chief variants of the plain machine in modern machine technology are synchronous-induction machines, pole-change stator windings, double-cage rotors, and most recently the provision of variable-frequency drives by use of semiconductor rectifiers and inverters. Nevertheless, 80% of the world's industrial a.c. motors (apart from fractional-kilowatt machines) are plain cage motors working on constant-frequency supplies.

8.2 *Action of the ideal induction machine*

The essentials of a 2-pole induction machine are shown in Fig. 8.1. Polyphase windings (not shown) on the stator produce a travelling sine-distributed m.m.f. which, acting on the short and uniform airgap, develops a correspond-

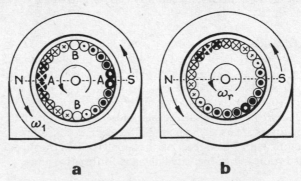

a   **b**

8.1 Elements of the induction machine.

ing travelling wave of flux of peak value Φ_m, as described in Sect. 3.14. The diagram indicates the axis of the gap flux at a given instant, and its direction of travel. The rotor carries a set of uniformly distributed full-pitch turns, each closed on itself.

In diagram (*a*), turn AA links no flux but has a maximum rate of change of flux linkage and so is the seat of the maximum induced e.m.f.; turn BB links maximum flux but the rate of change is momentarily zero and there is no e.m.f. The current that flows in each rotor turn at any instant depends upon the e.m.f. and the impedance: assuming this to be purely resistive, the current pattern, shown in (*a*), is identical with that of the e.m.f. But in fact each turn also has inductance, so that the alternating rotor currents are phase-delayed to give the pattern in (*b*). In each case the current distribution around the rotor gap surface is sinusoidal, and the condition equates to that in Fig. 1.10. As a consequence interaction mechanical forces are developed on the rotor conductors, urging them to move in the same direction as that of the stator field. The resultant torque in (*b*) is clearly less than in (*a*).

The behaviour of the machine is intimately related (i) to the *synchronous speed* and (ii) to the *slip*. The former is that of the travelling-wave airgap field, the latter is the fractional amount by which the angular speed of rotation of the rotor differs from synchronous speed.

Synchronous Speed: For a 2-pole machine the synchronous angular speed ω_s is the same as the angular frequency $\omega_1 = 2\pi f_1$ of the stator supply. In a machine with p pole-pairs, $\omega_s = \omega_1/p$.

Slip: If the angular speed of rotation of the rotor is ω_r, then the per-unit slip is $s = (\omega_s - \omega_r)/\omega_s$.

The rotor winding is isolated: it has no connected supply, and its power is transferred to it by means of the mutual gap flux. The amount of this power is primarily a function of the slip.

Let the rotor be at rest, a condition of *standstill*, with $\omega_r = 0$ and therefore $s = 1$. Both stator and rotor are at rest, and the angular frequency ω_2 of the rotor e.m.f.s and currents is the same as ω_1, the frequency of the stator e.m.f.s and currents. Taking the resistance and leakage reactance of a representative rotor phase to be r_2 and x_2, and its e.m.f. to be E_2, then the rotor current is $I_2 = E_2/(r_2 + jx_2)$, lagging E_2 by the phase angle $\arctan(x_2/r_2)$. The rotor m.m.f. pattern travels around the gap in synchronism with that of the travelling stator field. Interaction torque is developed in the direction of the stator field. The stator takes a current to balance the rotor m.m.f., just as does the primary of a static transformer to balance the secondary m.m.f., for the application of a supply voltage to the stator windings demands the presence of an airgap magnetic field and an approximate m.m.f. balance between stator and rotor. If we could ignore the stator leakage impedance, then this balance would be exact, as in the ideal transformer.

Now let the rotor rotate under torque at a speed ω_r in the direction of the travelling-wave field, with a *slip* between 1 and 0. The rate of change of linkage in the rotor winding falls, and the e.m.f. is reduced to sE_2 at the lower frequency $s\omega_1 = \omega_2 = \omega_1 - \omega_r$. The leakage reactance is also reduced, and the leakage impedance is now $r_2 + jsx_2$, so that the rotor current may not be much less than it was at standstill; moreover, its angle of lag is smaller and in consequence the torque is enhanced. The rotor m.m.f. pattern remains in synchronism with that of the stator, for it moves at angular speed ω_2 with respect to the rotor surface and is carried round bodily at ω_r to give a total angular rate of travel $\omega_2 + \omega_r = \omega_1$. As before, any current in the rotor calls for a balancing current in the stator.

If the rotor runs at *synchronous speed*, with $\omega_r = \omega_1$ and $s = 0$, the rotor windings move in an unchanging flux. Thus the rotor e.m.f. is $sE_2 = 0$; there is no current nor torque. The machine can develop torque only if there is a rotor e.m.f. and current, i.e. if there is a finite slip.

Operating Conditions

When a phase voltage V_1 is applied to the stator windings it is balanced (ignoring resistance) by the e.m.f. E_1 induced by a magnetic flux. Only part of this flux is mutual with the rotor and able to transfer energy thereto across the gap: the remainder is leakage flux proportional to the stator current. The greater the stator current, the smaller that proportion of the total flux which is mutual, so that Φ_m decreases with the load. In the ideal machine we shall disregard the stator leakage impedance, and assume that the mutual flux Φ_m is constant. For further simplicity, the machine is taken as having a basic 2-pole structure.

We shall consider the action of this ideal 2-pole machine in three operating modes, namely as a *motor* (with s between 1 and 0), as a *generator* (with s negative, i.e. with the rotor running at a speed higher than the synchronous), and as a *brake* (with s greater than unity, the rotor running backwards against the direction of the stator field). The assumptions, closely comparable to those adopted for the ideal transformer, are:

(i)   The stator and rotor windings are similar in distribution and have the same number of turns per phase.
(ii)   The stator applied phase voltage V_1 has a constant magnitude and frequency.
(iii)   The gap flux Φ_m is constant and travels at synchronous angular speed ω_1.
(iv)   The only loss is in the rotor resistance r_2.
(v)   The magnetizing m.m.f. is vanishingly small and the stator winding has negligible leakage impedance.

Although the assumptions imply a rather considerable simplification, the idealized machine can still present some significant characteristics.
Magnetization: The rotor winding is short-circuited on itself and has no external connection, the current being induced in it by the gap flux. As a result, the stator is always responsible for maintaining the magnetization by accepting from the supply the appropriate reactive power (here neglected). The stator applied voltage V_1 is balanced by an equal e.m.f. E_1 induced by the flux Φ_m, which also induces into the rotor winding the e.m.f. $E_2 = E_1$ at standstill and $sE_2 = sE_1$ at slip s. Any rotor current I_2 must be counterbalanced by an equal and opposite stator current $I_1 = -I_2$ to preserve m.m.f. equality

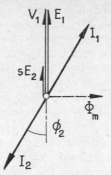

8.2 Ideal induction machine: phasor diagram.

and so maintain Φ_m. These conditions are represented in the phasor diagram of Fig. 8.2. The stator current, with reference to V_1, represents an input to the machine; in a sense, the rotor current is an 'output' with reference to sE_2.

Rotor Current: For any slip s, the rotor current is determined by the e.m.f. sE_2 and the appropriate leakage impedance $z_{2s} = r_2 + jsx_2$, whence the stator current balancing I_2 is

$$I_1 = -I_2 = \frac{sE_2}{z_{2s}} = \frac{sE_2}{\sqrt{[r_2{}^2 + (sx_2)^2]}} = \frac{E_2}{\sqrt{[(r_2/s)^2 + x_2{}^2]}} \qquad (8.1)$$

which has a lagging phase angle $\phi_2 = \arctan(sx_2/r_2)$. The equivalent circuit for the penultimate expression is shown in (*a*) of Fig. 8.3, and for the last in (*b*). The currents in (*a*) and (*b*) are, of course, the same; but in (*b*) the voltage $E_2 = E_1$ is, with the stated assumptions, a constant. Further, the active power conditions in the two equivalent circuits differ, for in (*a*) it is $I_2{}^2 r_2$ and in

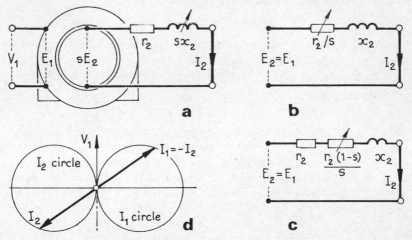

8.3 Ideal induction machine: rotor circuit.

(b) it is $I_2{}^2 r_2/s$. This difference is significant, and can be clarified by writing r_2/s as the series combination of r_2 and $r_2 (1-s)/s$, as in (c). With $I_2{}^2$ the former component gives the rotor I^2R loss, the latter gives a power that must correspond to that *converted* into mechanical power.

Current Locus: Suppose that $E_2 = E_1 = V_1 = 100$ V, and that the rotor circuit leakage impedance is made up of $r_2 = 0.1$ Ω and $x_2 = 0.5$ Ω. Working with the primary balancing current, then for a slip $s = 0.05$ p.u. we have $r_2/s = 2.0$ Ω, and $I_1 = 48$ A lagging E_1 by $\phi_1 = \arctan(0.5/2.0) = 14°$. Similarly for $s = 0.2$ p.u., $I_1 = 141$ A lagging by $79°$. For a negative slip of $s = -0.05$ p.u. the current is again 48 A but the phase angle is $\phi_2 = 166°$: this means that the active component is reversed, and the machine is delivering active power to the input terminals as a generator. For synchronous speed $s = 0$ the current is zero, while for infinite slip, forward or backward, the current is $E_1/x_2 = 200$ A at a phase angle of $90°$. If all these current phasors are plotted, the locus of their extremities is found to be a circle, Fig. 8.3(d).

Power: From Fig. 8.2, the stator active power input is given by

$$P_1 = V_1 I_1 \cos \phi_2 = E_1(-I_2) \cos \phi_2 = E_2(-I_2) \cos \phi_2 = P_2 \tag{8.2}$$

for as the ideal stator has no loss, the whole of its input power P_1 is transferred to the rotor by transformer action. Of the rotor input P_2, part is I^2R loss in the rotor circuit,

$$I_2{}^2 r_2 = (sE_2)I_2 \cos \phi_2 = sP_2$$

and the remainder is the power converted into the mechanical form:

$$P_m = (1-s)P_2$$

The important conclusion (general, and not dependent upon the assumption of constant flux) is that, of the power P_2 delivered across the gap to the rotor the fraction s becomes I^2R loss and the fraction $(1-s)$ appears as mechanical power (including windage and friction). Thus

$$P_2 : P_m : I^2R = 1 : (1-s) : s \tag{8.3}$$

A small slip, i.e. a running speed close to synchronous speed, is clearly advantageous. The rotor efficiency is $P_m/P_2 = (1-s)$.

Torque: As for a slip s the rotor speed is $\omega_r = \omega_1(1-s)$, the mechanical torque is

$$M = P_m/\omega_r = P_2(1-s)/\omega_1(1-s) = P_2/\omega_1 \tag{8.4}$$

The torque is thus proportional to the rotor power input regardless of slip. The power P_2 may be referred to as 'the torque in synchronous power' because it is the power that the actual torque would develop were it associated with synchronous speed ω_1 rather than the actual speed. The torque is obtained in terms of rotor quantities and the slip by using eqs.(8.1) and (8.2) and writing $\cos \phi_2$ as $r_2/\sqrt{[r_2{}^2 + (sx_2)^2]}$ to give

$$M = \frac{1}{\omega_1} \frac{sE_2{}^2 r_2}{[r_2{}^2 + (sx_2)^2]} = \frac{s\alpha}{s^2 + \alpha^2} K \tag{8.5}$$

where $K = E_2^2/\omega_1 x_2$ and $\alpha = r_2/x_2$. For a given ideal machine the torque therefore varies with slip and with the ratio of the rotor resistance to its standstill leakage reactance.

Eq.(8.5) gives $s^2 = \alpha^2$ for the condition of *maximum torque*, so that $s = \pm\alpha$ gives the information that there is a peak torque $M_{max} = K\alpha^2/2\alpha^2 = \frac{1}{2}K$ at each of two speeds, one below and one above synchronous speed.

The factor K includes the term $1/x_2$, so that the magnitude of the maximum torque is raised if the standstill rotor reactance is reduced: this is obvious because the rotor current, on which the torque depends, will be larger if the rotor circuit impedance is smaller, the effect being to enlarge the diameter of the current locus in Fig. 8.3(d).

The torque/slip and torque/speed relations for three values of $\alpha = r_2/x_2$ are drawn out in Fig. 8.4 for a machine in which the standstill leakage reactance

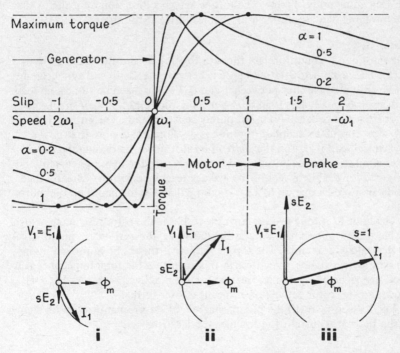

8.4 Ideal induction machine: torque/slip relations and phasor diagrams.

x_2 is fixed, but the resistance r_2 can be varied. In all cases the maximum torque is the same, the variation of resistance only changing the slip at which this torque is developed.

For most normal operating conditions the machine works near synchronous speed with a small slip (up to about 0.05 p.u.). In eq.(8.5) we can then legitimately neglect s^2 and express the torque as $M \simeq s(K/\alpha)$. For a given machine (K/α) is fixed, and the torque is directly proportional to the slip, giving a speed/torque characteristic in which the speed falls only slightly with increase of the torque. This is generally referred to as a 'constant-speed characteristic' resembling that of a d.c. shunt machine.

Generator Mode

The phasor diagram for stator quantities is shown at (i) in Fig. 8.4. The speed is hypersynchronous, the slip is negative (and usually small), the rotor e.m.f.s and currents have such direction as to demand active power *output* from the stator terminals. But magnetization is still dependent on the stator winding accepting reactive power for this purpose from the electrical source, so that the induction generator can only operate when connected to a live and synchronous a.c. system. If a *lagging* reactive power *input* is equated with a *leading* reactive power *output,* then the generator can be described as operating with a lending power factor. The torque acts in a direction opposite to that of the travelling field, requiring a mechanical drive at the shaft.

Motor Mode

The rotor e.m.f. conditions are those in Fig. 8.1 (*a*) and the rotor currents as in (*b*). The machine operates at speeds between standstill and synchronous speed, with positive slips between 1 and 0. Let the machine run on no load, with a slip typically less than 0.01 p.u. The rotor e.m.f. is very small, the rotor circuit impedance is almost purely resistive, and a current sufficient to develop a torque to maintain rotation is developed. Suppose that now a mechanical load is put on the shaft: the rotor slows, increasing the slip. The rotor e.m.f. rises in magnitude and frequency, producing more current and torque so long as the effect of the leakage reactance permits. Greater load results in more torque up to a maximum pull-out value and then the motor stalls.

The limit to torque production at low slips is the rotor resistance, which should therefore be small. At starting, however, the current is limited largely by the leakage reactance. The torque may nevertheless be improved by increasing the resistance, for although this augments the total impedance it reduces the angle of lag and results in more rotor active power. As eq.(8.4) shows, greater rotor active power means greater torque.

Fig. 8.4 shows at (ii) the phasor diagram of stator quantities. With change of slip the stator current follows the circular locus.

Brake Mode

If the induction machine is driven mechanically in a backward direction (i.e. opposite to the travelling-wave field) it still produces a forward torque and therefore acts as a brake, absorbing the mechanical power in rotor loss. The effect can be very useful in bringing a motor rapidly to rest. Suppose a motor is running normally at a small slip s, and the direction of the field is suddenly reversed. The rotor now has in effect a slip of $(2 - s)$ and enters the brake mode. The speed falls towards zero and the machine can be brought to standstill. It will not run up thereafter in the reversed direction if the supply to the stator is interrupted. This method of stopping a motor is termed 'plugging'.

The brake-mode conditions are indicated by the phasor diagram (iii) in Fig. 8.4.

8.3 *Cage and slip-ring rotor windings*

The two types of rotor winding in common use for induction machines are the cage and the slip-ring windings.

Cage Rotor

The essential features of the multiphase cage winding were discussed in Sect. 3.14. In practice it consists of a set of uniformly spaced bars accommodated in the rotor slots and connected at each end to conducting rings. The winding, which is electrically closed on itself as is required for induction machine operation, can be considered to be a series of full-pitch 'turns' formed by pairs of conductor bars a pole-pitch apart, joined together in a closed loop by the end-conductors. The assumed sinusoidal space distribution of bar currents in Fig. 8.5 shows that the end-rings carry the summation of the bar currents and

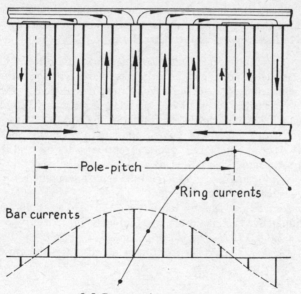

8.5 Currents in cage rotor.

affect the resistance of each 'turn'. Generally the effect is small, and the currents in the bars are assumed to retain a sine distribution.

Slip-ring Rotor

A slip-ring rotor carries a normal 3- or 6-ph winding, connected in star or delta and terminated on three slip-rings, which are short-circuited when the machine is in normal operation. A 2-ph winding is an acceptable but unusual alternative.

Comparison

The cage and slip-ring rotors are thus electrically equivalent as long as attention is confined to the fundamental sine-waves of voltage, current, flux, etc. We may proceed on the conclusion that the operation of a machine will be the same whichever type of rotor winding it has.

The practical reasons for the choice between cage and slip-ring windings are based in most cases on the following considerations:

(*a*)   the cage is permanently closed and its electrical characteristics are fixed, whereas the slip-ring winding permits of the variation of the electrical characteristics by the inclusion of external circuits via the slip-rings; and

(*b*)   the cage is adaptable to any number of poles, whereas the slip-ring winding has to be made for one (or possibly two) definite numbers of poles.

Consideration (*a*) is most cogent from the viewpoint of starting the motor against large load torques, where it is advantageous to be able to increase the resistance of the rotor circuit by the inclusion of a rheostat connected across the slip-rings; while (*b*) is of importance in connection with speed control.

The cage is always preferred, as it is a more robust and much cheaper winding. Some of its starting-torque disadvantages can be overcome by use of the double-cage or the deep-bar cage construction.

EXAMPLE 8.1: A 2-pole 50 Hz induction machine has a rotor of diameter 0.20 m and core length 0.12 m. The polyphase stator winding maintains in the gap a sine-distributed travelling wave of flux of peak density 0.54 T. The rotor winding comprises a cage of 33 bars, each of resistance 120 $\mu\Omega$ and leakage inductance 2.50 μH, including the effect of the end-rings. At a given load the rotor runs with a slip of 0.064 corresponding to an angular speed of 294 rad/s (46.8 rev/s). Find (i) the peak current per rotor bar and the rotor I^2R loss; (ii) the rotor m.m.f. per pole; (iii) the load angle between the gap-flux axis and the axis of the rotor m.m.f.; (iv) the electromagnetic torque developed; and (v) the mechanical power developed.

(i)   *Rotor bar current*: The slip velocity is $314 - 294 = 20$ rad/s or $20 \times 0.1 =$ 2.0 m/s peripherally. The maximum e.m.f. induced in a bar is $e_{bm} = Blu =$ $0.54 \times 0.12 \times 2.0 = 0.13$ V. The bar impedance is $z_b = (120 + j20 \times 2.5) =$ $(120 + j50) = 130 \angle 23° \mu\Omega$. The maximum bar current is therefore

$$i_{bm} = e_{bm}/z_b = 0.13/130 \times 10^{-6} = 1\ 000 \text{ A}$$

and the total rotor I^2R loss is $\frac{1}{2}(1\ 000^2 \times 120 \times 10^{-6} \times 33) = 1\ 980$ W

(ii)   *Rotor m.m.f.*: The peak linear current density is $A_2 = 1\ 000 \times 33/\pi \times$ $0.2 = 52\ 500$ A/m. The discussion in Sect. 1.11 shows that the rotor m.m.f. is

$$F_2 = \tfrac{1}{2}A_2 D = \tfrac{1}{2} \times 52\ 500 \times 0.2 = 5\ 250 \text{ A-t/pole}$$

(iii)   *Load angle*: If the rotor winding has only resistance, the peak current would occur in the bars at the positions of maximum gap flux density, and the rotor m.m.f. axis would be at right-angles to the gap flux axis. With the actual phase-lag of $23°$, the load angle is increased to $90° + 23° = 113°$: compare Fig. 8.1(b).

(iv) *Torque*: Using eq. (1.10),

$$M = -\tfrac{1}{2} \pi \times 0.20 \times 0.12 \times 0.54 \times 52\,500 \times \sin 113° = 98.5 \text{ N-m}$$

(v) *Power*: This is

$$P_m = M\omega_r = 98.5 \times 294 = 29\,000 \text{ W}$$

which can be checked from the rotor $I^2 R$ loss, using $P_m [s/(1-s)]$ = 29 000 (0.064/0.936) = 1 980 W. The total rotor input is $P_2 = 29\,000 + 1\,980 = 30\,980$ W and the efficiency (neglecting stator and mechanical losses) is 0.936 p.u.

8.4 Practical induction machine

In a practical induction machine the gap reluctance demands a magnetizing m.m.f., so that the stator current has to balance the rotor current and in ad-

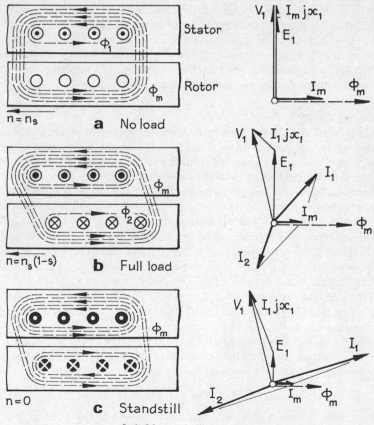

8.6 Magnetic flux patterns.

dition provide for magnetizing current and core loss. Again, the presence of stator winding inductance and leakage reactance makes the assumption of constant gap flux untenable.

Flux Changes

A simple idea of the flux variation can be gained from the three conditions of motor operation in Fig. 8.6. In (a), for no load with negligible rotor current, most of the flux is mutual (Φ_m) but a small proportion (Φ_1), perhaps 2–3% of the total, does not link the rotor. The total stator flux is $\Phi_m + \Phi_1$, corresponding to the two components E_1 and $I_m x_1$ of the applied voltage V_1, where I_m is the magnetizing current and the volt drop in stator-winding resistance is ignored. At full load (b) the mutual flux, now smaller, induces e.m.f.s and currents in the rotor, producing rotor leakage flux Φ_2 and requiring a balancing current component in the stator which increases the stator leakage. As V_1 has to provide for an increased stator leakage reactance drop $I_1 x_1$, the mutual flux component E_1 is less and Φ_m falls appreciably from its no-load value. At standstill (c) the rotor e.m.f. and current are large, and so must be the stator current. Much of the stator flux is now leakage, and the mutual flux has dropped to about one-half of the no-load level.

These effects, together with resistance, are taken into account by including in the equivalent circuit the stator phase resistance and leakage reactance. The magnetization requires that the stator and rotor m.m.f.s combine to produce the magnetizing and core-loss requirements.

Equivalent Circuit

This is constructed for one representative stator and rotor phase, and is developed from the elementary circuits in Fig. 8.3. The stator resistance r_1 and leakage reactance x_1 are separated from the common magnetic circuit, and the magnetizing and loss currents are considered as flowing through the shunt branches r_m and x_m. The equivalent circuit then appears as in Fig.8.7(a), the ideal transformer developing in the rotor the e.m.f. sE_2, which circulates I_2 through the resistance r_2 and the leakage reactance sx_2.

The rotor power is represented by the $I^2 R$ loss in r_2. A very useful modification is to use the rotor circuit (c) of Fig. 8.3 in order to include the mechanical power conversion: then the rotor e.m.f. is changed to E_2 and the rotor impedance parameters to r_2/s and x_2. Dividing r_2/s into r_2 and $r_2(1-s)/s$ then includes the converted power $I_2{}^2 r_2(1-s)/s = P_m$ explicitly. The final simplifying step is to assume the actual representative rotor phase winding to be replaced as in Fig. 8.7(b) by one having the same number of effective turns as the stator: this is the *equivalent rotor referred to the stator,* with the referred quantities such that the power, both active and reactive, in the referred rotor remain unaltered. Indicating referred quantities by a prime, then

$$E_2' = E_1, \quad I_2'^2 r_2' = I_2{}^2 r_2 \quad \text{and} \quad I_2'^2 x_2' = I_2{}^2 x_2$$

The conversions are evaluated in a manner analogous to that used for the secondary of a transformer, as follows.

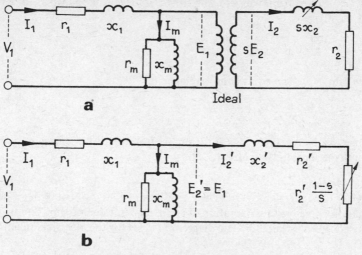

8.7 Equivalent circuits.

Slip-ring Rotor Winding: Here it is necessary only to write

$$E_2' = k_t E_2, \quad r_2' = k_t^2 r_2 \quad \text{and} \quad x_2' = k_t^2 x_2 \tag{8.6}$$

where $k_t = (K_{w1} N_1 / K_{w2} N_2)$ using the appropriate numbers of turns per phase in series (N) and winding factors (K_w).

Cage Rotor Winding: Eqs.(3.24) and (3.27) give the fundamental m.m.f.s respectively of a 3-ph winding carrying a current I_2' in N_1 turns per phase in series, and of a $2p$-pole cage winding in S_2 slots with an r.m.s. current I_b per bar: the expressions are

$$F_1 = 1.35 K_{w1} N_1 I_2' / p \quad \text{and} \quad F_2 = S_2 I_b / \sqrt{2}\pi p$$

For m.m.f. equivalence, $F_1 = F_2$ whence

$$I_b = I_2'(6 K_{w1} N_1 / S_2) \tag{8.7}$$

To derive the equivalent resistance r_2' we must first find the $I^2 R$ loss in the cage winding. If r_b is the resistance per bar, the total loss in the S_2 bars is $S_2 I_b^2 r_b$. It can be seen from the current distribution diagram in Fig. 8.5 that the end-ring current has a maximum made up of the sum of the instantaneous bar currents in a half pole-pitch, and therefore an r.m.s. value $I_c = I_b(S_2/2p\pi)$. The total $I^2 R$ loss for the S_2 bars each of resistance r_b and the two end-rings each of resistance r_c (taken around the periphery) can be equated to the total $I^2 R$ loss in a 3-ph winding carrying I_2':

$$I_b^2 [S_2 r_b + 2(S_2/2p\pi)^2 r_c] = 3 I_2'^2 r_2'$$

for a 3-ph stator, from which, using eq.(8.7), the stator equivalent of a cage-rotor resistance is

$$r_2' = \frac{12K_{w1}^2 N_1^2}{S_2}\left[r_b + \frac{0.1\,S_2}{2p^2}\,r_c\right] \tag{8.8}$$

In the same way, the equivalent reactance is

$$x_2' = \frac{12K_{w1}^2 N_1^2}{S_2}\,x_2 \tag{8.9}$$

where x_2 is the reactance of a rotor phase. It is usual to calculate only the slot leakage separately; see Sect. 2.9.

Phasor Diagram
Either of the equivalent circuits in Fig. 8.7 can be used to construct a phasor diagram for steady-state operation at some chosen value of slip. The two forms are given in Fig. 8.8. The diagram in (*a*) is for the electrical quantities

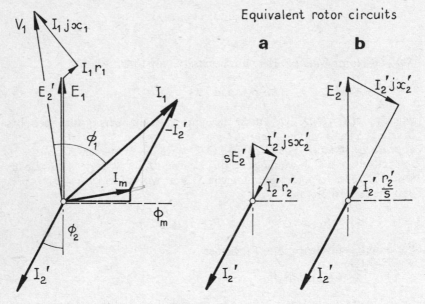

8.8 Phasor diagrams.

only, while (*b*) includes the total rotor power. The stator current comprises the components $-I_2'$ to balance the rotor m.m.f., and the magnetizing and core-loss components that make up the current I_m needed to magnetize the machine. The stator voltage V_1 accounts for the stator leakage impedance volt drops $I_1 r_1$ and $I_1 jx_1$, leaving the e.m.f. E_1 associated with the gap flux. In the rotor, I_2' is derived in (*a*) from sE_2' divided by the true rotor leakage impedance $(r_2' + jsx_2')$, whereas in (*b*) it is E_2' divided by $[(r_2'/s) + jx_2']$.

Approximate Equivalent Circuit

The equivalent circuits in Fig. 8.7 enable the characteristics of the induction machine to be evaluated for steady-state conditions by the ordinary processes of a.c. network solution. Some useful approximations can be made for normal operating conditions with small positive or negative slip, and for no-load and short-circuit conditions.

No Load: On no load the slip is so small that r_2'/s becomes very high, making the rotor in effect almost open-circuited, reducing the equivalent circuit to that in Fig. 8.9(a). The magnetizing impedance preponderates, being so much

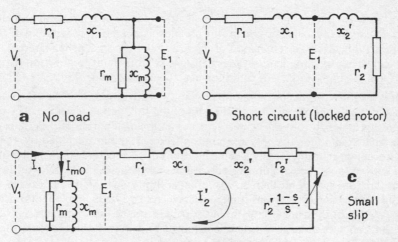

a No load   **b** Short circuit (locked rotor)

         c Small slip

8.9 Equivalent-circuit approximations.

greater than r_1 and x_1 that $E_1 \simeq V_1$. The loss is $V_1{}^2/r_m$ very nearly. However, there is a further small loss, in friction and windage, that demands a torque and therefore a rotor current and a balancing stator current so that the actual no-load current is I_0. A test of the machine on no load therefore gives, as stator input, the core and mechanical loss together with an almost negligible I^2R loss.

Short Circuit: In this condition the stator is excited (at normal, or for large machines, at reduced voltage) and the *rotor held at rest* ('locked'). The conditions in Fig. 8.9(b) show that, as E_1 is only about one-half of its no-load value and the rotor current is large, the magnetizing current can be neglected. The stator current is limited by the combined resistance $R_1 = r_1 + r_2'$ and the combined reactance $X_1 = x_1 + x_2'$. The active power input with the rotor locked is therefore the effective I^2R loss with its associated stray loss.

Operation with small slip: So long as the slip is small, the machine can be regarded as running with an almost constant gap flux, and the magnetizing branches r_m and x_m can be transferred to the primary terminals as in Fig. 8.9(c). From the network viewpoint this is a considerable simplification, and it is common where operating conditions permit to use this 'approximate' equivalent circuit for the purposes of easy calculation.

Limitations

There are numerous defects in the equivalent circuit. The impedance parameters are not constant. Apart from variation of ohmic resistance with temperature, the parameters may be expected to change with the loading conditions as follows:

r_1 and r_2: tooth saturation alters the eddy-loss coefficient (Sect. 3.2) and consequently the effective resistance.

x_1 and x_2: tooth and core saturation affect the leakage flux per unit current and therefore the effective inductive reactance.

r_m and x_m: the magnetizing reactance is dependent in magnitude on the saturation of the magnetic circuit; the core loss is not a simple function of the gap flux, as it is a summation of the loss in the ferromagnetic materials, in which the flux density varies in intensity from point to point, so that only at one fixed e.m.f. condition can r_m give the proper loss.

It is seen that the major cause of uncertainty is the saturation. The effect of saturation is to limit the flux in the iron of the flux paths concerned. The usual process in design is to calculate reactances on the assumption of infinite permeability, and then to apply a saturation factor. It is a formidable task, for an accurate solution for any given conditions of current and slip would require a knowledge of the path dimensions and the magnetic characteristics of the material at every point therein. The usual arbitrary division of the leakage flux into convenient regions sometimes involves two such components occupying in part the same iron path; but superposition of the components is not valid. A number of attempts have been made to deal with this problem, such as that devised by Chalmers and Dodgson [8].

8.5 *Steady-state theory*

The behaviour of the machine is investigated by use of the equivalent circuit, it being assumed that proper values can be assigned to the parameters to suit the operating conditions. Operation is considered in the steady state with constant stator applied voltage and frequency, and all values are per phase.

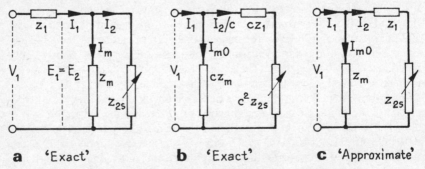

a 'Exact'   **b** 'Exact'   **c** 'Approximate'

8.10 Exact and approximate equivalent circuits.

In order to include the converted power explicitly, the equivalent circuit is based on Fig. 8.7(*b*). To simplify the analysis, the network of Fig. 8.10(*a*) is taken with the assumption that the stator/rotor effective turns ratio is 1/1; in using the results for general calculation it is only necessary to remember that z_2, r_2 and x_2 have to be read as $z_2{}', r_2{}'$ and $x_2{}'$. We seek the currents I_1, I_2 and I_m, and the torque, for any given slip s.

Stator

The stator current is $I_1 = I_m + I_2 = (E_1/z_m) + (E_1/z_{2s})$, and the applied voltage is

$$V_1 = E_1 + I_1 z_1 = E_1 + (I_m + I_2)z_1$$
$$= E_1[1 + (z_1/z_m) + (z_1/z_{2s})]$$

Now z_1 is small compared with z_m, but both are inductive and have impedance angles of the same order. It is convenient to write

$$c = 1 + (z_1/z_m) = c \angle \gamma \tag{8.10}$$

a complex number slightly greater than unity (e.g., 1.1) with a small angle γ, usually negative. Then

$$V_1 = E_1[c + (z_1/z_{2s})]$$

Curves typical of the variation of E_1 and I_1 with slip are shown in Fig. 8.11.

Rotor

It is now possible to obtain I_2 in terms of V_1:

$$I_2 = \frac{E_1}{z_{2s}} = \frac{V_1}{z_1 + cz_{2s}} = \frac{V_1}{(r_1 + jx_1) + c[(r_2/s) + jx_2]}$$
$$= \frac{V_1}{(r_1 + cr_2/s) + j(x_1 + cx_2)} \tag{8.11}$$

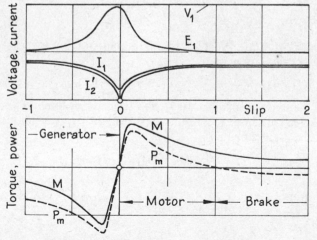

8.11 Induction machine characteristics.

Any circuit of impedance $r + jx$ in which one component (here the reactance) is constant while the other (resistance) is varied has a straight-line impedance locus, and in consequence the locus of its phasor current at constant voltage is a circular arc [Ref. 1]. The scalar magnitude of the rotor current (taking c as a *pure numeric*) is

$$I_2 = \frac{V_1}{\sqrt{[(r_1 + cr_2/s)^2 + (x_1 + cx_2)^2]}}$$

(8.12)

Torque

The total rotor input power per phase is $P_2 = I_2{}^2 r_2/s$ and the torque is P_2/ω_s. It therefore follows from eq.(8.3) that the torque is

$$M = \frac{1}{\omega_s} \cdot \frac{V_1{}^2 (r_2/s)}{(r_1 + cr_2/s)^2 + (x_1 + cx_2)^2}.$$

(8.13)

Curves typical of the torque (and rotor input power) are plotted in Fig. 8.11. Solving $dM/ds = 0$ for maximum torque gives the slip for this torque as

$$s_m = \pm \frac{cr_2}{\sqrt{[r_1{}^2 + (x_1 + cx_2)^2]}}$$

(8.14)

The value of the maximum torque is therefore

$$M_m = \frac{1}{\omega_s} \cdot \frac{V_1{}^2/2c}{\pm r_1 + \sqrt{[r_1{}^2 + (x_1 + cx_2)^2]}}$$

(8.15)

Magnetizing Current

At zero slip the rotor current vanishes and the magnetizing current is $I_{mo} = V_1/(z_1 + z_m) = V_1/cz_m$. When, for finite slips, rotor current flows and the e.m.f. falls to $E_1 = V_1 - I_1 z_1$, the magnetizing current reduces to I_m. Let the stator current $I_1 = I_m + I_2$ be written instead as $I_1 = I_{mo} + I_a$: then

$$I_a = I_2 + I_m - I_{mo} = I_2 + (E_1/z_m) - (V_1/cz_m)$$

Solving for I_a with $E_1 = V_1 - (I_{mo} + I_a)z_1$, we obtain (after some algebra) the result that $I_a = I_2/c$, whence

$$I_m = I_{mo} + I_a - I_2 = I_{mo} - I_a(c - 1) = I_{mo} - I_a(z_1/z_m)$$

As I_a has the same circular locus as I_2, the magnetizing current I_m for a slip s is the value I_{mo} at slip zero less a phasor that itself has a circular locus. In terms of the rotor current

$$I_m = I_{mo} - I_2[(c - 1)/c] = I_{mo} - I_2[z_1/(z_1 + z_m)]$$

(8.1⨍)

Stator Current

This is $I_1 = I_m + I_2$, and using eq.(8.16)

$$I_1 = \frac{V_1}{cz_m} - I_2 \frac{c-1}{c} + I_2 = \frac{1}{c}\left[\frac{V_1}{z_m} + I_2\right]$$

(8.17)

where $c = 1 + (z_1/z_m)$ from eq.(8.10), and I_2 is from eq.(8.11). Eq.(8.17) can be expressed by the alternative equivalent circuit in Fig. 8.10(*b*), which may offer some advantage in computing performance from known parameters.

A comparison of these results with those in Sect. 8.2 show that the stator impedance has an important influence on the behaviour of the machine. Generally the stator and rotor equivalent resistance and reactance values are comparable and the stator quantities cannot be neglected without serious error.

Within the limitations imposed by the variation of resistance and leakage reactance with saturation and frequency, eqs.(8.11)–(8.17) can be used to calculate machine performance.

EXAMPLE 8.2: A 36 kW, 415 V, 50 Hz, 3-ph, 10-pole, delta-connected cage machine has the following impedance parameters referred to a stator phase at normal operating temperature:

$$r_1 = 0.19 \ \Omega \qquad r_2 = 0.29 \ \Omega \qquad r_m = 143 \ \Omega$$

$$x_1 = 1.12 \ \Omega \qquad x_2 = 1.12 \ \Omega \qquad x_m = 16.8 \ \Omega$$

The performance of the machine is required for the following operating conditions:

   (i)   as a generator with a slip $s = -0.026$ p.u.;
  (ii)   at synchronous speed, $s = 0$;
 (iii)   as a motor at rated output, for which $s = 0.026$ p.u.;
 (iv)   as a motor at a slip corresponding to pull-out torque;
  (v)   at standstill, $s = 1.0$ p.u.;
 (vi)   as a brake with a slip of $s = 2.0$ p.u.;
(vii)   at infinite slip, $s = \infty$.

For condition (iv), eq.(8.14) gives $s_m = 0.133$ p.u. The calculations are set out in the Table, using the 'exact' equivalent circuit, Fig. 8.10(*a*), the phase voltage $V_1 = 415 \angle 0°$ V as reference phasor, and the following data:

$$z_1 \ = 0.19 + j \, 1.2 = 1.14 \angle 80.4° \ \Omega$$

$$z_{2s} \ = 0.29/s + j1.12 \ \Omega$$

$$z_m \ = 1/[(1/r_m) - j(1/x_m)] = 1.95 + j \, 16.6 = 16.7 \angle 83.3° \ \Omega$$

$$c \ = 1 + (z_1/z_m) = 1.07 \angle -0.2° \approx 1.07$$

$$\omega_s \ = 2 \, \pi \, 50/5 = 62.8 \ \text{rad/s}$$

The three-phase values in the Table include an estimated mechanical loss in windage and friction: this has to be supplied to the machine by the mechanical drive when the operating mode is generation or braking, but is deducted from the converted power P_m for motor-mode conditions.

The seven values of s are marked on the stator current locus in Fig. 8.12. The locus of $I_1 = I_m + I_2$ is seen to be circular, as also is that of the magnetizing current I_m, the details of which are more clearly shown in the enlarged part of the diagram, drawn to 5 times the scale of the main diagram. In com-

EXAMPLE 8.2: Performance of 36 kW 415 V 50 Hz 10-pole Induction Machine

Condition		Generator	Syn. speed	Motor	Max. torque	Standstill	Brake	Inf. slip
[1] Slip s	p.u.	−0.026	0.0	0.026	0.133	1.0	2.0	∞
[2] r_2/s	Ω	−11.2	∞	11.2	2.18	0.29	0.145	0
[3] z_2s	Ω	−11.2 + j1.12	∞	11.2 + j1.12	2.18 + j1.12	0.29 + j1.12	0.145 + j1.12	j1.12
[4] cz_2s	Ω	−12.0 + j1.2	∞	12.0 + j1.2	2.34 + j1.2	0.31 + j1.2	0.16 + j1.2	j1.2
[5] $z_1 + cz_2s$	Ω	−11.8 + j2.3	∞	12.2 + j2.3	2.53 + j2.3	0.50 + j2.3	0.35 + j2.3	0.19 + j2.3
[6] $z_2s/(z_1 + cz_2s)$	–	0.92 ∠ 5.6°	0.93 ∠ 0°	0.91 ∠ −5.3°	0.71 ∠ −15°	0.49 ∠ −2.3°	0.48 ∠ 1.2°	0.48 ∠ 4.7°
[7] $E_1 = V_1 \times$ [6]	V	382 ∠ 5.6°	386 ∠ 0°	378 ∠ −5.3°	296 ∠ −15°	203 ∠ −2.3°	199 ∠ 1.2°	199 ∠ 4.7°
[8] $I_m = E_1 \div z_m$	A	4.8 − j22.4	2.7 − j23.0	0.6 − j22.6	−2.8 − j17.6	0.9 − j12.1	1.6 − j11.8	2.3 − j11.7
[9] $I_2 = E_1 \div$ [3]	A	−33.7 − j6.7	0 − j0	32.6 − j6.2	88.2 − j81.2	37 − j171	26.4 − j175	14.6 − j177
[10] $I_2{}^2r_2$	kW	0.34	0.0	0.32	4.20	8.86	9.00	9.20
[11] P_2	kW	−13.04	0.0	12.52	31.50	8.86	4.52	0.0
[12] Pm	kW	−13.68	0.0	12.20	27.30	0.0	−4.48	−9.20
[13] $M = P_2/\omega_s$	N-m	−208	0.0	199	500	141	72	0.0
[14] $I_1 = I_2 + I_m$	A	−28.9 − j29.1 = 41 ∠ −135°	2.7 − j23.0 = 23.2 ∠ −83°	33.2 − j28.9 = 44 ∠ −41°	85 − j99 = 131 ∠ −49°	38 − j183 = 186 ∠ −78°	28 − j187 = 188 ∠ −82°	17 − j189 = 190 ∠ −85°
[15] $\cos \phi_1$	p.u.	0.71	0.12	0.76	0.66	0.20	0.15	0.09
[16] P_1	kW	−12.0	1.1	13.8	35.5	16.7	11.6	7.0
Three-phase values:								
Gross mechanical power $3Pm$	kW	−41.04	0.0	36.60	81.90	0.0	−13.44	−27.6
Mechanical loss	kW	0.64	0.62	0.60	0.40	0.0	0.63	–
Net mech. power	kW	−41.7	0.0	36.0	81.5	0.0	−14.1	–
Efficiency	p.u.	0.86	0.0	0.87	0.76	0.0	–	–

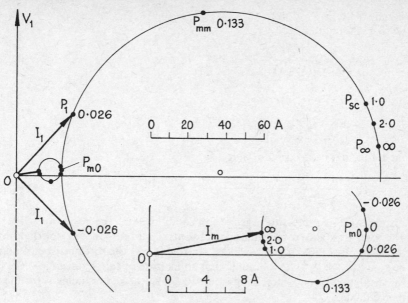

8.12 Circle diagram: 36 kW machine.

puting the tabulated results it is easily possible to introduce current-dependent variations in the resistance and reactance parameters; the locus diagrams will then not be precise circles.

8.6 *Approximate theory*

Use of the equivalent circuit (*c*) of Fig. 8.10 yields simpler expressions from which generally informative – but approximate – inferences can be made. In effect, the magnetizing current is taken as constant, and the conversion branch of the network becomes simply z_1 and z_{2s} in series with a constant phase voltage V_1 applied. The complex number *c* is now unity, and the following expressions can readily be derived.

Rotor Current

This is

$$I_2 = \frac{E_1}{z_{2s}} = \frac{V_1}{z_1 + z_{2s}} = \frac{V_1}{(r_1 + r_2/s) + jX_1} \tag{8.18}$$

where $X_1 = x_1 + x_2$ is the total combined stator and rotor leakage reactance at supply frequency referred to the stator winding. The scalar value of the current is

$$I_2 = \frac{V_1}{\sqrt{[(r_1 + r_2/s)^2 + X_1{}^2]}} \tag{8.19}$$

This has a circular locus.

Torque

Eq.(8.13) simplifies to

$$M = \frac{1}{\omega_s} \cdot \frac{V_1{}^2(r_2/s)}{(r_1 + r_2/s)^2 + X_1{}^2}$$

(8.20)

For the slip

$$s_m = \pm \frac{r_2}{\sqrt{[r_1{}^2 + X_1{}^2]}}$$

the torque has its maximum value

$$M_m = \frac{1}{\omega_s} \cdot \frac{V_1{}^2/2}{\pm r_1 + \sqrt{(r_1{}^2 + X_1{}^2)}}$$

(8.21)

The negative sign refers to negative slip conditions (i.e. the generator mode). The *rotor* resistance affects the speed for maximum torque, but not the torque itself. To obtain maximum torque at starting ($s = 1$) the rotor resistance must be $r_2 = \sqrt{(r_1{}^2 + X_1{}^2)}$. It is notable that an increase of *stator* resistance affects the maximum torque, increasing it for the generator and reducing it for the motor mode:

$$\frac{M_m \text{ (gen)}}{M_m(\text{motor})} = \frac{r_1 + \sqrt{(r_1{}^2 + X_1{}^2)}}{-r_1 + \sqrt{(r_1{}^2 + X_1{}^2)}} \simeq \frac{X_1 + r_1}{X_1 - r_1}$$

and the approximation holds if r_1 is small compared with X_1 as is normally the case. If $X_1 = 7r_1$, the torque ratio becomes 1.33. An attempt to control an overspeeding motor by connecting resistance into the stator lines may produce very high peak torques.

Losses and Efficiency

The obvious I^2R loss is $I_1{}^2 r_1 + I_2{}^2 r_2$ per phase, and is load-dependent. In addition there are various 'stray' losses (Sect. 8.12) which may also be taken as load-dependent. Certain losses are taken to sum to a constant, at least in the normal working range of speed: these are friction, windage and core losses. Per phase, the efficiency, or ratio [output/output + losses], is given approximately for an electromagnetically converted power P_m and mechanical loss p by

$$\eta = \frac{P_m - p}{(P_m - p) + \{I_2{}^2(r_1 + r_2) + p_s\} + [I_{mo}{}^2 (r_1 + r_m) + p]}$$

where p_s is the stray loss, and the core loss is represented by $I_{mo}{}^2 r_m$. The term in square brackets is taken as roughly constant so that maximum efficiency can be assessed as that for which the two sets of bracketed terms are equal. The variable and 'constant' losses of a machine on test can be inferred from its input on no load and with locked rotor.

EXAMPLE 8.3: Evaluate the performance of the machine of Example 8.2, using the 'approximate' equivalent circuit, for the same seven operating conditions.

The primary data are the same as in Example 8.3, but the magnetizing current is now

$$I_{mo} = V_1 [(1/r_m) - j(1/x_m)] = 2.9 - j24.7 \text{ A}$$

and from eq.(8.21) the slip for maximum torque is $s_m = 0.13$ p.u. The computed results are tabulated below: the items are numbered where possible in accordance with the Table for Example 8.2.

EXAMPLE 8.3: Performance of 36 kW 415 V 50 Hz
Induction Machine

Condition			Generator	Syn. speed	Motor
[1] Slip s	p.u.		−0.026	0.0	0.026
[3] z_{2s}	Ω		−11.2 + j1.12	∞	11.2 + j1.12
[5] $z_1 + z_{2s}$	Ω		−11.0 + j2.24	∞	11.4 + j2.24
[9] $I_2 = V_1 \div$ [5]	A		−36.3 + j7.4	0	35.0 − j6.9
[10] $I_2{}^2 r_2$	kW		0.40	0	0.37
[11] P_2	kW		−15.3	0	14.2
[12] P_m	kW		−15.7	0	13.8
[13] M	N-m		−244	0	225
[14] $I_1 = I_2 + I_m$	A		−33.4 − j32.1	2.9 − j 24.7	37.9 − j31.6
[15] $\cos \phi_1$	p.u.		0.72	0.12	0.77
[16] P_1	kW		−13.9	1.2	15.7
Output	kW		−16.0	0.0	13.6
Efficiency	p.u.		0.87	0	0.87

Condition		Max. torque	Standstill	Brake	Inf. slip
[1]	p.u.	0.130	1.0	2.0	∞
[3]	Ω	2.24 + j1.12	0.29 + j1.12	0.15 + j1.12	j1.12
[5]	Ω	2.43 + j2.24	0.48 + j2.24	0.34 + j2.24	0.19 + j2.24
[9]	A	92 − j184	38 − j178	33 − j182	16 − j185
[10]	kW	4.50	9.60	9.95	10.55
[11]	kW	35.1	9.60	5.00	0.0
[12]	kW	30.6	0.0	−4.95	−10.55
[13]	N-m	560	153	79	0
[14]	A	95 − j109	41 − j202	36 − j207	19 − j210
[15]	p.u.	0.66	0.20	0.17	0.09
[16]	kW	39.4	17.0	15.0	7.9
Output kW		30.5	0.0	0.0	0.0
Efficiency		0.78	0.0	−	−

There are divergences between the calculated figures in Examples 8.2 and 8.3 as a result of their different circuit bases. The stator and rotor current phasors again have a circular locus, but the magnetizing-current circle has shrunk to a single point.

8.7 *Current diagrams*

Current-locus diagrams for the induction machine are less common than they were before the advent of digital computation, but they are still useful for clarifying test results and for making comparisons of performance. Graphical niceties are out of place, however, because of the variability of impedance parameters, so that often a simplified diagram can give all that is necessary.

In general, r_1, x_1, r_2 and x_2 will be less for large than for small currents, with the reactances decreasing probably faster than the resistances. It may therefore be expected that the short-circuit current will be greater and of (probably) higher power factor than as estimated from values of r_1, x_1, r_2 and x_2 that are quite suitable for currents within the load range. It is evident that no single circle will be accurate over the whole range, and the diagram should be drawn with reference to the use to which it is put. If the full-load conditions are required, and maximum torque, the diagram should be drawn for characteristics applicable to the full-load current. It is not generally possible to do more than estimate the changes at very large short-circuit currents.

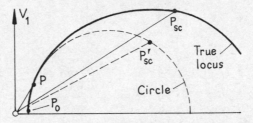

8.13 Practical stator-current locus.

That the true locus tends to an elliptical shape is shown by Fig. 8.13, drawn from test figures on a 415 V, 3-ph, 50 Hz, 6-pole, 1.2 kW machine with a short-circuit current small enough to be readily measured at normal voltage. The short-circuit test figures were:

Measured			Calculated			
Line voltage	Line current	Total power	Z_1	R_1	X_1	$\cos \phi$
V	A	kW	Ω	Ω	Ω	p.u.
63	2.5*	0.124	44	19.8	39.5	0.45
95	3.8	0.28	44	19.8	39.4	0.46
240	10.6	2.32	39	20.5	33.2	0.53
415	20.0	8.79	36	21.9	28.7	0.61

* Full-load current

From Fig. 8.13 it is clear that for finding currents in the normal load range the dotted circle, drawn from the no-load test and the short-circuit test at normal rated current, is adequate. But this circle will give erroneous results for starting current and torque.

'Exact' Current Locus Diagram

Using the theory developed in Sect. 8.5, the current locus can be drawn from a few known points, e.g. those marked in Fig. 8.12:

$$OP_{mo}: I_1 = I_{mo} = \frac{1}{c} \cdot \frac{V_1}{z_m} \quad \text{for the current at } s = 0;$$

$$OP_1, OP_{mm}, OP_{sc}, OP_\infty: I_1 = \frac{1}{c}\left[\frac{V_1}{z_m} + I_2\right]$$

where I_2 is given by eq.(8.11) with values of s corresponding respectively to some load (preferably full load), maximum torque, short-circuit $s = 1$, and infinite slip $s = \infty$. As all these points lie on a circle the centre is readily found by drawing a few chords.

The magnetizing current circle is very small. It can be drawn from the stator current circle by applying the analysis in Sect. 8.5, whence from eqs. (8.17) and (8.10)

$$I_1 = I_m + I_2 = I_{mo} + I_a = I_{mo} + I_2/c$$

For the phasor $OP_{sc} = I_{sc}$ of the stator current on short circuit. Fig. 8.14, $OP_{mo} = I_{mo}$ and $P_{mo}P_{sc} = I_a$. Draw a line $P_{sc}G$ of length cI_a and displaced in direction by the angle γ in c. Then $P_{sc}G = I_2$ and consequently OG must be I_m for the short-circuit condition. The same kind of construction can be used for any load point such as P_1, and as the angle γ applies in each case, all lines $P_{sc}G, P_1H \ldots$ must pass through the same focal point F on the stator current circle. Completing chords $FP_{mo}, P_{mo}H, HG \ldots$ enables the centre of the magnetizing current circle to be found.

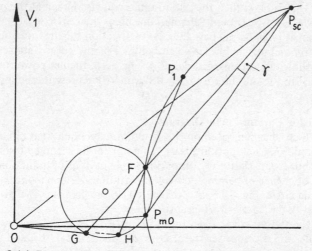

8.14 Geometrical construction for magnetizing current.

A short approximate method is to draw the line $P_{mo}P_{sc}$; bisect the angle $OP_{sc}P_{mo}$; and project the bisector to G, where the length $P_{sc}G = \sqrt{[(OP_{sc}) \cdot (P_{mo}P_{sc})]}$.

Diagram from Test Data

Although it is possible to obtain graphically a range of data from the current diagram, such information is limited in accuracy. A practical diagram therefore omits slip lines, loss lines and efficiency scales and aims to give a straightforward indication of full-load stator and rotor currents, full-load power factor, and pull-out torque, starting from test data or design figures. One of several such diagrams is given in Fig. 8.15, drawn to scale for one phase and normal rated voltage:

OP_0 $= I_0$, the no-load current at phase angle ϕ_0;

OA $= I_{0r}$, the reactive (pure magnetizing) component of I_0;

AP_0 $= I_{0a}$, the active component of I_0, associated with both stator core and mechanical losses per phase;

OB $= I_{sci} \simeq V_1/(x_1 + x_2')$, the 'ideal' short-circuit current;

AB = diameter of current-locus circle, of centre M;

OP_{sc} $= I_{sc}$, the stator short-circuit (standstill) current at phase angle $\phi_{sc} \simeq \arctan[(x_1 + x_2')/(r_1 + r_2')]$;

$P_{sc}D$ $= I_{sc} \cos \phi_{sc}$, the active component of I_{sc}, proportional to the short-circuit I^2R loss and divided at C in the proportion $P_{sc}C/CD = I_2'^2 r_2'/I_1^2 r_1$;

P_0C = torque line, extended to give the infinite-slip point P_∞;

P_0P_{sc} = output-power line.

Radii from M, perpendicular respectively to the torque and power lines, locate the points M_m and P_m: then the maximum torque is M_mG and the maximum output power is P_mN. Except for small machines the centre M of the circle can be taken as lying on the base line. A justification for the location of the output-power and torque lines is given below. For any given stator current $I_1 = OP$ as shown in the enlarged detail of Fig. 8.15, the net power output is proportional to PR and the torque to PS, while PT represents the input power.

'Approximate' Current Locus Diagram

A current locus diagram based on the 'approximate' equivalent circuit, with further simplifying assumptions, is of value not for its accuracy (which is doubtful) but for its clarity in comparing the behaviour of induction machines in respect of change of parameters and operation in various modes.

The main circle, Fig. 8.16, is constructed on the diameter P_0Y given by $V_1/(x_1 + x_2')$, the 'ideal' short-circuit current.

No-load: The stator active power input per phase is $P_0 = V_1I_0 \cos \phi_0 = I_{0a} = P_0A$ to scale, and it comprises stator core loss, mechanical loss, and a small rotor power to maintain rotation (taken as negligible). The reactive component

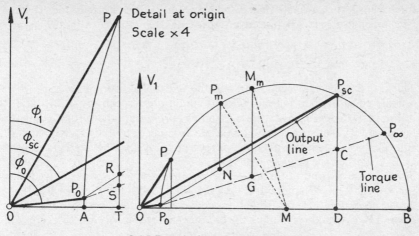

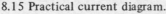

8.15 Practical current diagram.

of the stator no-load current is I_{or} = OA to scale, taken as constant. On load, the rotor speed falls and with it the mechanical loss; at the same time there is a rising rotor core loss. It is assumed that the sum of the mechanical and core losses is constant and that $P_0 = P_0 A = QT = KD = YB$ to scale.

Short-circuit: The stator current is I_{sc} = OP$_{sc}$ and the stator active power is $P_{sc} = V_1 I_{sc} \cos \phi_{sc} = P_{sc}D$ to scale, of which KD has been taken as the 'constant' loss. The remainder, $P_{sc}K$, must be the total stator and rotor I^2R loss per phase. Straight lines $P_0 P_{sc}$ and $P_0 K$ then contain the appropriate I^2R loss

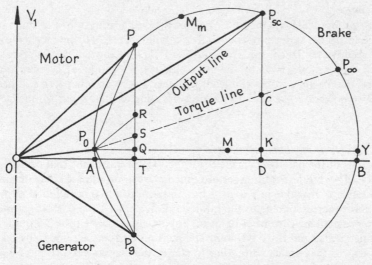

8.16 Approximate current diagram.

for any load, such as that represented by point P in the motor mode. The line PT, perpendicular to AB, has the intercept RQ: the ratio

$$\frac{RQ}{P_{sc}K} = \frac{P_0Q}{P_0K} = \frac{P_0P(P_0P/P_0Y)}{P_0P_{sc}(P_0P_{sc}/P_0Y)} = \frac{(P_0P)^2}{(P_0P_{sc})^2} = \frac{I_2{}^2}{I_{2sc}{}^2}$$

which, taking the equivalent rotor and the stator phase currents as very roughly equal for load and short-circuit conditions respectively, is the ratio of total I^2R losses for these cases. As active currents are the vertical components, and PT therefore represents the stator input active power, RQ the I^2R and QT the constant loss, then the output power is PR. The line P_0P_{sc} can then be labelled the *output-power line*.

Torque: The net output torque can be regarded as the net power output plus the rotor I^2R loss. The total I^2R loss, RQ for a load P and $P_{sc}K$ for short-circuit conditions, can be divided respectively in the same proportions RS/SQ and $P_{sc}C/CK$ corresponding to the rotor/stator I^2R loss ratio (or, approximately, the rotor/stator phase resistance ratio r_2'/r_1) by the line P_0C. Then the intercepts PS and $P_{sc}C$ represent torque (in synchronous power to the power scale), and the line P_0C can be called the *torque line*.

Infinite Slip: At $s = \infty$ there is no rotor current and consequently no rotor I^2R loss. Projecting P_0C to P_∞ therefore gives the corresponding condition. The part of the circle between P_{sc} and P_∞ then gives the brake mode.

Motor Mode: The circular locus between P_0 through P to P_{sc} is for operation of the machine as a motor. The stator current $I_1 = OP$ for a required output power PR or torque PS can then be located, and its power factor $\cos \phi_1$ found. The efficiency is PR/PT and the slip is RS/PS (but it would be idle to measure these from the diagram). The pull-out torque (or maximum output power) can be obtained from the point M_m (or P_m) where a radius from the centre M, drawn perpendicular to the torque line (or the power line), cuts the circumference.

Generator Mode: This is the lower part of the circle from P_0 to P_∞. For a load point P_g the stator power output as a generator is P_gT, the mechanical input to drive the machine is P_gR, and the I^2R and 'constant' losses are, as for the motor mode, given by the intercepts RS, SQ and QT. Regarded as an output, the machine works at a leading power factor: in other terms, it has an active power *output* but still requires a lagging reactive *input* for magnetization.

8.8 *Effects of machine parameters*
Variations of the parameters have the effects on performance that can be illustrated with reference to a set of comparative approximate circle diagrams. The 'standard' machine in Fig. 8.17 has, in terms of 1.0 p.u. full-load rating as a motor (corresponding to the intercept P_1R) and a 1.0 p.u. stator current OP_1, a magnetizing current of 0.3 p.u., a short-circuit current of 2.5 p.u., and equal stator and rotor I^2R loss. The per-unit currents are not to be taken as typical, as they are chosen rather to clarify the diagrams (i)–(iv) in Fig. 8.17. The lettering on all diagrams corresponds to that of Fig. 8.16.

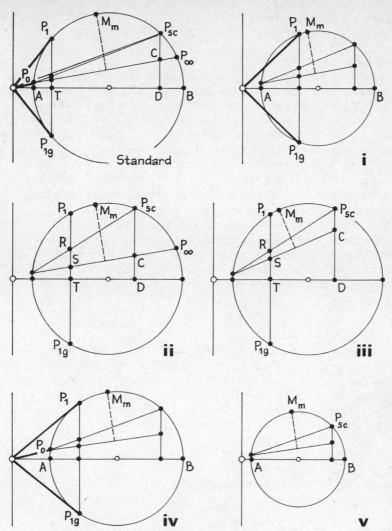

8.17 Effect of varying machine parameters.

Leakage Reactance

It is not necessary to consider the leakage reactances of stator and rotor separately. In diagram (i) the total reactance is about 1.3 times that of the 'standard' machine, and it becomes difficult to accommodate the required rated output because the diameter of the circle, **AB**, is reduced to about 0.75 of its original size. The pull-out torque is of the same order as the full-load torque so that the machine as a motor has little reserve for overload. Comparable restrictions are placed on the performance of the machine as a generator.

The leakage reactance plays a dominant role in the capability of the machine, but not very much can be done in design to reduce it.

Resistance

Increase of the *rotor* resistance is the most important practical variation, as it has a direct effect on the torque. Diagram (ii), for a rotor phase resistance twice that of the 'standard', shows a substantial increase in the torque at starting as a motor, but the full-load efficiency ($P_1 R/P_1 T$ for the motor mode, and $P_{1g}T/P_{1g}R$ for the generator mode) is naturally reduced. The brake mode is extended and the braking torque is increased. Use is made of rotor resistance in the torque control of slip-ring machines by connecting a rheostat across the slip-rings (which for normal running are short-circuited). If, however, the higher rotor resistance is built into the machine (as when deep-bar rotor conductors are used) there may be cooling difficulties.

Doubling the *stator* resistance, as in (iii), lowers the efficiency without improving the low-speed or brake torque. In the generator mode there is a substantial increase in the driving torque demand. The additional stator I^2R loss may cause difficulties in cooling unless the added resistance is external, an unusual arrangement producing low-speed current limiting as the only advantage.

Magnetizing Reactance

Although the teeth and cores of the magnetic structure are semi-saturated to economize in material, the airgap between the stator and rotor presents the major portion of the magnetic-circuit reluctance. As the machine has to be magnetized from the a.c. supply system, the gap length is kept short (of the order of 1 mm) compatible with the limitations imposed by mechanical clearance, shaft deflection and out-of-balance magnetic pull. The shaft must be very stiff to preserve centrality of the rotor with the stator.

A long airgap yields a diagram like that in (iv). The full-load stator current is increased and its power factor lowered. The approximate equivalent circuit obscures the fact that excessive magnetizing current restricts the diameter AB of the current circle.

Voltage

Voltage reduction seriously reduces performance. A common case is that of starting a motor by star-delta switch. Diagram (v) is drawn for this case, the working voltage per phase in star connection being $1/\sqrt{3}$ of normal, reducing currents by this factor and powers by the factor squared, i.e. to one-third of normal.

8.9   *Motor performance in the steady state*

On no load with the rotor windings short-circuited, the machines runs at a speed very close to the synchronous speed at which the stator travelling-wave field revolves. The rotor current is very small, because only the loss torque (for windage and friction) has to be developed. The stator takes a corresponding active current which, with the core-loss current, is the active no-load current component. The magnetizing current is much larger than the active current, so that the no-load power factor is low: e.g. 0.1 or (for small motors) 0.25.

If a load torque is applied to the shaft, the rotor tends to slow down. The rotor frequency is very low, the inductive reactance has little effect, and the rotor current is nearly in phase with the rotor e.m.f.: i.e. it is substantially a torque-producing current. The slip increases so as to produce enough rotor current and torque; and, very nearly, the load torque and slip are proportional. Thus both the slip/torque and speed/torque relations are rectilinear. The increase in rotor current necessitates a rise in the corresponding (active) component of stator current, and the power factor rises.

On full load the slip may be about 0.05 p.u. in a 2 kW motor down to around 0.01 p.u. in a 1 MW machine, and the full-load stator current is about three times the no-load value. The power factor rises to a value between 0.8 and 0.9. The efficiency, which is of course zero at no load, rises to a maximum at or (usually) below full load. This maximum occurs roughly where the I^2R losses in stator and rotor sum to an amount equal to the no-load 'constant' losses. At higher loads, the I^2R losses increase more rapidly than the output, so that the efficiency slowly falls.

For moderate overloads the stator current and slip increase roughly in proportion to the load, while the power factor, as the circle diagram indicates, remains sensibly unchanged. For larger overloads both current and slip increase

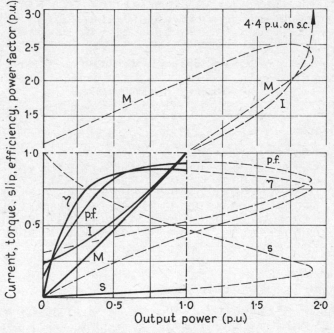

8.18 Motor performance curves.

more rapidly, until at maximum load (about 100% overload) the current may
be about three times its full-load figure, at a power factor in the neighbourhood of 0.7. At this stage a small increase in overload will cause the motor to
stop suddenly. The current will rise to e.g. five times rated full-load value, the
torque will be small, and the power factor lower, 0.2–0.4. The machine will
not restart until the load torque has been reduced below that available at starting. The rating of a motor must be reckoned in relation to its pull-out or stalling (maximum) torque, a usual relation of maximum to full-load torque
being 2:1.

A typical set of characteristic curves over the complete range is given in
Fig. 8.18.

The efficiency and power factor at rated load depend on the size and
application, and particularly on the nominal speed $n_1 = f/p$ (r/s) or 60 f/p
(r/min) for a machine with p pole-pairs and a supply of frequency f. Because
of less effective use of material and greater leakage inductance a low-speed
machine is heavier, less efficient and more reactive than a fast machine. The
figures below compare the data for a 5 kW machine at various speeds, and a
range of cage and slip-ring machines.

5 kW 50 Hz 3-ph Cage Motor

$2p =$	2	4	6	8	12	16
Efficiency p.u.	0.84	0.85	0.84	0.83	0.81	0.79
Power factor p.u.	0.90	0.86	0.81	0.74	0.72	0.70
Relative mass	0.85	1.0	1.3	1.6	2.4	4.4
Relative cost	0.98	1.0	1.25	1.45	2.0	3.1

Cage and Slip-ring 50 Hz Motors

Rating	Efficiency/power factor, p.u.					
	Cage motors			Slip-ring motors		
kW	$2p = 4$	8	12	4	8	12
1	0.72/0.75	—	—	—	—	—
5	0.84/0.86	0.83/0.74	0.81/0.70	0.82/0.83	0.81/0.67	0.79/0.66
10	0.86/0.87	0.84/0.78	0.82/0.71	0.84/0.84	0.83/0.74	0.80/0.73
20	0.88/0.89	0.85/0.83	0.83/0.75	0.87/0.88	0.85/0.80	0.82/0.77
50	0.90/0.90	0.89/0.85	0.87/0.82	0.89/0.90	0.88/0.83	0.86/0.81
100	0.91/0.92	0.90/0.89	0.89/0.84	0.91/0.92	0.89/0.87	0.88/0.83
1 000	0.93/0.94	0.93/0.93	0.92/0.91	0.93/0.94	0.92/0.91	0.91/0.90

Types and Applications
Cage Motors: The four main classes are:

1. Motors with normal starting torque and current, having cages with low
resistance and reactance, low full-load slip, good efficiency and power factor,
and high pull-out torque.

2. Motors with normal starting torque and low starting current, having large rotor reactance but the same slip and efficiency as (1). The power factor and pull-out torque are less.

3. Motors with high starting torque and low starting current, using deep-bar or double-cage rotors. A 2 p.u. starting torque with moderate current is obtained on full voltage. The efficiency, power factor and pull-out torque are lower than for (1). The motors are useful for starting against load (e.g. reciprocating compressors).

4. Motors with high full-load slip, using a comparatively high-resistance cage, with large starting torque, low starting current and low efficiency. Used for drives with heavy starting but light running duty. If employed for loads with rapidly fluctuating torques (e.g. punch presses) they may be attached to flywheels for load-peak equalization.

Slip-ring Motors. Suitable for heavy, frequent starting and accelerating duty-cycles. A high starting torque is obtained with a low starting current at high power factor. The rotor losses are mainly in the external resistance, easing the problem of rotor cooling.

Change-pole Motors. Cage machines with double or single stator windings to develop two (occasionally three) pole numbers and corresponding synchronous speeds.

Variable-speed Motors. Machines with additional equipment to give a wide and continuous speed range. Modern forms employ solid-state devices (thyristors or, for low ratings, transistors) to derive a variable-frequency motor input from a constant-frequency main supply by cycloconverter or rectifier/inverter circuits. Sophisticated electronic sensing and control systems can give accurate speed-holding and adjustment over a wide range, with braking and reversal.

Gear-motors are high-speed motors with integral unit gear construction giving output shaft speeds of 10–500 r/min. The motor is generally a 4-pole 1 500 r/min machine. If the service is heavy, the gear must be very robustly designed. Gear-motors may be applied to conveyors, agitators, low-speed fans, blowers and screens.

Special machines include: submersible motors, usually 2-pole machines with long small-diameter vertical rotors, the liquid passing through the rotor itself; motors with solid steel rotors, employed for intermittent starting of other machinery; 2-ph motors for servomechanisms in which one phase is supplied at constant voltage with the other phase provided with a quadrature voltage of varying magnitude and polarity; and selsyns for servo error-detection or control.

Linear motors utilize the induction principle in linear in place of rotary form, for short-throw actuators, crane traversing, and high-speed railway traction.

8.10 *Motor starting*

To accelerate a motor from rest to operating speed requires an input of energy to supply (i) the losses and (ii) the kinetic energy of the motor and its load.

The I^2R loss alone is equal to the stored kinetic energy, an important consideration in the frequent starting of high-inertia loads. In the design and choice of a drive motor it is necessary to take account of the frequency of the starting duty, the duration of each start, and the starting torque demanded. For large machines there may be limitations imposed by the supply authority on the magnitude of starting-current peaks.

The torque per phase is given by $M = I_2{}^2 r_2/s \simeq kI_1{}^2/s$. Comparing the torques M_n for rated current I_n and slip s_n and M_s for a starting current I_s at slip unity, the ratio is

$$M_s/M_n = (I_s/I_n)^2 s_n \qquad\qquad\qquad (8.22)$$

If a motor is started 'direct-on-line', i.e. switched on to a normal voltage supply, the starting current is $I_s = I_{sc}$. For a machine with $I_{sc} = 5\,I_n$ and $s_n = 0.04$ p.u., the starting torque is $M_s = 5^2 \times 0.04 = 1.0\,M_n$, the machine developing full-load torque at starting with five times full-load current.

The starting torque of the induction motor is increased by augmenting its natural rotor resistance. This is readily feasible with a slip-ring motor; the rotor winding of a plain cage machine, however, is not accessible, and although a high-resistance cage would improve the starting torque, there would be undue rotor I^2R loss on load. A starter may therefore either increase the rotor resistance, or act only to limit the starting current by some form of reduction of the voltage and, in consequence, also of the torque. The starting methods are the following. *Slip-ring:* external rotor resistance or impedance. *Cage:* direct-on-line, stator impedance, autotransformer, star/delta switch, clutch.

Slip-ring Motors

Maximum torque at starting is achieved if the rotor resistance per phase (including an external rheostat) is roughly equal to the total leakage reactance, and at the same time the stator current is reduced and its power factor raised. The slip-rings may be connected to a liquid resistor, an iron-cored inductor, or a metallic rheostat which is cut out in steps during run-up.

Liquid Resistor. In the arrangement of Fig. 8.36, the electrodes are raised or lowered to change the effective resistance. For small motors the electrodes can be fixed: the high initial starting current causes vapour to be formed at the electrode surfaces to give a high resistance, an effect that diminishes as the rotor current falls with rise in speed.

Inductor. For high rotor frequencies (as at starting or with 'plug' braking), eddy loss in the inductor core provides a high effective resistance, which diminishes as the frequency falls during run-up.

Rheostat. The starting current depends on the torque required and on the thermal rating of the rheostat. The lower the torque, the greater the number of resistance steps required, calling for more contactors. The slip-rings are usually short-circuited for normal-speed running.

Where the number of resistance steps is small, the starting rheostats may be mounted on the rotor shaft, and arranged to be cut out by a centrifugal switch. The use of slip-rings and external switchgear is avoided, but the device is complicated if frequent starting is required (and consequently much heat developed), and if more than two resistance steps are needed. A motor fitted with a centrifugal switch will start itself automatically if stalled.

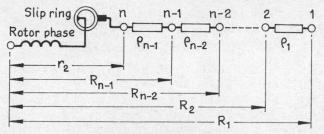

8.19 Rotor resistance starter.

Fig. 8.19 shows a rotor phase of resistance r_2 and one branch of the starter rheostat with sections of resistance $\rho_1, \rho_2 \ldots$ such that the total resistance on steps $1, 2 \ldots$ is $R_1, R_2 \ldots$ As each step is completed, the current becomes I_{max}, the speed rises reducing the current, and when a lower limit I_{min}, is reached a rheostat section ρ is cut out. Consequently, taking the upper and lower current limits to apply throughout the start, and the phase resistance at the start of a step to be R_p when the slip is s_p, falling to s_q just before the next step is made, we have

$$I_{max} = \frac{V_1{}'}{\sqrt{[(R_p/s_p)^2 + X^2]}} \quad \text{and} \quad I_{min} = \frac{V_1{}'}{\sqrt{[(R_p/s_q)^2 + X^2]}}$$

where X is the total motor leakage reactance and $V_1{}'$ is the stator applied voltage, both referred to the rotor. These expressions give

$$R_1/s_1 = R_2/s_2 = \ldots r_2/s_n \quad \text{and} \quad R_1/s_2 = R_2/s_3 = \ldots = R_{n-1}/s_n$$

taking the slip for normal running on the given load to be s_n. Combining the two sets of equalities and writing $s_1 = 1$ yields

$$s_2/s_1 = s_3/s_2 = \ldots = R_2/R_1 = R_3/R_2 = \ldots = r_2/R_{n-1} = \gamma$$

whence $R_1 = r_2/s_n, R_2 = \gamma R_1, R_3 = \gamma R_2$, etc. Further, $\rho_1 = R_1 - R_2, \rho_2 = R_2 - R_3 = \gamma \rho_1$ etc.

Fig. 8.20 shows a typical arrangement. Stator contactors are provided for main switching (with overcurrent protection) and for forward and reverse rotation by interchange of a pair of stator lines. Rotor rheostat sections are short-circuited by standard 3-pole contactors in a controlled sequence.

Limitation of the starting current to constant peaks is rarely applied in practice as it requires too many contactors. The number can be reduced by cutting sections from the rotor phases in sequence instead of simultaneously, provided that the current unbalances are not prolonged by a slow start.

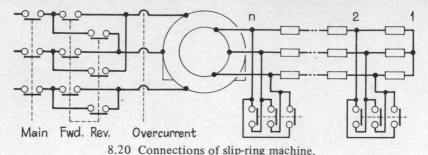

8.20 Connections of slip-ring machine.

The rotors of machines of like rating, but driving separate shafts, can be connected to a common rheostat to maintain equal speeds. If one rotor lags, its e.m.f. is phase-advanced and it takes a greater current. The action may fail for low slips because the circulating current, though large, has a low p.f. and may not develop an adequate alignment torque.

Rotor Voltage. The choice is arbitrary. For small machines there is little difficulty in keeping it below 500 V, an upper limit if starters and switchgear are hand-operated. With large machines it is necessary to limit the current to avoid difficulties with slip-rings, brushes and short-circuiting gear, and rotor voltages of 1—2 kV may be chosen.

Saturistor Control. If a 'hard' magnetic material such as Alnico is used as part of the core of an inductor excited by a coil carrying an alternating current, the effective resistance and inductance of the coil are low provided that the excursions of a.c. magnetic field intensity do not exceed the coercive force of the material. At higher excitations the hysteresis causes a marked rise in loss, and therefore in effective coil resistance and impedance. With a fully developed hysteresis loss, which for Alnico is typically 16 W/cm^3, further rise in the power dissipation is due mainly to the coil resistance, so that the overall effective resistance declines with increasing current. The impedance and power of such an inductor vary with the exciting current in a way indicated in Fig. 8.21(a). At maximum resistance the power factor of the exciting coil is about 0.75, and is independent of frequency because the hysteresis loss is itself proportional to the frequency. The construction of the inductor usually takes the form of Alnico blocks between normal laminated-steel yokes, as the hard material is most readily produced in simple shapes. The operating characteristics can (if needed in a closed-loop control system) be adjusted by superposing a d.c. excitation in an additional control winding.

Such an inductor, called a *saturistor* (or saturable resistor), can replace each of the star-connected branches of a starting rheostat across the slip-rings of an induction motor to provide automatically an almost constant starting torque of, say, 1.5 times full-load torque over the starting range of speed, with a current of less than twice normal, as shown in Fig. 8.21(b). The characteristics are almost ideal for several classes of mechanical load. A suitable quantity of Alnico, or other similar hard magnetic material, is included in each inductor in accordance with the effective resistance/current relation required. It is usually desirable to short-circuit the saturistor bank when the machine attains

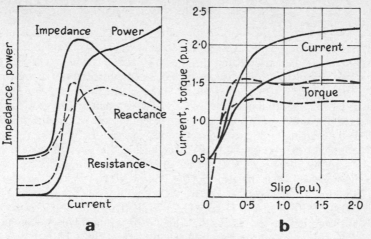

8.21 Saturistor control.

full working speed in order to limit the heating of the inductor windings
and to preserve the full pull-out torque level. The control must be arranged to
re-insert the saturistors during speed change and in the event of undervoltage
conditions, in order to limit the subsequent current inrush. If starting is likely
to be delayed, a heat sink must be provided to avoid excessive temperature-
rise of the Alnico; and if the machine is to run as a brake it may be necessary
to shunt the saturistors by a rheostat. If the saturistors can be mounted on the
rotor itself, elimination of slip-rings and brushes is made possible.

Cage Motors

As a cage rotor winding is electrically isolated, the starting current can (apart
from special devices) be reduced only by lowering the voltage applied to the
stator. If the voltage be reduced from normal to the fraction x, the no-load
and short-circuit currents will be changed roughly by the same proportion;
but the torque and output for a given slip become the fraction x^2 of normal,
as discussed in connection with diagram (v) of Fig. 8.17.

Direct Switching. A cage motor may be switched 'direct-on-line', i.e. on to
normal voltage, momentarily taking several times full-load current at a low
power factor. The motor may be damaged by overheating if the start is delayed by excessive load torque or supply-voltage depression. Direct switching
may be the subject of Supply Authority regulations, which may be expressed
as the permissible ratio of starting kVA to rated kW; the following are
typical:

Range of rating, kW:	1–6	6–40	40–250	250–500	500–1 500	1 500–4 000
Maximum ratio kVA/kW:	10	9	8	7.7	7.4	7.2

From eq.(8.22) the ratio of starting to full-load torque (at which the slip is s_n) is $M_s/M_n = (I_{sc}/I_n)^2 s_n$.

EXAMPLE 8.4: A 3-phase, 3.3 kV, 50 Hz, 10-pole, star-connected induction motor has a no-load magnetizing current of 40 A and a core loss of 30 kW. The stator and referred rotor standstill leakage impedances are respectively $(0.18 + j1.6)$ and $(0.40 + j1.6)$ Ω/ph. With the motor supplied from 3.3 kV busbars having a short-circuit level of 27 MVA, estimate (i) the gross torque, stator current and power factor when the machine runs with a slip of 0.03 p.u., and (ii) the starting torque with direct-on-line switching.

The rated phase voltage is 1.90 kV. The busbar short-circuit current level is $27 \times 10^3/3 \times 1.90 = 4.75$ kA, and the effective system impedance (assumed to be purely reactive) is therefore $1.90/4.75 = j0.40$ Ω.

(i) *Running*. At the slip $s = 0.03$ p.u., the total impedance is $Z_1 = [(0.18 + 0.40/0.03) + j(1.6 + 1.6)] = (13.5 + j3.2)$ Ω/ph. This is nearly 35 times as great as the system impedance so that the motor terminal voltage can be taken as 3.3 kV. Using the approximate equivalent circuit, Fig. 8.10(c):

$$I_2 = 1\,900/(13.5 + j3.2) = (134 - j32) = 138 \text{ A}$$
$$P_2 = 3 \times 138^2 \times 0.40/0.03 = 762 \text{ kW}$$
$$P_m = 762(1 - 0.03) = 740 \text{ kW}$$
$$M_e = 762/62.8 = 740/60.8 = 12.1 \text{ kN-m}$$
$$I_0 = (10/1.90) - j40 = (5 - j40) \text{ A}$$
$$I_1 = (5 - j40) + (134 - j32) = (139 - j70) = 156\, \angle - 27° \text{ A}$$
$$\text{p.f.} = \cos(-27°) = 0.89$$

(ii) *Starting*. The rotor current is large enough for the magnetizing branch current to be neglected. The series impedance of the motor is $Z_1 = [(0.18 + 0.40) + j(1.6 + 1.6)] = (0.58 + j3.2)$ Ω/ph; this is increased to $(0.58 + j3.6) = 3.65$ Ω/ph because of the system reactance. Then

$$I_2 = 1\,900/3.65 = 520 \text{ A}$$
$$M_s = 3 \times 520^2 \times 0.40/62.8 = 5.2 \text{ kN-m}$$

With zero system reactance the current would be 585 A and the torque 6.5 kN-m, an increase of 25%.

Stator Impedance Starting. The inclusion of a resistor or an inductive reactor in each of the stator input lines reduces the stator terminal voltage to a fraction x of normal, the fraction approaching unity as the speed of the motor approaches normal. The initial starting current is $I_s = xI_{sc}$ and the starting torque is $M_s = x^2 M_{sc}$. The ratio of starting to full-load torque is $M_s/M_n = x^2 (I_{sc}/I_n)^2 s_n$, or the fraction x^2 of that obtainable with direct switching. The method may be used for the smooth starting of small machines such as those driving centrifugal oil purifiers; but the star-delta switch may be cheaper and may give a better starting torque.

Auto-transformer Starting. Suppose that the auto-transformer in Fig. 8.22 is used to reduce the stator applied voltage to the fraction x of normal. Then the ratio of starting to full-load torque is $M_s/M_n = x^2 (I_{sc}/I_n)^2 s_n$ for a normal-voltage short-circuit current I_{sc}, the same as for the stator-impedance method.

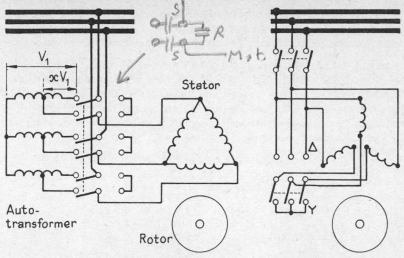

8.22 Autotransformer starter.   8.23 Star/delta starter.

But here the transformer primary current is x times the secondary current and therefore $x^2 I_{sc}$. For the same *line* current, the starting torque is $1/x$ times as great as that for the stator-impedance method. Further tappings can provide voltage steps during starting, the auto-transformer being switched out after the start is completed. The transformer needs only a short-time current rating but its leakage impedance should be small to avoid undue limitation of the starting current.

Star-delta Switching. For this method, Fig. 8.23, a motor must be built to run normally with a delta-connected stator winding. For starting the windings are connected in star, reducing the phase voltages to $1/\sqrt{3} = 0.58$ of normal. The starting current per phase is $I_s = 0.58 I_{sc}$, the line current is $0.33 I_{sc}$, the starting torque is 0.33 of its short-circuit value, and the ratio

$$M_s/M_n = \tfrac{1}{3}(I_{sc}/I_n)^2 \, s_n$$

The method is effective if the starting torque is adequate, and cheap so long as the supply voltage does not exceed about 3 kV.

Solid-state Starting. For loads with a rising torque/speed relation, a stepless start at a low voltage between 0.25 and 0.7 p.u. (i.e., a torque between 0.06 and 0.5 of the d.o.l. value) can be obtained by anti-parallel thyristors with phase control (Sect. 8.16) in each supply line, either (i) directly, to give voltage control by appropriate triggering, or (ii) across the secondary of a gapped-core inductor with its primary in the supply line, to give variable-impedance control.

Clutch Starting. A centrifugal clutch coupling has an inner part attached to the motor shaft and an outer part connected to the load. Spring-loaded blocks on the inner member come into contact with the outer member under centrifugal force to lock the driving and driven parts together. The blocks may be coated with friction-brake material, or may be metal running in oil. A comparable method utilizes a hydraulic torque-converter or 'fluid coupling'. Both methods prevent stalling and permit the motor to be rated on running rather than starting conditions, but involve additional cost and maintenance.

EXAMPLE 8.5: A small cage motor has a short-circuit current of 6 times full-load current, and a full-load slip of 0.05 p.u. Find in terms of normal values the motor and line starting currents and torque for (a) direct switching; (b) stator resistance and (c) auto-transformer starting with the motor current limited to 2.0 p.u.; and (d) star-delta starting. (e) What auto-transformer ratio gives full-load torque at starting?

The results are tabulated, using the various expressions already found:

Per-unit of normal value	(a)	(b)	(c)	(d)
Motor phase voltage	1.0	0.33	0.33	0.58
current	6.0	2.0	2.0	3.5
Line current	6.0	2.0	0.67	2.0
Torque	1.8	0.2	0.2	0.6

The torque is the same in (b) and (c) for the imposed condition of twice full-load motor current, but the *line* current for the auto-transformer start is only one-third of that for the rheostat.

(e) For full-load torque at starting, $M_s/M_n = (I_s/I_n)^2 \, s_n = 1$, from which $I_s = I_n/\sqrt{s_n} = 4.46 \, I_n = 0.75 \, I_{sc}$. Thus a 75% tap is required. The line current is $(0.75)^2 = 0.56$ times full-load current.

Starting-current peaks
In the foregoing the short-duration switching transients are ignored: they are discussed in Sect. 8.21. The 'steady-state' starting current does not fall appreciably until the motor has attained over two-thirds of normal speed, and it is desirable to delay the sequence of transitions to the normal running connection. Thus in Fig. 8.24, for star-delta starting, the transition should not be

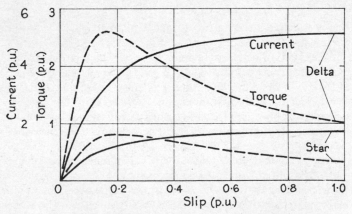

8.24 Torque and current in star/delta starting.

made until the speed is within 20% of synchronous; otherwise there will be
a peak line current considerably greater than full load — greater in fact than
the standstill current in star.

Any starting method that involves a momentary disconnection of the main
supply may cause severe current peaks. Stator excitation of the magnetic
field is withdrawn, but the short-circuited rotor develops induced currents
that hold the original field at the position on the rotor surface that it had
at the moment of interruption. This field, 'frozen' to the rotor and carried
round by it, induces e.m.f.s of reducing magnitude in the open-circuited stator
winding, and (because of the slip) of time phase anything between coincidence
and opposition to the supply voltages when they are re-applied in delta. In
the former condition there will be large current surges in all phases. In fast
switching, with interruption between 0.1 and 0.3 s, the star to delta transition should preferably be such that the phase winding connected across terminals RN of a supply of sequence RYB is switched in delta to RB; similarly
YN to YR and BN to BY. Then the applied line phasor voltages are retarded
by 30°, compensating to some extent the effects of rotor retardation. The
modified star-delta transition of Fig. 8.25 avoids current surges by retaining

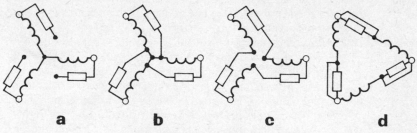

a   **b**   **c**   **d**

8.25 Modified star/delta switching.

the line connections without interruption: (*a*) the motor is started in star;
(*b*) resistors are paralleled with the phase windings leaving the motor unaffected; (*c*) the star point is opened putting the windings in delta with series resistance; and (*d*) the resistors are short-circuited.

Starters and Contactors
The functions of a 'starter' include stopping and possibly heavy-current
braking duty, and protection of the motor from overcurrent and undervoltage
conditions. Overcurrent relays are thermally operated by fixed heaters, associated with the individual phase currents, that deflect bimetal strips. Large
deflections, or those differing between the respective phases, operate to trip
the contactor for protection against sustained overcurrent or single-phasing,
and may initiate an alarm. Protection from excessive winding temperatures
is secured by thermistors embedded in the windings, one for each phase section. The thermistors have a nonlinear resistance/temperature characteristic
which is used to unbalance a bridge network if the winding temperature rises
excessively.

Contactors. Three standardized categories of a.c. contactor are concerned with motors: their designations (BS 775) are

AC4: for cage motors where starting, reversing, plugging and 'inching' duty is required and where very high peak currents may occur.
AC3: for cage motors involving starting and switching off with the motor still running, and so not normally required to break the starting or the stalling current.
AC2: for slip-ring motors in which current levels are limited.

In addition to its category designation, a contactor starter is given a class of intermittent duty which defines the number of operations per hour that it can deal with while still complying with the thermal, electrical and mechanical specifications.

8.11 *Harmonic effects*

With certain numbers of poles and of stator and rotor slots in cage motors, peculiar and deleterious behaviour may be observed when the machine is started. For example, with the number S_1 of stator slots equal to the number S_2 of rotor slots, the machine may refuse to start at all, a phenomenon termed *cogging*. With other ratios of S_2/S_1 the motor may exhibit a tendency to run stably at a sub-normal speed (e.g. one-seventh): this is called *crawling*. In some cases excessive *vibration* may be set up at sub-synchronous speeds, due to pulsations of mechanical force on the teeth, generating *noise* (e.g. a medium-pitched howl). The efficiency of a machine is inferior to that predicted by simple theory because of *load* (stray) *loss*. All these phenomena originate in the very complicated nature of the instantaneous gap flux distribution, which is strongly affected by the slot openings as these are usually wider than the length of the radial airgap. Slip-ring machines are similarly affected, but the essential difference in the present context is that a cage winding, being in effect a multiphase winding, can circulate a current under any harmonic e.m.f. produced by the gap flux, except that for which the wavelength is equal to the pitch of the bars.

In Sects. 8.4—8.8 the performance of an induction machine is discussed in terms of simple 'main' and 'leakage' fluxes, assuming the machine to be mechanically, magnetically and electrically balanced, with both time-varying and space-distributed quantities strictly sinusoidal. This is basically valid, but the second-order effects of present concern, which are of importance in design and performance, depend on the precise gap-flux distribution, i.e. on its harmonic content.

Gap-flux Harmonics

The m.m.f. of a uniformly distributed stator winding (Sect. 3.14) contains a sine-distributed fundamental and a family of space harmonic components of order $n = 6m \pm 1$, where m is a positive integer. The harmonics form waves of m.m.f. travelling at speeds $1/n$ of that of the fundamental, in the same direction if $n = 6m + 1$ (i.e. 7th, 13th), in reverse if $n = 6m - 1$ (e.g. 5th, 11th).

For a similar 3-ph rotor winding (also carrying sinusoidal phase currents) there is a comparable fundamental and harmonic family of m.m.f.s. Now the gap flux results from the stator and rotor m.m.f.s in combination, and as rotor movement continually alters the relative orientations of the stator and rotor phase windings, the resultant harmonic m.m.f.s, and the gap flux harmonics produced by them, change cyclically. A further effect is a cyclic change in the stator-rotor mutual inductance.

Such harmonics alone would have minor effect on a slip-ring machine, and even less with a cage machine. But they are greatly intensified by slotting, which not only introduces 'steps' in the m.m.f. wave, Fig. 3.29(b), and further harmonics, but also modulates the gap flux as, at a given radial plane, the rotor and stator teeth move in and out of register. High-frequency components, responsible for load loss and noise, are generated.

Binns and Schmid [111] list the following harmonic components developed in a cage machine. *Stator:* (i) stator m.m.f.; (ii) rotor m.m.f.; (iii) slot-ripple modulated by (i); (iv) saturation. *Rotor:* (v) slip-frequency currents due to the main-flux fundamental; (vi) components associated with (i); (vii) slot-ripple due to stator teeth; (viii) saturation; (ix) currents induced by harmonic fluxes. The chief loss-producing harmonics are those in (iii), but winding asymmetry and non-uniform gap-length are aggravating factors.

Supply-voltage Harmonics
The effects of non-sinusoidal supply voltages, encountered more particularly in inverter-fed machines, are discussed in Sect. 12.3.

Cogging and Crawling
If the stator and rotor slot numbers, S_1 and S_2, are equal or have an integral ratio, the slot-harmonic fluxes give rise to strong alignment forces when the machine is at rest, and these may exceed the tangential accelerating torque to cause the machine to *cog*.

Fifth- and seventh-order harmonic fluxes may be deemed as produced by sets of additional poles superimposed on the fundamental poles. They generate rotor e.m.f.s, currents and torques of the same general torque/slip shape as that of the fundamental, but related to 'synchronous' speeds respectively 1/5 backwards and 1/7 forwards, as indicated in Fig. 8.26(a). The resultant torque/speed curve, combined with the fundamental, shows a marked 'saddle'. If large, the 7th harmonic torque may prevent the speed from standstill rising above about 1/7 of normal. The motor therefore *crawls*. With certain slot ratios it is possible for a stator and a rotor m.m.f. harmonic to rotate together at some subsynchronous speed to develop a synchronizing torque at that speed, as in Fig. 8.26(b), again with a tendency to crawl.
Harmonic Induction Torques. In a slotted m-phase stator winding, the e.m.f. distribution factors of harmonics of order $n = 6Ag' \pm 1$ (with $A = 0, 1, 2 \ldots$ and g' slots per pole and phase) are the same as those for the fundamental. A 4-pole motor with 24 stator slots ($S_1 = 24$, $g_1' = 2$) will thus encourage 11th and 13th harmonics, and torque 'saddles' may be observable at +1/13 and

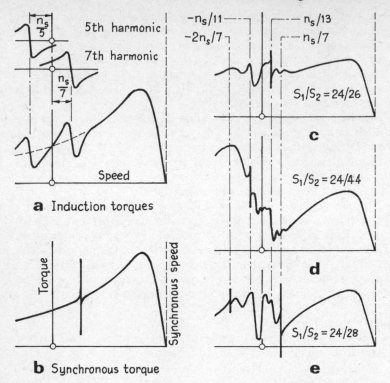

a Induction torques

b Synchronous torque

8.26 Harmonic torque effects.
    (*a*) Harmonic asynchronous (induction) torques.
    (*b*) Harmonic synchronous torques.
    (*c, d, e*) Torque/speed curves of 50 Hz 4-pole motor with various
        slottings.

$-1/11$ of the synchronous speed. If the half-wave-length of a stator-produced harmonic is equal to the rotor bar pitch, the harmonic torque is intensified: e.g. with $S_1 = 24, S_2 = 44$ the 11th harmonic has 44 half-waves in the gap each similarly located with respect to a rotor bar. The result is a strong 11th harmonic torque.

Harmonic Synchronous Torques. The rotor slotting produces space harmonics of order $(S_2/p) \pm 1$, the plus sign referring to travel in the direction of rotor rotation. Consider a 4-pole motor with $S_1 = 24, S_2 = 18$: the stator m.m.f. has a reversed 11th and a forward 13th, the rotor a reversed 13th and a forward 15th harmonic. For a fundamental synchronous speed n_s the 'synchronous' speed of the 13th harmonic is $+n_s/13$ and $-(n_s - n)/13$ relative to stator and rotor respectively when the speed of rotation is n. The two 13th harmonics are in step when $+n_s/13 = n - (n_s - n)/13$, i.e. when $n = n_s/7$. Thus there is a torque discontinuity at $n_s/7$ produced not by the 7th but by the 13th harmonics. The torque/speed curves in (*c*), (*d*) and (*e*) of Fig. 8.26 exhibit both induction and synchronous harmonic torques. The two are distinguished in that the operating speed changes slightly for the former but is constant for the latter, for a small variation of the shaft load.

Slotting. Alger [32] has analyzed the effects of slotting on the cogging and crawling phenomena in terms of gap-permeance waves, the permeance at any point in the gap being multiplied by the resultant m.m.f. there and the harmonic content extracted. The most inportant term is a function of rotor position angle θ and angular speed ω_r given by $\cos\left[(S_2 - S_1)\,\theta - S_2\,\omega_r t\right]$. Binns [33] criticizes the permeance-wave method as giving a poor approximation where m.m.f. and permeance harmonics of comparable order are concerned, because the flux paths across the gap, assumed by Alger to be radial, are not actually so. Binns uses an energy method based on zigzag leakage as well as gap permeance variation.

Many rules for slotting have been suggested, the recommendations depending to some extent on the analytical assumptions made. For example, cogging torques can be eliminated by making $(S_1 - S_2)/2p$ integral, S_1/S_2 fractional, and S_2 not widely different from S_1; asynchronous harmonic torques are limited if S_2 does not exceed $1.25\,S_1$; to limit synchronous harmonic torques, S_2 should not be $6px$ or $6px \pm 2p$, where x is a positive integer; slot harmonics are reduced if S_2 is not made equal to $S_1 \pm 2p$ or $S_1 \pm p$ or $\frac{1}{2}S_1 \pm p$. The cases illustrated in Fig. 8.26 abrogate one or other of these rules.

Noise

Noise results from mechanical vibration in a machine and its supporting structure, and directly from the coolant, especially if it stimulates natural structural frequencies.

Mechanical Forces. Dynamic out-of-balance of the rotor generates vibration at a frequency determined by the speed of rotation, and in 50 Hz machines with p pole-pairs will have a frequency of $50/p$, to which the ear is relatively insensitive. Bearings, particularly of the ball or roller types, produce noise by the rolling or sliding motion between the fixed and moving surfaces.

Magnetic Forces. These are important. The magnetic flux crossing the gap varies rapidly with time and relative position owing to the fluctuation of slot currents and tooth configurations, with the result that forces of magnetic attraction across the gap are subject to change and so generate vibration, especially in machines with few poles in which the variations are more concentrated and the shafts less stiff. Eccentricity of the airgap surfaces, whether of the stator or the rotor, can cause strong vibratory forces of magnetic origin. Of the wide frequency band of forces, those of low frequency are the more important in their ability to transfer vibration to the structure, while those in the range 100–10 000 Hz produce directly radiated noise. The teeth respond to varying forces by a 'cantilever' vibration. The core responds as a rather complex thick cylinder; a long core may develop vibrations as a cylinder with travelling waves of mechanical displacement around it.

Aerodynamic Noise. This results from rapid changes in pressure of the coolant gas which radiate directly into the gas stream or excite structural vibrations. The causes of pressure fluctuation are: (i) Variation in the resistance to gas-flow brought about by the movement of rotor slots and ducts past the stator gap surface. (ii) 'Air beating' by fan blades. (iii) Vortices in the wake of objects in the gas stream, which break up against other objects to develop aeolian tones. (iv) The flow of coolant through ducts having abrupt changes of cross-section, generating turbulence and random pressure fluctuations.

The methods of reducing noise, recommended by Ellison and Moore [34], are by reducing the known causes and by modifying coolant transmission paths. Noise of magnetic origin can be reduced only by lowering the gap flux density (which is uneconomic), but some mitigation is achieved by elimination of flux space harmonics. It is desirable to avoid any asymmetry, to keep the structural members stiff, to give careful attention to fan design, and to improve stream-line flow of the cooling gas (but this conflicts with the efficacy of thermal pick-up from the cooling surfaces). In some cases it may be helpful to employ anti-vibration mountings.

All parasitic fields are natural noise-producers. High tooth saturations, and zigzag leakage with unsuitable slot numbers, generate magnetic forces which move around the rotor as the slot openings fall in and out of register. It is usually considered that skewing of the rotor (or, occasionally, of the stator) slots can reduce magnetic force fluctuations and so reduce noise.

Skewed Slots

Some effects of skewing are discussed in Sect. 3.13, and Fig. 3.28 indicates how the aim of skewing − the reduction of slot harmonics − is achieved. However, there are some undesirable effects. The direction of the current in a rotor bar is no longer axial, but has a peripheral component proportional to the sine of the angle of skew which magnetizes the stator and rotor cores axially, superimposing a flux that disturbs the symmetry of the gap flux-density distribution and introducing some degree of further saturation. There are also the effects of increase in leakage reactance and stray loss.

It is shown by Binns and Dye [35] that the cogging torque of any cage machine can be made zero by a skew of a rotor slot-pitch or a multiple thereof, although this may not be the optimum value on other grounds. For example, the disturbance to the magnetic symmetry gives rise to a gap energy sensitive to rotor axial position, so that in consequence there is an axial force on the rotor of magnitude proportional to the fractional difference between the gap densities at the two ends of the machine. If the two densities are respectively B' and B'', and the mean density is B (with all densities in r.m.s.) the axial force is

$$f_a = \tfrac{1}{2}\pi D l_g (B' - B'') B / \mu_0$$

for a machine of rotor diameter D and radial gap length l_g. In a typical case B' and B'' may differ by 20% and the force may be some score newtons.

The beneficial effect of skewing can be seriously impaired as a consequence of imperfect insulation between rotor bars and the slots in which they lie, a problem of particular difficulty with cast-aluminium cages.

Load Losses

Friction, windage, core and I^2R losses can be estimated from no-load and

short-circuit tests (Sect. 8.22). The measured loss on load is invariably greater, by the amount of what is called the 'stray' load loss, made up chiefly of high-frequency tooth loss and harmonic rotor I^2R. Varying fluxes in any conducting and/or magnetic parts of the teeth, cores, windings, shaft and enclosing structure will generate loss, but the chief locations are the following:
Fundamental-frequency: (i) Eddy-currents in stator conductors. (ii) Eddy currents in the end regions. (iii) Displacement of stator and rotor fundamental m.m.f. waves produced by skewing.
High-frequency: (iv) Harmonics in the rotor due to stator m.m.f harmonics, and (v) in the stator due to rotor m.m.f. harmonics.

It is necessary to distinguish between *no-load* and *load* stray losses. The former occur as a result of permeance variations producing harmonic fluxes and tooth-pulsation effects due to relative tooth positions on no load; the latter are the additional (and more important) effects arising from leakage fluxes on load, which may reduce the full-load efficiency by 1% or considerably more. The stray load-loss is particularly evident if the slip is large; it may double the torque developed in the braking mode and lead to serious errors in the estimation of accelerating and braking performance.

Alger [32] and Christofides [36] have given analytic treatments of several forms of stray loss. The physical concepts are, however, far from simple, and the superposition of arbitrary 'components' of leakage flux is highly complicated by saturation.

Losses associated with self-leakage fluxes (i.e. not mutual to stator and rotor windings) contribute nothing to primary energy transfer and fall logically under the heading of stray loss. But gap leakage flux (in contrast to slot and overhang leakage) is partly mutual as it crosses the gap: components of gap leakage can and do produce torques, and cause saddles in the torque/speed relation. In this connection, the tooth-pulsation fluxes induce harmonic currents in a cage winding leading to I^2R losses dependent on the winding impedance, and these losses rise with increasing difference between stator and rotor slot numbers. Moreover, the impedance of the cage to inducing e.m.f.s depends on its impedance to the core in view of the phase relation of some of the e.m.f.s, especially with skewed slots. Consequently the use of a suitable slot ratio and the insulation of the bars in their slots can substantially reduce the loss. But cage bars are rarely insulated because of structural difficulties.

Several tests [37] have been devised for measuring stray loss, in particular the so-called *reversed-rotation* test (Sect. 8.22). The tests are not suitable for large machines, and in any case cannot precisely measure the losses sought. In respect of the declaration of efficiency in motor specifications, Schwarz [39] concludes that making a nominal allowance for stray load loss is at present the only reasonable and practicable method.

8.12 *High-torque cage motors*
Compared with a slip-ring machine of like rating, the cage motor has the following advantages:

(i) About one-half of the rotor conductor material and a smaller inherent I^2R loss.

(ii) Smaller rotor leakage reactance, and in consequence a better power factor and a higher pull-out torque (the diameter of the current circle depends on the reactance).

(iii) Higher efficiency, so long as the cage resistance does not have to be augmented to obtain adequate starting torque.

(iv) No slip-rings, brushgear, short-circuiting devices, rotor terminals or starting rheostats, and consequently a lower cost.

The disadvantages are almost entirely connected with starting. The most usual standard low-resistance cage machine gives 0.3–0.45 p.u. torque with 2.5–3.5 p.u. current on a start at 60% voltage. If the starting conditions are onerous, some form of high-torque cage machine may be suitable. The methods are: Deep-bar cages, using skin effect; or Double-cage construction.

Deep-bar Cage Machines
The behaviour of a deep bar of rectangular section is discussed in Sect. 3.2, giving eq.(3.6) and the characteristics in Fig. 3.3. But these are based on the assumption of infinite core and tooth permeability, a condition that does not exist at starting on normal voltage. Chalmers and Dodgson [8] discuss various attempts to take account of saturation (generally by taking a rectilinear B/H relation or even a fixed saturation density such as 2 T), and give a refined method of their own that can be programmed for computation. If saturation is neglected, the predicted short-circuit current, power factor and torque are liable to considerable error, with the current underestimated and the torque overestimated e.g. by 20%. With strong saturation in the region of the tooth tips, the leakage flux is less concentrated there, the bar current is consequently less constricted, and the effective bar resistance falls, producing the difference between predicted and actual results.

Some typical bar shapes are shown in Fig. 8.27: (i) is the simple rectangular bar requiring some means of retaining it, (ii) is the wedge bar, and (iii) and (iv) are versions of the T-bar which is preferred by some makers. The behaviour of the full-depth T-bar (iii) is less predictable than the recessed form (iv), largely because of the difficulty in assessing tooth-tip saturation. The bar (v) is fixed tightly by a steel key to provide rigidity.

The starting current being confined chiefly to the outer part of the bar,

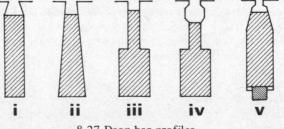

8.27 Deep-bar profiles.

this part rapidly heats, but the thermal conductivity spreads the heat to the lower part. The T-bar has the mechanical advantage that the shoulders keep the bar more tightly in the slot: vibration and thermal cycling (leading to expansion and contraction) can be very damaging to a slack bar and its insulation (if any).

Double-cage Machines
In these the rotor carries two concentric and separate cage windings, outer and inner, arranged in a slot arrangement that may take a variety of forms, Fig. 8.28. The outer cage has a high resistance together with the low reactance

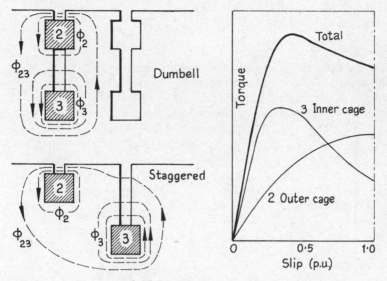

8.28 Double-cage machine.

of a normal single cage, while the inner cage has a low resistance and the high reactance obtained as a result of the long, narrow, slitted lip. The slits are necessary to prevent the inner cage from being 'missed' by the main flux, for if the inner bars were buried in the rotor iron, the gap flux would link only the outer cage.

By suitable choice of the cage resistances, the number and breadth of the slot openings and the depth of the inner cage, it is possible to control the shape of the torque/speed curve so as to obtain, within certain limits, any desired torque characteristic. To a first (and rather rough) approximation it is possible to consider the cages as producing quite separate torques, their sum giving the total torque of the motor, as in Fig. 8.28. A wide range of operational characteristics is obtained by suitably modifying the individual cage resistances and leakage reactances. This flexibility is achieved at some cost. In comparison with a plain cage motor, the pull-out torque is smaller because one cage produces its maximum torque at a speed different from that of the other cage. The high-resistance cage bears the burden of most of the loss during starting; with too frequent starts it may burn out. The additional inner-cage leakage lowers the full-load power factor, and the high outer-cage resistance increases the full-load loss and impairs the efficiency. However, as a normal single-cage machine with a small rotor I^2R loss could not develop the same starting torque, the energy efficiency of the double-cage motor may be greater for the given duty.

Diecast Rotors. The double-cage construction is sometimes used with diecast aluminium rotor windings. During the casting process the upper and lower parts 2 and 3 of the 'dumbell' are connected by an aluminium fillet which provides electrical contact between cages along their length. However, analysis and experiment both show that the assumption of separate cages is a reasonable approximation.

Performance

Taking the two cages as magnetically as well as electrically separate, the performance of a machine can be estimated by drawing two circle diagrams and adding corresponding rotor current phasors to obtain the effective total. If in Fig. 8.29 points P_2 and P_3 correspond to outer- and inner-cage currents for

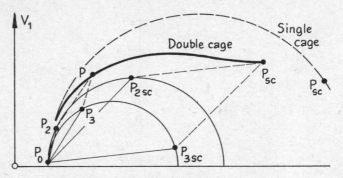

8.29 Current locus of double-cage motor.

a given slip, their phasor sum gives P, a point on the resultant current locus from P_0. As the outer cage has the lesser reactance its circle has the greater diameter; and the short-circuit point P_{2sc} is near to the crest of the circle since the high resistance contributes the major part of the starting torque. The inner-cage circle is smaller and its short-circuit point P_{3sc} is lower. As both the resistance and leakage reactance of the double cage is greater, the combined rotor current must always lie within the circle of the corresponding plain motor. The torque of the double-cage machine can be obtained by drawing torque lines for each circle separately, and combinining the two values for a given slip.

The allocation of resistance and leakage reactance to the two cages is intricate, and depends on the torque/speed characteristic required. The more nearly the inner cage approaches a normal single-cage construction, the bigger is its circular locus, as in (*a*) of Fig. 8.30, and the more nearly will the pull-out torque approach that of the plain single-cage motor; but at the same time the starting current will increase and the starting torque reduce as a result of the greater parallelling effect of the two cage resistances. At the other extreme is the use of a large inner-cage reactance and an adjustment of the outer-cage resistance to achieve a high starting torque, Fig. 8.30(*b*); but the pull-out

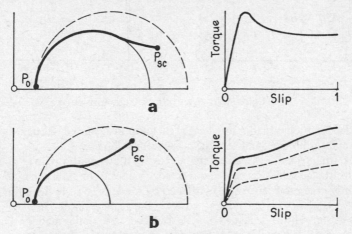

8.30 Current loci and torque/slip curves of double-cage motor.

torque is now but little greater than the full-load value. This characteristic, combined with stator rheostat control as indicated by the dotted torque/slip relations, is applicable to cranes. It gives a controllable hoisting torque together with inherent overspeed limitation for lowering.

The Table below gives a comparison between typical slip-ring and cage motors of comparable rating.

Per-unit values	Slip-ring	Single-cage	Double-cage* D	S
Efficiency, full-load	0.88	0.90	0.90	0.90
half-load	0.86	0.89	0.87	0.89
Power factor, full-load	0.85	0.92	0.86	0.88
half-load	0.67	0.82	0.76	0.82
Slip, full-load	0.04	0.028	0.021	0.022
Starting current, d.o.l.	–	6.0	5.6	5.5
star-delta	–	2.0	1.9	1.8
Starting torque, d.o.l.	–	1.3	2.1	2.1
star-delta	–	0.43	0.7	0.7

* Slotting: D, dumbell; S, staggered

EXAMPLE 8.6: A 22.5 kW, 400 V, 50 Hz, delta-connected double-cage motor with staggered slotting has the equivalent circuit per phase in Fig. 8.31 with

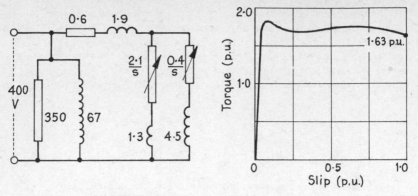

8.31 Equivalent circuit and torque/slip curves.

impedances in ohms and mutual cage reactance disregarded. Obtain the current locus and torque line, and plot the gross torque/slip curve.

Referring to Fig. 8.32, co-ordinate axes are drawn with origin at point O_{23}, with resistance values vertical and reactance values horizontal. The locus of the outer-cage impedance $z_2{}'$ is obtained by drawing to scale from the point for $s = 1$ (that is, $r_2{}' = 2.1\ \Omega$, $x_2{}' = 1.3\ \Omega$) the vertical line $z_2{}'$ marked off in terms of slip. The impedance locus $z_3{}'$ is drawn in a similar manner. Each impedance locus is now inverted about O_{23} to give the circular loci $y_2{}'$ and $y_3{}'$, with centres respectively at M_2 and M_3, a suitable common admittance scale being chosen to give a convenient size of circle. On the admittance loci, the s values are marked in by inversion from the impedance lines. Now, for each chosen slip, $y_2{}'$ is added to $y_3{}'$ to give the dotted locus $y_{23}{}'$ which, re-inverted about O_{23}, results in the locus of $z_{23}{}'$, the combined impedance of the inner and outer cages in parallel. The stator impedance is now added by shifting the origin by $0.6\ \Omega$ to scale downwards, and $1.9\ \Omega$ horizontally to the left, locating the new origin O_{123}. Impedances measured from O_{123} are for the whole machine excluding the magnetizing impedance. The corresponding admittance y_{123} is the inversion of $z_{23}{}'$ with respect to O_{123}. The final origin O is located from O_{123} by shifting to scale by $1/r_m$ and $1/x_m$. Then O is the origin of the terminal admittance, and the locus y_{123} becomes also the locus of the total stator current $I_1 = V_1 y_{123}$, where V_1 is the phase voltage. The *torque line* is found by calculating the stator I^2R loss at a number of points, and setting up an intercept from the horizontal base-line equal to scale to the active component of current corresponding thereto. Then, very nearly, the remainder of the active current (and the corresponding active power) is the rotor input and (to another scale) torque. Thus at standstill the stator current is 114 A, the loss per phase is $114^2 \times 0.6 = 7\ 800$ W, the active current component is $7\ 800/400 = 19.5$ A; and this, set off vertically beneath the unity-slip point, locates the point on the torque line. The torque/slip curve, Fig. 8.31, can be drawn by measurement from the stator current locus marked in slip values; it shows a comparatively uniform torque over the speed range during the starting operation averaging about 1.7 p.u. of full-load torque. At standstill,

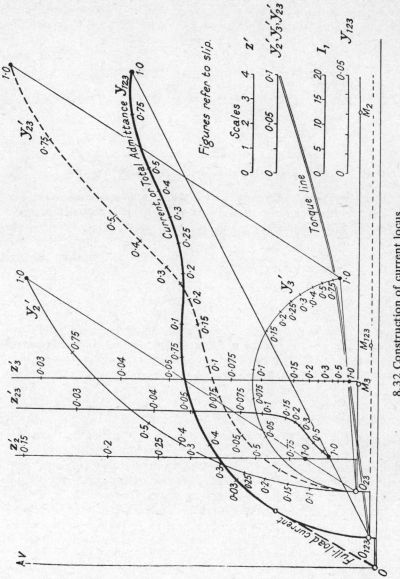

8.32 Construction of current locus.

1.63 p.u. torque is obtained with 4.75 p.u. current. The pull-out torque is about 1.8 p.u.

'Exact' Equivalent Circuit. From the leakage-flux patterns in Fig. 8.28 it is clear that the outer cage has a small flux Φ_2 and the inner cage a much greater flux Φ_3; and there is a common flux Φ_{23} endowing the two cages with mutual inductance. It may therefore not be valid to assume the cages to form separate rotor circuits, particularly if the rotor has dumbell slotting.

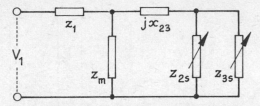

8.33 Double-cage motor with cage mutual reactance.

Modifying the equivalent circuit of Fig. 8.10(a) to accommodate two cages with mutual reactance gives the network in Fig. 8.33. The parallel branches z_{2s} and z_{3s} together give, at any slip s, a combined impedance $r_e/s + jx_e$ where

$$r_e = \frac{r_2 r_3 (r_2 + r_3) + s^2 (r_2 x_3{}^2 + r_3 x_2{}^2)}{(r_2 + r_3)^2 + s^2 (x_2 + x_3)^2} \tag{8.23}$$

$$x_e = \frac{r_2{}^2 x_3 + r_3{}^2 x_2 + s^2 x_2 x_3 (x_2 + x_3)}{(r_2 + r_3)^2 + s^2 (x_2 + x_3)^2} + x_{23} \tag{8.24}$$

These are rather complicated functions of slip, but if we take r_{e0} and x_{e0} as their values at $s = 0$ and define a ratio m for the increase in overall resistance of r_e divided by the decrease of x_e compared with r_{e0} and x_{e0}, for a given slip s, then

$$m = \frac{\Delta r}{\Delta x} = \frac{r_e - r_{e0}}{x_{e0} - x_e} = \frac{r_2 + r_3}{x_2 + x_3}$$

which is independent of slip and serves as a useful parameter for dealing with the allocation of cage impedances for securing the operating characteristics required. Alger [32] deals with several approximate analyses in terms of m, variations of which permit adjustment of the torque/slip curve, and of r_{e0} and x_{e0} on which the small-slip characteristics depend. Chalmers and Mulki [40] use the same parameters in a design-synthesis procedure in which the cage resistances and reactances are related to the shape of the torque/slip curve, the upper limits of short-circuit current and rotor $I^2 R$ (or rated slip), and the lower limits of starting and pullout torques and full-load power factor.

Eqs.(8.23) and (8.24) show that, for small slips, the double cage can be represented by a single cage with the approximate parameters

$$\frac{r_e}{s} = \frac{1}{s} \cdot \frac{r_2 r_3}{r_2 + r_3} \quad \text{and} \quad x_e = \frac{r_2{}^2 x_3}{(r_2 + r_3)^2} + x_{23}$$

in which x_2 has been neglected. The former is simply the parallel combination of r_2/s and r_3/s: these determine the full-speed characteristics. Calculation of the short-circuit conditions at $s = 1$ gives the starting torque, and inspection will show roughly the type of torque/slip curve to be expected.

8.13 *Motor speed control*
The plain induction motor in normal operation has inherently a nearly constant speed. Certain applications, however, call for two or more speeds, and in others a continuously variable speed range is a necessary feature. Two-speed motors, for example, are commonly required for boiler forced-draught fan drives; ships with a.c. propulsion need reversing and low-speed manoeuvring facilities; and cranes must have speeds adjusted to suit the loads to be lifted and lowered. In process plants such as paper-making and steel-rolling it may be vital to achieve closely accurate speed control of a number of associated motors to 0.01% or better, a requirement obtainable only by closed-loop control.

Basic Methods
The speed n of a motor with p pole-pairs, connected to a supply of frequency f and running with a slip s with respect to its synchronous speed $n_s = f/p$, is

$$n = (f/p)(1 - s)$$

showing that the three basic methods by which the speed can be varied are:
Slip: The slip s for a given torque can be varied in cage motors by changing the supply voltage, and in slip-ring motors by including external rheostats into the rotor circuits, or by applying to them slip-frequency voltages from auxiliary plant in order to convert the slip power into a useful form.
Poles. The synchronous speed f/p can be changed by altering the number of pole-pairs. Change-pole motors are specially wound for this purpose, usually for two speeds but occasionally for three or more.
Frequency. The speed can be varied over a wide range by stepless control of the supply frequency f. Solid-state frequency-changers are generally used to obtain the necessary supply.

Alternative Methods
There are several special methods of occasional usefulness. A 3-ph motor with a 1-ph rotor winding can be made to run stably at one-half of normal synchronous speed. A motor with stator and rotor fed at different frequencies can operate at synchronous speeds corresponding to the sum and difference frequencies. If the essential simplicity of the induction machine is given up, a wide speed range can be obtained by providing the rotor with a commutator.

8.14 *Slip control*
The efficacy of slip control by voltage reduction is affected by the torque/speed characteristic of the load. For a given slip, the motor torque is proportional to the square of the applied stator voltage. Fig. 8.34(*a*) shows typical motor characteristics for applied voltages of 1.0, 0.71 and 0.5 p.u., and three

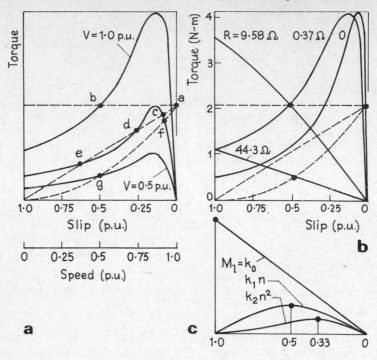

8.34 Slip control.

load torques M_l, viz (i) constant ($M_l = k_0$), (ii) proportional to speed ($M_l = k_1 n$) and (iii) proportional to speed-squared ($M_l = k_2 n^2$). Combining motor and load torque characteristics shows that for (i) there will usually be a single stable speed a and an unstable speed b, and that the motor cannot operate at all in either of the two low-voltage conditions. For (ii) there may be two stable speeds c and e, separated by an unstable speed d. For (iii) there is a stable speed f at 0.71 p.u. voltage, and an unstable one g for 0.5 p.u. Thus if the slip is to be controlled by voltage reduction, the motor must be suitably designed for its load and the voltage reduction properly chosen. Incidentally, it is possible to stabilize speeds b, d and g by providing a feedback loop control.

Slip Power. With substantial speed changes, high levels of slip power are concerned. Eqs.(8.3) and (8.4) show that a given torque M implies a corresponding active power input P_2 to the rotor regardless of speed, and that the fraction s of P_2 appears in the rotor circuits as slip power. With cage rotors the slip power is the rotor $I^2 R$ loss, but with slip-ring rotors most of the slip power is transferred to external devices. Let the torque demand at speed n be proportional to n^x; then the corresponding power is $P_m = k n^{x+1}$. The rotor input power is $P_2 = P_m/(1 - s)$ and the slip power is sP_2. In terms of the power demand $P_{mm} = k n_s^{x+1}$ at synchronous speed, the slip-power coefficient

$$g_s = sP_2/P_{mm} = (n/n_s)^{x+1} [s/(1 - s)] = s(1 - s)^x$$

expresses the relative power (including rotor internal I^2R) that has to be dealt with for a given slip. Putting $x = 0$, 1 and 2 gives respectively the three representative torque/speed characteristics (i), (ii) and (iii) in Fig. 8.34(a), and the corresponding slip powers are shown in (c). For (i) the slip power is proportional to slip, and comprises the whole rotor input at standstill. For (ii) the slip power is a maximum of $0.25P_2$ at $s = 0.5$; and for (iii), typical of a fan drive, the peak slip power is $0.15 P_2$ at $s = 0.33$.

Voltage Control

Stator voltage reduction by series resistors or inductors, though wasteful and current-dependent, can be applied to cage motors of rating 1 kW or less. Machines with star-delta starters may be run economically by switching to the star connection (reducing the phase voltage to 0.58 p.u. and the pull-out torque to one-third) provided that the load characteristic is suitable. For large machines a more sophisticated method is to adjust the stator voltage by interposing inverse parallel diodes with thyristors, or by bidirectional thyristors. Erlicki, Ben Uri and Wallach [41] describe the *burst-firing* method, i.e. the application of normal voltage for a few periods on and off in continuous succession, using thyristors for the switching function. Applied to the stator, burst-firing induces transients and increases loss, and is suitable only for small cage motors. It is more successful when applied to a slip-ring rotor, as the gap-field does not then collapse during the periods of quench.

Rotor Resistance Control

In the normal operating region the torque and slip are related to an approximation by $M = ks/r_2$, so that the slip for a given torque is proportional to the rotor-circuit resistance. Connection of external rheostats across the slip-rings (as for starting) therefore lowers the speed at the cost of a greater dissipation of slip power as heat, so that the overall efficiency must always be less than $(1 - s)$. Further, there is a greater change of speed with load. In practice the method is used for occasional minor speed reductions as in mine-ventilating fan drives, and also for 'load equalization'.

A simple analysis can be based on the approximate theory of Sect. 8.6 with the stator resistance r_1 neglected and the rotor-winding resistance r_2 augmented by an external rheostat R such that $R_2 = r_2 + R$. Then with all phase quantities referred to the stator, eq.(8.20) gives

$$M\omega_s = P_2 = V_1^2(R_2/s)/[(R_2/s)^2 + X_1^2]$$

for the active power input to the rotor. The maxima M_m and P_{2m} occur for a slip $s = s_m = R_2/X_1$ for which $M_m\omega_s = P_{2m} = V_1^2/2X_1$, whence for a given fraction q of maximum torque

$$q = \frac{M}{M_m} = \frac{P_2}{P_{2m}} = \frac{2X_1(R_2/s)}{(R_2/s)^2 + X_1^2}$$

Rearrangement yields a quadratic equation which solves to

$$R_2 = s(X_1/q)[1 \pm \sqrt{(1 - q^2)}]$$

Of the two values of R_2 only one is realizable. The other gives an unstable operating point on the negative-slope region of the torque/speed curve, or is less than r_2 implying that R is negative. Fig. 8.34(*b*) shows the method applied to the following Example.

EXAMPLE 8.7: A 375 kW 3.0 kV 50 Hz 3-phase 10-pole slip-ring induction motor with star/star connection has a rotor resistance $r_2 = 0.39\ \Omega$/ph and a total leakage reactance $X_1 = 5.75\ \Omega$/ph (both referred to the stator). The stator/rotor effective turns ratio is 4.65 and the full-load slip is 0.022 p.u. Ignoring stator resistance and mechanical loss, estimate the external rotor resistance, the slip power and the efficiency for a speed of one-half synchronous when the load has (i) constant full-load torque, (ii) a torque of full-load value at normal speed but is proportional to speed-squared.

Synchronous speed is 62.8 rad/s. Per phase, the full-load output is 125 kW, the full-load torque is $125/(62.8 \times 0.978) = 2.04$ kN-m and the pull-out torque is $1730^2/(2 \times 5.75 \times 62.8) = 4.15$ kN-m.

(i) *Constant Torque.* Here $q = 2.04/4.15 = 0.49$ and $q^2 = 0.24$. Then

$$R_2 = 0.50(5.75/0.49)\,[1 \pm \sqrt{(1 - 0.24)}] = 10.97 \text{ (or } 0.76)\ \Omega$$

Thus $R = R_2 - r_2 = 9.58\,(\text{or } 0.37)\ \Omega$ of which only the former gives a stable operating point. The slip power $sP_2 = 0.5(125/0.978) = 63.9$ kW/ph, of which 55.8 kW is dissipated in R. The mechanical output is $125(0.5/0.978) = 63.9$ kW, equal to the rotor slip power, and with other losses ignored the efficiency is 0.5 p.u. The actual resistance in rotor terms is found by dividing R by the square of the turns ratio, giving $9.58/4.65^2 = 0.44\ \Omega$.

(ii) *Torque proportional to speed-squared.* At one-half synchronous speed the torque is $2.04(0.50/0.978)^2 = 0.53$ kN-m; whence $q = 0.53/4.15 = 0.128$, and $q^2 = 0.016$. Then

$$R_2 = 44.7 \text{ (or } 0.18)\ \Omega \quad \text{and} \quad R = 44.3 \text{ (or } -0.21)\ \Omega$$

showing that the alternative solution is not possible. The slip power is 16.6 kW, nearly all dissipated in R, and the efficiency is again 0.5 p.u. In rotor terms $R = 2.05\ \Omega$.

The motor characteristics in Fig. 8.34(*b*) are marked with the stator equivalents of the external rotor resistance R, that marked $R = 0$ being for a short-circuited slip-ring condition.

Slip Regulator. A motor driving a load (such as a rolling mill) subject to sudden load peaks may be fitted with rotor rheostats and a shaft-mounted flywheel. On sudden load application the speed falls, part of the torque demand being met by release of kinetic energy from the flywheel. With fall of load the motor accelerates, restoring energy to the flywheel. The equalization obtained by a contactor-controlled variable rotor rheostat is shown in Fig. 8.35(*a*), which also shows the less effective result when the rheostat is fixed. A liquid rheostat, Fig. 8.36, may be used, the electrode position being set by a 'torque motor' which exerts a static torque proportional to the main motor current to increase the resistance (and so the slip) with rise of load.

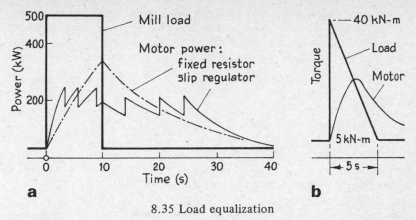

a

b

8.35 Load equalization

More elaborate methods employ a servo-motor with sensors responding to both torque and speed.

EXAMPLE 8.8: A 1 000 kW 50 Hz 10-pole induction motor developing full-load torque at 573 rev/min (60 rad/s) is provided with a flywheel, the total stored kinetic energy at 60 rad/s being 9.0 MJ. The fixed rotor external resistance is such that the slip for full-load torque is increased to 0.10 p.u. The load cycle, Fig. 8.35(b), comprises an idling torque M_0 = 5.0 kN-m, an instantaneous peak M_1 = 40.0 kN-m, and thereafter a uniform decrease to M_0 in 5.0 s. Find the motor torque/time relation.

Rated full-load torque is 16.7 kN-m. If this is developed at slip 0.10 p.u., and if the motor torque is assumed proportional to slip, then $M = ks$ where k = 167 kN-m per unit slip. With the load pulse applied at t = 0, the load torque is represented by $M_l = (40 - \alpha t)$ where α = 7.0 kN-m/s. The system

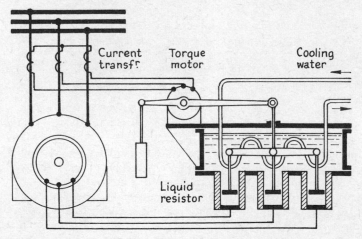

8.36 Slip regulator.

inertia is $J = 9.0 \times 10^6/\frac{1}{2} \times 60^2 = 5\,000$ kg-m^2 and the synchronous speed is $\omega_s = 62.8$ rad/s. The motor torque M balances the mechanical load and inertia torques M_l and $J(d\omega_r/dt)$ at any angular speed ω_r. Writing $\omega_r = \omega_s(1-s)$, then $d\omega_r/dt = -\omega_s(ds/dt) = -(\omega_s/k)(dM/dt)$, and the balance equation is

$$(dM/dt) + (k/J\omega_s)M = (k/J\omega_s)(M_1 - \alpha t)$$

which solves to

$$M = [M_1 - M_0 + \alpha(J\omega_s/k)][1 - \exp(-k/J\omega_s)t] + M_0 - \alpha t$$

$$= 52 - 49 \exp(-t/1.88) - 7t$$

over the interval $t = 0$ to $t = 5$ s. The peak motor torque occurs at $t = 2.45$ s, and the motor torque/time curve is given in Fig. 8.35(*b*).

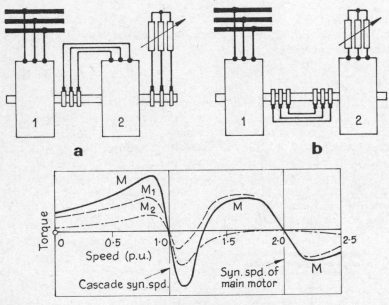

8.37 Cascade connection.

Cascading

This arrangement, Fig. 8.37, is one of the earliest developed to make effective use of *slip power*, which is supplied by the main motor to an auxiliary induction machine mechanically coupled, direct or through gearing, to the main-motor shaft. The slip power may be fed to the auxiliary stator as in (*a*), its rotor having slip-rings and starting rheostats or a plain cage winding. Alternatively the two rotors may be interconnected as in (*b*).

Let p_1, p_2 be the respective pole-pair numbers of the main and auxiliary motors, and f_1 the supply frequency: then the synchronous speed of the main motor is $n_1 = f_1/p_1$. Suppose n to be operating speed of the set on no load. In connection (*a*), the frequency in the main rotor and auxiliary stator is $f_2 =$

$(n_1 - n)p_1$ and the synchronous speed of the auxiliary machine is $n_2 = f_2/p_2$. Taking the no-load slip of the auxiliary rotor to be zero, then the set runs at $n = n_2 = n_1[p_1/(p_1 + p_2)]$. This is for unidirectional torques, but if the main and auxiliary torques oppose, the no-load speed is $n = n_1[p_1/(p_1 - p_2)]$. When the set is loaded, the speed drops fractionally. Electrical power to the main stator is passed (except for stator loss) to its rotor, where the power divides into (i) a part proportional to n converted into mechanical power, and (ii) the slip power proportional to $(n_1 - n)$ transferred to the auxiliary machine which employs it in mechanical power output and I^2R loss. The mechanical outputs are in the ratio $p_1 : p_2$. Fig. 8.37 shows a group of typical torque, power and current characteristics for a set with $p_1 = p_2$. Cascade operation has inherently low power factor, efficiency and pull-out torque, but it has been applied in a.c. ship-propulsion systems, where low manoeuvring speeds are important.

Variable-speed Slip-power-recovery Schemes
These enable slip-ring induction motors to operate at continuously variable speeds efficiently and economically. There are two groups:
Mechanical Recovery. Slip power from the slip rings is converted into mechanical power to supplement the output of the main motor.
Electrical Recovery. Slip power is frequency-converted and returned to the main supply.

Early examples of the former are those due to Kramer (1906), and of the latter the Scherbius (1906) and Clymer (1942) systems. All were based on the use of auxiliary machines (d.c. and a.c. commutator motors, synchronous motors and converters, variable-ratio transformers, etc.). In modern variable-speed sets, solid-state rectifiers and inverters can replace most of these.

Simplified examples of mechanical and electrical slip-power recovery are shown in Fig. 8.38. In both the control of slip is by field control of the d.c. machine. Decreasing the field reduces the armature e.m.f., allowing increased currents to flow in the slip-ring rotor, which accelerates to a speed determined by the mechanical load. A small difference between the slip-ring and d.c. armature voltages (referred back through the rectifier) then circulates the load current. In a comparable manner, the speed of the main motor falls if the d.c. excitation is raised. In the control characteristics shown, the operating speeds correspond at no load approximately to the intersections of the slip-ring and referred d.c. armature voltages: on load there is a speed drop of 0.05 p.u. or less. The minimum operating (or *base*) speed is shown in each case as about 0.5 p.u. of synchronous speed.

Electrical recovery applied to a constant-torque load implies that the power input to the induction machine is constant regardless of its speed (but most of the slip power is returned to the supply). With mechanical recovery, the torque contribution of the d.c. motor reduces the mechanical load taken by the induction motor: the set then takes a constant power input for a drive with torque inversely proportional to speed (i.e. constant mechanical power). In each case the efficiency is reasonably good, being lowest at base speed. The mechanical recovery is simpler, while the electrical method is well adapted

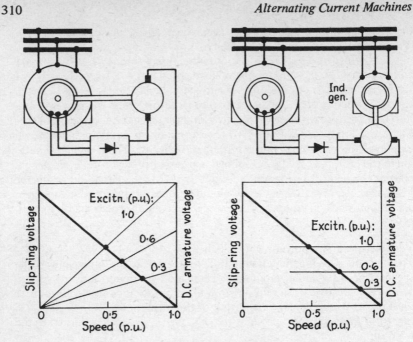

8.38 Slip-power recovery.

to high-speed drives unsuitable for a direct-coupled d.c. machine. Both can be used for driving centrifugal pumps (e.g. for sewage and drainage) and mine-ventilating fans.

The use of the d.c. machine is avoided in the modification in Fig. 8.39, where the slip-frequency currents are rectified in a diode bridge network; the

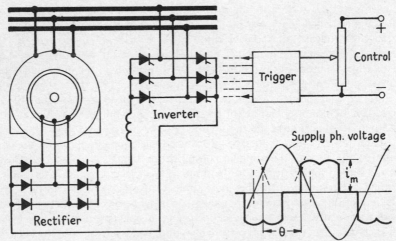

8.39 Slip-power recovery.

unidirectional output current is smoothed and passed on to a 3-ph line-commutated inverter at a rate depending on the supply voltage, the rectified direct voltage and the thyristor firing angle. The inverted current has a fixed wave-form and a constant conduction angle of $2\pi/3$ rad. The onset of conduction with respect to the phase-voltage zero is controlled by the firing angle θ. The harmonic content of the inverted current is substantially invariable: in terms of a peak current i_m of unity, the magnitudes of the fundamental, 5th, 7th, 11th and 13th harmonics are 1.05, 0.24, 0.12, 0.10 and 0.07. As power flow through the rectifier is unidirectional, only subsynchronous speeds are obtainable.

In rectifier/inverter recovery, part of the active power input to the motor passes through the rotor back to the supply, but the lagging reactive power is absorbed within the motor and rectifier. The bridge rectifier is, in effect, a loss-free lagging reactive power device, a frequency changer and a harmonic-current generator, feeding the rotor with reactive power to magnetize the gap at each harmonic frequency, while drawing from the rotor the equivalent in fundamental reactive power. The fundamental of the inverted current leads the phase voltage by an angle equal to the supplement of the firing angle and is therefore a source of leading reactive power (or a sink of lagging reactive power), and the harmonics cause 'distortion' power loss. As regards compensation, the provision of stator capacitors intensifies the harmonics, but capacitors across the rotor terminals can markedly decrease the leakage reactance of the motor and raise the peak torque.

The usual range of inverter firing angle for wide speed regulation down to standstill is $90°-160°$, allowing a safe interval for commutation. The ratio (active/reactive) power of the inverter varies from zero at a $90°$ firing angle corresponding to maximum speed up to 0.9 lagging for $160°$, the low-speed condition.

8.15 *Pole-changing*

A cage motor with two (or more) separate stator windings of different pole numbers can run at two (or more) integrally related speeds. Certain change-pole motors have a single winding, the connections of which can be changed to give different pole numbers; but separate windings, appropriately matched to the load conditions, may be preferred for such loads as lifts and hoists where a speed change must be made without losing control of the motor. Separate windings are, however, difficult to accommodate.

The coil pitch of a stator winding is naturally fixed, but its electrical span depends on the number of poles. A coil pitch one-eighth of the circumference provides full-pitch coils for an 8-pole connection, two-thirds for a 6-pole and one-half for a 4-pole connection. Too narrow a span is to be avoided, and for a pole-change winding for a 2/1 speed ratio a possible arrangement is a coil span of $1\frac{1}{3}$ pole-pitch for the larger and $\frac{2}{3}$ for the smaller pole number, in each case with a coil-span factor of 0.87. The spans may in practice be nearer to 1 and $\frac{1}{2}$ (coil-span factors 1.0 and 0.71) in order to avoid excessive leakage reactance in the lower-speed connection.

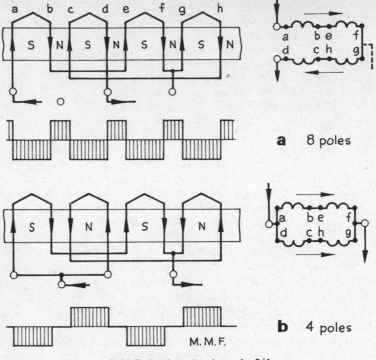

8.40 Pole-changing in ratio 2/1.

Pole Ratio 2/1

The essential method is given in Fig. 8.40, only one phase being shown. For the higher pole number the coils are in series (*a*). For the lower number (*b*) they are in series-parallel, giving the m.m.f. patterns corresponding respectively to 8 and 4 poles. It is common to use a double-layer winding, with a 120° spread for the higher number. An important design consideration is to limit space harmonics, which may produce a tendency to 'crawl'.

Pole Ratio 3/1

A method due to Rawcliffe and Jayawant [42] consists in making three sub-groups of each phase winding. Fig. 8.41 shows (*a*) two pole-pitches of a normal 3-ph winding, and (*b*) its subdivision into what amounts to nine phases. If subgroups a_1, $-b_3$, a_2 are connected in series they form a normal phase of a 2-pole machine (the other phases are b_1, $-c_3$, b_2 and c_1, $-a_3$, c_2). If now the subgroup a_1, a_2, a_3 is series connected it forms one phase of a 6-pole winding, the *b* and *c* subgroups being similarly series connected. In each case the windings have an electrical spread of 60° for all pole numbers. For the same supply voltage the 6-pole arrangement gives a rather low gap flux density, and to increase it one-third of each phase may be omitted in the 2-pole connection.

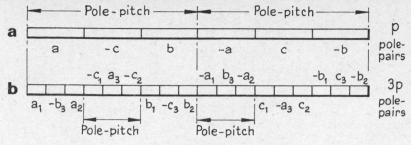

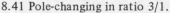

8.41 Pole-changing in ratio 3/1.

Operating Conditions

Referring throughout to the stator winding, let subscripts 1 and 2 imply conditions with the smaller and larger pole-numbers p_1 and p_2. The ratio of gap flux densities B and that of the pull-out torques M are

$$\frac{B_1}{B_2} = \frac{N_2 K_{w2} p_1}{N_1 K_{w1} p_2} \cdot \frac{V_1}{V_2} \quad \text{and} \quad \frac{M_1}{M_2} = \frac{(B_1 K_{w1})^2}{(B_2 K_{w2})^2} \cdot \frac{p_2}{p_1}$$

where N, K_w and V refer to the turns in series per phase, the winding factor and the phase voltage. The relative maximum output powers are $P_1/P_2 = (M_1/M_2)(p_2/p_1)$. Neglecting saturation, the magnetizing and short-circuit current ratios are

$$\frac{I_{m1}}{I_{m2}} = \frac{B_1 p_1}{B_2 p_2} \cdot \frac{N_2 K_{w2}}{N_1 K_{w1}} \quad \text{and} \quad \frac{I_{sc1}}{I_{sc2}} = \frac{N_2^{\,2}}{N_1^{\,2}} \cdot \frac{V_1}{V_2}$$

for the latter ignoring change in overhang leakage. If the same phase voltage is applied for each connection, then $V_2 = V_1$. The ratios are only approximate, the magnetizing-current ratio being 10–20% greater and the short-circuit current ratio 10–15% smaller than as given, because of the changed m.m.f. pattern and the different conditions of saturation.

The choice of winding arrangement depends on the load. Two cases are considered; others (e.g. for fan ventilation drives) may sometimes be more apposite.

Constant Torque: If the same torque range is required for both speeds (i.e. power proportional to speed), the windings are best utilized if

$$B_2 \simeq B_1 \quad \text{or} \quad N_2 K_{w2} V_1 / N_1 K_{w1} V_2 \simeq p_2/p_1$$

and the ratio of the pull-out torques is then

$$M_1/M_2 \simeq (K_{w1}/K_{w2})^2 (p_2/p_1)$$

Constant Power: For torque inversely proportional to speed, the ratio $B_1/B_2 = p_1/p_2$ gives the same overload capability on each speed and the same current density in the windings.

Fig. 8.42 shows one phase of a double-layer winding arranged for 2:1 pole-changing. The coil-span is ½ for p_1 (high speed) and full-pitch for p_2,

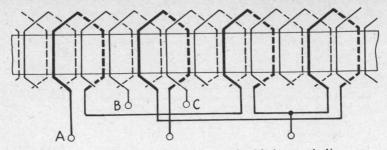

8.42 Pole-changing in ratio 2/1 with double-layer winding.

and the phase-spreads are inherently 60° and 120° (elec.). The winding factor
is $K_w = k_d k_e$, whence for p_1 it is $K_{w1} = 0.96 \times 0.71 = 0.68$, and for $p_2 = 2p_1$
it is $K_{w2} = 0.83 \times 1.0 = 0.83$. If the phase coil groups are in series, and the
terminal connection is delta for p_1 and star for p_2, then

$$N_1/N_2 = 1 \quad V_1/V_2 = 1/\sqrt 3 \quad B_1/B_2 = 0.71 \quad M_1/M_2 = 1.5$$
$$I_{m1}/I_{m2} = 0.65 \quad I_{sc1}/I_{sc2} = 1.7 \quad P_1/P_2 = 3.0$$

so that 'constant-torque' conditions are roughly achieved. For a 'constant-
power' condition the winding connections could be parallel-star/series-delta,
giving

$$N_1/N_2 = 1/2 \quad V_1/V_2 = 1/\sqrt 3 \quad B_1/B_2 = 0.71 \quad M_1/M_2 = 0.68$$
$$I_{m1}/I_{m2} = 0.9 \quad I_{sc1}/I_{sc2} = 1.7 \quad P_1/P_2 = 1.4$$

Common connections for $p_2 = 2p_1$ are parallel-star/star and parallel-star/delta,
which require only six external stator terminals and for which the flux
densities do not differ by more than 40 and 20% respectively. On pole-
changing the stator phase-sequence is reversed to preserve the same direction
of rotation.

Voltage Control. Combining this with pole-changing can give more speed
availability in certain cases. It can readily be inferred from Fig. 8.34(*a*) that
if the number of poles could be doubled (and the synchronous speed halved)
there may be several more running speeds.

Slip-ring Rotors. Cage rotors can respond to any stator pole number if properly designed, but slip-ring windings must be pole-changed to correspond to
the stator polarity. Windings can be devised to act as conventional windings
for one pole number (and so available for rheostat control) and as a permanent short-circuit to the other.

Pole-amplitude modulation

To have separate stator windings for pole-changing is wasteful of conductor
material, requires deeper slots and a larger frame size, and involves greater I^2R
loss and leakage reactance. The method of *pole-amplitude modulation* (so
called by analogy with the balanced modulator in telecommunications in
which the carrier wave is suppressed and there are two side-frequencies)

enables several pole-combinations to be economically obtained with a single winding and without increase in frame size. Modulation of the m.m.f. of a p-pole-pair gap field yields two superposed fields of pole-pairs $p_1 = p - k$ and $p_2 = p + k$, where k is an integer. Then by suitable 3-ph connections one of these is eliminated, leaving the other. The term 'modulation' refers to reversal or suppression of the current in parts of the winding. In Fig. 8.43, the m.m.f.

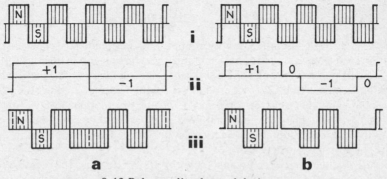

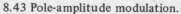

8.43 Pole-amplitude modulation.

in rectangular form of one phase of a 3-ph 8-pole winding is shown in (i), while (ii) represents two kinds of modulation. In (a) the modulation is in fact a reversal of current in the right-hand half of the winding; in (b) there is the additional feature of the suppression of the m.m.f. in certain poles (i.e. the relevant conductors are disconnected). The result is shown in (iii), and in each case the m.m.f. is a combination of 6-pole and 10-pole fields. Thus with $k = 1$ and the basic $p = 4$ for the unmodulated winding we obtain $p_1 = p - k = 3$ and $p_2 = p + k = 5$.

When these 1-ph patterns are allied to those of the other two phases, the resultant can be a 3-ph field of one or other pole-number; a given mechanical spacing has different electrical spacings for different pole-numbers, and can be made such as to favour one of the pole-numbers and suppress the other.

Let the angular phase spacing for the basic $2p$-pole winding be $r(2\pi/3)$ rad where r is any integer other than 3 or a multiple of 3. After modulation to $p \pm k$ the electrical spacing becomes $[(p \pm k)/p] r(2\pi/3)$. One of the modulation products is suppressed if its phase spacing is a multiple of 2π rad, say $2m\pi$; that is

$$\frac{p \pm k}{p} r \frac{2\pi}{3} = 2m\pi \quad \text{whence} \quad \frac{m}{r} = \frac{1}{3}\left(1 \pm \frac{k}{p}\right)$$

To suppress $p_1 = p - k$ it is necessary for $m/r = \frac{1}{3}(1 - k/p)$; the angular phase spacing for p_2 is derived from $(1 + k/p) = 2 - 3m/r$ so that

$$\frac{p + k}{p} r \frac{2\pi}{3} = \left(2 - \frac{3m}{r}\right) r \frac{2\pi}{3} = (2r - 3m) \frac{2\pi}{3}$$

So long as r is not a multiple of 3, the spacing is $\pm 2\pi/3$, $\pm 4\pi/3$, $\pm 6\pi/3$ ...
and so correct for developing p_2. Similarly, suppressing p_2 develops p_1. With
$r = 1, 2, 4, 5 \ldots$ and $m = 0, 1, 2 \ldots$ the numbers of pole-pairs obtainable for
pole-changing are

$$k = 1: \quad 1/2, \; 4/5, \; 7/8, \qquad \ldots (3m + 1)/(3m + 2)$$

$$k = 2: \quad 2/4, \; 5/7, \; 8/10 \qquad \ldots (3m + 2)/(3m + 4)$$

$$k = 3: \quad 3/6, 12/15, 21/24 \; \ldots (9m + 3)/(9m + 6)$$

Harmonics. A 3-ph m.m.f. has only an approximately sinusoidal space distribution, and when modulated by a rectangle (or shortened rectangle), itself
equivalent to a fundamental and a series of odd-order harmonics, the result is
rich in harmonic m.m.f.s. Each produces a component modulation and a corresponding number of space harmonics (some forward, others backward)
capable of developing 'spurious' torques. Such undesirables could be reduced
if sinusoidal modulation were practicable, but only an approximation in the
form of a stepped wave is feasible.

Overall Modulation. In the original paper by Rawcliffe et al. [43] simple reversal was the modulation method, considered in respect of the m.m.f. or current
distributions of individual phase windings. Later developments [44] dealt
with total modulation, i.e. the effect of applying a space modulation wave to
the current distribution of a complete winding. The modulation of a p-pole-
pair basic winding by a $(p + q)$-pole-pair wave is equivalent to modulation of
the phase windings individually by separate waves each of $(p - q)$ pole-pairs,
and conversely. The outcome was the development of a wide-ratio two-speed
single-winding machine as described by Fong [45]. Considerable ingenuity
has been expended on the technology of p.a.m. machines, which are commercially available for close or wide ratios of two or of three speeds, all
from a single basic 3-ph winding.

Generally, p.a.m. windings have coil groups comprising different numbers
of coils, or even different numbers of coil groups per phase. The design technique lies in the arrangement of the individual coils, the choice of the phase
to which each coil should belong, the relative current directions, and whether
these have to be reversed for pole-changing. The three component phase windings may each be unbalanced, but by the method of *symmetrization*, described
by Fong and Broadway [112], they may in combination be equivalent to a
balanced 3-ph winding. An algebraic-trigonometric logic has been devised to
simplify and extend the commercial range of p.a.m. motors for close or
wide ratios. Two pole-numbers, $2p_1$ and $2p_2$, are achieved by simple six-
terminal series/parallel switching. The phase windings with the terminals
A, B and C have intermediate tappings, a, b and c. For both star and delta
connection the pole-numbers are obtained as follows:

$2p_1$: supply to abc with ABC joined.
$2p_2$: supply to ABC with abc isolated.

Wide ratios such as 4/16 and 16/40 can be had, and 6/16 and 6/20 (but not
6/18). Wide ratios suit compressors, hoists and machine-tools.

8.16 Frequency control

Straightforward frequency control of propulsion-motor speeds is sometimes employed on shipboard, because the generator frequency and voltage are readily variable. Portable industrial tools such as drills and grinders may be fed from high-frequency sources (e.g. 110/220 V at 180 Hz and 125/250 V at 200 Hz) to drive 2-pole induction motors at 10 800 and 12 000 r/min having an active power/mass ratio as high as 1 kW/kg.

Efficient conversion plant has been developed to enable cage motors to be operated at speeds adjustable over a wide range, usually incorporating closed-loop control of the speed setting. In modern practice the frequency-changer is a solid-state device, able to develop the required frequency and voltage from a d.c. source or, more usually, from a fixed-frequency a.c. mains supply. Converters employ thyristors or, for low-power motor ratings, power transistors.

Static Frequency Conversion

Thyristors for frequency conversion operate in a switching mode in which they either conduct or 'block'. The thyristor trigger electrode (or 'gate') can be pulsed with a suitable voltage to initiate conduction, but thereafter loses control: it cannot cut off the current. Restoration to the blocking mode, a process of *commutation,* requires the thyristor current to be reduced to zero by some circuit property and a reverse bias to be applied between cathode and anode for several microseconds. If the thyristor is fed from an a.c. supply, the natural reversal of the alternating voltage each half-period can be exploited to secure both of these requirements, giving *natural* (or phase) *commutation.* This technique is applied in the cycloconverter. In the more important d.c.-link converter, a primary a.c. supply is first rectified to d.c., which is then inverted to give an a.c. output of controllable frequency. As the inverter thyristors are d.c. fed, natural commutation is not feasible. Auxiliary circuits for *forced commutation,* usually employing charged capacitors, must therefore be provided. For triggering the conduction mode, a single pulse of $150-300$ μs duration or a train of 25 μs pulses with a mark/space ratio of $1/5$ may be applied to the gate.

Cycloconverter

The simplest 3-ph/3-ph cycloconverter requires 18 thyristors to generate an output frequency necessarily lower than that of the primary supply, for if it exceeds about one-third of the input frequency the output waveform becomes intolerably distorted. The essential connections are given in Fig. 8.44. The thyristors are sequentially switched to 'fabricate' the output waveform from segments of the supply voltage, the switching rate and sequence being determined by the control circuitry (not shown in detail) which develops and distributes the low-power firing pulses. A typical waveform is illustrated for phase A showing the piecemeal development of a low secondary output frequency f_2 from a primary frequency f_1. The control of voltage is by an advance or delay of the firing pulses. Centre-tapped interphase inductors are required to

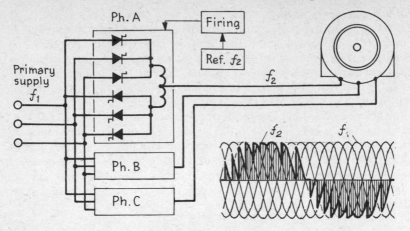

8.44   Cycloconverter.

smooth transfer from one thyristor to the next. The cycloconverter is able to
change the frequency in either direction so that it can be applied to the regenerative braking of induction motors; but generally its limited frequency
range restricts its use to traction (e.g. for developing $16\frac{2}{3}$ Hz from a 50 Hz
primary supply for railways with low-frequency 1-ph series-motor drives),
and to steel-rolling and comparable drives with many motors that must have
group speed control.

D.C.-link Converter
If the primary a.c. supply is first rectified in a diode or thyristor bridge and
the resulting direct voltage filtered, the d.c. supply can be fed to a static inverter to change it into a.c. (usually 3-ph) power at any other frequency within
a wide range (e.g. 5–150 Hz). The inverter power thyristors are triggered in
sequence so that a rectangular or stepped output waveform is generated. The
control circuity comprises an adjustable-frequency reference oscillator with
logic elements to form and distribute firing pulses in the proper sequence and
at the required rate, and as the inverter has a d.c. input its output frequency
limit is determined by the turn-on and forced-commutation turn-off times
of its constituent thyristors. For small equipments rated up to a few kilovolt-

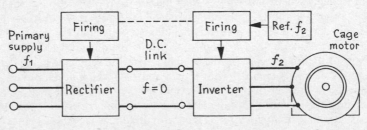

8.45   D.C. link converter.

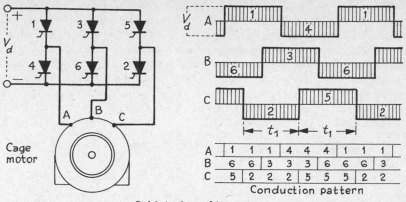

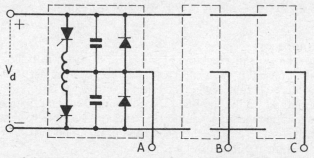

8.46 Action of inverter.

amperes, power transistors may be substituted for thyristors, with the advantages of higher switching rates, simpler commutation, and sinusoidal outputs at frequencies up to 400 Hz.

The frequency control of the speed of an induction motor generally involves also a concomitant control of its voltage, for in order to preserve a constant peak flux density in the gap and an invariable condition of core saturation, the voltage must be roughly proportional to the frequency. The d.c.-link arrangement in Fig. 8.45 may be realized in two main forms. In (i) the rectifier is a diode bridge developing an approximately constant link voltage; in (ii) a controlled thyristor rectifier supplies the link with a variable voltage. Details of some practical forms of d.c.-link converter are described in Sect. 12.2. In each case the output frequency to the motor is determined by the firing pattern of the inverter thyristors. More elaborate methods may be used if the speed accuracy required is such as to demand closed-loop control.

The basic inverter action can be seen from the simplified diagram in Fig. 8.46. Let thyristor 1 be on and 4 be off: this condition is maintained for a time interval t_1, at the end of which 1 is turned off and 4 on for a further time t_1. The two rectangular half-waves form a complete period of frequency.

8.47 Three-phase inverter with forced commutation.

1/2 t_1 supplied to phase A of the motor. Similar waves with appropriate timing are applied to phases B and C. The motor inductance delays current decay, so that the thyristors must be paralleled by diodes as in Fig. 8.47, which also shows the forced-commutation capacitors. When accelerating the motor from rest, the inverter operates at minimum frequency (e.g. 5 Hz) initially, the frequency being raised at a rate appropriate to the inertial and load torque demand by feedback control of the reference oscillator frequency, which in turn controls the firing pulse timing. Fig. 8.48(*a*) shows typical oscillograms of the motor line voltage and current.

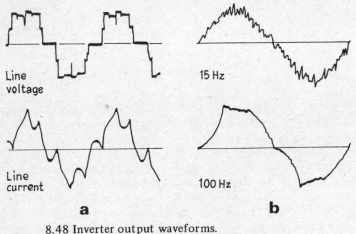

8.48 Inverter output waveforms.
(*a*) Thyristor inverter   (*b*) Chopped d.c.

For small motors it is possible to obtain a near-sinusoidal inverter output from a d.c. link by commutating each phase of the machine rapidly across the positive and negative terminals, the ratio of the durations of the opposing-polarity pulses determining the voltage level. Fig. 8.48(*b*) gives typical waveforms for lower and higher speeds. In the latter case the limit of sine-voltage forming by the chopping process is exceeded, and the supply unit then operates with the thyristors switched only once per period.

Motor Performance
Harmonics. The switching function of a diode or thyristor rectifier distorts the current waveform from the primary a.c. supply system, particularly if the system impedance is appreciable. The effect is to develop disturbing voltage dips and to impose on the current a strong harmonic content. With installations of large rating the supply authority may require supply to be taken at a higher voltage (e.g. 33 kV), or the installation of harmonic filters, or the use of a 24-pulse system of rectification.

The harmonic content of the output of a frequency-changer adds to motor losses. Compared with sinusoidal operation, the losses in a cage motor fed from an inverter are augmented by: (i) I^2R arising from harmonic currents and by increased skin effect in windings; (ii) high-frequency core loss; (iii) loss due to skew leakage flux, which may be large at harmonic frequencies; (iv) loss in the end windings; (v) load (stray) loss arising from space-harmonic m.m.f.s excited by time-harmonic currents. Investigations of the additional losses have been undertaken by Chalmers and Sarkar [46] and by McLean [70].

Harmonic voltages, and the currents produced by them in the frequency-dependent motor leakage impedance, develop forward- or backward-directed harmonic torques due to the interaction at various speeds of pairs of gap-flux and current components. With k harmonic currents forming travelling-wave fields, there are k^2 harmonic torques, comprising one fundamental oscillating torque, $(k - 1)$ forward or backward torques, and $(k^2 - k)$ hunting torques. According to Largiarder [47] the last-named may be an appreciable fraction (e.g. $7\frac{1}{2}\%$) of the main driving torque. Hunting occurs when gap-flux harmonics and rotor currents, moving at different peripheral speeds, interact to generate a torque oscillation that makes no *mean* contribution to the useful torque. At the low-speed end of the control range the motor may exhibit a hunting torque modulation at six times working frequency.

Speed. Inverter drives give a wide range of operating speed, continuously variable from standstill to 9 000 r/min for a 2-pole motor. Maximum speeds of 4 500 r/min or less are more economically provided if the motor is a 4- or 6-pole machine. The speeds quoted are for a maximum inverter output frequency of 150 Hz; higher values can be obtained if the inverter is equipped with fast-switching thyristors. The parameters of interest in control are the rotor resistance and the load inertia. Lawrenson and Stephenson [48] show that when a machine is reversed by dropping the inverter frequency, the speed falls more slowly than the instantaneous synchronous speed: the slip is therefore negative and the machine regenerates. Inverter-fed motors running at constant speed are analysed by Ward [49] using 2-reaction methods.

Constant-flux Operation. Eq.(3.19) shows that the phase e.m.f. of an a.c. machine is proportional to the product of flux and frequency. In variable-frequency speed control it is necessary to avoid excessive magnetic circuit saturation on low frequencies by maintaining a phase voltage approximately proportional to the operating frequency. With such a relation, a motor can provide full rated torque over the whole speed range. Let the e.m.f., slip and leakage reactance be respectively E_0, s_0 and x_0 at the 'normal' angular frequency ω_0, and E, s and $x = x_0(\omega/\omega_0)$ for a different frequency ω. Then with $E/\omega = E_0/\omega_0$ to preserve a constant gap-flux density, it is shown in Sect. 12.3 that the gross torques are given respectively by

$$K \frac{\omega_0 s_0 r}{r^2 + s_0^2 x_0^2} \quad \text{and} \quad K \frac{\omega s r}{r^2 + s^2 [x_0(\omega/\omega_0)]^2}$$

These torques are the same, and take the same rotor current, if $s\omega = s_0\omega_0$.

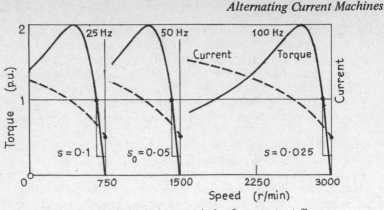

8.49   Torque/speed characteristics for constant flux.

The torque/speed and current/speed relations for any operating speed are therefore closely related, as indicated in Fig. 8.49 for a motor operated on 25, 50 and 100 Hz. In this simple analysis no account is taken of changes in mechanical and core loss, or in effective resistance, or in magnetizing current or saturation, resulting from the change in frequency. To obtain a constant E/ω it is necessary to relate the motor voltage V to the operating frequency f by $V = a + bf$.

8.17 *Power-factor adjustment*
Because of its a.c. excitation, the full-load p.f. of an induction motor rarely exceeds 0.9, and to avoid tariff penalties it may pay to raise the p.f., by (i) connecting capacitors across the stator terminals or (ii) using the synchronous-induction machine. Formerly (iii) phase-advancers or (iv) special phase-compensated machines were employed. The broad effects of these methods are shown in Fig. 8.50 in terms of the modifications that they impose on the normal circle diagram. In (i) the circle is shifted bodily to the left by the amount of the constant leading current of the capacitor; in (ii) the normal operation is synchronous, the machine having a d.c. rotor excitation derived from a rotating exciter or a static rectifier; in (iii) and (iv) the operation is modified by injecting a capacitance effect into the rotor circuit or into the magnetizing branch of the equivalent circuit.

Static Capacitors
If the motor operates at full load with input active, reactive and apparent powers P, Q and S, and therefore at a power factor $\cos \phi = P/S$, the raising of the power factor to $\cos \phi'$ (for which $Q' = P \tan \phi'$) requires a leading reactive power $Q - Q' = P(\tan \phi - \tan \phi')$ and the capacitance required is $C = (Q - Q')/\omega V_1^2$. The capacitor may be made smaller by raising its operating voltage, e.g. by an autotransformer, if the motor supply voltage is low. Where adequate compensation is provided for full-load power-factor improvement, the system is overcompensated (though not, as a rule, significantly) at low levels of load. The switching requires care to avoid the possibility of momen-

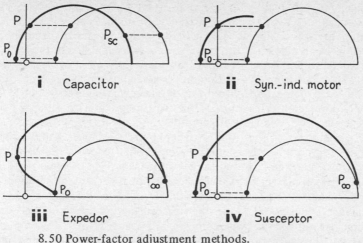

i Capacitor   **ii** Syn.-ind. motor

iii Expedor   **iv** Susceptor

8.50 Power-factor adjustment methods.
(i) Capacitor. (ii) Synchronous-induction motor.
(iii) Expedor. (iv) Susceptor.

tary self-excitation of the motor as an induction generator, a condition that might produce a stator overvoltage condition. The capacitance of the phase-advancing capacitor bank may have to be limited to a value that makes self-excitation impossible.

Synchronous-induction motor
This machine is described in Chap. 10. A noteworthy feature is the large leading current taken at small loads, and the machine can be useful for compensating the power factor of a group of uncompensated motors.

Phase-advancers
A phase-advancer is a device connected into the rotor circuit of the induction motor, but mechanically external to it. The two basic devices can be classified as: (i) *Expedor*, developing a voltage that is a function of the rotor current and has some phase relation thereto; it generates or absorbs an e.m.f., equivalent to an impedance rise or fall. (ii) *Susceptor*, developing a voltage that is a function of the rotor open-circuit e.m.f. and has some phase relation thereto; it affects the magnetizing current of the motor, and consequently its effective magnetic-circuit susceptance.

In either case, the basic method consists in so adding to the rotor induced e.m.f. that the rotor current is advanced in phase, and the consequent m.m.f. reaction causes the stator current to be correspondingly advanced. The method requires the use of a slip-ring rotor and the injection thereby of voltages of suitable magnitude, phase-displacement and slip frequency from a separate phase-advancer machine. The principal types of advancer are
Expedor: Kapp vibrator, Leblanc advancer, Scherbius advancer.
Susceptor: frequency converter.

Kapp Vibrator. A small d.c. armature is connected in series with each slip-ring connection to carry the slip-frequency rotor phase current. The d.c. armature develops a slip-frequency vibratory torque with a permanent or separately-excited field system, and an e.m.f. corresponding to the instantaneous armature speed. On account of the inertia, the e.m.f. has a 'capacitive' effect. The vibrator must be cut out and replaced by a rheostat for starting.

Leblanc and Scherbius Advancers. Slip-frequency currents are supplied to a 3-phase commutator armature. If this armature is at rest, a rotating field of slip frequency is developed and a counter-e.m.f. of slip frequency appears at the commutator brushes. If now the armature is rotated by an auxiliary motor at a speed above that corresponding to slip frequency it develops (still at slip frequency) e.m.f.s proportional to the main rotor currents and of leading phase-displacement. The advancer is not applied to the slip-rings until the main motor has been started and is up to its operating speed.

Frequency Converter. This takes the form of an armature with d.c. type winding, commutator and slip-rings, connected mechanically to the main motor. Three-phase brushgear on the commutator is connected to the main-motor rotor circuit, and an e.m.f. of slip frequency is injected by energizing the advancer armature through its slip-rings from the main a.c. supply. Rotation of the armature at the speed of the main motor results in slip-frequency e.m.f.s being tapped off the commutator, of phase dependent on the brush-rocker position.

Successful attempts have been made to incorporate phase-advancing devices into small induction motors to avoid the complication of the additional machine. These, however, as well as the types briefly described above, are no longer in common use.

8.18   *Braking*

Simple friction brakes may serve for hoists, lifts and cranes. Electrical methods are adopted for sophisticated drives, especially where precision in braking time or load positioning is called for, but they may impose severe duty cycles and involve both thermal and mechanical stresses.

Regenerative Braking

The motor, overdriven by the load into a negative-slip hypersynchronous speed, becomes an induction generator. In a slip-ring machine, the braking can be maintained by a moderate *rotor* external resistance and still return useful energy to the supply. Braking fails when the speed falls to synchronous speed. The braking range is, however, extended if the stator has a pole-change winding, in which case the rotor winding is normally of the cage type.

Plugging

Braking is obtained by reversal of the stator connections while the motor is running, so reversing the direction of the travelling-wave airgap field. The slip is then greater than unity and the machine develops a braking (i.e. a reversed) torque, corresponding to the region between P_{sc} and P_∞ in the circle diagram of Fig. 8.16. The stator and rotor currents are large. Cage motors of about 20 kW are plugged direct, using the star connection if a star-delta switch is provided; larger motors require stator resistors. Slip-ring motors employ rotor resistance for current limitation. Problems of thermal rating may arise for frequent braking duty, and contactors may need frequent servicing.

Insofar as torque transients (Sect. 8.20) can be ignored, or have a duration small compared with the braking time, the essentials of plugging can be based on the steady-state torque/speed characteristics of the motor and its load, Fig. 8.51, together with the inertia of the drive. In the diagram, (*a*) shows two

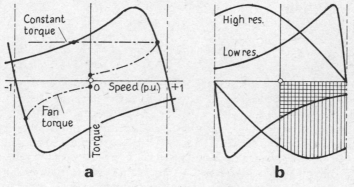

8.51 Torques during plugging.

typical load torques, one for a fan and the other for a hoist or elevator; (*b*) shows that the power available for plugging can be increased by augmenting the *rotor* resistance. Simultaneous reversal of the stator connections and inclusion of rotor resistance in a slip-ring machine is fully practicable, but the same effect can be achieved by including saturistors in the rotor winding. Alternatively, a deep-bar rotor cage may give a convenient rise in effective rotor resistance following the plug-switching operation as the slip frequency is suddenly increased from s to $(2-s)$ times ω_1. Shepherd [54] examines the characteristics and requirements for rapid plugging (including the effect of control by saturistors) and shows that there is an optimum effective resistance.

EXAMPLE 8.9: A 420 V 50 Hz 3-phase star-connected 6-pole induction motor having stator and referred-rotor leakage impedances both given by $(1.0 + j2.0)$ Ω/ph, drives a pure-inertia load at 960 r/min. The total inertia of the drive is $J = 3.0$ kg-m^2. Estimate the time to bring the drive to standstill by plugging. Neglect friction, windage and load torques other than inertial, and the effect of transient switching torques.

Immediately after switching the speed is given by $\omega_r = \omega_s(1-s)$ and the angular acceleration by $\alpha = d\omega_r/dt = -\omega_s(ds/dt) = M/J$, where M is the motor torque over the range $s_1 = 1.96$ p.u. initially to $s_2 = 1.0$ at rest. Then d$t =$

$-\omega_s(J/M)\,\mathrm{d}s$. Taking the torque per phase as the approximation in eq.(8.20), then

$$\mathrm{d}t = -\frac{J\omega_s{}^2}{3V_1{}^2}\left[\frac{(r_1 + r_2/s)^2 + X_1{}^2}{r_2/s}\right]\mathrm{d}s$$

Integrating the term in square brackets gives

$$[(r_1{}^2 + X_1{}^2)/2r_2]\,(s_2{}^2 - s_1{}^2) + 2r_1(s_2 - s_1) + r_2\ln(s_2/s_1)$$

$$= -24.1 - 1.9 - 0.7 = -26.7$$

with $r_1 = r_2 = 1.0$ and $X_1 = (x_1 + x_2) = 4.0\ \Omega$; then the total time is

$$t = 3.0 \times 105^2 \times 26.7/420^2 = 5.0\ \mathrm{s}$$

in which $\omega_s = 2\pi \times 1\,000/60 = 105$ rad/s.

Two-speed Plugging

In some circumstances, loads driven by pole-change motors may be retarded by switching from the high-speed to the low-speed connection. A braking torque is developed in this range. Fig. 8.52(a) indicates that during run-down the torque changes sign twice in rapid succession, passing through a strong reverse peak. Such a duty calls for a robust design, with taper-shaft couplings, welded instead of keyed cores, well-braced stator windings and reinforced cage-rotor bars and end-rings. It is to be noted that inclusion of resistance in series with the *stator* windings is deleterious, for it actually increases the peak reverse torque: this is shown by the circle diagrams (b) in Fig. 8.52.

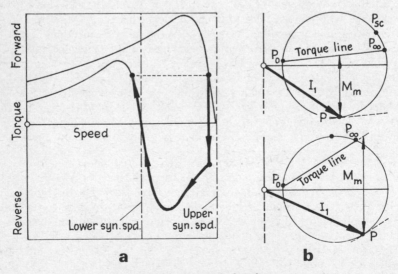

8.52 Two-speed plugging.

Motors required to plug high-inertia loads may have their own inertia minimized by low speed and small diameter, a combination which gives a high inherent leakage reactance. The magnetizing current tends to be large because the facility of an easily repaired stator winding necessitates the use of open slots.

D.C. Injection Braking
In this method the stator windings, immediately after their disconnection from the 3-phase supply, are excited with direct current from a rectifier and (in the case of slip-ring machines) resistance is introduced into the rotor circuits. The d.c. excitation establishes a *stationary* gap flux, the magnitude and position of which depends on the resultant combined m.m.f. of the d.c. stator and induced rotor currents. A rotor e.m.f. proportional to the flux and to the speed is generated in the rotor, the braking effect being the result of the I^2R loss. A typical connection diagram is given in Fig. 8.53; the

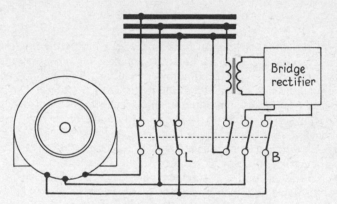

8.53 D.C. injection braking.

machine operates as a motor with contactors L closed, while with L open and B closed a direct current is fed through two stator phases, the third being left on open circuit. Alternative connections (such as those in Fig. 10.44) are possible.

Approximate Analysis. Rotation of the rotor at synchronous speed in the stationary airgap field gives conditions similar to those in an a.c.-fed motor at standstill; at rest, with no induced rotor e.m.f., the condition is like that of an a.c. machine at synchronous speed. The operation is therefore described by an equivalent circuit whose parameters have the factor $S = (1 - s)$, where s is the slip, and fed with a constant applied current, I_1 in Fig. 8.54(a), instead of a constant applied voltage. Ignoring saturation and rotor leakage, the rotor impedance is simply r_2 and the equivalent circuit has the phasor relation $I_m = I_1 - I_2$ and closely the scalar relation $I_m^2 = (I_1^2 - I_2^2)$. The flux due to I_m generates a rotor e.m.f. $E_2 = SI_m x_m$, where x_m is the magnetizing reactance chosen to relate the magnetizing current to the rotor e.m.f.: this leads to the

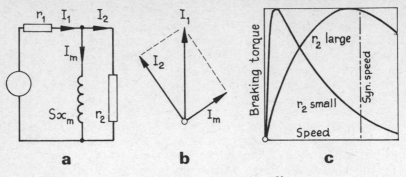

8.54 D.C. injection braking effects.

equivalent circuit in (a). Resolving Sx_m and r_2 into their series equivalent gives the resistance term

$$r_{2s} = r_2(Sx_m)^2/[r_2{}^2 + (Sx_m)^2]$$

from which the rotor I^2R loss per phase for a speed having the fraction S of synchronous speed is $I_1{}^2 r_{2S}$. The braking torque corresponds to a synchronous power per phase of $I_1{}^2 r_{2S}/S$. Maximum torque occurs for $S = r_2/x_m$ and depends only on the stator current and the magnetizing reactance. The braking-torque/*speed* relation resembles that of the torque/*slip* for a normal induction motor, as shown in Fig. 8.54(c). For dynamic braking the average torque is of importance, and it can be shown to be greatest for $r_2 = \frac{1}{2}x_m$.

Effect of Saturation. The net magnetizing current is greatest at zero speed, for then there is no opposing rotor current, so that the airgap flux is a maximum and saturation conditions are inevitable. The value of x_m will be low at standstill and will rise with increase of speed. The difficulty in applying the approximate analysis is consequently the strong effect of saturation on x_m. Krishnamurthy [52] has investigated the gap-flux distribution for a machine excited as in Fig. 8.53 for various speeds, and some of these patterns are given in Fig. 8.55 to bring out the variations in flux level, position and waveform. Clearly the speed and braking effort is a complex function of the saturation. Butler [53], using the equivalent circuit of Fig. 8.54(a) but with rotor leakage reactance included, has derived operating characteristics with

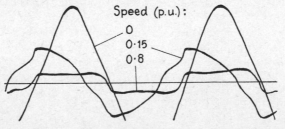

Speed (p.u.):
0
0·15
0·8

8.55 Gap-flux distribution for constant exciting currents.

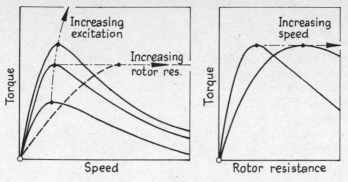

8.56 D.C. injection braking characteristics.

saturation taken into account. He shows that a severe torque limitation results from saturation but that for most practical applications it is permissible to ignore rotor leakage.

Characteristics. Typical performance curves are given in Fig. 8.56. Injection braking takes little power from the supply and gives smooth braking torque, useful for mine winders and high-inertia loads. Its advantage over plugging is the absence of a reverse-rotation airgap field (and therefore no tendency for the machine to run backwards) and a lower rotor I^2R loss. Dubey and De [55] give methods of evaluating the performance of a d.c. braking system with saturistors on the rotor, and show that the braking performance can be optimized by a suitable combination on the rotor of resistors and saturistors.

Capacitor Braking

A motor on load, disconnected from the supply and reconnected to a bank of capacitors, will self-excite as an induction generator (Sect. 8.20). An alternating voltage is built up in the stator windings until balance between the capacitive and saturating magnetizing reactance obtains. The load inertia drives the machine with a slip negative with respect to the 'synchronous' speed of self-excitation, and the speed falls. At the same time the magnetizing current also falls, reducing the saturation level and increasing the magnetizing reactance, so lowering the self-excitation frequency. By proper choice of capacitance, the 'synchronous' speed can be maintained below the rotor speed (i.e. with negative slip) down to low speeds. The method is of use where positive braking to standstill is not required, the load stopping the machine in the time allowable. Resistors may be used in the stator circuit in series with the capacitor bank to control the torque. Where braking to rest is a requirement, the magnetic field built up by self excitation is exploited by short-circuiting the stator windings through resistors for the final stage. If the load inertia is high, however, a d.c. injection scheme may be used instead.

Capacitor braking, alone or combined with d.c. or magnetic braking, has the advantages of high braking torque, lower loss, small energy demand from the supply, and quiet operation.

Braking Conditions

In terms of synchronous power, the motor torque is the rotor input P_2, while the rotor I^2R loss is sP_2. The ratio of I^2R loss to torque is therefore simply the slip s. The average loss between any two speeds corresponding to slips s_1 and s_2 is therefore $P_2(s_1 + s_2)/2$. The time t for the speed change depends on the inertia of the drive system, and the energy expended in the change is $P_2 t(s_1 + s_2)/2$ which, for given conditions, is proportional to $(s_1 + s_2)$. In acceleration from standstill $(s_1 = 1)$ to normal running $(s_2 \simeq 0)$, that part of the rotor input dissipated electrically is proportional to 1. During retardation by plugging the initial condition is approximately $s_1 = 2$, and at standstill $s_2 = 1$: the loss is thus proportional to 3. If acceleration or braking is by variation of the stator supply frequency, as in the inverter or capacitor methods, the synchronous speed is itself changed and the slip remains small so that the loss is proportional to, say, 0.2. The motor duty therefore depends not only on the inertia but also on the method of braking. Because of the time taken to stop them, high-inertia loads may be unsuitable for plugging.

Precision methods of braking can be very effective. Conveyors and multi-motor drills stopped every 8 s, machine-tools brought to rest with a positional accuracy of ± 0.025 mm, and a 600 Hz motor running near 36 000 r/min brought to rest in $3\frac{1}{2}$ s are impressive examples.

8.19   *Unbalanced operation*

Unbalanced conditions of operation may arise accidentally, or may be deliberately introduced, by asymmetrical supply voltages or inequalities of phase impedance.

Voltage Asymmetry

The performance of a 3-ph induction motor is modified if its stator phase voltages differ in magnitude or phase-displacement, or both. Such voltages can be considered as the superposition of symmetrical positive-, negative- and zero-sequence components, each of which (disregarding saturation nonlinearities) produces corresponding currents in the windings. The p.p.s. components develop a 'normal' travelling-wave m.m.f. in accordance with the phase sequence and direction of rotation; the n.p.s. components develop an m.m.f. travelling in the reverse direction. The effect of z.p.s. components, usually small, resembles 1-ph excitation of the machine with three times the number of poles.

As regards p.p.s. and n.p.s. components, the motor behaves as the combination of two machines, one with a phase voltage V_+ and slip s, the other with V_- and $(2 - s)$. The resultant mechanical power per phase is

$$P_m = I_{2+}{}^2 r_2 \left[(1 - s)/s \right] - I_{2-}{}^2 r_2 \left[(1 - s)/(2 - s) \right]$$

where the currents are determined by the respective sequence voltages and impedances. The resultant torque is $P_m/\omega_s(1 - s)$. The torque/slip curves in (a) of Fig. 8.57 are for a large degree of asymmetry. Minor asymmetries do not greatly affect the torque, but *any* asymmetry increases the I^2R loss. *Speed Control and Braking.* Asymmetric-voltage control has been applied to

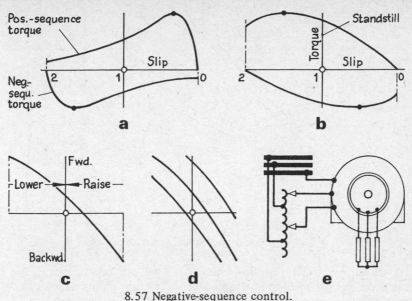

8.57 Negative-sequence control.

cranes. If a slip-ring motor is heavily loaded on rotor resistance, the torque/slip relations (*a*) in Fig. 8.57 are changed to those in (*b*), and the net torque approaches a linear relation with slip (*c*). The control enables a crane to develop a stable *upward* torque to restrain a load while it is being lowered. One method of achieving the family of characteristics in (*d*) is to connect an auto-transformer across two supply lines as in (*e*). If the tap positions can be crossed, it is possible to control from full forward symmetry to full backward symmetry through a single-phase condition when the two tappings coincide.

Single-phasing. This term refers to operation with one stator terminal accidentally disconnected (as by a blown fuse or a faulty contact on the 'run' side of a star-delta switch), a common cause of winding burnout. Fig. 8.58 compares the performance of a 3-ph motor under normal and single-phasing conditions. Clearly, if single-phasing occurs, a loaded motor may stall, because the disconnection of a stator terminal introduces a n.p.s. torque.

Single-phase Braking. If the stator windings of a 3-ph motor with a synchronous speed n_s are connected in series or parallel to a 1-ph supply so that all three currents are co-phasal, the distribution of the gap m.m.f. produces three times the normal number of poles, all pulsating at supply frequency. A rotor running at a speed higher than $n_s/3$ is therefore in a generating mode and is retarded. A machine with a high-resistance rotor can be braked down to standstill, a method sometimes used for the light-duty braking of machine-tools.

Winding Unbalance

A 3-ph motor with a 1-ph rotor winding has two stable operating speeds,

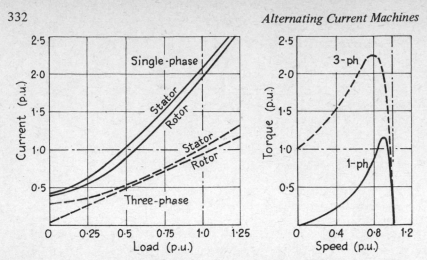

8.58 Normal and single-phasing performance.

respectively approaching synchronous and one-half synchronous speed. A comparable effect is observed when a 3-ph rotor has a winding unbalance, deliberate or resulting from a winding fault or a difference in the external impedances connected to the slip-rings.

The p.p.s. currents set up in the stator by a symmetrical p.p.s. 3-ph voltage form a normal travelling wave of m.m.f. of synchronous speed n_s with respect to the windings, and the corresponding gap flux induces p.p.s. stator and rotor e.m.f.s. The rotor e.m.f. circulates currents of slip frequency, but in consequence of phase unbalance these can themselves be resolved into p.p.s. and n.p.s. components. The latter travel relative to the rotor at a speed $-sn_s$, and therefore at $(-n_s + 2n)$ relative to the stator. The stator reacts to the rotor n.p.s. gap flux by induced currents of frequency $(1 - 2s)f$. The result is a reversed interaction torque which imposes a dip in the net torque/speed characteristic at one-half of synchronous speed. If the rotor phase impedance unbalance is large, the dip is deep enough to permit of stable half-speed operation. Several further phenomena occur. Negative-sequence currents are injected into the supply system, and pairs of p.p.s. and n.p.s. flux-current combinations develop pulsating torques and consequent vibration (except at half speed); the p.p.s. and n.p.s. flux axes move into and out of coincidence, producing strong cyclic magnetic saturation.

If half-speed running is to be an intended operating condition, the rotor unbalance should include a short-circuited phase winding, and saturation should be mitigated by use of an over-size frame. Transfer of the operating speed from one-half to full is difficult, and so is reduction from full to half speed unless the load torque is sufficient.

8.20   *Induction generator*
Operation of an induction machine in the generator mode, with negative slip, has been discussed in Sects. 8.1–8.8. There are two excitation conditions of interest.

External Excitation

To determine its voltage and frequency, and to supply its lagging reactive magnetizing power, an induction generator is connected to an energized power network with synchronous machinery. The action can then be inferred in simple terms from Fig. 8.17: the active power generated is typically TP_{1g}, the mechanical drive power is $P_{1g}R$, and the network supplies the reactive power OT. The torque/slip characteristic lies in the negative-slip region of Fig. 8.11. If the prime-mover driving torque exceeds the peak electrical torque, the speed rises and at some high value of negative slip the generating effect vanishes.

The machine can utilize waste heat of process steam in chemical works and the power from small remote hydro plants. To minimize transients and heavy motoring currents, the set is brought up to synchronous speed before the stator windings are connected to a 3-phase network, remote starting and connection being readily possible. The induction generator needs little auxiliary equipment, it can run without hunting and its rotor can carry a robust cage winding or even be a solid cylinder. On external short circuit the excitation collapses, giving some automatic protection. The prime-mover need not be governed, although the generator must then be able to sustain overspeeds up to 2 or 3 p.u. The drawbacks are the need for a.c. magnetization, the moderate efficiency and (for mechanical reasons) the short air-gap.

Self-excitation

With the magnetizing reactive power provided by a capacitor bank, an external a.c. supply is not needed; but the operating frequency and the generated

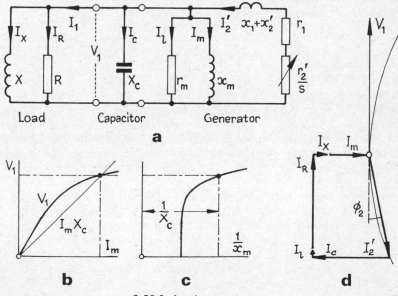

8.59 Induction generator.

voltage are affected by the speed, load and capacitor rating. The capacitor must also supply any lagging reactive power demand.

Provided that the rotor has an adequate remanent field, the machine will self-excite. The operating conditions can then be described in terms of the approximate equivalent circuit (*a*) of Fig. 8.59. On *no load*, $I_c = V_1 \omega C = V_1/X_c$ must be equal the magnetizing current $I_m = V_1/x_m$. But V_1 is a function of I_m so that stable operation requires the line $I_m X_c$ in (*b*) to cut the voltage/excitation characteristic. If, as in (*c*), the terminal voltage V_1 is plotted as a function of the magnetizing susceptance $1/x_m$, the operating point is fixed by the corresponding capacitive susceptance $\omega C = 1/X_c$, and this settles the frequency. Under *load* conditions, the generated power $V_1 I_2 \cos \phi_2$ provides for the load power in R and the loss in r_m. The reactive currents must sum to zero, i.e.

$$V_1/X + V_1/x_m + I_2 \sin \phi_2 = V_1/X_c$$

which determines the capacitance for a given load and voltage. The several currents in network (*a*) are shown in phasor diagram (*d*).

EXAMPLE 8.10: On rated supply, a 3.0 kV 50 Hz star-connected induction machine has the phase-current components: ideal-short-circuit $V_1/(x_1 + x_2)$, 750 A; core-loss, 10 A. The magnetizing current is given by

Voltage (line) kV:	1.45	2.5	3.0	3.4
Current A:	20	40	60	90

For a frequency 50 Hz, estimate the star-connected capacitor values (i) just to initiate self-excitation, (ii) to generate 3.0 kV line, and (iii) to provide a load current $(125 - j20)$ A/ph.

Working per phase, the magnetizing susceptance is found:

Phase voltage V_1	kV:	0.84	1.44	1.73	1.96
Magnetizing current I_m	A:	20	40	60	90
Susceptance I_m/V_1	mS:	24	28	35	46

(i) The minimum capacitive susceptance for self-excitation is 24 mS, so that $C = 24 \times 10^{-3}/314 = 76\ \mu$F/ph. (ii) Using the method of Fig. 8.59, interpolation gives 43 mS for 1.73 kV/ph (3.0 kV line), whence $C = 137\ \mu$F/ph. (iii) The rotor current circle is drawn with diameter 750 A to scale; then $I_m = 60$ A, $I_l = 10$ A, $I_R = 125$ A, $I_X = 20$ A. The active current is $125 + 10 = 135$ A, and with $I_X + I_m = 80$ A the capacitor current is 104 A, requiring $C = 190\ \mu$F/ph.

Performance. The voltage regulation of a self-excited induction generator may be large, and the terminal voltage waveform is likely to be distorted because of the need for a saturated condition to stabilize the excitation. The machine has been used in aircraft as a self-starting motor and brushless generator: for this application the driving speed for constant frequency must rise with increase of load, and because of the inevitably wide variations in magnetization characteristic it may be necessary, in machines to run in parallel, for star-connected saturable inductors to be fitted.

Accidental Self-excitation. Dangerous voltages may occur with induction generators working over long transmission lines if synchronous machines at the far end become disconnected and the line capacitance excites the induction machine. Motors with power-factor correction capacitors may generate if the capacitors remain connected to the running machine when for some reason the supply is cut off.

Braking. Self-excitation can be used for braking, as described in Sect. 8.18. On disconnection of the stator from the supply and its subsequent connection to a capacitor bank, the stator voltages are self-excited to develop a travelling-wave field, and for a generating mode the 'synchronous' speed of the field must be lower than the speed of rotor rotation. The drive power is drawn from the inertia energy and the speed consequently falls. The slip (i.e. the difference between the momentary field and shaft speeds) remains negative because, with reducing current, the effective magnetizing inductance rises — perhaps to 10 times its saturated value. But eventually the inductance stabilizes and further speed drop brings the machine into a motor mode.

Equivalent Circuit

The equivalent circuit for any given self-excited frequency ω_e can be expressed in terms of the parameters r_1, r_2, x_1, x_2, x_m and X_c all referred to the *normal* angular frequency ω_1 by using a factor $a = \omega_e/\omega_1$. If the shaft speed is expressed as the factor b of normal synchronous speed, then $a - b = s$, the slip (which is negative). The equivalent circuit for frequency ω_e can then be drawn as in Fig. 8.60, in which normal-frequency values of reactance are

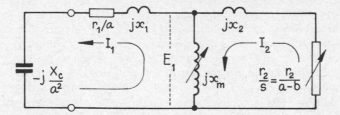

8.60 Self-excited induction generator.

retained while the stator and rotor resistance and the capacitive reactance are suitably modified. As there is no externally applied voltage, the mesh equations for currents I_1 and I_2 are

$$0 = I_1 [(r_1/a) + j(x_1 - X_c/a^2)] + (I_1 - I_2) jx_m$$

$$0 = I_2 [r_2/(a-b) + jx_2] + (I_2 - I_1) jx_m$$

From these, eliminating I_1 and I_2 and separately equating the 'real' and 'imaginary' parts to zero, we have

$$\frac{x_m}{x_1} = \frac{(r_1 x_2/r_2 x_1)[(b-a)/a] + (X_c/a^2 x_1) - 1}{1 - (r_1/r_2)[(b-a)/a]}$$

Knowing x_m/x_1 and therefore x_m, for chosen values of a, b and X_c, the value of E_1 can be obtained from the magnetization characteristic and the rotor current derived. The braking torque per phase is then $I_2{}^2 r_2/(b-a)$.

8.21   *Transients*

The steady-state stator and rotor currents and the travelling-wave gap flux that exist in an induction machine running at a fixed speed with a constant unidirectional torque have to be established from the instant when the stator of the inert machine is switched on to the supply. Magnetic energy has to be supplied, as has the kinetic energy of the rotating system and the various losses, as well as the energy absorbed by the load in raising it to the final steady-state condition. The energizing process is complicated by the reaction of cores to the rapid rise of flux density, the skin effect in conductors, and the mechanical deformations of motor shaft and structure as quickly varying torques are developed. The difference between the actual and the steady-state conditions for given speed and load are bridged by transient currents and fluxes, which appear instantaneously and thereafter decay exponentially at rates determined by the parameters and time-constants of the several electric circuits and the instantaneous speed of the machine. The term *switching* is applied to transients arising as a result of the energizing of an inert machine.

Transients are also generated when one steady state is changed to another, as when *re-switching*. If a running machine has its load suddenly altered, or if it is disconnected from the supply for a brief period and then reconnected, transient currents and torques appear. These transients are sometimes of great severity, as when a motor is plugged or dynamically braked, or subjected to supply-voltage interruptions especially where there are capacitor banks across the stator terminals. Considerable torque and current transients may occur even when motor terminal connections are changed during run-up by a star-delta switch. Very large induction machines, when switched, re-switched or subject to the voltage fluctuation resulting from a power-system fault, may react strongly on the system by drawing or supplying transient power.

Constant-linkage theorem. Physical appreciation of the transient condition is afforded by the *theorem of constant linkage,* which states that in a closed circuit of zero resistance, the algebraic sum of its magnetic linkages remains constant. No circuit (superconduction apart) is devoid of resistance, but the linkage nevertheless remains instantaneously constant immediately after a sudden change because it takes time for stored magnetic energy to be altered. a process approximating to a time-exponential rise or decay. When, for example, a 3-phase voltage is suddenly applied to the stator terminals of an inert induction machine, phase fluxes begin to build up at rates proportional to the instantaneous voltages (which are inevitably unequal). The stationary rotor develops corresponding opposition currents that initially maintain the inert condition of zero linkage. If the rotor moves, these currents may be carried round, but rapidly decay. On disconnecting a running motor, the stator currents are quenched; but as the short-circuited motor maintains its linkage by a unidirectional current, a trapped or 'frozen' flux is carried round with the rotor. The trapped flux decays, but induces rotational e.m.f.s in the stator windings. The transient currents and torques that occur on re-switching depend on the stator e.m.f.s so induced, and if these have phase-opposition to the supply voltages at the instant of reconnection, the transient currents may be severe and the transient torque may have a negative (retarding) peak.

Analysis

Prediction of transient performance is difficult. The differential equations that must be used for adequate modelling of an induction machine are complicated, and if there is a change of speed during the process of a transient, the equations are (in the mathematical sense) nonlinear, quite apart from inevitable physical nonlinearity resulting from deep-bar effects and magnetic saturation.

For transient analysis, Wood and his co-workers [56] use the 2-axis model (Sect. 7.5). Using the impedance matrix of Fig. 7.14, the voltage equations for the stator 2-axis windings 1D and 1Q, and the corresponding rotor windings 2D and 2Q, are:

$$1D: \quad v_{1d} = r_1 i_{1d} + (L_m + L_1)p i_{1d} + L_m p i_{2d}$$

$$1Q: \quad v_{1q} = r_1 i_{1q} + (L_m + L_1)p i_{1q} + L_m p i_{2q}$$

$$2D: \quad v_{2d} = r_2 i_{2d} + (L_m + L_2)p i_{2d} + L_m p i_{1d}$$
$$+ (L_m + L_2)\omega_r i_{2q} + L_m \omega_r i_{1q}$$

$$2Q: \quad v_{2q} = r_2 i_{2q} + (L_m + L_2)p i_{2q} + L_m p i_{1q}$$
$$- (L_m + L_2)\omega_r i_{2d} - L_m \omega_r i_{1d}$$

The 2-axis transient equivalent voltages for the case to be considered are then imposed and the equations solved, by iteration if there is a change in the rotor speed ω_r. In many cases the rotor circuits are short-circuited and have no externally applied voltage, so that $v_{2d} = v_{2q} = 0$. The validity of the method can be shown by presuming steady-state conditions with a constant slip s. Then p is replaced by $j\omega_1$ and ω_r by $\omega_s(1-s)$ where ω_s is related to ω_1 by the number of stator pole-pairs. Combining ω_1 with the inductances to give reactances, and inserting the condition of short-circuited rotor windings, then there is no difference between the d- and q-axis windings except one of time-phase. Consequently it is possible to write $v_{1d} = v_1$, $v_{1q} = -jv_1$, $i_{1d} = i_1$, $i_{1q} = -ji_1$, $i_{2d} = i_2$ and $i_{2q} = -ji_2$. Then

$$1D, 1Q: \quad v_1 = [r_1 + j(x_m + x_1)] i_1 + jx_m i_2$$

$$2D, 2Q: \quad 0 = [r_2 + j(x_m + x_2)] i_2 + jx_m i_1$$
$$- (1-s)[(x_m + x_2)i_2 + x_m i_1]$$

These express the behaviour of the steady-state equivalent circuit, and the d- and q-axis contributions to the torque are equal.

There are alternative analytical methods. Smith and Sriharan [57] apply instantaneous symmetrical components; Enslin, Kaplan and Davies [58] apply the differential equations of coupled circuits using parameters derived from indirect tests and show that the transient stator and rotor current components each consist of a pair of terms with complex angular frequencies. Torque components arise from the consequent flux and m.m.f. distribution waves defined by amplitude, decrement and angular speed of rotation. Unidirectional synchronous torques are developed by the interaction of flux and m.m.f. waves of the same angular speed, pulsating torques by those of different speeds, with all stator-flux/rotor-m.m.f. and rotor-flux/stator-m.m.f. products accounted. There exist nine decaying transient torques superimposed on the steady-state torque.

Switching

Starting-torque transients are affected by the instant of connection of the first phase to close, and the delay (if any) in closure of the other two phases. The initial peak torque may also be affected by residual flux left after a previous duty cycle. Assuming the machine to remain at rest, Slater and Wood [56] give Fig. 8.61(a) as a typical torque/time pattern with simultaneous switching

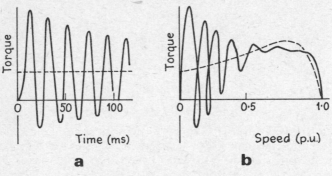

8.61 Starting transients.

on all phases, the main component being a damped alternating torque. An approximate expression for the total peak torque at starting in terms of the steady-state locked-rotor torque is

$$M_p \simeq 1 + \sqrt{[k(1 + \omega_1{}^2\tau^2)]}$$

where $\tau = (L_1 + L_2)/(r_1 + r_2)$ is a time-constant derived from the stator and rotor leakage inductance and resistance, and k is a factor to account for delay between the application of the first and last line voltages; $k = 1$ for simultaneous switching.

Transient torques persist for several periods of the supply frequency. In the *slow* start of a motor with a large inertia load, the effects may die away before the machine has gathered speed. The *fast* run-up of an unloaded machine may, in contrast, exhibit strong torque pulsations, Fig. 8.61(b), the torque reversals producing the characteristic 'loops'. The steady-state torque/speed curve is given for comparison. Both curves in Fig. 8.61 are for sinusoidal applied voltage; additional and complicated effects are likely if the supply voltages are derived from thyristor inverters.

Re-switching

When a running motor is disconnected from the supply and then reconnected, a transient torque is developed, primarily as a result of the e.m.f.s induced into the stator phase windings by the gap flux of the unidirectional decaying rotor currents, as already described. The phase angle of these e.m.f.s with respect to the re-applied supply voltage is significant, and the amplitude and sign of the first peak of the transient torque are closely dependent on the rotor speed and the duration of the interruption. In the worst case the first peak may reach 15 p.u. of full-load torque. The corresponding transient peaks of stator current are typically 5–7 p.u. of full-load current.

Plugging, dynamic braking and busbar transfer are examples of fast re-switching in which re-connection may be made with appreciable e.m.f.s existing in the stator winding. In *plugging* the re-connected supply has the reversed phase-sequence; even at reduced voltage the transient torques and currents are severe, e.g. with 15 p.u. peak torque and 20 p.u. current in terms of rated values. Plugging is a drastic operation, complicated by the effects of eddy currents and saturation.

Marked transient effects occur when *terminal capacitors* are used for power-factor correction, self-excitation of an induction generator or capacitance braking. When a motor supply is interrupted with the capacitors left across the stator terminals, the capacitors tend to maintain the gap flux and the stator may build up an overvoltage in spite of the drop in rotor speed. On re-connection with unfavourable conditions of e.m.f. and supply voltage phasing, the resulting current and torque transients can be severe. In the case of an induction generator, the effects of reconnection are intensified by the rise in speed due to the prime-mover.

Transients during *d.c. injection braking* are normally small, and are more likely to be caused by short-circuit of the d.c. supply than by the injected voltage. Consequently they become less significant the longer the duration of disconnection.

An example of a re-switching transient is given in Fig. 8.62. The machine under full-load conditions is disconnected from rated voltage for 93 ms (4.65 periods of a 50 Hz supply) and then re-connected. The torque wave is a computed electromagnetic gross torque.

Electromechanical Transients

Run-up and Overshoot. During the direct-on-line start and run-up of an unloaded cage motor, the measured torque/speed characteristic may exhibit a depressed peak-torque, and be observed to reach a supersynchronous speed overshoot before settling to the small steady-state positive slip. Both effects are due to the action of the short-circuited rotor in opposing rapid changes of flux linkage. As the rotor approaches synchronous speed, the rate of decay

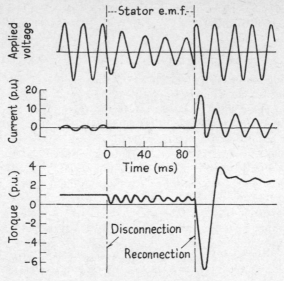

8.62 Reconnection transients

of its currents may be slow enough to give the temporary effect of d.c.-excited poles, which consequently develop a synchronising torque, like that of a synchronous motor with a retarded load angle. The machine therefore 'hunts' about synchronous speed for one or two oscillations before the rotor torque adjusts to the steady-state condition.

Change of Load. Reduced low-slip damping and transient synchronizing torques are also exhibited when a running motor is subjected to sudden changes of load. During the rapid retardation following a large load application, the trapped rotor flux combines with the main synchronously rotating field to produce synchronizing torque which reverses cyclically as the speed drops and develops wide resultant fluctuations of motor torque, the peaks decaying as the rotor currents settle to the new condition.

Fluctuating Load. If a load, such as a reciprocating compressor, imposes a slip and torque oscillation, cyclically generated rotor fluxes are trapped, of magnitude dependent on the load-oscillation frequency and the short-circuit time-constant. They produce synchronizing torques, the elastic nature of which in combination with the load inertia of the set gives a natural oscillation frequency. If this is close to the fluctuation frequency of the load (or to one of its harmonics) the slip-swing may be augmented by resonance, causing a strong pulsation of the stator-current waveform envelope. Analysis of such cases, with examples, is given by Middlemiss [109].

Faults

System Faults. Where these reduce the terminal voltage of a loaded induction motor the speed of the machine will fall. In the post-fault voltage-recovery period, large motors make heavy reactive power demands that may depress the system voltage still further and inhibit re-acceleration. The nature of the torque/speed characteristic of the load is important; if, for example, load torque is proportional to speed, recovery of normal operating conditions is likely to be more quickly attained.

Short Circuit. When a system short-circuit fault occurs at or near the terminals of a large induction motor, the machine makes a significant contribution to the system fault current, particularly if it is running on light load. The short-circuit current is always more severe for a 1-phase than for a 3-phase fault. The most important quantity is the first peak current as it determines the 'make' rating of the protective circuit-breaker. In general the short-circuit current has a d.c. component of slow decay, and an a.c. component. The latter is larger than the direct-on-line starting current and may reach 10−20 times full-load current; it can be represented by two exponential terms related to the d-axis transient and subtransient reactances and time-constants, viz.

$$x = x_1 + x_2 x_m/(x_2 + x_m) \quad \text{and} \quad \tau = (1/\omega r_2)[x_2 + x_1 x_m/(x_1 + x_m)]$$

In cage rotors the resistance r_2 and leakage inductance L_2 may vary considerably with rotor frequency. It is adequate for r_2 and $x_2 = \omega L_2$ to use values for rated frequency for the subtransient x_d'' and τ_d'', and d.c. values for the transient x_d' and τ_d' terms. Then for a pre-fault terminal voltage V_1 and a short circuit near to the motor terminals, the r.m.s. current at time t is approximately

$$I = V_1 \left[\left(\frac{1}{x_d''} - \frac{1}{x_d'} \right) \exp \left(-\frac{t}{\tau_d''} \right) + \frac{1}{x_d'} \exp \left(-\frac{t}{\tau_d'} \right) \right]$$

To a reasonable approximation the current can be estimated on a basis of the subtransient reactance alone.

8.22 *Testing*

Few stock motors undergo a comprehensive test, but new types call for more detailed testing to confirm design figures. On large machines the stator is magnetically checked for imperfections, the rotor for balance and (for slip-ring machines required to run for long periods at high slip) an underspeed load test with slip-power recovery.

D.C. Resistance

Accessible windings are checked by a d.c. method, the conductor temperature being recorded for calculating the resistance at specified operating temperature.

Open-circuit (Voltage Ratio)

This, for slip-ring machines, consists in applying rated voltage and frequency to the stator windings and measuring the voltage between slip-rings as a check for balance. The measured phase voltages V_1 and V_2 do not give the effective turns-ratio because of stator leakage; but if V_2 is applied to the rotor and the stator voltage is then v_1, the effective turns-ratio is close to $\frac{1}{2}(V_1 + v_1)/V_2$.

Run-up

A starting test may be applied to cage motors to confirm that they are capable of starting against a specified load torque and inertia without crawling or developing noise.

No-Load

This gives core, pulsation, friction and windage losses, magnetizing current and no-load power factor. It reveals mechanical out-of-balance and faulty connections. The stator windings are connected to a supply of rated voltage and frequency, with instruments to measure voltage, current, power and slip. After starting, the rotor (if it is of the slip-ring type) is short-circuited as for normal operation. Power and current are read at, say 1.2 p.u. of normal voltage and the test repeated at lower voltages down to the value at which the stator

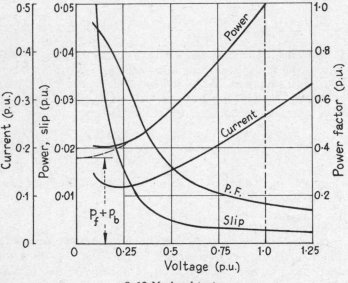

8.63 No-load test.

current begins to rise. Fig. 8.63 gives a plot of typical results. At normal voltage the no-load current I_0 is between 0.2 and 0.4 p.u. of full-load current and the power factor is low as the reactive (magnetizing) component I_{0r} of I_0 predominates. With reduction of voltage, the flux falls almost in proportion and the power curve is roughly parabolic as the core loss is proportional to the square of the flux density; the power factor rises as the active current component I_{0a} increases to satisfy the mechanical loss. At about 0.2 p.u. voltage the magnetizing current and core loss are small, but the mechanical loss calls for a larger active current component and the slip rises to permit its circulation. The power is therefore almost entirely for mechanical loss, and if the power P_0 is extrapolated to zero voltage (if necessary by plotting the straight-line relation of P_0 to V_1^2) the intercept is the mechanical loss p_b for brush friction (if there are slip-rings with no brush-raising device) and p_f for windage and bearings. The power actually includes a small I^2R loss, for which, in small machines, a correction can be made.

Locked-rotor (Short-circuit)

The rotor, short-circuited as for a running condition, is held stationary, and the stator is supplied at a low voltage of normal frequency. The position in which the rotor is clamped may affect the stator current: if so, a mean position is found, or the rotor permitted to revolve very slowly during the test. The applied stator voltage is raised in steps, with readings of current and power, until the current reaches not more than twice normal. The temperature must be taken for each reading so that the conditions can be normalized. Typical results are shown in Fig. 8.64, plotted to a base of stator current. The

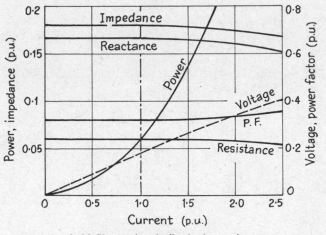

8.64 Short-circuit (locked-rotor) test.

power curve is nearly a parabola, as the power input is almost entirely I^2R loss. For, say, 0.2 p.u. voltage, the flux density is about one-tenth of its normal value and the core loss is consequently very small.

The leakage impedance is seen to fall with higher currents as a result of saturation, and the resistance may drop a little for the same reason. For deep-bar and double-cage machines, the effective rotor standstill leakage reactance increases at normal speed above its actual standstill value due to the redistribution of the rotor current. For accurate performance calculations the motor reactance may therefore be re-determined at e.g. 5 Hz for a 50 Hz machine.

Performance

In the assessment of efficiency by loss-summation, BS 269, the losses to be taken into account are:

Fixed loss: (*a*) core loss; (*b*) bearing friction; (*c*) total windage; (*d*) brush friction.

Direct load loss: I^2R loss (e) in stator winding, (f) in rotor winding, (g) brush loss.

Load loss: stray loss (h) in core, (j) in conductors.

Losses (a), (b), (c) and (d) are found by deducting the small no-load I^2R loss from the total measured loss in the no-load test. Losses (e) and (f) for slip-ring motors are calculated from the currents (obtained from the circle diagram) and resistances of the windings. For cage rotors, loss (f) with the small part of (h) and (j) appertaining to the rotor, is determined from the expression —

$$\text{(output + windage + friction)} \times s/(1-s)$$

Loss (g) for all brushes together is taken as the slip-ring current $\times$ 1 V. The stray load loss is covered by deducting $\frac{1}{2}\%$ from the full-load efficiency, and *pro rata*; or it may be estimated on the basis of a reverse-rotation test.

The circle diagram is an essential feature of the method, and it gives also the full-load current, power factor and slip, and an indication of the pull-out torque.

EXAMPLE 8.11: The following test data refer to a 22.5 kW, 500 V, 3-phase, delta-connected, 50 Hz, 4-pole cage motor:

No Load

Line voltage	V:	564	515	480	320	160	75
Line current	A:	9.8	8.6	7.8	5.4	4.5	7.9
Power, W1	kW:	−2.05	−1.55	−1.22	−0.25	0.28	0.36
W2	kW:	3.67	3.05	2.62	1.40	0.69	0.60
W1 + W2	kW:	1.62	1.50	1.40	1.15	0.97	0.96

Locked Rotor

Line voltage	V:	60	100	140
Line current	A:	18.6	32.0	44.8
Power, W1	kW:	−0.27	−0.75	−1.50
W2	kW:	0.83	2.35	4.65
W1 + W2	kW:	0.55	1.60	3.15

The figures above are corrected to 75 °C. The stator resistance (hot) is 0.68 Ω/ph. Separate the principal losses, and find the full-load current, power factor, efficiency and slip, the pull-out torque, and the maximum power output.

From the plot (a) of no-load readings in Fig. 8.65, the no-load values at normal voltage are $V_1 = 500$ V, $I_0 = 8.2$ A, $P_0 = 1.45$ kW, whence cos $\phi_0 = 0.202$ and $\phi_0 = -78.4°$. By extrapolation, the friction and windage loss is $p_f = 0.90$ kW. The no-load I^2R loss for the three phases, calculated from the stator phase resistance, is 0.05 kW. The core loss (using the nomenclature given in a later paragraph on loss-separation) is $p_{e1} + p_{h1} + p_p = 1.45 - 0.90 - 0.05 = 0.50$ kW.

Plotting the locked-rotor readings, (b), and taking values for full-load current current (estimated at 32 A), we have $V_1 = 100$ V, $I_{sc} = 32$ A, $P_{sc} = 1.60$ kW, and therefore $\cos \phi_{sc} = 0.29$, $\phi_{sc} = -73.1°$. The corresponding values *pro rata* for normal voltage are $V_1 = 500$ V, $I_{sc} = 160$ A, $P_{sc} = 40$ kW, $\cos \phi_{sc} = 0.29$; the stator I^2R loss is then 17.4 kW, so that the rotor I^2R loss is $40.0 - 17.4 = 22.6$ kW.

The circle diagram, Fig. 8.65(c), is drawn for convenience with line values and a current scale of 10 A per unit length. One current unit vertically therefore represents $\sqrt{3} \times 500 \times 10 \times 10^{-3} = 8.66$ kW. For a full load of 22.5 kW a vertical intercept of $22.5/8.66 = 2.60$ units is necessary between the full-load point P and the out-put line $P_0 P_{sc}$. This fixes P and gives the stator current $OP = 32$ A to scale at a power factor of 0.906. The full-load losses and input are therefore

	kW
Friction and windage	0.90
Core	0.50
Stator I^2R	0.70
Rotor I^2R	0.90
Output	22.5
Input	25.5

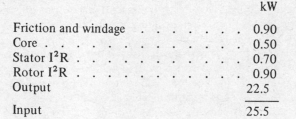

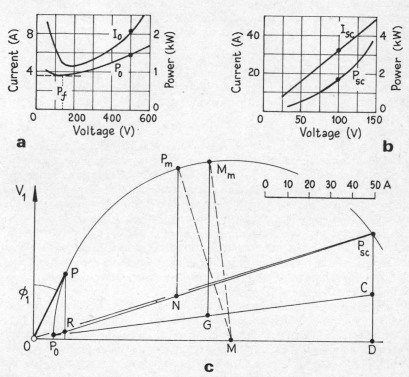

c

8.65 Test results and circle diagram.

The efficiency is 22.5/25.5 = 0.88 p.u. (less ½% for stray loss). The full-load slip is 0.90/24.3 = 0.037 p.u. Maximum power (given by P_mN = 5.75 units) is 5.75 x 8.66 = 49.8 kW. The torque line is P_0C, where C divides $P_{sc}D$ in the proportion 22.6/17.4 of the short-circuit rotor/stator I^2R loss. The maximum torque is 2.5 times the full-load torque. The starting torque $P_{sc}C$ on full voltage approximates to the full-load torque, but saturation makes this estimate unreliable.

Temperature-rise

A true test of the temperature-rise can be achieved only by direct loading by means of a brake or a generator. With two identical machines mechanically coupled, a load test can be arranged by supplying one at normal voltage and frequency from a mains supply, the other from an auxiliary source of lower frequency so that it operates in a generator mode at a negative slip. A check is made with the roles interchanged.

For machines of large rating, or with vertical shaft, or with an inaccessible shaft extension, a variable-frequency test provides an alternative. The machine is run up on no load, and a second voltage of variable frequency and magnitude is then introduced. The result is a cyclic acceleration and retardation with the machine operating alternately in motor and generator modes. The fluctuating current depends on the difference frequency and the motor inertia, and can be adjusted to an r.m.s. average corresponding to full-load current.

Insulation

Insulation is tested by high-voltage ohmmeter and by the application of a specified overvoltage (e.g. from a test transformer) between open-circuited phase windings, and between windings and earth.

Loss-separation

Apart from the I^2R loss in accessible windings, which can be calculated, the losses in an induction machine include: p_b, the brush-friction loss; p_e, the eddy-current loss in the core steel; p_f, the mechanical losses in bearing friction and windage; p_h, the hysteresis loss; and p_p, the pulsation loss. The friction losses require no comment. Using the subscripts 1 and 2 for stator and rotor respectively, the stator core loss is $p_{e1} + p_{h1}$ at normal flux. At standstill the rotor core loss is $p_{e2} + p_{h2}$, when the rotor is open-circuited. At a slip s, the loss becomes $s^2p_{e2} + sp_{h2}$, since the former is proportional to the square and the latter to the first power of the frequency. The eddy-current loss corresponds to a torque proportional to the slip, but the hysteresis loss corresponds to a constant torque, since sp_{h2} is the product of the slip s and a constant power, or the slip-speed sn_1 and a constant torque. The rotating field produced by the stator may be conceived as magnetizing the rotor, producing corresponding rotor poles. The hysteretic effect is, however, to retard the rotor poles behind the stator ones, so that a continuous attraction is set up between the stator and induced rotor poles. This *hysteretic torque* urges the machine in the direction of the stator rotating field. The power supply for the loss in rotor hysteresis must naturally be given to the stator in association with the main rotating field: but the loss is employed usefully in producing hysteretic torque which tends to raise the speed.

The pulsation loss is a high-frequency tooth loss in stator and rotor, having the frequency $S_2 n$ in the former and $S_1 n$ in the latter, and produced by variations of gap reluctance as the tooth tips pass each other. The loss depends on the relative slot-pitches in a manner not easily susceptible to calculation.

No-load Test with Rotor Driven: Suppose a *slip-ring* machine to be driven with open-circuited rotor and stator excited from a normal-voltage, normal-frequency supply. As the rotor speed is varied, áll the losses will vary with the exception of the stator loss ($p_{e1} + p_{h1}$). Since the friction losses are supplied by the drive, the stator electrical input corrected for I^2R loss consists of

$$(p_{e1} + p_{h1}) + (p_{e2} + p_{h2})$$

at standstill, and

$$(p_{e1} + p_{h1}) + s(sp_{e2} + p_{h2})$$

at slip s, for stator and rotor core losses. The rotor losses produce a forward torque, and therefore power, to drive the machine, and this must come also from the stator. From eq.(8.3) the stator takes in addition

$$(1 - s)(sp_{e2} + p_{h2}).$$

At synchronous speed the torque due to rotor eddy-currents is zero, but that due to rotor hysteresis may have any value between the limits $\pm p_{h2}$, depending on the position of the rotor with respect to the stator rotating poles. The electrical power supplied to the stator therefore changes discontinuously by $2p_{h2}$ between speeds just below and just above synchronism.

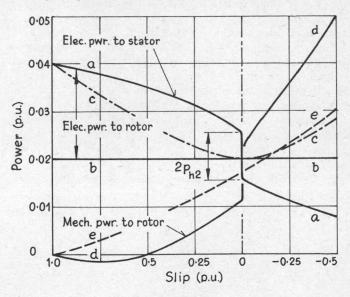

8.66 Loss estimation.

Fig. 8.66 shows loss/slip curves typical of the conditions described, viz. the motor driven by external means with its rotor open-circuited and the stator energized at normal voltage and frequency. Curve *a* represents the stator input; and curve *b* the constant stator loss ($p_{e1} + p_{h1}$). The intercept between *a* and *b* is the loss given electrically to the rotor ($sp_{e2} + p_{h2}$). The part $s(sp_{e2} + p_{h2})$ which is lost in the rotor is the intercept between *b* and *c*: the remainder between *c* and *a* is the part developed by the rotor as a forward driving power. The net mechanical power supplied to or by the rotor at various speeds is given by curve *d*. This is obtained by comparing the mechanical power needed to drive the rotor with the stator first excited then unexcited. The difference indicates what mechanical power the rotor develops from the electrical power it receives from the stator. The intercept between *d* and the base-line is the mechanical power that must be externally supplied to turn the rotor. Curve *e* is the sum of *d* and the intercept between *c* and *a*. Since *d* does not include purely mechanical losses, then *e* (which is the total mechanical power supplied from drive and stator together) is the pulsation loss which causes an effect resembling friction.

Subdivision: The following seven tests can be used to separate the several losses discussed above.

1. Measure the power to the stator at normal voltage and frequency with the rotor open-circuited and at rest. The stator input P_1 corrected for I^2R loss is the standstill core loss corresponding to curve *a* in Fig. 8.66.

2. Measure the power to the stator with the rotor short-circuited and running on no load: corrected for I^2R loss this gives the stator core and pulsation losses and the friction losses, P_2. The rotor eddy-current loss is very small; its hysteresis loss is almost wholly returned as a driving torque, providing part of the friction loss.

3. Measure as in 2 but with the brushes raised, P_3.

4. With the machine running as in 2, the rotor circuit is suddenly opened and the stator input measured immediately thereafter. The pulsation and friction losses cause the rotor to slow down.

5. Apply to the rotor the voltage that produces $\frac{1}{2}(V_1 + v_1)$ in the stator, as described above under 'Open-circuit (Voltage-ratio)'. Measure the rotor input with the machine running on no load with the stator short-circuited, to give the conditions in 2 but inverted. The corrected rotor input is P_5.

6. With the machine running as in 5, open-circuit the stator. The rotor input falls to P_6, corresponding to test 4.

7. Obtain the friction losses from a no-load test or a retardation test, to give P_7.

From these power observations (corrected in each case for I^2R loss)

$$p_b = P_2 - P_3 \qquad\qquad p_{h1} = P_6 + \tfrac{1}{2}(P_2 - P_5 - P_1)$$

$$p_f = P_7 - p_b \qquad\qquad p_{h2} = P_4 + \tfrac{1}{2}(P_5 - P_1 - P_2)$$

$$p_{e1} = P_1 - P_6 \qquad\qquad p_p = \tfrac{1}{2}(P_5 + P_2 - P_1) - P_7$$

$$p_{e2} = P_1 - P_4$$

The method, based on small differences, is not easy, but with care yields useful results.

Load Loss

The American Standard method [59] for assessing the 'stray' load loss of a single large cage machine comprises the two following tests.

Rotor-removed Test: A measurement of stator eddy-current and end-region fundamental-frequency loss components is obtained by circulating full-load current in the stator windings with the rotor removed from the machine. It is assumed that the end-region flux conditions are unaffected by the absence of the rotor, but this may not in fact be so especially with 2- and 4-pole machines with high electric loadings. The corrected fundamental-frequency stray loss is p_{1s}.

Reversed-rotation Test: This consists in driving the machine mechanically backwards at synchronous speed ($s = 2$) with the stator supplied at a voltage sufficient to circulate full-load current I_1; the rotor current then has full-load value I_2, and its equivalent I^2R loss for m representative phases is $mI_2{}^2r_2$. One-half of this, $mI_2{}^2r_2/s = mI_2{}^2r_2/2$, is provided by the stator and the other half by the mechanical drive. The total input to the stator is therefore

$$p_1 = mI_1{}^2r_1 + \tfrac{1}{2}mI_2{}^2r_2 + p_{1s}$$

while on the assumption that the high-frequency stray load loss p_{2s} is entirely supplied mechanically the total drive power is

$$p_m = p_f + \tfrac{1}{2}mI_2{}^2r_2 + p_{2s}$$

From these two measurable powers is derived the expression

$$p_{2s} = p_m - p_f - (p_1 - mI_1{}^2r_1 - p_{1s})$$

which is the mechanical power less the stator input corrected for its I^2R loss and fundamental-frequency load loss.

Chalmers and Williamson [38] discuss the validity of the method. They conclude that reasonable accuracy is obtainable in some cases in spite of the three assumptions made, namely that the tooth-frequency flux pulsations (i) are the main source of h.f. load loss, (ii) are supplied mechanically, and (iii) are the same for both directions of full-speed rotation. The errors can, however, be large where stator-winding pitch leads to high belt-leakage flux and loss, and where there are large sub-harmonic fluxes due to irregular stator windings.

Factory Testing

The testing of induction machines during and after manufacture can present special problems, particularly when customers consider it important to test machines under full-load conditions. These problems are considered in Chapter 12, Sect. 12.5.

9

Induction Machines:
Construction and Design

9.1 *Constructional features*

The essential features are: a laminated stator core carrying a poly-phase winding; a laminated rotor core carrying either a cage or a polyphase winding, the latter with shaft-mounted slip-rings; a stiff shaft to preserve the very short air-gap; a frame to form the stator housing and carry the bearings; and a cooling system.

Cores

The stator and rotor cores are built up of thin sheets of special core steel with good magnetic properties. The laminae are insulated from one another by varnish, paper or sprayed china clay films. In the smaller sizes the stator plates form complete annular rings, but for larger machines they are assemblies of segmental stampings. Typical methods of securing segmented stator cores are shown in Fig. 9.1. The gap surfaces of stator and rotor have suitable slots

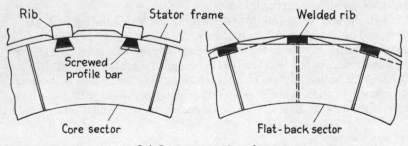

9.1 Sector stampings for stator.

punched out, of shape dependent upon the kind of winding chosen. Long cores are provided with spacer plates at intervals to provide radial cooling ducts.

In small motors the rotor core is mounted directly on the shaft, to which it is keyed, and clamped between a shoulder on the shaft and a shrink ring, with the slots skewed. For large machines the rotor is carried on a fabricated spider.

350

Frame

The frame may be die-cast or fabricated. Machines up to about 50 kW rating may have frames die-cast in a strong silicon-aluminium alloy, sometimes with the stator core cast in. The end-covers carry the bearing housings, and the method allows of ready thickening of the wall in regions of higher mechanical stress. The surface finish requires no machining.

The advantage of fabrication for large machines is its adaptability to new designs and modifications. The main constructional difficulty is the avoidance of distortion when the component steel plates are welded together. Essentially the frame is a short cylinder, or a box, with end plates and axial ribs. When the inner rib surfaces have been machined and the ends turned where necessary for the end covers, the frame is ready to receive the stator core, assembled and wound externally. Some fabricated elements are shown in Fig. 9.2. At (*a*) the

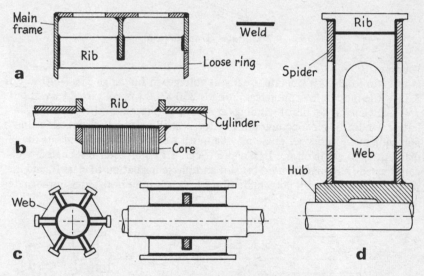

9.2 Fabricated constructions.

core length has an additional centre plate for rigidity; at (*b*) the core is compressed between end-plates; the rotor spider in (*c*) comprises a shaft with arms and stiffeners; the large diameter rotor spider (*d*) is built up from a hub, arms and ribs, the latter machined to take the rotor core.

Machines in the range of 250 − 10 000 kW are often built on a modular basis to fit a variety of specifications. The motor frame is a fabricated rectangular enclosure, Fig. 9.3, containing the stator core and provided with end plates with bearings to carry the rotor. Openings *a* can be sealed by plates, fitted with louvres or connected to air ducting. The top is covered by a structure *b* adapted to requirements of drip- or flame-proofing, or of total enclosure with a heat-exchanger.

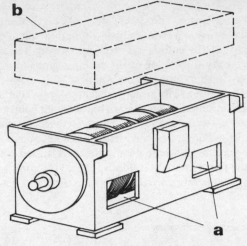

9.3 Modular construction.

Windings

Motors up to 250 kW operate usually at voltages in the range 380–440 V. For
3.3 kV the minimum economical rating is 250–300 kW, and for 11 kV about
750 kW. General details of windings are given in Chapter 3; the double-layer
lattice winding is most common for stators. Small motors with few slots and
requiring many turns per phase may be mush wound. Modern insulants for
diamond coils are in classes E, B and F (Sect. 3.1). Polyester foil coated with
compressed fibre may be used for slot and phase insulation to class E, and
plastic foil backed with polyamide fibres for class F, in both cases impregnated
with a class F varnish.

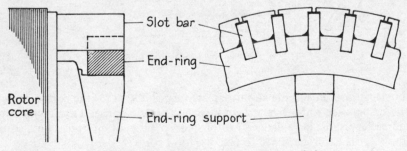

9.4 Connection of deep rotor bars to end-ring.

Cage rotor windings comprise copper or aluminium bars in slots (usually
uninsulated), welded or brazed to the end-rings. Fig. 9.4 shows a method of
connecting deep bars to end-rings: the jointing, as the most likely site of failure, requires care. It is also necessary to keep the bars tight in the slots, for
with mechanical vibration and thermal cycling a loose bar is quickly damaged.

Much of the difficulty can be avoided in small cage machines by integral casting of aluminium bars and end-rings. The stacked rotor core is used to form part of the die in a gravity-fed process, or by a centrifugal or pressure technique. The end moulds form the rings with fins as simple fan blading. A silicon alloy (6–12% Si) is employed, as pure aluminium does not cast well.

Where slip-rings are required, the rotor winding is of the conventional 3-ph type, a wave winding being common for its convenient adaptability. The choice of voltage is usually free; its standstill open-circuit slip-ring level is typically 100 V in small machines, and up to about 1 kV for large machines.

Enclosure and cooling

Besides the usual types of enclosure (Sect. 7.6) a flameproof construction is available for Division 1 Areas (mines, petroleum plants, etc., with hazardous atmospheres). This is a totally-enclosed construction with all joints flanged so that any flame generated by an internal explosion will be cooled in its passage through the flange so as to be externally innocuous. Machines without such flame-proofing are used in Division 2 Areas where flammable gas is not normally present.

Surface-cooled 3-phase motors for low and medium outputs are commonly made with finned housings. With higher outputs this simple method is inadequate, and the machines may be self-ventilated with two separate air circuits and tubular heat-exchanger. The inner circuit takes up the heat from the windings and cores, and thence to the cooler where the outer air circuit passes the heat to ambient; the two circuits are not in direct air contact. There is a fan for each circuit. The construction is suitable up to about 5 MW. At higher ratings the outer circuit may be water-cooled if there is a convenient supply.

Shafts and bearings

The shaft of an induction machine is short and stiff to maintain the necessarily short airgap. The use of ball or roller bearings makes accurate centring of the rotor readily possible. Where massive rotors demand journal bearings, the self-aligning spherical-seated type is useful.

Slip-rings may be pressed together on a body of reinforced thermo-setting resin carried on a mild-steel hub. Brush-lifting and short-circuiting gear is not generally fitted to modern motors. In small machines the rings may be located on the shaft between the rotor core and the bearing, or outboard. In the latter case the shaft has to be bored to take the slip-ring leads through the bearing.

A drawing in Fig. 9.5 shows the general arrangement of a large slip-ring motor. A small cage motor is shown in Fig. 9.7.

9.2 Main dimensions

Apart from some special machines, the manufacture of small and medium-rated induction motors is concentrated into a series of standard frames to cover a wide range by adjustment of rotor diameter and length, as indicated in Sect. 7.6. The synchronous speeds available with 50 Hz machines are 3 000, 1 500, 750, 600 . . . r/min, and 3 600, 1 800, 900 . . . r/min for 60 Hz supply. All these can be produced with a limited number of frames. The frame houses a

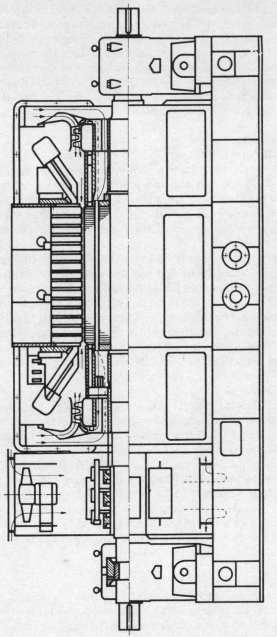

9.5 Slip-ring induction motor: 3 700 kW, 11 kV, 50 Hz, variable speed.

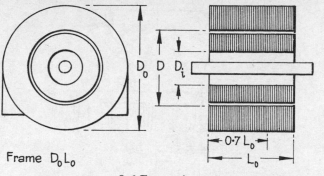

Frame $D_0 L_0$

9.6 Frame size.

stator of given outside diameter D_0, together with its bearings, end-covers etc., and a maximum active length L_o, Fig. 9.6. A variation of rating is possible by alternative core lengths down (say) to $0.7 L_o$, and the stator bore can be increased for multipolar designs.

The rating of a frame depends not only on its dimensions and the shaft speed, but also on the specific magnetic and electric loadings, the type of enclosure, load characteristics, and the starting-stopping duty. A demand for a low speed, such as 600 r/min, can at normal frequency be met only by a multipolar construction, which is uneconomic for machines of small rating.

The manufacture of small motors is highly competitive. The aim is a range of standardized cores, frames, cooling systems and insulation with the facility for economically meeting minor variations in specifications. A basic choice is that of the ratio of the stator and rotor outside diameters and the slotting, on which the saturation levels and I^2R loss depend. A further requirement is that the stalling torque should be about twice the full-load torque, a ratio that depends on the total leakage reactance: the latter is proportional to the leakage permeance and the square of the number of stator turns per phase, and it cannot be changed without some sacrifice elsewhere, such as in a variation of slot depth.

Economy demands that as much as possible of a range design be standardized, and 4-pole motors up to 250 kW may have completely standardized core punching dimensions, consequent upon which the design process is virtually reduced to specifying an appropriate winding from a standard list.

The following Examples give typical design data. The first is set out as a 'hand' calculation to indicate the application of design formulae, and represents the end product of a series of trials.

EXAMPLE 9.1: 5 kW, 415 V, 3-ph, 50 Hz, 4-pole cage motor for star-delta starting, standard duty cycle S1 (Sect. 4.8), to BS 3939. See Fig. 9.7.
Main dimensions. The appropriate frame has $D_o = 240$ mm, $L_o = 90$ mm, foot-mounted with a shaft-centre height of 132 mm. The power factor and efficiency on rated load are estimated each to be 0.83 p.u. The rated input is therefore $S = 5\,000/(0.83 \times 0.83) = 7\,260$ VA. The value of k in eq.(1.16) is $1.11\pi^2 K_w =$

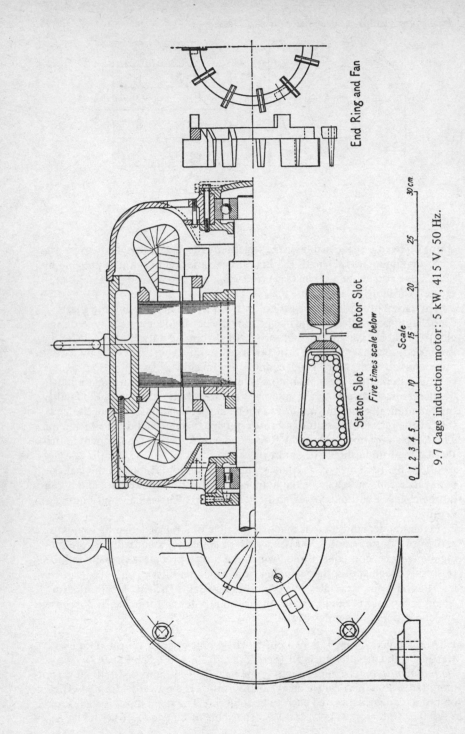

End Ring and Fan

Stator Slot   Rotor Slot
Five times scale below

0 1 2 3 4 5   10   15   20   25   30 cm

Scale

9.7 Cage induction motor: 5 kW, 415 V, 50 Hz.

11 K_w, and with full-pitch stator winding $K_w = 0.955$. The synchronous speed is $n = 25$ r/s. The specific magnetic and electric loadings are taken as $B = 0.48$ T and $A = 27\,000$ A/m. The main dimensions are obtained from

$$D^2 l = 7\,260/(11 \times 0.955 \times 0.48 \times 27\,000 \times 25) = 0.0021 \text{ m}^3$$

Taking the full available core length, then $l = L_o = 0.90$ m and hence $D = 0.15$ m. As the core length is too short to require cooling ducts, the net iron length is $l_i = 0.9l = 0.081$ m. The main airgap dimensions are thus

$$D = 150 \text{ mm}, \quad Y = 118 \text{ mm}, \quad l = 90 \text{ mm}, \quad l_i = 81 \text{ mm}, \quad l_g = 0.40 \text{ mm}$$

The radial gap is the minimum economically obtainable without surface grinding or special care.

Stator winding. The no-load flux is based on an e.m.f. equal to the supply voltage of 415 V/ph in delta connection. Then $\Phi_m = 0.48 \times 0.118 \times 0.09 = 5.1$ mWb. The number of turns per phase is

$$N_1 = 415/(4.44 \times 0.955 \times 50 \times 5.1 \times 10^{-3}) = 384$$

With $S_1 = 36$ slots giving $g_1' = 3$ slots per pole and phase, then there are $z_1 = 64$ conductors per slot. The estimated full-load current is $I_1 = 7\,260/(3 \times 415) = 5.8$ A. Taking a current density $J_1 = 4.9$ A/mm^2 the required conductor area is $a_1 = 1.18$ mm^2, obtained by enamelled round wire of diameter 1.22 mm (bare), 1.40 mm (insulated). The stator slots are shaped to permit of parallel-sided teeth each 6 mm wide. Allowing 0.5 mm slot liner and 1 mm slack, the slot depth including a 3 mm slot opening and a 1 mm lip, is 24 mm. The mush-type winding has a mean-turn length of 345 mm, and a total length per phase of 265 m. The phase resistance at 75 °C is

$$r_1 = 0.021 \times 265/1.18 = 4.7 \ \Omega$$

The total stator I^{2}R loss is $3 \times 5.8^2 \times 4.7 = 475$ W.

Rotor winding. A suitable cage is contained in $S_2 = 30$ slots, $g_2 = 7\frac{1}{2}$ slots/ph, slot-pitch $y_{s2} = 15.6$ mm, and a skew of one slot-pitch. An estimate of the rotor equivalent phase current is $I_1 \cos \phi_1 = 4.8$ A. From eq.(8.7) the r.m.s. bar and end-ring currents are

$$I_b = 4.8(6 \times 0.955 \times 384/30) = 350 \text{ A}$$
$$I_c = 350 \times 30/4\pi = 835 \text{ A}.$$

The rotor cooling is good, and the current density may be $J_2 = 7.6$ A/mm^2. The required bar area is $a_2 = 350/7.6 = 46$ mm^2. The bar dimensions are chosen with regard to the mechanical strength and saturation of the teeth. Slots 10.5 mm deep and 6.5 mm wide can house conductors 8.75 mm $\times$ 5.75 mm with rounded corners and light insulation. The bar length is 120 mm and its resistance at 75 °C is 0.055 mΩ; the I^{2}R loss for 30 bars is 200 W. The end-rings are cast with fan blades, and have an area of about 120 mm^2 and a mean diameter of 117 mm. The resistance around each is 0.064 mΩ (hot), and for the two rings the I^2R loss is 90 W. The total I^2R loss in the cage is 290 W. The rotor resistance expressed as a stator phase equivalent is

$$r_2' = 290/(3 \times 4.8^2) = 4.3 \ \Omega/\text{ph}.$$

No-load current. The effective gap length $l_g' = k_g l_g$ is found by applying eq. (2.8) in Sect. 2.5:

$$k_g = \frac{13.1}{13.1 - (0.74 \times 3.0)} \cdot \frac{15.6}{15.6 - (0.40 \times 1.0)} = 1.24$$

so that $l_g' = 1.24 \times 0.40 = 0.50$ mm. The effective gap area per pole is $Yl = 0.0106 \ \text{m}^2$. The cores are constructed from No. 42 grade 0.4 mm plates. The Table shows the estimation of the magnetizing m.m.f., based on the procedure given in Sect. 2.5. The magnetizing component of the no-load current from eq.(2.10) is

$$I_{om} = 530 \times 2/(1.17 \times 384 \times 0.955) = 2.47 \ \text{A/ph} = 4.3 \ \text{A/line}$$

The core loss, taken as confined to the stator core and teeth, is estimated from the curves in Fig. 2.5(b):

Core: peak density, 1.50 T; specific loss, 15 W/kg; mass, 8.75 kg; loss, 132 W.
Teeth: estimated peak density, 1.75 T; specific loss, 20 W/kg; mass, 3.15 kg; loss, 63 W.

The total core loss is 200 W. Including 60 W for mechanical loss, the total no-load loss is 260 W, requiring an active component $I_{0a} = 260/3 \times 415 = 0.21$ A/ph. The no-load current is therefore

$$I_0 = \sqrt{(2.47^2 + 0.21^2)} \simeq 2.5 \ \text{A/ph} = 4.3 \ \text{A/line}$$

at a power factor $\cos \phi_0 = 0.085$. The magnetizing reactance per phase is $x_m = 415/2.47 = 168 \ \Omega$.

Magnetization: $\Phi_m = 5.1$ mWb

a, area (m^2); l, length (m); $B_{30} = 1.36 \ \Phi_m/a$; $B_c = \Phi_m/2a$

Part	a	l	B_{30}	B_c	H	F
Stator core	0.0017	0.060	–	1.50	2 400	144
teeth	0.0044	0.024	1.59	--	4 200	100
Airgap	0.0106	$0.50/10^3$	0.65	–	520 000	260
Rotor teeth($\frac{1}{3}$)	0.0047	0.0105	1.49	–	2 100	20
core	0.0024	0.030	–	1.06	210	6

| | | | | | Total: | 530 A-t |

Short-circuit current. The specific slot permeances are

$$\lambda_{s1} = \frac{18}{3 \times 9} + \frac{2}{8} + \frac{2 \times 2}{8 + 3} + \frac{1}{3} = 1.61 \quad \text{and} \quad \lambda_{s2} = 1.65$$

Referring the rotor value to stator terms requires multiplication by $(K_{w1}{}^2 S_1)/(K_{w2}{}^2 S_2) = 0.955^2 \times 36/30 = 1.094$, giving 1.81. The total slot permeance coefficient is $1.61 + 1.81 = 3.42$. Then eq.(2.28) gives the *slot leakage reactance* in stator terms:

$$x_s = [4\pi f \mu_0 N_1{}^2/pg_1{}']L_s\lambda_s = 19.4 \times 0.09 \times 3.42 = 6.0 \,\Omega$$

For the overhang leakage, eq.(2.27) gives $L_0\lambda_0 = 0.34$ and the *overhang leakage reactance* in stator terms:

$$x_o = 19.4 \times 0.34 = 6.6 \,\Omega$$

From eq.(2.25) the *zigzag leakage reactance is*

$$x_z = \frac{5}{6} 168 \left[\frac{1}{9^2} + \frac{1}{7.5^2} \right] = 4.2 \,\Omega$$

With a cage rotor there is negligible differential leakage. The total leakage impedance per phase referred to the stator winding is given by

$$R_1 = r_1 + r_2{}' = 9.0 \,\Omega; \quad X_1 = x_s + x_o + x_z = 16.8 \,\Omega; \quad Z_1 = 19.0 \,\Omega$$

The short-circuit current is $I_{sc} = 415/19.0 = 21.8$ A/ph $= 38.0$ A/line at a power factor $\cos \phi_{sc} = 0.47$.

Performance. The current diagram or the equivalent circuit gives

$$I_1 = 5.7 \text{ A}, \quad I_2{}' = 4.6 \text{ A}, \quad \cos \phi_1 = 0.83$$

for the full-load phase values, confirming the design figures. The full-load efficiency is calculated as follows

		W
Stator I^2R:	$3 \times 5.7^2 \times 4.7$	458
Rotor I^2R:	$3 \times 4.5^2 \times 4.3$	262
Core loss:		200
Mechanical loss:		60
Total loss:		980
Output:		5 000
Input:		5 980

The *efficiency* is $5\,000/5\,980 = 0.835$, reduced by $\frac{1}{2}\%$ for stray load loss to 0.83 p.u.

The full-load *slip,* from the ratio of rotor I^2R loss to total rotor input, is

$$s = 262/(262 + 60 + 5\,000) = 262/\,5\,322 = 0.049 \text{ p.u.}$$

The *starting torque* for direct switching in delta is given by the current diagram or by the equivalent circuit as 1.16 p.u., which is reduced by star-delta switching to 0.39 p.u. It is in practice somewhat greater because saturation reduces the leakage reactance and eddy currents raise the rotor effective resistance at stand-still. The *pull-out torque* is 2.0 p.u.

Cooling. The rotor end-rings are bare and carry fans, and the stator overhang is well ventilated. From the drawing in Fig. 9.7 the cooling surface across which air is fanned is approximately a cylinder formed by the inner faces of the overhang and core, and the annular faces formed by the stator coil ends. The back of the stator also dissipates. The inner area is estimated as 0.17 m². On account of the rotor peripheral speed (u = 11.75 m/s) the inner surfaces are more effective than the back of the core by the factor $(1 + 0.1\ u)$ = 2.2, making the equivalent surface 0.17 x 2.2 = 0.375 m². The back of the stator has a surface of area 0.068 m², so that the total is 0.44 m²: it has to dissipate about 660 W of I^2R and core loss. Assuming from Sect. 4.3 a coefficient c = 0.03, the temperature-rise is

$$\theta_m = 0.03 \times 660/0.44 = 45\ ^\circ C$$

The rotor loss is 260 W, and it has ample cooling surface: its temperature-rise is about 35 °C, well within the specified limits and with a margin for the rapid heating during a start.

EXAMPLE 9.2: Outline data are given below for two slip-ring motors differing from the normal industrial range: the slotting and main dimensions are set out in Fig. 9.8.

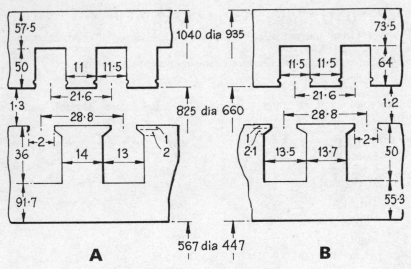

A   567 dia 447   B

9.8 Slot and core details.

Motor A: 270 kW, 3.0 kV, 3-ph, 50 Hz, 10-pole slip-ring motor to drive a centrifugal water pump, rated with reduced specific magnetic and electric loadings for a 30 °C temperature-rise in ambient air at 55 °C and 100% humidity.
Motor B: 75 kW, 3.0 kV, 3-ph, 50 Hz, 8-pole slip-ring motor, flame-proof for driving mine-haulage machinery, rated for a 50 °C temperature-rise; the restricted ventilation enforces low I^2R loss, especially in the rotor.

		A	B
Rating			
Output	kW	270	75
Power factor	p.u.	0.84	0.86
Efficiency	p.u.	0.93	0.94
Input	kVA	346	93
Line/phase current	A	66.6	17.9
Main Dimensions			
Magnetic loading	T	0.39	0.225
Electric loading	kA/m	30.9	14.9
Stator bore	m	0.825	0.66
Gross bore length	m	0.40	0.50
Ducts, no./width	mm	5/10	nil
Net core length	m	0.35	0.50
Iron length	m	0.32	0.45
Radial gap length	mm	1.30	1.20
Pole pitch	m	0.26	0.26
Stator			
Winding/connection		2-layer diamond/star	
No-load flux per pole	mWb	40.0	29.1
Turns per phase	—	200	286
Number of slots	—	120	96
Distribution factor	—	0.958	0.958
Coil-span factor	—	1.0	0.991
Winding factor	—	0.958	0.950
Slot pitch	mm	21.6	21.6
Conductors per slot	—	2 x 5	2 x 9
Conductor size	mm	6.5 x 3	6.5 x 2.2
area	mm^2	19.0	14.0
Length of mean cond.	m	0.963	1.032
Resistance	Ω/ph	0.426	0.886
Current density	A/mm^2	3.50	1.27
Total f.l. $I^2 R$ loss	kW	5.70	0.84
Copper mass	kg	220	250
Rotor			
Winding/connection	—	2-layer bar wave/star	
Slip-ring o.c. voltage	V	650	490
Turns per phase	—	45	48
Number of slots	—	90	72
Winding factor	—	0.960	0.960
Slot pitch	mm	28.8	28.8
Conductors per slot	—	2(2 x 3)	2 x 2
Conductor size	mm	13 x 3	20 x 5
area	mm^2	77	98.5
Length of mean cond.	m	0.862	0.963
Resistance	mΩ/ph	21.2	19.7

Table continued . . .

		A	B
Current density	A/mm²	3.32	2.42
Total f.l. I²R loss	kW	4.00	0.49
Copper mass	kg	175	265

No-load current

M.M.F. per pole		A-t	1 100	500
Magnetizing	current	A	24.6	6.3
	reactance	Ω	70.5	275
Core loss		kW	5.7	2.2
No-load	loss	kW	8.6	3.1
	current	A	24.6	6.3
	power factor	p.u.	0.067	0.095

Short-circuit current

Slot reactance	Ω	2.26	9.85
Overhang reactance	Ω	1.56	3.75
Zigzag reactance	Ω	1.13	4.40
Differential reactance	Ω/ph	0.30	0.90
Total reactance	Ω/ph	5.25	18.9
Total resistance	Ω/ph	0.85	1.57
Total impedance	Ω/ph	5.3	19.0
Short-circuit current	A	327	91
power factor	p.u.	0.16	0.083

Performance

Full-load losses:			
stator I²R	kW	5.7	0.84
rotor I²R	kW	4.0	0.49
brush	kW	0	0.14
core	kW	5.7	2.2
mechanical	kW	2.9	0.9
load (stray)	kW	0.9	0.33
total	kW	19.2	4.9
Output	kW	270.0	75.0
Input	kW	289.2	79.9
Efficiency	p.u.	0.93	0.94
Full-load rotor input	kW	276.9	76.5
slip	p.u.	0.0145	0.0061
speed	r/min	591	745
Pull-out torque	p.u.	2.70	2.87

9.3 Design

Traditional formulae and methods are inadequate for modern economic design to extended limits and greater loadings, and for the use of digital computers.

The earliest application of digital computation was simply as a slide-rule. A later step (taken first in induction machine design in 1956) was a form of progressive synthesis, the designer making initial estimates of the independent variables and intervening to modify them in the light of the computed performance. This method is still valuable for the investigation and refinement of specific details; but with fast modern computers it may be expensive to interrupt calculations.

In the design of small induction motors, which have completely standardized coreplates, synthesis is considerably simplified, for many independent variables become constants. The chief independent variables are the stator turns per phase and the core-length, and their determination can be optimized on a cost basis. This means an organized procedure for scanning the area of possible designs until a minimum of the cost function is found.

Only in the case of transformer design (Sect. 6.9) have full synthetic-design programs been so far possible, as a consequence of the relatively few variables compared with those essential in a rotating machine. Here the synthesis must include analysis and subsequent iterative processes, i.e. design analysis (the calculation of performance for any current set of independent variables) and direct-search or gradient iterations to find maxima or minima. Gradient methods require a knowledge of the derivatives of the appropriate function, which carry considerable information regarding the function but are often very difficult to formulate. If these derivatives cannot be stated, direct-search methods have to be used: such a condition frequently arises because some variables (such as slot numbers) are discrete, not continuous.

Formulation

For digital computation, the many variables in the traditional 'hand' design (Sect. 9.2) must be reduced by changing them into fewer new ones with suitable interrelations. Many such formulations are in use: one is the following.

Given the core length, outside diameter, current and flux densities and the specification, the method is (i) to cast the basic equations for starting and pull-out torque, starting apparent power and full-load slip into functions of three variables, the total leakage permeance Λ and the ratios G_1 and G_2 of the stator and referred rotor phase resistances to the *total* leakage reactance, so that $G_1 = r_1/X_1$ and $G_2 = r_2'/X_1$; (ii) to eliminate slot dimensions from the equations for Λ, G_1 and G_2 by using expressions for slot shape and employing tooth and core flux densities in terms of their levels compared with the gap density. For example, eq.(8.15) for the maximum torque can be expressed as

$$M_m = 0.075 \frac{[1.05 B_g(D - l_g)lK_w]^2}{p\Lambda [G_1 + \sqrt{(1 + G_1)^2}]}$$

G_1 is obtained from $r_1 = \rho_1 L_{mt} N_1/a_1$ and $X_1 = 2\pi f N_1^2 \Lambda$. The mean length of stator turn is based on the core-length l and pole-pitch $\pi D/2p$ in a form such as $L_{mt} = (2l + 1.6\pi D/p)$. The conductor area per slot $a_1 = k_{s1}a_{s1}S_1/6N_1$ uses the gross slot area a_{s1}, the space factor k_{s1} and the number S_1 of stator slots. Then

$$G_1 = \frac{0.95 \, \rho_1 (2l + 5D/p)}{k_{s1} a_{s1} S_1 f \Lambda}$$

A rather more complicated expression gives G_2. Often only a few slot combinations are acceptable for reasons of harmonic torques and noise, and the choice is further limited if the same core plates have to serve for several pole-numbers. It may be possible to fix S_1 and S_2 for a given frame. The calculation of Λ is in any case expressed in terms of a series of slot dimensions, based on expressions such as those in eqs.(2.22 – 27). The slot dimensions also settle the tooth and core areas for a given frame. The effects of changes in various parameters can be computed in the search for an optimum design. The Table shows typical directions of change in slot area, permeance, G-values and the ratio J_1/J_2 of stator/rotor current density that result from an increase in the stator bore D, core length l, and the ratios b_1 and b_2 of the stator and rotor core or tooth flux density to the gap density:

Increase in:		b_1	b_2	D	l
Effect on:					
Gross slot area	a_{s1}	up	0	down	0
	a_{s2}	0	up	up	0
Permeance Λ		0	0	0	up
Ratio	G_1	down	0	up	0
	G_2	0	down	down	0
Current density J_1/J_2		down	up	up	0

A design synthesis on this basis gives main dimensions to satisfy specified torque and starting apparent-power requirements. The program is run to arrive at a minimum number of punching forms to cover each frame, and some economic compromise struck for each required rating. It may be found that three sets of punchings per frame are needed: one for a 2-pole machine, in which the core behind the slots has to carry a large flux; one for 4- and 6-pole machines; and one for motors with 8 or more poles. For each set and frame the program is run with D and l independently varied to enable the compromise dimensioning to be selected. A motor can then be designed using the pull-out torque as a criterion. For a given pair of D and l values only one gap flux density gives the required pull-out torque and the core and tooth densities follow, as also does the stator current density. The cooling has, of course, always to be checked.

Programs
Design synthesis programs are not easy to devise. The problem is discussed by Chalmers and Bennington [60], who outline a method for large non-standard-

ized cage motors that co-ordinates a number of suitable techniques. The program routines are:

Passive: Routines that release data (e.g. tool sizes) to the program or which carry out certain design analyses.

Performance: Routines associated with the calculation of equivalent-circuit parameters and calculate performance therefrom.

Active: Program organizing the search for solutions.

Function: Holds a series of function subroutines such as those for calculating reactance, those stating magnetic *B/H* relations and those for dealing with saturation, slot-openings and end-winding dimensions.

Input and Output: Reads in data, checks its validity and transfers control to a 'steering' routine; each time control is referred back, more data are read in. An update routine records the state of all output parameters at the end of a total cycle. An output routine organizes the specific output routines, and prints out the design and performance information.

Monitor: This consists of two sets of orders, (i) printing information concerning the course of calculation of a particular design to indicate the manner in which the design is obtained, to facilitate detailed improvements; and (ii) trapping errors.

Such a program may take 80 s of computer time but produces several alternative designs in the process.

Inverter-fed Motors

Economic design depends on standardization. The individual matching of a motor to its inverter conflicts with the low-cost advantage of a standard machine in that a wide range of operating frequencies is concerned and a significant harmonic content is present. The leakage inductance should be low with c.s. inverter feed to reduce voltage peaks, but high for v.s. inverters to limit harmonic currents. The problem is discussed by De Buck [132] for motors accelerated by a controlled rise in frequency: he suggests the adoption of a standard leakage reactance of about 0.3 p.u. at nominal frequency (50 or 60 Hz), with series inductors between the inverter and motor terminals where necessary to suit the inverter characteristics. Stator windings must have restricted conductor and slot depths to limit eddy-current loss (Sect. 3.2) at high frequencies; and as rotor cages develop additional loss at harmonic frequencies, consideration may have to be given to subdivided bars (Fig. 3.2), or even to a coil winding. A rotor of either normal or modified form, dimensionally interchangeable, can then be selected as necessary. Skewed slotting is undesirable for the reasons set out in Sect. 3.13, and is in any case unnecessary as parasitic torques present no problem.

10
Synchronous Machines: Theory and Performance

10.1 *Type and construction*

Large a.c. power networks operating at constant frequency (e.g. 50 or 60 Hz) rely almost exclusively on synchronous *generators* for the provision of electrical energy, and may have synchronous *compensators* at key points for reactive-power control. Private, stand-by and peak-load plant with diesel or gas-turbine prime-movers also have synchronous generators. Non-land-based synchronous plant is found on oil rigs (with a generating capacity of 50 MVA upward), on large aircraft (with hydraulically driven generators working at 400 Hz), and on ships for variable-frequency supply to synchronous propeller motors. Synchronous *motors* provide constant-speed industrial drives with the possibility of power-factor correction, but are not often built in small ratings for which the induction motor is cheaper.

The speed n_s (in r/s) of a synchronous machine is related to the frequency f_1 and the number of pole-pairs p by $n_s = f_1/p$. With a 2-pole construction this means 3 000 and 3 600 r/min for 50 and 60 Hz respectively. The use of a rotating d.c. field system is almost universal, as it permits the a.c. windings to be placed on the stator where they are more readily braced against electromagnetic force and insulated for high voltage. The rotor field construction may be with *salient poles,* or be *cylindrical* with no polar protuberances.

Generators. The prime-mover speed has a profound influence on the constructional form, and in all large units the limiting feature is the centrifugal force on the rotor. Steam-turbine-driven machines (*turbo-generators*) must run at high speed, and have 2-pole rotors of solid forged steel, of diameter limited to about 1.2 m and of axial length several metres. Generators driven by water power (*hydro-generators*) are built for a wide range of comparatively low turbine speeds, and are axially short, but have large diameters to accommodate the many salient poles.

Compensators. Except at start-up, these are not called upon to develop or absorb mechanical power, and are built with cylindrical or salient-pole rotors and a minimum of mechanical loss.

Motors. Both constructional forms are used, in accordance with the synchronous speed required. As a plain synchronous motor develops a torque only at synchronous speed, provision must be made for it to start in some other way, usually by fitting the rotor with a cage winding to develop an induction torque.

10.2 *Action of the ideal synchronous machine*

The essential action of a polyphase synchronous machine is most simply introduced in terms of an *ideal* cylindrical-rotor machine connected to *infinite busbars*. The ideal machine has stator windings of negligible resistance and leakage inductance, a uniform airgap, a magnetic circuit of high permeability and devoid of saturation, and balanced loading. The infinite busbar system to which the machine is connected produces a balanced 3-phase voltage of constant magnitude and frequency, and can deliver or absorb both active and reactive power without limit.

The machine in Fig. 10.1 has stator windings (not shown) connected to the

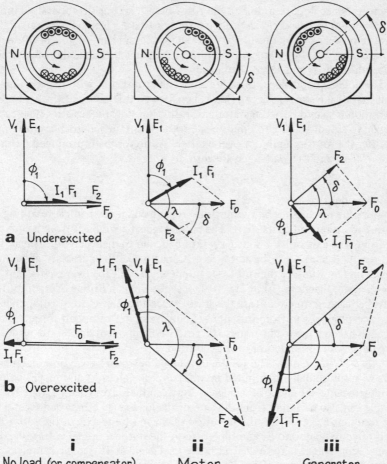

10.1 Ideal synchronous machine: phasor diagrams.

infinite busbar, and a rotor carrying a d.c. exciting winding. To work at all the rotor must rotate at synchronous speed. No torque can be developed if the rotor is unexcited, but the stator must take in lagging reactive power to magnetize the machine to a gap flux per pole of Φ_m in order that a stator e.m.f. E_1 can be induced to balance the applied voltage V_1. If the rotor is now given a small d.c. excitation, it takes over part of the task of exciting the magnetic circuit, reducing the demand of the stator for magnetizing power. This is the condition of *under-excitation*, (a) in Fig. 10.1

With the machine on *no load* (zero mechanical load, and therefore also zero active electrical power), the conditions in (i) apply, and the phasor diagram can be drawn using the representative-phase concept of Sect. 7.1 and incorporating the travelling-wave m.m.f. F_1 of the stator and the rotating m.m.f. F_2 of the rotor. As there is no torque, the m.m.f. axes remain in alignment with a zero torque angle, and the two m.m.f.s combine to give the resultant F_0 necessary to produce Φ_m. If, as in (b), the rotor m.m.f. is considerably raised into the condition of *overexcitation*, the stator must take a leading, demagnetizing current in order that the resultant F_0 shall remain unchanged and still generate Φ_m and $E_1 = V_1$. (For clarity in the phasor diagrams the currents correspond in length to the m.m.f.s that they produce.)

The *no-load* conditions, then, are these: the rotor runs with its axis aligned with that of the stator; the gap flux is constant so that always $E_1 = V_1$; the magnetization is shared between stator and rotor; the stator takes a reactive current, lagging to aid, or leading to oppose, the rotor m.m.f. in accordance with whether the rotor is under- or over-excited. As no torque is produced, both the torque angle and the load angle are zero.

Motor mode

Suppose the machine to be running on no load, and a mechanical retarding torque to be applied to its shaft. The rotor begins to fall back from its coaxial position. The axis of its m.m.f. F_2 is retarded, and to maintain normal gap flux and resultant m.m.f. F_0, the stator m.m.f. must make the necessary adjustments. Depending on whether the rotor is (a) underexcited or (b) overexcited, the phasor/space diagrams (ii) in Fig. 10.1 describe the conditions of stability that must be reached. In either case the stator takes an active current component, thus accepting power from the supply and developing a forward torque on the rotor to balance the load torque. The reactive component acts, as on no load, to compensate for under- or over-excitation.

Torque and load angles. The torque developed between two cylindrical members with sinusoidally distributed m.m.f.s of peak values F_1 and F_2 per pole is proportional to the sine of the angle λ between their axes (Sect. 1.11). The torque can also be found from the sine of the load angle δ between F_2 and the resultant F_0. As F_2 has the direction of the magnetic axis of the rotor, and F_0 is directly related to E_1 and therefore to the stator terminal voltage V_1, it is more convenient here to deal with the *load angle* δ. This angle is indicated in the phasor diagrams of Fig. 10.1. It will be observed that δ is negative for the motor mode.

The synchronous machine thus acts as a motor by a backward adjustment of its load angle. Greater mechanical load calls for an increase in negative load angle. But this cannot go on without limit, for if δ exceeds $-\frac{1}{2}\pi$ rad (elec.) the torque falls below its maximum and the rotor (by 'falling out of step') will desynchronize and stop.

Diagrams (a) and (b) for the motor are drawn with the same load angle. Overexcitation results in greater active current and therefore torque, and the maximum power capability is markedly raised.

Generator mode

Suppose again that the machine is running on no load, but that now a prime mover connected to its shaft provides a driving torque in the direction of the stator travelling-wave field. The rotor moves forward, developing a load angle δ of sign opposite to that for the motor mode. The relative position of the axis of F_2 makes necessary a change in F_1 to provide the constant resultant F_0. The conditions are shown in Fig. 10.1 (iii), with (a) for underexcitation and (b) for overexcitation. The active component of the stator current reverses, so delivering power into the supply and developing a counter torque to balance the driving torque. The machine thus generates. Again the reactive component of I_1 must provide proper compensation, and this affects the power factor at which the machine works. With respect to output, underexcitation causes ϕ_1 to be a leading angle (meaning that there is an input of lagging reactive magnetizing power from the supply); overexcitation has the opposite effect. For the same load angle δ, greater excitation means greater power, as it does for the motor mode.

Compensator (phase-modifier) mode

A synchronous machine designed to run unloaded, with its shaft connected neither to a mechanical load nor to a prime mover, can operate only in accordance with the no-load conditions (i) of Fig. 10.1. Variation of the rotor excitation causes the machine to take a purely reactive stator current, lagging if underexcited and leading if overexcited. It therefore behaves at its stator terminals as if it were a polyphase inductor or capacitor, changing from one to the other in accordance solely with the rotor excitation. Such a machine, called a *synchronous compensator*, can be used to control the voltage regulation of transmission-line systems by modifying the resultant receiving-end load. The limit to the capacitive effect is the level of rotor excitation, while the inductive effect, as it demands underexcitation, cannot be extended beyond the excitation level below which the machine becomes unstable (tends to lose synchronism).

Operating conditions

The same machine, without change in its stator connections or its direction of rotation, can operate as motor or generator solely by accommodating its load angle to the mechanical conditions imposed on its shaft. Varying the rotor excitation then controls the power factor of the stator current, but not its power component, except that the *limit* of active power and torque is determined by the excitation. With mechanical load or drive beyond this limit the machine is unable to develop a balancing electromagnetic torque, and falls or rises out of synchronism

As synchronous speed is the only stable operating condition in which sustained torque can be developed, the machine has no torque when it is at rest. It has to be started and brought up to synchronous speed by some auxiliary means.

E.M.F diagrams

The gap flux Φ_m which generates E_1 is excited by F_1 and F_2 in combination. However, in the ideal cylindrical machine, the absence of saturation makes it valid to imagine that F_1 and F_2, acting across the uniform gap length, excite *separate* sinusoidally distributed gap fluxes Φ_a and Φ_t which superpose to give Φ_m as resultant. The two flux components can then be considered to induce distinguishable e.m.f.s E_a and E_t in the stator windings, having the phasor sum E_1. Consider the diagram (b)(ii) for the over-excited motor in Fig. 10.1: if this be drawn with e.m.f.s instead of m.m.f.s, then $E_1 = V_1$ is drawn leading $\pi/2$ rad on F_0, and therefore E_a and E_t bear the same relation respectively to F_1 and F_2. The e.m.f. E_a due to the stator's own m.m.f. and flux is in quadrature with the phasor of current I_1 and is proportional to I_1. The e.m.f. diagram then appears as in Fig. 10.2(a), and it can be related to the equivalent circuit (b), in which E_a can be taken as produced by the stator current in an 'armature-reaction reactance' external to a 'machine' in which only E_t due to rotor excitation is generated.

Reaction reactance. The stator m.m.f. and its reaction on the rotor m.m.f. to give the resultant m.m.f. is real enough: its representation by an 'external' inductive reactance is fictitious, but convenient. Although it somewhat obscures the physical action and applies only to steady-state conditions, by its use we can derive direct expressions for power and torque in terms of measurable quantities — the stator terminal voltage, the rotor exciting current, and the load angle. The (armature-) *reaction reactance* $x_a = E_a/I_1$, a measure of the stator m.m.f. per unit current, is readily calculable from the stator winding details and the gap dimensions, or it can be obtained directly from the open- and short-circuit characteristics of the machine.

Open-circuit characteristic. The o.c.c. relates the phase e.m.f. of the stator on open circuit to the rotor excitation when the machine is driven at synchronous speed. The measured phase e.m.f. E_{oc} then directly determines E_t for any given rotor excitation.

Short-circuit characteristic. The s.c.c. is found by driving the machine (preferably at synchronous speed, but in an ideal machine the actual speed — provided that it is not zero — does not matter) with the stator phases short-circuited. The conditions are then that $V_1 = E_1 = 0$, so that there is no resultant gap excitation, and in consequence the stator and rotor m.m.f.s just balance out with $F_1 = F_2$.

Open- and short-circuit characteristics are shown in Fig. 10.2(c). On the basis of the equivalent circuit (b), the argument is that E_{oc} for a given excitation is E_t, and that therefore I_{sc} for the same rotor excitation is such that $E_a = E_t$; whence the reaction reactance must be $x_a = E_{oc}/I_{sc}$.

Operating characteristics

The phasor diagram in Fig. 10.2(d) is for the same conditions as in (a), namely an overexcited motor mode. The input current is shown resolved into the active component $I_{1a} = I_1 \cos\phi_1$ and the leading reactive component $I_{1r} = I_1 \sin\phi_1$. The corresponding components of reaction reactance voltage can be seen by inspection to have the scalar magnitudes

$$I_{1a}x_a = -E_t \sin\delta \quad \text{and} \quad I_{1r}x_a = V_1 - E_t \cos\delta$$

Multiplying through by V_1 gives the input active and reactive electrical powers per phase:

$$P_1 = -(V_1/x_a)\, E_t \sin\delta \tag{10.1}$$

$$Q_1 = (V_1/x_a)\, [E_t \cos\delta - V_1] \tag{10.2}$$

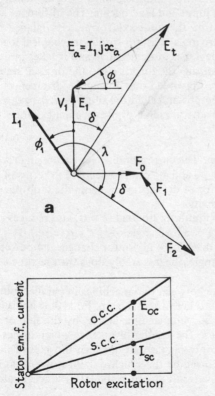

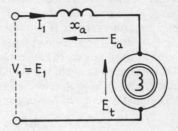

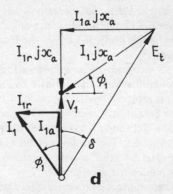

10.2 Ideal synchronous machine: equivalent circuit.

These relate the powers per phase to the rotor excitation level and the load angle for a machine of given reaction reactance connected to an infinite busbar of voltage V_1. As the ideal machine has no losses, P_1 is also the mechanical power per phase, and if divided by the synchronous speed ω_s gives the torque. For a motor the load angle δ is negative, giving P_1 as a positive input; for a generator with a positive load angle, the input P_1 is negative, meaning an electrical output. A positive value for Q_1 means that the reactive power input is leading.

The relations in eqs.(10.1) and (10.2) are illustrated in Fig. 10.3 for a machine with a reaction reactance $x_a = 1.25$ p.u. The curves of active and reactive input powers are drawn for excitations E_t expressed in per-unit values, where 1.0 p.u. excitation is that which gives $E_t = V_1$. For any given power P_1 and excitation E_t there is a particular load angle δ and an associated leading or lagging reactive power Q_1.

Synchronizing power and torque. A synchronous machine tends to maintain itself in synchronism with the supply. Suppose a machine, running on a steady load P_1 with a load angle δ, to begin to rise or fall in speed: then the load angle begins to diverge from the balance condition and the input power also changes, so that a power differential between input and load occurs. The difference acts against the angular disturbance to return the rotor axis towards its balance condition; for example, if a motor driving a constant-torque mechanical load begins to slow down, its load angle becomes more negative and it takes in more electrical power, the excess restoring the rotor axis to the original state. The machine will remain synchronized provided that an angular divergence brings about an appropriate change in power, called the *synchronizing power* and defined in terms of power per electrical radian of displacement: hence

$$p_s = \mathrm{d}P_1/\mathrm{d}\delta = (V_1/x_a)E_t \cos \delta \qquad\qquad (10.3)$$

is the synchronizing power per radian. The corresponding synchronizing torque is $M_s = p_s/\omega_s$. More practically, both p_s and M_s can be specified in terms of power or torque per mechanical degree of displacement, using the multiplier $p(\pi/180)$ for a machine with p pole-pairs.

The synchronizing power is a maximum for no load ($\delta = 0$) and reduces as the machine is more heavily loaded. For a load angle of $\pm\frac{1}{2}\pi$ rad (elec.) it vanishes altogether, because angular divergence no longer changes the power. Up to this limit, the larger the load angle, the less stably does the machine operate.

Synchronizing power can circulate between two machines in parallel (but not connected to an infinite busbar) to keep them in step. Eq.(10.3) holds if V_1 is replaced by the E_t value of the second machine, and x_a by the sum of the two individual reaction reactances. The angular divergence which brings the synchronizing power and torque into play is that between the axes of the two rotors. The 'faster' machine is slowed by the synchronizing power that it circulates to the 'slower' machine, as a result of which the latter is accelerated to reduce the divergence.

Hunting. Acceleration or retardation of a machine by synchronizing torque imparts to the rotating system an angular acceleration and a change of kinetic

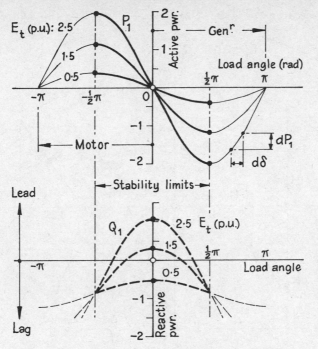

10.3 Active and reactive power/load-angle relations.

energy. A machine will therefore over-swing beyond the equilibrium angle, build up a synchronizing torque in reverse, and continue to hunt (i.e. to support variations of speed about the mean synchronous speed) unless the kinetic swing energy can be dissipated, as it must be in a practical machine. As the machine has a synchronizing torque proportional to divergence, the analogue of an 'elastic' property, its natural undamped frequency of hunting is given by

$$\omega_0 = \sqrt{(M_s/J)} \tag{10.4}$$

when the moment of inertia of the rotating parts is J.

Voltage regulation. For a generator, the *regulation* is the voltage rise at the terminals when a given load is thrown off, the machine being disconnected from the infinite busbar but the speed and excitation remaining unchanged. The rise is the scalar difference between E_t and V_1 and the regulation is defined as

$$\epsilon = (E_t - V_1)/V_1 \text{ p.u.} \tag{10.5}$$

The regulation depends on the load and power factor. It may be zero for a leading power factor load, or even be negative. Generally synchronous generators operate on lagging loads and have a large positive regulation.

Electrical load diagram. This shows all possible operating conditions for a

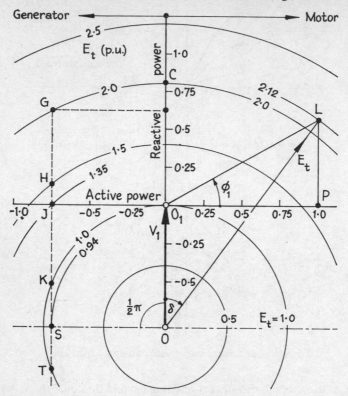

10.4 Ideal synchronous machine: electrical load diagram. The scales are in
per-unit values for a reaction reactance of 1.25 p.u.

machine on infinite busbars. It is based on the phasor diagram (*d*) of Fig. 10.2:
let this represent full-load conditions. Then the electrical load diagram, Fig.
10.4, is built round it with the constant busbar voltage V_1 represented by OO_1,
E_t by OL and $I_1 j x_a$ by LO_1, with PO_1 and LP as its active and reactive components. As PO_1 is for full-load active current and so also for 1.0 p.u. input
active power P_1, this line can be scaled in per-unit active power with positive
values for motor and negative for generator action. The same scaling can be
used for the vertical axis through OO_1 to give the per-unit reactive power Q_1,
with positive for leading and negative for lagging inputs. Load point L then
gives direct scale values of P_1 and Q_1 for the operating condition, which requires E_t = OL to scale. Constant values of E_t appear as circles struck from the
origin O. The angle PO_1 L is the terminal phase angle between current and
voltage (in this case leading, because the machine is operating in the motor
mode with overexcitation). The angle O_1 OL is the load angle, in this case
negative.

The diagram can give the essential operating features for any specified steady-
state condition. Point G, for example, corresponds to inputs of $P_1 = -0.75$ p.u.

and $Q_1 = +0.60$ p.u.: it represents working in the generator mode with *outputs* $-P_1 = +0.75$ p.u. and $-Q_1 = -0.60$ p.u., obtained with an excitation $E_t = OG = 2.0$ p.u. (i.e. twice that corresponding to $E_t = V_1$). If the excitation is reduced, the prime-mover power remaining unchanged, the operating point travels along the dotted line GHJ . . ., the reactive power changing to compensate. At point S the load angle is $\frac{1}{2}\pi$ rad, corresponding to maximum power with an excitation of $E_t = 0.94$ p.u. With higher excitations the operating point can lie in a region where $\delta > \frac{1}{2}\pi$ rad, but it is not stable. The horizontal line OS, the maximum-power line, can also be called the *steady-state stability limit.*

The electrical load diagram gives solutions of eqs.(10.1) and (10.2). It makes evident that a synchronous machine working on infinite busbars can operate as a motor or compensator or generator, respectively at a negative, zero or positive load angle, depending upon the mechanical power given to it or demanded from it. The level of the excitation determines the operating power factor (and to a lesser extent the load angle), and the greater the excitation the greater is the maximum power capability of the machine.

V-curves. These, called so because of their shape, relate input current I_1 to excitation E_t for specified constant active powers. They can be constructed direct from the electrical load diagram. For example, the operating conditions for a generator, with its per-unit reaction reactance the same as that in Fig. 10.4 and delivering 0.75 p.u. active power, are given by the dotted line GHJKS. The stator phase current is proportional to O_1G, O_1H . . . for excitations of 2.0, 1.5 . . . p.u., so that plots of the current/excitation relations can readily be drawn. Fig. 10.5 shows V-curves obtained in this way from Fig. 10.4. As

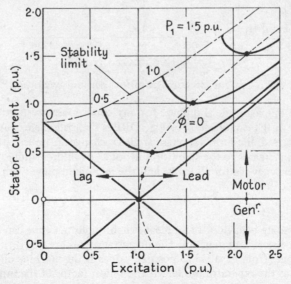

10.5 V-curves.
The scales are in per-unit values for a reaction reactance of 1.25 p.u.

the electrical load diagram is symmetrical with respect to generator and motor operation, the V-curves apply to both.

Constant active load, variable excitation. This condition implies a constant shaft torque (driving or driven) and a constant active current component I_{1a}. Change in excitation affects only the reactive current component I_{1r}, which acts to make up the deficiency in rotor excitation for underexcited conditions, or to oppose the excess for overexcitation. As a result, the terminal power factor is changed, and can be made leading or lagging at will. Excitation also affects the load angle, which reduces as the excitation is raised. Further, there are maximum and minimum excitations for a given active power; synchronous running is lost if the excitation is too low, while the maximum is not normally reached in a normal machine because of the very large stator and rotor currents involved. A typical characteristic is given by the dotted line GHJKS in Fig. 10.4.

Constant excitation, variable active load. Triangle OLO_1 in Fig. 10.4 is composed by the scalars E_t, $I_1 x_a$ and V_1. If all three are divided by x_a they yield the triangle RPO in Fig. 10.6 in which OP = I_1. The circular locus of P gives

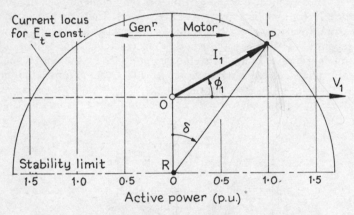

10.6 Current circle diagram for constant excitation.

the current in magnitude and phase for any condition corresponding to a given fixed value of the excitation E_t. Angle ORP is the load angle δ. As the active load is increased, the terminal phase angle ϕ_1 changes from lead to lag through zero, and δ increases to the stability limit of $\pm\frac{1}{2}\pi$ rad. For underexcitation the input current may lag for all loads, and the maximum power will obviously be reduced.

Summary
The steady-state behaviour of an ideal synchronous machine can be inferred from its electrical load diagram. The important features are (i) the load angle, which changes from negative to positive values as the machine changes from the motor to the generator mode; (ii) the power factor of the input current, which varies in accordance with the requirement that the combined m.m.f.s F_1 and F_2 of stator and rotor shall yield across the gap a resultant F_0 to mag-

netize the machine to a flux that generates the e.m.f. E_1 balancing the applied voltage V_1.

EXAMPLE 10.1: An ideal 1.5 MW, 50 Hz, 3-ph, 8-pole synchronous machine with a reaction reactance of 1.25 p.u. runs on 6.6 kV infinite busbars. The moment of inertia of the rotor and its mechanical attachments is 3 100 kg-m^2. The machine operates

(*a*) on no load with 1.0 p.u. excitation;
(*b*) as a compensator with 2.0 p.u. excitation;
(*c*) as a motor on full load (1.0 p.u.) at power factor 0.87 leading;
(*d*) as a generator on 0.75 p.u. active power at an output power factor of 0.78 lagging.

For each case estimate the load angle, synchronizing power and torque, and the natural frequency of hunting oscillation; for case (*d*) find the voltage regulation.

The electrical load diagram in Fig. 10.4 is drawn for $x_a = 1.25$ p.u.; the power/load-angle curves in Fig. 10.3 and the V-curves in Fig. 10.5 also apply here. Using per-unit values based on the full rating (1.5 MW) enables calculations to be made for the whole machine direct instead of per phase.

(*a*) *No Load*: The no-load operating point with $E_t = 1.0$ p.u. is O_1 in Fig. 10.4. The stator current and the load angle are both zero. From eq.(10.3) the synchronizing power is

$$p_s = (1.0/1.25)1.0 = 0.80 \text{ p.u./rad(e.)} = 1.20 \text{ MW/rad(e.)}.$$

As 1 rad(m.) = 4 rad(e.), then also

$$p_s = 0.80 \times 4 = 3.2 \text{ p.u./rad(m.)} = 0.056 \text{ p.u./deg(m.)}$$

The synchronous speed is $\omega_s = 2\pi 50/4 = 78.6$ rad(m)/s so that the corresponding synchronizing torque is

$$M_s = 3.2 \times 1.5 \times 10^3/78.5 = 61 \text{ kN-m/rad(m.)}$$

Applying eq.(10.4), the natural angular frequency of hunting oscillation is

$$\omega_0 = \sqrt{(61\ 000/3\ 100)} = 4.44 \text{ rad(m.)/s}$$

that is, an oscillation frequency $f_0 = 0.71$ Hz, or one complete swing in 1.4 s.

(*b*) *Compensator*: Raising E_t to 2.0 p.u. shifts the operating point to C. The load angle is still zero (there being no active power concerned) but the machine takes a reactive input of 0.80 p.u. = 1.2 Mvar leading as a synchronous capacitor. The values required are

$$p_s = (1.0/1.25)2.0 = 1.60 \text{ p.u./rad(e.)} = 6.4 \text{ p.u./rad(m.)}$$
$$M_s = 6.4 \times 1.5 \times 10^3/78.5 = 122 \text{ kN-m/rad(m.)}$$
$$\omega_0 = \sqrt{(122\ 000/3\ 100)} = 6.28 \text{ rad/s}, f_0 = 1.0 \text{ Hz}.$$

(c) *Motor*: The operating point is L, with 1.0 p.u. active power (1.5 MW), 0.57 p.u. leading reactive power (0.85 Mvar) and 1.15 p.u. apparent power (1.72 MVA). The necessary excitation is readily calculated, or measured from the load diagram: it is E_t = 2.12 p.u. The load angle is $-36° = -0.63$ rad(e.). The synchronizing power and torque are

$$p_s = (1.0/1.25) \times 2.12 \times \cos(-36°) \times 4 = 5.49 \text{ p.u./rad(m.)}$$
$$M_s = 5.49 \times 1.5 \times 10^3/78.5 = 105 \text{ kN-m/rad(m.)}$$

and the natural frequency is ω_0 = 5.8 rad(m.)/s, f_0 = 0.92 Hz.

(*d*) *Generator*: The conditions are given by G, with OG = E_t = 2.0 p.u., and inputs of -0.75 p.u. active and $+0.60$ p.u. reactive. The output of the machine as a generator is therefore $+0.75$ p.u. = 1.12MW and -0.60 = 0.9 Mvar lagging. By calculation or measurement, $\delta = +28° = +0.49$ rad(e.). Then

$$p_s = 5.64 \text{ p.u./rad(m.)}, \quad M_s = 108 \text{ kN-m/rad(m.)},$$
$$\omega_0 = 5.9 \text{ rad(m.)/s}, \quad f_0 = 0.94 \text{ Hz}$$

The regulation is $\epsilon = (E_t - V_1)/V_1 = 1.0$ p.u., because obviously the open-circuit terminal voltage is double the busbar voltage.

10.3 *Practical synchronous machines*
The assumptions, explicit and implicit, made in dealing with the ideal machine in Sect. 10.2 are the following:

(i) a symmetrical cylindrical construction in which the stator and rotor m.m.f.s and their resultant are sinusoidally distributed;

(ii) an unsaturated magnetic circuit of high permeability, and a uniform radial gap providing the whole reluctance and producing a sine-distributed gap flux under all conditions;

(iii) negligible electrical, magnetic and mechanical losses, and no magnetic leakage;

(iv) purely steady-state operating conditions.

A practical machine may have salient poles, and even if its construction is cylindrical it still has some degree of effective polarization; it therefore has a non-sinusoidal distribution of gap flux. The windings are distributed in slots, and give rise to space harmonics of gap-flux density so that the induced e.m.f. contains time harmonics besides a fundamental. The machine has conduction, core and mechanical losses producing heat, and it has a significant level of leakage flux. Its operating conditions include not only the steady state, but also sudden load change, hunting, starting, and the effects of faults both within the machine structure and in the electrical network into which it is connected. Finally, saturation of the magnetic circuit varies widely with load and excitation and markedly influences the behaviour of the machine.

Losses
Core losses occur in the yoke, core, tooth and end-winding regions of the magnetic circuit, the stator (and, in cylindrical machines, the rotor) teeth being normally worked at high levels of saturation. I^2R and eddy-current losses occur in the stator windings, and the rotor excitation power requirement may be very large in machines of high rating. Load (stray) losses are caused by alternating

leakage fluxes that appear, when the machine is loaded, in pole-faces and end regions. Mechanical windage and friction loss may be considerable with machines of large diameter or high speed; and in heavy generators the bearing friction may demand forced cooling to avoid excessive temperature-rise in the lubricant.

Electrical and magnetic losses have to be dissipated by a cooling system, sometimes elaborate, in order to limit the thermal stress imposed on the winding insulation. In turbo-generators the structural temperature-rise may cause significant dimensional changes that have to be accommodated by permitting cyclic expansion and contraction.

Leakage flux

Rotor leakage flux adds to the saturation conditions of the rotating part of the magnetic circuit. Self-flux linking the stator slot and overhang conductors, besides contributing to eddy-current and stray losses, endows the stator phase windings with leakage reactance x and a corresponding volt-drop I_1x. It is particularly important during transients as it forms the chief limitation on current growth.

Saturation

Magnetic saturation considerably affects the performance of a synchronous machine in respect of its excitation, regulation and circuit impedances. By introducing a strong non-linearity it makes a linear analysis, like that in Sect. 10.2, almost impossible. The saturation phenomenon is a function of the magnetic material concerned and its working flux density, and is therefore a local effect that varies from point to point in space and (for varying flux densities) in time. For a machine its mass overall effect is normally taken as represented by the open-circuit characteristic.

Open-circuit and short-circuit characteristics. The *open-circuit characteristic* or o.c.c. relates the terminal voltage on open circuit at normal speed to the amount of field excitation. Since under these conditions the terminal voltage measured is the induced e.m.f., which depends on the total flux Φ_m linking the stator winding, the o.c.c. is a measure of the saturation in the magnetic circuit. At low values of field excitation the o.c.c. is rectilinear: the iron parts of the magnetic circuit have only a small magnetic loading and are highly permeable, so that the chief reluctance in the circuit is that of the gap. For a density of, say, 0.8 T a laminated ferromagnetic material requires only about 100 A-t/m, but an air path requires 640 000 A-t/m. The part of the excitation devoted to the gap is defined by the air-line. Fig. 10.7 shows an o.c.c. typical of a large turbo-generator: comparison with the o.c.c. in Fig. 10.2(c) indicates the effect of saturation. For high flux densities the ferromagnetic parts suffer a considerable decline in permeability, and the o.c.c. for the upper range of field excitation has a comparatively small slope. At normal terminal voltage, corresponding to 1.0 p.u. field excitation, about 20% of the exciting m.m.f. is expended on the gap, whereas at 1.25 p.u. voltage the gap m.m.f. has risen to 1.0 p.u. and the iron m.m.f. to 0.7 p.u. giving a total of 1.7 p.u. for the whole

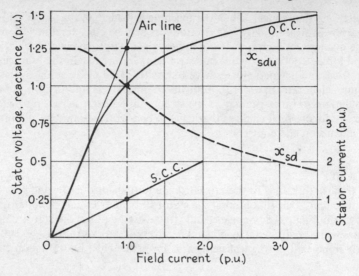

10.7 Open- and short-circuit characteristics.

magnetic circuit. At higher saturations the excitation expended on the iron paths exceeds that on the gap.

The *short-circuit characteristics* (s.c.c.) relates principally the 'armature reaction' to the field excitation. If there were no leakage reactance nor resistance, the short-circuit current in the stator would furnish the a.c. equivalent of the d.c. rotor excitation. The field excitation has to exceed this equivalent value by the amount necessary to produce the small flux that generates the e.m.f. for circulating the short-circuit current through the resistance and leakage reactance. So long as this flux is small enough the s.c.c. remains a rectilinear relation between armature and field currents. At high values of short-circuit current, however, the field current is high and the consequent leakage produces saturation in the poles after which a given increase in the exciting current produces less increase in the stator current. In most machines the curvature of the s.c.c. is only slight, and starts well above full-load current; thus a single point together with the origin suffices to locate the s.c.c.

Synchronous reactance. As just observed, in an actual machine the excitation on short circuit has not only to balance the stator current, but must also generate the voltage drop in resistance and stator leakage reactance. Thus the equivalent circuit for the ideal machine, Fig. 10.2(b), must be replaced by that in Fig. 10.8(a), and the ideal phasor diagram by Fig. 10.8(b). The winding resistance r and the true leakage reactance x are additions to the ideal machine, and now E_1 differs from V_1 by the volt-drops due to I_1 in r and x. Alternatively, V_1 and E_t can be related more directly by combining x and the reaction reactance x_a to give an effective total reactance $x_s = x_a + x$ called the *synchronous reactance.* If further we include r, the total impedance is the *synchronous impedance, z_s.*

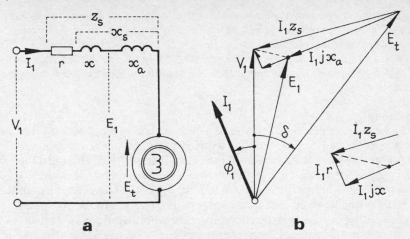

a **b**

10.8 Practical synchronous machine: equivalent circuit and phasor diagram.

For a given field current, the terminal e.m.f. E_{OC} on open circuit and the current I_{SC} on short circuit are obtained respectively from the o.c.c. and s.c.c. The effective internal (synchronous) impedance is $E_{OC}/I_{SC} = z_s = r + jx_s$, $= r + j(x + x_a)$, with x_a dependent on saturation. In a large machine, x and x_a are typically 0.1 and 1.0 p.u., but r may be only about 0.01 p.u.; then z_s is numerically almost the same as x_s, and the volt drop $I_1 r$ can normally be omitted from the phasor diagram.

In constructing Fig. 10.8(*b*) for a given load condition (i.e., specified V_1, I_1 and ϕ_1) the approximate method of Fig. 10.9 can be used. First the synchronous reactance x_s (the dotted curve) is obtained from the o.c.c. and s.c.c. Next, E_1 is found from V_1 by adding the phasors $I_1 r$ and $I_1 jx$, to give point E on the o.c.c. Compared with the unsaturated air line, OE represents an equivalent saturation characteristic for the specified load, and CD is the appropriate value of x_s. The phasor diagram can now be completed to give E_t and point B. The required field current is OF.

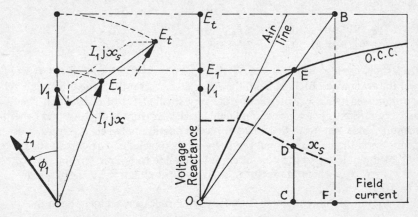

10.9 Effect of saturation.

Load angle. For strict consistency with the concept of load angle employed for the ideal machine, the angle between E_t and E_1 should be taken. However, it is far more convenient to adopt as a definition the angle between E_t and V_1 for a practical machine, as indicated in Fig. 10.8(*b*). In effect, this is to consider the synchronous reactance rather than the reaction reactance as the operative quantity.

Saliency

In a salient-pole machine, and also to a minor extent in a cylindrical-rotor machine, the gap flux produced by a distributed m.m.f. depends upon the orientation of the m.m.f. axis with respect to that of the saliency, and the gap reluctance presented to it. The very different effects of orientation have been discussed before, in Sect. 3.14 and Fig. 3.32. The component F_{1d} of the (assumed) sinusoidally distributed stator m.m.f. directed in the pole or direct axis operates over a magnetic circuit identical with that of the rotor field system; but the quadrature-axis component F_{1q} is applied across the interpolar space, producing a smaller flux per ampere and a markedly different distribution.

The two-reaction theory considers direct- and quadrature-axis reactions separately, assigning different reaction reactances to each so that the appropriate axis synchronous reactances are

$$x_{sd} = x + x_{ad} \quad \text{and} \quad x_{sq} = x + x_{aq}$$

for the direct and quadrature axes respectively. The q-axis reactance is smaller than the direct-axis reactance because a given m.m.f. gives rise to a smaller gap flux.

It is necessary to distinguish the active and reactive components of I_1 from its d- and q-axis components. The former are related to the applied voltage V_1 through the terminal phase angle ϕ_1, the latter to the magnetic axis of the rotor. In Fig. 10.2(*a*) it is seen that E_t is drawn at right-angles to the rotor magnetic axis given by F_2; hence the d-axis component is at right-angles to E_t. The two sets of components are shown in Fig. 10.10. When δ is the load angle and ϕ_1 is the terminal phase angle, then

$$
\begin{array}{lll}
I_{1a} = I_1 \cos \phi_1 & I_{1r} = I_1 \sin \phi_1 & I_1 = \sqrt{(I_{1a}{}^2 + I_{1r}{}^2)} \\
I_{1q} = I_1 \cos(\phi_1 - \delta) & I_{1d} = I_1 \sin(\phi_1 - \delta) & I_1 = \sqrt{(I_{1q}{}^2 + I_{1d}{}^2)}
\end{array}
$$

On a short-circuit test $\delta = 0$ and ϕ_1 is very nearly $-\frac{1}{2}\pi$ rad, so that $I_1 = I_{1r} = I_{1d}$ and the reactance obtained from the o.c.c. and s.c.c. ratio is the direct-axis synchronous reactance x_{sd} (neglecting the very small effect of resistance). The q-axis synchronous reactance is obtainable in design from flux plots, and experimentally by a 'slip test', Sect. 10.15. As the relation between x_{sq} and x_{sd} depends upon the constructional geometry of the machine, particularly the ratio b/Y of the pole-arc to the pole-pitch, it is possible to infer x_{sq} from the expression $x_{sq} = K_r x_{sd}$, where the factor K_r is given approximately by the graph in Fig. 10.10.

Values of direct- and quadrature-axis synchronous reactances lie in the

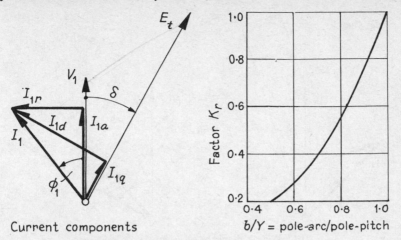

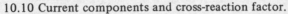

Current components b/Y = pole-arc/pole-pitch

10.10 Current components and cross-reaction factor.

range 0.8–1.6 p.u. and 0.65–1.0 p.u. respectively, the higher figures applying to synchronous compensators.

Other practical features
Harmonics are discussed in Chapter 3, and mechanical considerations in Chapter 7. Electrical features of generators and motors are dealt with in subsequent sections of this chapter. First, the steady-state action of the machine with saliency is taken up to show certain modifications that must be made to the theory of the ideal machine. The analysis of transient behaviour cannot be undertaken on a basis of steady-state theory: it requires the much more elaborate circuit model described in Chapter 7.

10.4 *Steady-state theory*
The theory of the practical synchronous machine differs from that of the ideal in respect (i) of the use of the synchronous reactance in place of the reaction reactance, (ii) of the inclusion, where its effect is significant, of stator-winding resistance, and (iii) of saliency. Further, the synchronous reactance is not a constant but varies with the conditions of magnetic-circuit saturation.

Phasor diagram
Motor mode. Phasor diagrams for a motor with a given active input power are given in Fig. 10.11, with (*a*) for overexcited and (*b*) for underexcited conditions and for known d- and q-axis synchronous reactances. (The volt drop $I_1 r$ in resistance has been omitted for clarity: it would be inserted as in Fig. 10.8(*b*) as a phasor cophasal with the current I_1.) Each reactance voltage is drawn in quadrature with its respective axis current: $I_{1q} x_{sq}$ is smaller than it would be in a machine without saliency (for which $x_{sq} = x_{sd} = x_s$) and as a result the load angle is also less. The salient-pole machine is magnetically 'stiffer' as the structural polarization contributes reluctance power by alignment effect.

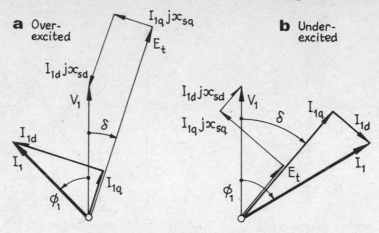

10.11 Phasor diagram of salient-pole motor.

Generator mode. A phasor diagram for an overexcited generator is given by Fig. 10.12(*a*).

In drawing these phasor diagrams, the d- and q-axis currents and reactance voltages are not known until the load angle between V_1 and E_t is fixed, but δ in turn depends upon the axis reactance voltages. A graphical construction to resolve the difficulty is given in Fig. 10.12(*b*) for the case of the overexcited

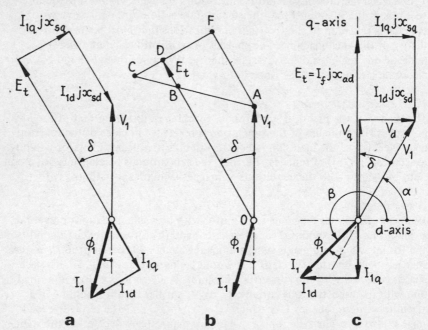

10.12 Phasor diagram of salient-pole generator.

generator. It is assumed that the terminal conditions (voltage, current and phase angle) and the two axis reactances are known: then V_1 and I_1 can be drawn to locate the point A. Draw AC in a direction perpendicular to I_1 and of length equal to $I_1 x_{sd}$. Locate point B such that BA/CA is the ratio x_{sq}/x_{sd}. Then the line OB has the required direction of E_t, and the magnitude of E_t is found by dropping the perpendicular CDF on OB, where CFA is a right-angle. Then OD is the phasor E_t, DF is $I_{1q} j x_{sq}$ and FA is $I_{1d} j x_{sd}$. The ratio BA/CA is the same as K_r in Fig. 10.10. If the saliency is negligible, then CA = $I_1 j x_s$ and OC = E_t, the load angle being increased as a result.

Circuit theory. The dynamic circuit theory of Sect. 7.5 can give the steady-state operation as a special case. For a 2-pole generator on steady balanced load and connected to a symmetrical 3-phase supply of angular frequency ω_1, the speed is $\omega_r = \omega_1$, the angle θ in Fig. 7.13 is $\omega_1 t$ reckoned from the instant that the axis of phase A coincides with the d-axis, and the field current i_f is constant. Let the applied voltage and input current of phase A be $v_a = v_m \cos(\omega_1 t + \alpha)$ and $i_a = i_m \cos(\omega_1 t + \beta)$; corresponding values for phase B and C are written by adding $(-2\pi/3)$ and $(-4\pi/3)$ respectively to the cosine terms. Applying eq. (7.20) gives

$$v_d = v_m \cos \alpha \qquad\qquad i_d = i_m \cos \beta$$
$$v_q = -v_m \sin \alpha \qquad\qquad i_q = -i_m \sin \beta$$

These, being independent of time, are *direct* currents and voltages.

The circuit equations are now written direct from the matrix. All terms in $p = d/dt$ are zero and there are no currents in KD or KQ. We can write $r_d = r_q = r$, and $L_d = L_q = L$ from symmetry, whence $\omega_1 L = x$, the leakage reactance. Further the reaction reactances x_{ad} and x_{aq} can be written for $\omega_1 L_{ad}$ and $\omega_1 L_{aq}$ respectively, and the axis synchronous reactances as $\omega_1 (L_{ad} + L) = x_{sd}$ and $\omega_1(L_{aq} + L) = x_{sq}$. In per-unit terms, the equations for the three active circuits F, D and Q reduce to

and
$$v_f = r_f i_f, \qquad\qquad v_d = r i_d + x_{sq} i_q$$
$$v_q = r i_q - x_{sd} i_d - x_{ad} i_f$$

The term $-x_{ad} i_f$ is a measure of the gap flux due to the field excitation and is consequently represented by e_t.

The results can be used to construct the complexor diagram, Fig. 10.12(c), in terms of r.m.s. values. The phasor applied voltage and input current for phase A are

$$V_1 = V_d + V_q = (\cos \alpha + j \sin \alpha)(v_m/\sqrt{2}) = (v_d - j v_q)/\sqrt{2}$$
$$I_1 = I_d + I_q = (\cos \beta + j \sin \beta)(i_m/\sqrt{2}) = (i_d - j i_q)/\sqrt{2}$$

There is no essential difference between (c) and (b) of Fig. 10.12.

In per-unit terms, the inductance coefficients in eq.(7.17) can be expressed as their corresponding reactances. Only currents i_f, i_d and i_q are concerned, so that the torque is

$$M = -x_{ad} i_f i_q - (x_{sd} - x_{sq}) i_d i_q$$

The last term is the reluctance torque, which vanishes in a non-salient machine, leaving only the torque due to the main flux $x_{ad} i_f$.

Operating characteristics

Because of the dependence of synchronous reactance on the degree of saturation and its high per-unit value in a normal machine compared with the stator winding resistance, the latter is neglected in dealing with performance characteristics except, of course, where losses and their thermal implications are concerned. Analytical expressions are greatly simplified in this way, and are quite adequate for the purpose.

Active power. The active power input to a *non-salient* machine is given by eq.(10.1) modified to include the leakage reactance x, i.e. with x_s substituted for x_a to give

$$P_1 = -\frac{V_1}{x_s} E_t \sin \delta \tag{10.6}$$

which is positive for the motor mode (δ negative). In a *salient* machine it is necessary to employ the axis synchronous reactances. The active power is $P_1 = V_1 I_1 \cos \phi_1$: Fig. 10.10 shows that we may write

$$I_1 \cos \phi_1 = I_{1q} \cos \delta - I_{1d} \sin \delta$$

while from the phasor diagram in Fig. 10.11(a) we have

$$I_{1q} x_{sq} = -V_1 \sin \delta \quad \text{and} \quad I_{1d} x_{sd} = E_t - V_1 \cos \delta$$

Making the substitutions then gives

$$P_1 = V_1 \left[-\frac{E_t}{x_{sd}} \sin \delta - \left(\frac{V_1}{x_{sq}} - \frac{V_1}{x_{sd}} \right) \sin \delta \cdot \cos \delta \right]$$

$$= -\frac{V_1}{x_{sd}} E_t \sin \delta - \frac{V_1^2 (x_{sd} - x_{sq})}{2 x_{sd} x_{sq}} \sin 2\delta \tag{10.7}$$

The first term has the form of eq.(10.6). The double-angle term, due to saliency, expresses the difference in the axis gap reluctances through the corresponding synchronous reactances; and if, as in a non-salient machine, there is no difference the term vanishes. It is of interest to observe that the saliency term is independent of E_t and therefore exists even when the machine is unexcited, for the presence of structural polarization adds an alignment component to the torque. Taking x_{sq} as typically equal to $0.6 x_{sd}$, then eq.(10.7) becomes

$$P_1 = -(V_1/x_{sd}) E_t \sin \delta - \tfrac{1}{3}(V_1^2/x_{sd}) \sin 2\delta$$

If, for example, $E_t = 2V_1$ and the load angle is small enough to equate the sine of the angle to the angle itself, then the active power is $P_1 = -(V_1^2/x_{sd})(2 + \tfrac{2}{3})\delta$, showing a 25% contribution made by saliency; and there is still an active power even if the excitation is zero.

Fig. 10.13 shows the non-saliency and saliency power components and their sum for a machine working on infinite busbars with $x_{ad} = 1.25$, $x = 0.15$ and

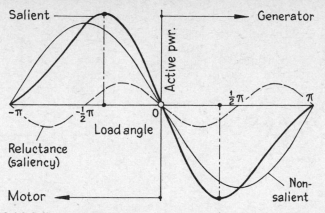

10.13 Salient-pole machine: active-power/load-angle relation.

therefore $x_{sd} = 1.40$ p.u. The q-axis synchronous reactance is taken to be $x_{sq} = 0.70$ p.u. For clarity the power/load-angle relation is graphed for the single excitation $E_t = 2.0$ p.u., for which eq.(10.7) gives

$$P_1 = -1.43 \sin \delta - 0.36 \sin 2\delta$$

for the per-unit active power input per phase. Maximum power occurs for $\delta = \pm 1.21$ rad (69°), compared with the $\frac{1}{2}\pi$ rad (90°) for a non-salient machine. *Synchronizing power and torque.* An additional term now appears in eq.(10.3) to to give

$$p_s = \frac{dP_1}{d\delta} = \frac{V_1}{x_{sd}} E_t \cos \delta + \frac{V_1^2 (x_{sd} - x_{sq})}{x_{sd} x_{sq}} \cos 2\delta \qquad (10.8)$$

for the synchronizing power per phase and per electrical radian of displacement. Expressed as per mechanical radian and divided by the synchronous angular speed, the result is the synchronizing torque.

Steady-state reactance
Power, stability and regulation problems of a synchronous machine can be solved only if the reactances are known, but these depend on the flux conditions and the degree of saturation.

Short-circuit ratio. The stator phase voltage is proportional to the product of the effective gap flux Φ_m and the turns per phase N_1. A large flux requires a bigger core area (which in turbo-type machines means a longer core as the diameter is limited by peripheral speed) but less conductor material. By contrast, with more turns a smaller core area is needed but the conductors must be deeper and have more cross-sectional area. The two shapes are indicated at (*a*) in Fig. 10.14. Now a fundamental thermal limit is the m.m.f. rating F_2 of the rotor. Part of F_2 opposes F_1 of the stator, part provides the gap-flux magnetization, F_0. With fewer stator turns, F_1 is reduced and the gap length may be ex-

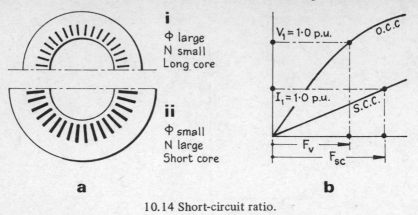

i
Φ large
N small
Long core

ii
Φ small
N large
Short core

a   **b**

10.14 Short-circuit ratio.

tended: with more stator turns the gap length must be reduced, and for a given load the load angle is greater and the machine is less 'stiff' and therefore less stable.

The relative flux/turns proportionality is embodied in the *short-circuit ratio* shown in Fig. 10.14(*b*):

$$r_{sc} = \frac{F_v}{F_{sc}} = \frac{\text{p.u. excitation for normal voltage on o.c.}}{\text{p.u. excitation for rated stator current on s.c.}}$$

Modern turbo-generators normally have a s.c.r. between 0.5 and 0.6, but this must be raised to 1.0—1.5 if the loading is likely to be capacitive (as with connection to long unloaded overhead transmission lines or extensive h.v. cable runs). The range 1.0—1.5 is also common for low-speed generators. If a generator works on a leading zero-power-factor load, its stator m.m.f. F_1 is direct magnetizing, causing self-excitation. To retain voltage control the machine must have positive rotor excitation: the gap must therefore be lengthened to increase F_v and the stator turns reduced in number to limit F_{sc}.

The short-circuit ratio is the reciprocal of the d-axis per-unit synchronous reactance if the latter is defined for normal voltage and rated current. But x_{sd} for a given load is affected by the saturation conditions, whereas r_{sc} is specific and uni-valued for a given machine.

Synchronous reactance. Defined from the o.c.c. and s.c.c. for any given excitation, x_{sd} is not a constant, as Fig. 10.7 clearly shows. To avoid ambiguity the *unsaturated synchronous reactance*, related to the gap alone, is defined as

$$x_{sdu} = \frac{\text{p.u. voltage on o.c.c. air-line}}{\text{p.u. stator current on s.c.c.}}$$

for any excitation, and it is a constant rather greater than the reciprocal of r_{sc}. It is the value normally quoted for modern machines.

Adjusted synchronous reactance. The choice of x_{sd} for a given machine and

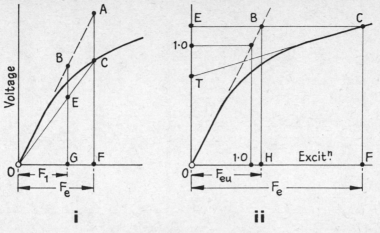

10.15 Synchronous reactance.

loading condition requires judgement. Two methods of graphical estimation are given in Fig. 10.15.

(i) Let OF = F_e be the excitation on the o.c.c. for the internal e.m.f. E_1 for a given load (i.e. the phasor addition of the terminal voltage and the internal voltage drops in resistance and in leakage reactance): then CF is the open-circuit voltage E_1, while AF is the value E_{1u} that would be generated were there no saturation. Join OC. Let OG = F_1 be the stator m.m.f.: then BG = E_{au} is the air-line reaction e.m.f. corresponding. The saturated value is taken to be EG = E_a, and the appropriate synchronous reactance is

$$x_{sd} = (E_a/I) + x = x_a + x$$

(ii) The o.c.c. is drawn with 1.0 p.u. excitation corresponding to 1.0 p.u. voltage on the air line. The internal e.m.f. E_1 is obtained as in (i). From it the o.c.c. excitation OF = F_e and the air-line excitation OH = F_{eu} are found. Then OF/OH = $F_e/F_{eu} = k_1$, a saturation factor. A tangent to the o.c.c. is drawn through C and produced to cut the ordinate axis in T, dividing EO in the ratio TO/ET = k_2. Then the synchronous reactance appropriate to the load condition is

$$x_{sd} = \frac{x_{sdu} - x}{\sqrt{[k_1 (1 + k_2)]}} + x$$

Potier reactance. The Potier reactance is the leakage reactance obtained in a particular way from a test on the machine at full-load current of power factor zero lagging. Such a test, requiring but little power, furnishes the excitation for short circuit and for normal rated voltage, in each case with full-load z.p.f. current.

Consider an unsaturated machine with negligible resistance. Its o.c.c. is the straight line in Fig. 10.16 (a). Point S on the abscissa axis is such that OS = F_{sc} is the field excitation producing full-load current on short circuit (which is

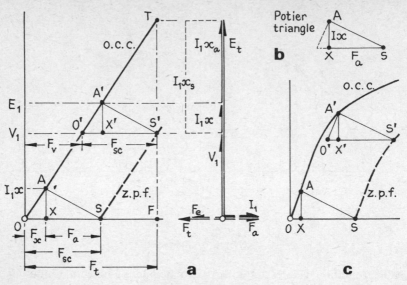

10.16 Potier reactance.

naturally at a phase angle of 90° lag). F_{sc} must be sufficient to counter the
direct-demagnetizing armature reaction F_a, also to circulate the current through
the leakage reactance, which requires an induced e.m.f. $AX = Ix$ and a
corresponding field excitation $OX = F_x$. Then $F_{sc} = F_a + F_x$. Now for a given
z.p.f. current both F_a and F_x are constant, and will apply equally when the
total excitation is increased to F_t: for F_t reduced by F_{sc} gives the remainder
F_v producing the air-line terminal voltage V_1. The triangle $O'A'S'$ is identical
with OAS. Thus AS and $A'S'$ are equal and parallel and the dotted line SS' becomes the voltage/excitation characteristic for full-load current at z.p.f. Then
$A'X' = AX$ is again the Ix drop in leakage reactance, giving the internal e.m.f.
E_1. On open circuit the terminal voltage for an excitation $OF = F_t$ would be
$TF = E_t$, so that $TS' = Ix_s$, the synchronous-reactance voltage drop on z.p.f.
load. These points are illustrated by the phasor diagram in Fig. 10.16(a).

Triangle AXS in (b) is the *Potier triangle.* Derived from the o.c.c. and the
short-circuit point for full-load current, it is assumed to apply equally to a
normal saturated case, shown in (c). Both the o.c.c. and the z.p.f. characteristics
are now curved, but the triangle OAS must everywhere fit between the two
characteristics. It follows that the z.p.f. curve should be the o.c.c. shifted bodily
by the distance AS = $A'S'$. However, points A' and S' do not quite correspond
to A and S, for although the gap flux is substantially the same, the greater total
field excitation increases the field leakage and causes more pole saturation, so
that the o.c.c. and z.p.f. curves have slightly different slopes. As a result $A'X'$
is found to be a little greater than AX, and the value of x obtained from it
(known as the *Poiter reactance*) exceeds the value based on pure leakage flux.

The Potier triangle can be determined from the o.c.c., the field excitation
for full-load current on short circuit, and the leakage reactance x. If x is not

known, a z.p.f. characteristic is needed. Two points suffice: S for short circuit, and S' for normal voltage. Then from S' the horizontal line S'O' = SO is drawn, and O'A' parallel to OA to cut the o.c.c.

Electrical load diagram

Non-salient machine. With resistance neglected the electrical load diagram is identical with that in Fig. 10.4 for the ideal machine. Its practical use is in the operation of large synchronous generators. In such machines the basic para-meters are the active power (which is settled by the output of the prime mover) and the power factor, chosen to suit the network loading conditions. The maxi-mum allowable stator and rotor currents must also be considered, as influenc-ing temperature-rise limits. Other factors include operation at leading power factors, and the general problem of synchronous stability.

To avoid undue complexity, saturation and resistance effects on the load diagram are ignored, and an unsaturated value of direct-axis synchronous react-ance selected. The machine is assumed to be connected to infinite busbars. With the various operating limits included, the diagram is called an *operating chart*, Fig. 10.17. Its basis is shown in (*a*). For a given terminal voltage V_1 and

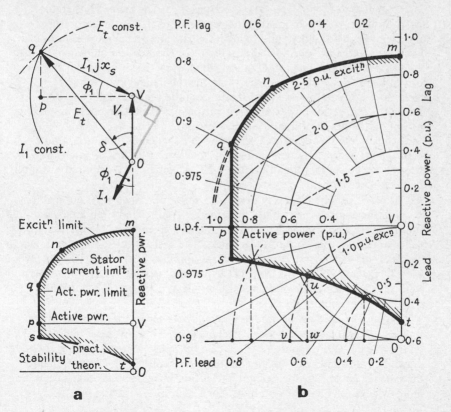

10.17 Operating chart for cylindrical-rotor generator.

current I_1 at a phase angle ϕ_1 (i.e. for a given output power) the e.m.f. E_t is obtained by adding the phasors V_1 and $I_1 j x_s$. For constant current (i.e. constant apparent power), $I_1 j x_s$ is constant and its locus is a circle centred on the point V. Constant excitation circles are struck from O. The load angle is that between E_t and V_1, the phase angle is contained between $I_1 j x_s$ and the horizontal line through V. Other features are based on Fig. 10.4, except that the diagram is drawn in terms of generator *output* and not a negative input. The operating chart in Fig. 10.17 is for a synchronous reactance of 1.67 p.u.: for zero excitation the current is $1.0/1.67 = 0.60$ p.u., so that the length OV corresponds to $Q_1 = 0.6$ p.u. reactive power, fixing both active and reactive power scales.

The working area encloses all operating points within the specified limits of rotor excitation, stator current, prime-mover power and stability as indicated in the capability diagram (*a*) of Fig. 10.17. The generator rating is taken as 1.0 p.u. apparent power at a p.f. of 0.9 lagging, and the maximum drive power as the corresponding active-power rating, i.e. 0.9 p.u. The former locates point *q* and the latter the line *qp* in (*b*). The stator-current limit line is the arc *qn* struck from centre V. At *n* the rotor-current limit becomes operative: it is here assumed that the rotor current must not exceed a value corresponding to $E_t = 2.5$ p.u. (i.e. 2.5 times V_1). The arc *nm* is therefore struck from centre O, completing the capability chart for the generator/ prime-mover set for unity and lagging p.f.s. The line *qp* cannot be continued downward to the theoretical stability limit, for the smallest increment of load would cause the machine to fall out of step. A practical condition is for the load to be restricted to 0.1 p.u. less than the theoretical limit for any excitation. Consider point *v* for 1.0 p.u. excitation in (*b*): reduce O*v* to O*w* by 0.1 p.u. active power, then *wu* cutting the excitation circle in *u* fixes a point on the practical stability limit curve *sut*.

The completed working area, shown with a shaded outline, is *mnqpsut*. A load point within this area defines the active, reactive and apparent powers, the current, power factor and excitation. The load angle, if required, is found by direct measurement.

Salient machine. The difficulty here is the differing values of d- and q-axis synchronous reactance. The method described by Walker [61] represents a simple modification of the non-salient operating chart, given known (unsaturated) values of the two synchronous reactances. The basis of the method is shown in Fig. 10.18(*a*) which (in terms of generator *output*) restates Fig. 10.12(*b*) where AC $= I_1 j x_{sd}$ and OBD $= E_t$. Draw O′C equal and parallel to OD: then O′ is a point on a saliency circle of diameter OQ $= V_1 [(x_{sd}/x_{sq}) - 1]$. Further, CO′ produced meets point Q.

Diagram (*a*) can now be used as a salient-pole generator operating chart, similar to that for the non-salient machine except that E_t is the intercept O′C on the line QO′C. For constant excitation the appropriate E_t locus is found by drawing rays from Q and marking on them constant length O′C external to the saliency circle. The loci are not circles but limacons, their shape becoming more evident for smaller excitations. Each such E_t locus has a point correspond-

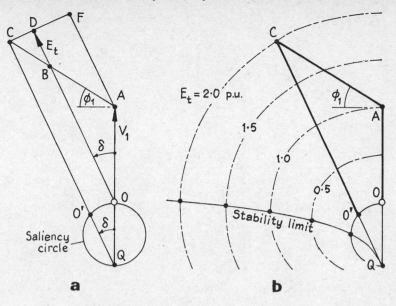

10.18 Basis of chart for salient-pole generator.

ing to maximum active power, and a curve drawn through these points forms the theoretical limit of stability, as indicated in Fig. 10.18(b).

The generator operating chart obtainable is given in Fig. 10.19. It is substantially the same as for the non-salient machine except in the low-excitation region. The machine is presumed to have a current rating limited by 1.0 p.u. active power at a power factor not lower than 0.9. Then mn represents the excitation limit imposed by rotor heating; nps is the prime-mover mechanical power limit; sq is imposed by stator heating; qu is the practical stability limit; and ut is a restriction imposed by the requirement that there shall always be a positive field excitation, for it is possible for a salient-pole machine to run without excitation — or even with a small reversed excitation — though under such conditions it is susceptible to de-synchronizing should small disturbances occur. *Motors and compensators.* The motor action only requires the chart to be reflected across the vertical axis of V_1 to give an operating chart having features in every way comparable with those for generation. In the case of a compensator, the chart reduces simply to the vertical axis.

10.5 *Steady-state performance*

Where a machine operates on an infinite busbar, the general features of its performance can readily be inferred from the operating chart. Some generators may, however, work individually or in parallel on isolated loads (e.g. on shipboard or in large aircraft).

Isolated generator

Consider an isolated synchronous generator, driven at constant speed and pro-

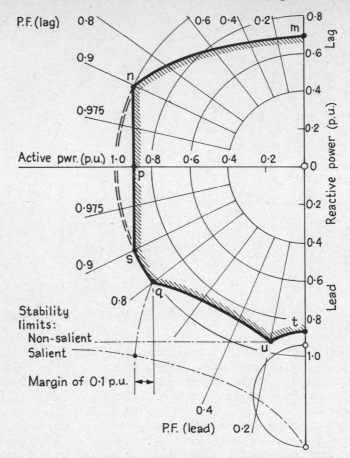

10.19 Operating chart for salient-pole generator.

vided with constant excitation. Disregarding stator resistance and saliency it can be inferred from the phasor diagram of the ideal machine that for loads of *unity power factor*, the terminal voltage is given by $V_1^2 + (I_1 x_s)^2 = E_t^2 =$ constant, and the V_1/I_1 relation, Fig. 10.20(a), is substantially an ellipse with semi-axes E_t and I_{sc}. The current I_{sc} is that which flows when the load resistance is reduced to zero, a condition giving $V_1 = 0$ and $I_{sc} = E_t/x_s$. For *zero power factor* loads, we have the scalar relation $V_1 = E_t \pm I_1 x_s$, giving the two straight lines in (a) for zero power factor leading and lagging. Loads of intermediate power factor give V_1/I_1 relations between these limits. If the terminal voltage V_1 is to be maintained constant, then the excitation will have to be adjusted to suit the load and power factor, as shown in Fig. 10.20(b); all unity and lagging-p.f. loads require increase of excitation with load current, but low leading-p.f. loads, for which the stator m.m.f. is direct-magnetizing, require the excitation to be reduced. Although variation in the conditions of satura-

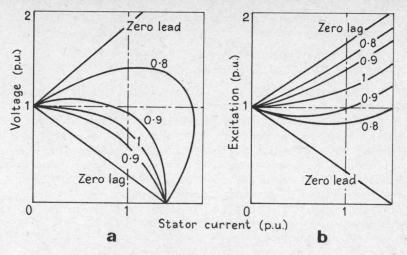

10.20 Generator load characteristics.
(*a*) Constant excitation. (*b*) Constant voltage.

tion modify x_s, the effect on the V_1/I_1 relations is not enough to invalidate the load characteristics in Fig. 10.20.

Voltage regulation. This, defined above in eq.(10.5), is an important characteristic affecting protection, voltage control and exciter rating. Apart from a direct test, the regulation can be estimated from design or low-load test data. The regulation of a salient-pole machine is not likely to differ much from that of a cylindrical-rotor machine with the same value of x_{sd}, and as no simple method of prediction is precise the same method can be applied to each form of structure. The imprecision results from the complex changes in saturation that occur in the machine from full load to no load. Four methods of prediction of the regulation are given in the following example. The general data needed are the o.c.c., the s.c.c., the leakage impedance, and the z.p.f. full-load current test to obtain the Potier reactance.

EXAMPLE 10.2: Find the regulation for full-load current (*a*) at p.f. 0.8 lagging and (*b*) at p.f. 0.8 leading for a 500 kVA, 3.3 kV, 3-phase, star-connected, 50 Hz salient-pole generator. Resistance, 0.02 p.u.; leakage reactance, 0.08 p.u.; ratio (pole-arc/pole-pitch), 0.67; reaction coefficient $K_r = 0.34$; o.c.c. and s.c.c., Fig. 10.7 with 1.0 p.u. voltage corresponding to 3.3 kV line, 1.0 p.u. excitation to 5 kA-t/pole, and full-load current on short-circuit circulated by a field excitation of 5 kA-t/pole. The short-circuit ratio is $r_{sc} = 1.0$, and the unsaturated synchronous reactance is $x_{sdu} = 1.25$ p.u.

1. *Simple M.M.F. method*: This identifies E_1 with V_1 an open-circuit as in an ideal machine, and builds up the m.m.f. triangle F_1, F_2, F_0 as in Fig. 10.1 (iii) for a given terminal voltage V_1 and load phase angle ϕ_1. The regulation so obtained is generally less than the true value

Taking $F_0 = 1.0$ p.u. and $F_1 = 1.0$ p.u., then for (*a*) $F_2 = 1.78$ p.u. giving

$E_t = 1.26$ p.u. from the o.c.c.; similarly for (b) $F_2 = 0.89$ and $E_t = 0.94$ p.u. The regulations are therefore

(a) $\epsilon = (1.26 - 1.0)/1.0 = 0.26$ p.u.
(b) $\epsilon = (0.94 - 1.0)/1.0 = -0.06$ p.u.

The negative sign indicates that the open-circuit voltage after throwing off the load is less than the terminal voltage on load.

2. *Synchronous impedance method.* This is based on the phasor diagram, Fig. 10.8(b), modified for a generator (i.e. with negative input current). There is considerable doubt as to the choice of the synchronous impedance. The value from the o.c.c./s.c.c. tests at maximum current may be taken, but even so the saturation conditions may be lower than for normal load. Generally, regulations greater than those actually occurring are given by the method.

At full-load current the synchronous impedance is almost identical with the synchronous reactance, i.e. $1.0/1.0 = 1.0$ p.u. Then for (a) $E_t = 1.80$ and for (b) $E_t = 0.90$ p.u., whence

(a) $\epsilon = 1.80 - 1.0 = 0.80$ p.u.
(b) $\epsilon = 0.90 - 1.0 = -0.10$ p.u.

3. *Adjusted synchronous impedance method.* The construction in Fig. 10.15(ii) is used. For full-load current at p.f. 0.8, taking into account the resistance and leakage reactance,

$$E_1 = 1.0 + (0.8 \pm j0.6)(0.02 + j0.08) = 1.06 \text{ or } 0.97 \text{ p.u.}$$

Then for (a): OF = 1.40 p.u., OH = 1.06 p.u., OF/FH = k_1 = 1.32, TO/ET = 1.51; x_{sdu} = 1.25; thence

$$x_{sd} = 0.10 + (1.25 - 0.10)/\sqrt{[1.21(1 + 0.76)]} = 0.87$$

and $E_t = 1.55$ p.u. For (b) $E_1 = 0.97$, OF = 1.18, OH = 0.97, $k_1 = 1.18/0.97 = 1.21$, TO/ET = 0.76, $x_{sd} = 0.87$, so that $E_t = 0.85$ p.u. The regulations are

(a) $\epsilon = 1.55 - 1.0 = 0.55$ p.u.
(b) $\epsilon = 0.85 - 1.0 = -0.15$ p.u.

4. *Reaction method*: In effect this is based on the phasor diagrams in Fig. 10.12, with the effect of resistance and leakage reactance included. The internal e.m.f. E_1 for the two cases, as already found in Method 3, is respectively 1.06 and 0.97 p.u. In (a), E_1 requires an excitation of 1.18 p.u., and the leakage impedance drop of 0.08 p.u. requires 0.07 p.u. excitation, both obtained from the o.c.c., so that the reaction m.m.f. is $1.0 - 0.07 = 0.93$ p.u.; AB = 0.34 AC, and F_2 is found to be 1.87 p.u. giving $E_t = 1.28$ p.u. from the o.c.c. In (b), $E_1 = 0.97$ and in a similar way $E_t = 0.94$ p.u. The regulations are

(a) $\epsilon = 1.28 - 1.0 = 0.28$ p.u.
(b) $\epsilon = 0.94 - 1.0 = -0.06$ p.u.

Summarizing the results: for (a) we obtain 0.26, 0.80, 0.55 and 0.28 p.u., while for (b) the values are -0.06, -0.10, -0.15 and -0.06. The very wide differences give emphasis to the inaccuracy of regulation prediction in practical machines by simple methods.

Two generators in parallel

When two synchronous generators are connected in parallel to a local load, they have an inherent tendency to remain in step by exchange of synchronizing power, the development of which depends on the fact that the stator impedance is predominantly reactive. If it were not, the machines could not operate stably in parallel. Consider identical machines, 1 and 2 in Fig. 10.21, in parallel

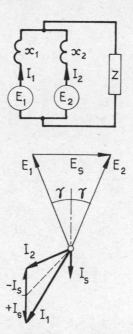

10.21 Two generators in parallel.

and with a common load: if the internal e.m.f.s, E_1 and E_2, diverge because of the acceleration of machine 1, the resulting phase-difference 2γ develops the voltage difference E_s that circulates a current I_s in the local circuit of the two stators through the two impedances x_1 and x_2 (neglecting resistance). The reactance causes I_s to have the phase of an active-power component: consequently machine 1 produces $P_s = E_1 I_s$ as a generator and supplies it to machine 2 as a motor. Within the limits of maximum power, therefore, the machines develop a mutual synchronizing power.

The argument does not, of course, require the machines to be identical, and in general they have different e.m.f.s, different impedances and different load/ speed regulations. The latter are imposed by the prime-mover governors, which are arranged so that a speed reduction is necessary for an increase in mechanical power. Both machines must run in sychronism, and their output powers are therefore determined by the respective governor setting and load/speed characteristic.

EXAMPLE 10.3: Two parallel-running synchronous generators have e.m.f.s of 1 000 V/ph, and their phase impedances are $z_1 = (0.1 + j2.0)$ and $z_2 = (0.2 + j3.2)$ Ω. They supply a common load of impedance $(2.0 + j1.0)$ Ω/ph. Find their terminal voltage, load currents, active power outputs and no-load circulating current for a phase divergence of 10 deg(e.). The governor characteristics are identical.

Taking $E_1 = (1\ 000 + j0)$ V, then $E_2 = 1\ 000(\cos 10° - j\sin 10°) = (985 - j174)$V. Applying the Millman (parallel-generator) theorem with a common load impedance Z, then

$$V\left[\frac{1}{Z} + \frac{1}{z_1} + \frac{1}{z_2}\right] = \frac{V}{Z_0} = \frac{E_1}{z_1} + \frac{E_2}{z_2}$$

gives the common terminal voltage V. The admittance summation yields $Z_0 = 0.905 \angle 66.3°$, and the right-hand side gives the total short-circuit current $I_{sc} = 808 \angle -90.7°$. Hence

$$V = I_{sc}Z_0 = 730 \angle -24.4° = (667 - j303)\ V$$

is the common terminal voltage. For the individual load currents,

$$I_1 = (E_1 - V)/z_1 = 224 \angle -44.9° = (160 - j159)\ A$$
$$I_2 = (E_2 - V)/z_2 = 106 \angle -64.3° = (46 - j96)\ A$$

The corresponding active powers per phase are

$$P_1 = 155\ kW \quad \text{and} \quad P_2 = 60\ kW$$

so that machine 1 is retarded and machine 2 advanced. With the machines operating in parallel on zero load, the circulating current is

$$I_s = (E_1 - E_2)/(z_1 + z_2) = (34 - j1)\ A$$

which is substantially an active-power current, corresponding to a generated power of 34 kW/ph in machine 1.

Unless the machines have an exceptionally large inertia, and respond only slowly to synchronizing power, the impedances z_1 and z_2 are not the synchronous impedances, but have lower (dynamic) values.

Generator on system busbars

If the system can, at the generator terminals, be considered as an infinite busbar, the generator characteristics are those derived from the operating chart of Fig. 10.17 or 10.19. If, however, the infinite busbar (or system reference point) is remote from the machine, it is necessary to include with the generator impedance an addition to take account of the transmission link.

Motor

The performance of large motors is subject to system effects, but small machines can generally be considered to operate on infinite busbars with the characteristics already discussed.

V-curves. The variation of stator current with field excitation on constant power form curves of V shape. Although V-curves apply equally to generators, their interest lies mainly with motor and compensator operation.

O-curves: The electrical active-power input per phase to a motor is $P_1 = V_1 I_1 \cos \phi_1$. The converted power (i.e. the mechanical power plus constant rotational loss) is $P_m = P_1 - I_1^2 r$, whence

$$I_1^2 - (V_1/r)I_1 \cos \phi_1 + P_m/r = 0$$

a quadratic showing that there are two possible current values for a given converted power, one with a lagging and the other with a leading power factor. The active current I_{1a} taken to develop a given P_m is greater for power factors less than unity because of the additional $I^2 R$ loss that then obtains. The phasor locus of I_1 is consequently a circle, for if I_1^2 is written as $(I_1 \cos \phi)^2 + (I_1 \sin \phi)^2$ in the expression above, then I_1 must lie on a circle struck from a centre M displaced by a distance $V_1/2r$ from the origin O of the phasor V_1, as shown in Fig. 10.22. The radius of the circle is $\sqrt{[(V_1/2r)^2 - (P_m/r)]}$. The construction of the mechanical load diagram (or O-curve diagram) is as follows. Let $OM = V_1/2r$ to scale: draw with M as centre a circle of radius OM. This circle represents the current locus for $P_m = 0$. Any smaller circle on centre M represents the current locus

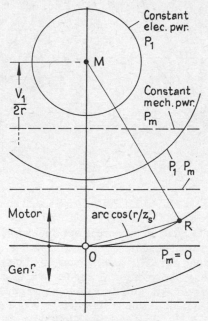

10.22 O-curves.

for some constant mechanical developed power P_m. For a load such that $P_m/r = (V_1/2r)^2$ there is a single value of current $I_1 = V_1/2r$ corresponding to the maximum possible power: the power circle has in fact shrunk to zero radius and become the point M. Such a condition, impracticable because of excessive heating and critical stability, is in fact the *maximum-power-transfer* condition commonly exploited in telecommunication circuits.

Fig. 10.22 is completed by the addition of the line OR, of length V_1/z_s and drawn at angle arc $\cos(r/z_s)$ to MO. Circles with R as centre give the loci of operating points with constant excitation. Current circles from M with radii greater than $V_1/2r$ serve for operation in the generator mode. The O-curves are based on constant resistance and synchronous reactance and are therefore of academic interest only.

EXAMPLE 10.4: The 500 kVA machine of Example 10.2 is rated as a motor with a full-load output of 375 kW. Plot O- and V-curves for full, half and zero mechanical power output. Resistance, 0.02 p.u. = 0.435 Ω/ph; leakage reactance, 0.08 p.u. = 1.74 Ω/ph; core loss, 8 kW, and friction and windage loss, 7 kW, assumed invariant.

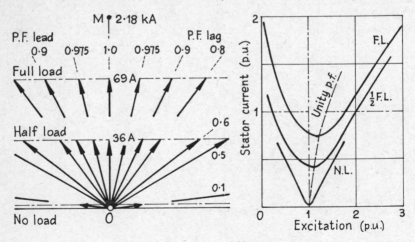

10.23 O-curves and V-curves.

The curves are shown in Fig. 10.23. Point M is located at a distance $V_1/2r = 2.18$ kA/ph from the origin O. As this is about 20 times full-load current, the curvature of the O-curves in the actual working range is so small that the circular arcs could justifiably be replaced by straight lines.

Full load: The output is 375 kW. The total converted power is 375 + 15 = 390 kW which, at unity power factor, would demand an active input current of 68 A. This in turn produces an I^2R loss of 6 kW, so that the actual input is about 396 kW and 69 A/ph. A similar iteration for a power factor of 0.50 gives an input of 375 + 15 + 28 = 418 kW, 836 kVA and 147 A, this current producing 28 kW of I^2R loss. The result is obtainable directly from the full-load O-curve.

Half load: The converted power is $\frac{1}{2}(375) + 15 = 202.5$ kW. At unity p.f. the input is 204 kW, 36 A/ph.

No load: The power input consists of losses only, so that the I^2R loss is important, particularly at low power factors.

The phase currents for various power factors are as follows:

Power factor		1.0	0.95	0.8	0.5	0.1	0.05
Current, A/ph:	full-load	69	73	88	147	–	–
	half-load	36	38	46	74	–	–
	no-load	2.6	2.7	3.3	5.3	28	85

Using the o.c.c. and s.c.c. and the phasor diagram for a salient-pole motor, the excitation for each load can be obtained and the V-curves constructed to give the result shown in Fig. 10.23.

10.6 Transient performance

The behaviour of a synchronous machine under conditions of transience may have a dominant effect on its application and may demand special attention in its design. For example, the possibility of a generator being suddenly short-circuited when running fully excited necessitates robust design of its end-windings to withstand the disruptive forces generated electromagnetically by the currents, which in a 500 MW turbo-generator may exceed 100 kN axially and 50 kN peripherally. These forces depend on the square of the current.

The phase-swinging of a machine resulting from switching operations, faults, or sudden changes of load is a typically occurring transient condition of an unpredictable nature, while the induction starting of a synchronous motor is a predictable transient important to the fulfilment of the motor's duty cycle.

The behaviour of a generator subject to a fault condition depends on (i) the location of the fault in the connected network and the impedance up to the fault, (ii) the phase windings involved, (iii) the instant in a period at which the fault occurs, (iv) the change in prime-mover speed, (v) the control responses of the governor, voltage regulator and protection systems, (vi) the generator loading and excitation, and (vii) the generator parameters, which are quite different from those appropriate to the steady state. As the differences are most evident under the conditions of a sudden terminal short circuit, such a fault is first discussed, assuming constant speed in (iv), neglect of the effects in (v) and a no-load condition in (vi).

10.7 Short circuit

The time immediately following a terminal short circuit is arbitrarily divided into three successive intervals:

1. A short time (e.g. one or two periods of the supply frequency) during

which the conditions are largely dependent upon the flux linking the stator and rotor windings at the instant of fault initiation.

2. A subsequent interval of transient decay of current amplitudes consequent upon damping and the rise of armature reaction.

3. A final period during which steady-short-circuit conditions obtain: a machine will normally be open-circuited before this period is reached.

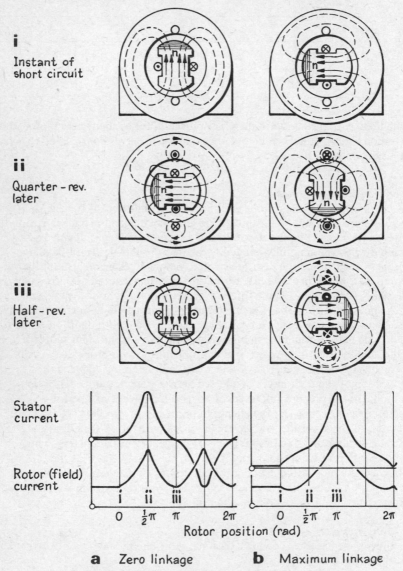

i
Instant of
short circuit

ii
Quarter - rev.
later

iii
Half - rev.
later

Stator
current

Rotor (field)
current

a   Zero linkage   **b**   Maximum linkage

10.24 Constant linkage: single-phase short circuit.

Initial conditions

A physical insight into the behaviour of a generator during time-interval (1) above can be based on the *theorem of constant linkage,* which states that the linkage of a loss-free inductive circuit, closed on itself, cannot be altered. On stator short circuit, the stator windings are closed by the fault, while the rotor exciting winding is closed through its exciter; and if resistance can initially be neglected, the linkage of each winding must remain at the level it had at the instant of short circuit.

Consider an elementary *1-phase* machine, Fig. 10.24, comprising a 2-pole field system and a concentrated stator winding. Let the stator, initially on open circuit, be suddenly short-circuited as at (*a*) when in a condition of zero linkage, instant (i). As the rotor rotates to a new position (ii) at right-angles to the first, it tends to establish full normal linkage in the stator winding: this the stator must oppose by a current in the direction shown, forcing the rotor field into a leakage path. Simultaneously the rotor current must increase to maintain its own flux constant despite the now increased reluctance of its magnetic circuit. At position (iii) the stator current is again zero and the rotor current sinks to normal. A similar argument for subsequent instants gives the current/position relation shown.

Suppose now that the position of the rotor at the instant of stator short circuit is that at (i) in Fig. 10.24(*b*). The stator has maximum linkage. Rotation of the rotor into position (ii) removes its linkage from the stator, which consequently assumes a current to produce its own, in leakage paths. In position (iii) the rotor attempts to produce full reversed flux in the stator, and is counteracted by the stator current rising to a high value to maintain unchanged its original linkage. The peak current is determined by the reluctance of the leakage paths. The rotor in position (iii) has its flux forced into a leakage path mainly inside the stator bore, so that the rotor current rises too. The current changes are shown in (*b*). In actuality, the rotor current variations take place partly in the rotor winding, partly in other closed paths available for current flow, such as the rotor body (if solid), wedges and end-covers, or damper windings.

The short-circuit current is seen to be a function of relative position of stator and rotor, and immediately after short circuit the currents can be assessed in terms of rotor angle and appropriate leakage inductances.

At the moment of short circuit in a *3-phase* machine, the flux linking the stator from the rotor is caught and 'frozen' to the stator, giving a stationary replica of the main-pole flux. For this purpose each phase will in general carry a d.c. component. The rotor maintains its own poles, and accordingly there are two systems of similar poles, one rotating with the rotor and the other fixed with the stator. Each time the rotor poles are in the position they occupied at the moment of short circuit the rotor (field) current is normal and all the stator phase currents zero (if the machine were previously unloaded). One half-period later the stator and rotor poles are again axially coincident, but are opposite in sense: yet constant linkage must be maintained in each, and current peaks result. Further, the stationary field held by the stator induces a normal-frequency e.m.f. in the rotor. The resultant a.c. component in the effective rotor field

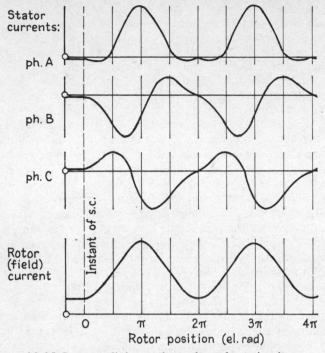

10.25 Constant linkage: three-phase short circuit.

circuit develops a fundamental-frequency flux, which due to rotation produces in the stator windings a *double-frequency* or *second-harmonic* current. Thus the general form of the transient phase current is a wave comprising fundamental, second-harmonic and d.c. components. The rotor current has a large fundamental-frequency fluctuation, part of which will flow in eddy-current paths. Fig. 10.25 gives curves of stator-phase and rotor-circuit currents exhibiting these effects.

Analysis of the currents in a simple ideal 2-pole machine when short-circuited can be based on the treatment in Sect. 7.5. The machine is reduced to the three circuits F, D and Q, with resistance neglected in accordance with the constant-linkage theorem. It is convenient to make $t = 0$ the instant of short circuit, at which the axis of phase A make the initial angle θ_0 with the d-axis: then $\theta = \omega_1 t + \theta_0$. Prior to zero time the stator axis voltages are $v_d = 0$ and $v_q = -x_{ad} i_{f0}$ for an exciting current i_{f0}. The effect of the short circuit is to reduce v_q instantaneously to zero, represented by the sudden application of a step-function voltage $- v_q 1$ to the q-axis terminals. The currents thereafter are superimposed on those that existed prior to $t = 0$. In fact there is a current only in F, so that the currents in D and Q will be the short-circuit currents, while that in F is the component superimposed on i_{f0}. The speed is taken as invariable, with ω_r equal throughout to ω_1 The only equations concerned are consequently (7.12), (7.15) and (7.16), omitting the resistances and all KD and KQ effects. Then writing inductance coefficients for convenience as reactances, we have

F: $\quad 0 = (1/\omega_1)[x_{ad} + x_f)\, p i_f + x_{ad}\, p i_d]$ (i)

D: $\quad 0 = (1/\omega_1)[(x_{ad} + x)\, p i_d + x_{ad}\, p i_f + \omega_1(x_{aq} + x) i_q\,]$ (ii)

Q: $-v_q 1 = (1/\omega_1)[(x_{aq} + x)\, p i_q - \omega_1 x_{ad} i_f - \omega_1(x_{ad} + x) i_d\,]$ (iii)

From (i), $i_f = -i_d[x_{ad}/(x_{ad} + x_f)]$ can be eliminated from (ii) and (iii). Then from (ii)

$$i_q = -\frac{1}{\omega_1} \cdot \frac{1}{x_{aq} + x}\left[\frac{x_{ad}\, x_f}{x_{ad} + x_f} + x\right] p i_d = -\frac{1}{\omega_1} \cdot \frac{x_d'}{x_{sq}} \cdot p i_d$$

where x_d', the term in square brackets, is called the *d-axis transient reactance*. Substituting for i_f and i_q in (iii) and simplifying then gives

$$i_d = \frac{\omega_1^2}{x_d'} \cdot \frac{1}{p^2 + \omega_1^2} \cdot v_q 1 \tag{10.9}$$

This solves by standard methods to give i_d as a function of time, and from it i_q:

$$i_d = (v_q/x_d')(1 - \cos \omega_1 t) \quad \text{and} \quad i_q = -(v_q/x_{sq}) \sin \omega_1 t$$

It remains to transform the axis currents into phase equivalents by eq.(7.19) with $\theta = \omega_1 t + \theta_0$. For phase A

$$i_a = v_q\left[\frac{1}{x_d'} \cos(\omega_1 t + \theta_0) - \frac{x_{sq} + x_d'}{2 x_d' x_{sq}} \cos \theta_0 - \frac{x_{sq} - x_d'}{2 x_d' x_{sq}} \cos(2\omega_1 t + \theta_0)\right]$$

$$\underset{\substack{\text{normal frequency} \\ \text{a.c. component}}}{} \quad \underset{\substack{\text{asymmetric (d.c.)} \\ \text{component}}}{} \quad \underset{\substack{\text{double-frequency} \\ \text{a.c. component}}}{} \tag{10.10}$$

And similarly for phases B and C. The transient field current is

$$i_f = v_q \frac{x_{ad}}{x_{ad}(x_f + x) + x_f x} [1 - \cos \omega_1 t]$$

These results resemble the curves in Fig. 10.25. Neglect of resistance implies that all current components persist without decay.

EXAMPLE 10.5: A 2-pole generator running on no load and excited to normal busbar voltage is dead-short-circuited on all three phases at the instant that the axes of the field and phase A are in coincidence. Find the current in phase A and in the field winding. Per-unit reactances: $x_{ad} = 1.50$, $x_{aq} = 0.60$, $x = 0.10$, $x_f = 0.13$ p.u. Resistances: all negligible.

Here $v_q = 1.0$ p.u. and $\theta_0 = 0$. The d-axis transient reactance is

$$x_d' = \frac{x_{ad}x_f}{x_{ad}+x_f} + x = 0.22 \text{ p.u.}$$

and $x_{sq} = x_{aq} + x = 0.70$ p.u. Using the solutions above we get

$$i_a = 4.5 \cos \omega_1 t - 3.0 - 1.5 \cos 2\omega_1 t$$

$$i_f = 4.2(1 - \cos \omega_1 t) + 1.0 = 5.2 - 4.2 \cos \omega_1 t$$

the unity being added to i_f for the initial value $i_{f0} = 1.0$ p.u. The values correspond to those plotted in Fig. 10.25.

Subsequent conditions

Eq.(10.9) above shows that, neglecting resistance, the normal-frequency component of the *initial* short-circuit current is limited by the transient reactance
x_d'. The *steady-state* short-circuit current, however, is limited by $x_{sd} = x_{ad} + x$.
In Example 10.5 we found $x_d' = 0.22$ p.u., whereas x_{sd} is given as 1.60 p.u.,
a much higher value. The high initial short-circuit current is determined by a
true leakage inductance, and as armature reaction takes a finite time to build
up the quite different flux pattern threading the various closed windings of the
machine, there is a time interval of several seconds between the initial high-
valued currents and the attainment of the much lower level of steady-short-
circuit conditions.

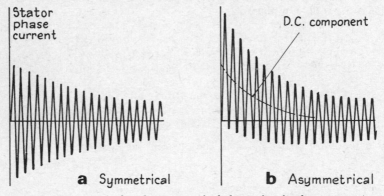

10.26 Symmetrical and asymmetrical short-circuit phase currents.

Fig. 10.26 shows two limiting patterns of phase short-circuit current. In (*a*)
the instant of short-circuit is such that there is no asymmetry — it corresponds
roughly to the conditions in Fig. 10.24(*a*) — while (*b*) is a case in which there
is a maximum d.c. component. The rapid attainment of symmetry and the
slower shrinking of the current envelope to the steady-short-circuit level each
depend upon an effective resistance/inductance ratio related to the type of
construction. Thus in the solid rotor body of a turbo-generator the eddy-

currents in rather ill-defined paths shunt the field winding resistance and contribute to the rate of energy loss, so damping both the asymmetry and the short-circuit current amplitude. In salient-pole machines the damping windings in the pole faces serve a similar purpose.

In brief, *3-phase* short-circuit currents are characterized by (i) a high initial value determined by an effective leakage inductance; (ii) some phase-current asymmetry depending upon the instant in the cycle at which the short circuit occurs; (iii) rapid attenuation of the d.c. components; (iv) slower attenuation of the a.c. component to a value determined by the synchronous reactance.

If a short circuit occurs between one phase terminal and earth with the other two phases unaffected, the *1-phase* short-circuit current develops a pulsating stator m.m.f. with an a.c. component that can be regarded as resolved into two equal and oppositely rotating component m.m.f.s. That component moving with the rotor gives rise to a demagnetizing effect, but the other reacts on the field winding with a double-frequency pulsation. The d.c. component of the phase current will, as in the 3-phase case, produce stationary magnetic poles reacting on the field circuit at normal frequency. The rotor current then exhibits (i) a general increase to balance the forward-rotating m.m.f. of the a.c. reaction, (ii) a double-frequency current due to the backward-rotating m.m.f., and (iii) a normal-frequency current resulting from the d.c. component, which vanishes as symmetry is approached (or is absent if there is no asymmetry).

Short-circuit current envelope. On a 3-phase dead-short-circuit, the a.c. components of the phase currents — that is, the currents that remain after the d.c. transient components have been extracted — occupy identical envelope shapes, of which Fig. 10.27 is typical. The envelope is substantially exponential in its variation with time, with a time-constant τ' of the order of 1−2 s. At the beginning, however, there is a much more rapid rate of decay with a time-constant τ'' of 0.2 s or less. Extrapolation of the main part of the envelope back to zero

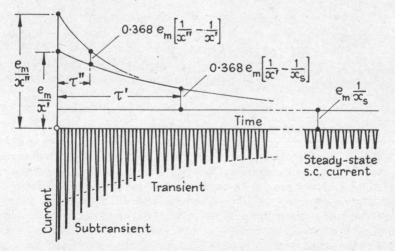

10.27 Analysis of symmetrical short-circuit current.

time would give a peak current of e_m/x': the actual current peak at zero time, given by e_m/x'', is in fact considerably greater. It is necessary to discuss the reason for the existence of the *sub-transient* reactance x'' and time-constant τ'', as well as of the *transient* values x' and τ'. The explanation is based on the flux paths enforced at various times during the run of the sudden short circuit. *Axis reactances.* For a terminal short circuit, all the effective reactances — subtransient, transient and synchronous — are d-axis quantities. If the short circuit occurs at a point remote from the generator terminals into the connected network, the phase angle may be less and quadrature-reaction effects may appear together with some increase in the resistance and inductance parameters concerned in the short-circuit paths. The general problem then requires for solution a knowledge of the subtransient, transient and synchronous reactances and time-constants for both d- and q-axes. Typical values are given in the Table, together with data for negative- and zero-sequence reactance. The n.p.s. reactance is a rough mean between x_d'' and x_q''; the z.p.s. reactance is small and is considerably affected by the stator winding coil-pitch and phase-spread.

Synchronous Machine Reactances (p.u.) and Time-constants (sec.)

Quantity		Turbo-generator	Hydro-generator with dampers	without dampers	Compensator
Synchronous: d-axis	x_{sd}	1.0–2.5	0.5–1.5	0.5–1.5	0.8–2.0
q-axis	x_{sq}	1.0–2.5	0.35–1.0	0.35–1.0	0.5–1.5
Armature leakage:	x	0.1–0.25	0.1–0.2	0.1–0.2	0.1–0.2
Subtransient: d-axis	x_d''	0.1–0.25	0.1–0.2	0.15–0.3	0.1–0.25
q-axis	x_q''	0.1–0.25	0.2–0.8	0.35–1.0	0.2–0.8
Transient: d-axis	x_d'	0.2–0.35	0.2–0.3	0.2–0.35	0.2–0.35
q-axis	x_q'	0.2–0.35	0.25–0.8	0.35–1.0	0.25–0.8
Sequence: n.p.s.	x_2	0.1–0.35	0.15–0.6	0.25–0.6	0.15–0.5
z.p.s.	x_0	0.01–01	0.04–0.2	0.04–0.2	0.03–0.2
Time-constant: d.c.	τ_a	0.1–0.2	0.1–0.2	0.1–0.2	0.1–0.2
subtransient	τ''	0.03–0.1	0.03–0.1	0.03–0.1	0.03–0.1
transient	τ'	1.0–1.5	1.5–2.0	1.5–2.0	1.5–2.5

A physical appreciation of the terms subtransient and transient can be had by consideration of the diagrams in Fig. 10.28, drawn for a salient-pole machine in which the damping circuits are more explicit than they are in a turbo-type machine. For the d-axis, (c) shows the *steady-state* condition in which the stator d-axis m.m.f. contributes flux that traverses the 'normal' magnetic circuit. But if the stator current appears suddenly as in the case of a short circuit, the closed circuits prevent a corresponding sudden change of linkage: an opposing current is induced into the damper winding, forcing the stator flux into the gap as in (a). With the exponential decay of the damper-winding current, the stator m.m.f. is able to force its flux more deeply into the pole (although against the opposi-

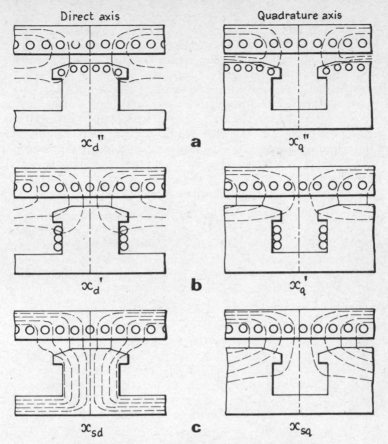

Direct axis   Quadrature axis

x_d'' **a** x_q''

x_d' **b** x_q'

x_{sd} **c** x_{sq}

10.28 Flux paths for salient-pole machine reactances.

tion of induced field current) as indicated in (b). Eventually the steady-state short-circuit condition (c) is reached. As the inductance of a winding is measured by the linkage per unit current, which is in turn dependent on the magnetic reluctance of the flux path, the effective stator inductance is a minimum in condition (a), and smaller in (b) than in (c). The inductance coefficients multiplied by ω_1 give the corresponding reactances x_d'', x_d' and x_{sd}. A similar but less pronounced effect occurs for q-axis stator currents. Because the flux path in (b) differs but little from that in (c), the q-axis transient reactance x_q' has a magnitude of the same order as that of the synchronous reactance x_{sq} in a machine with saliency.

In a turbo-generator there are no specific damper windings in the pole faces, but the solid rotor body and, to a limited extent, the metallic slot wedges provide eddy-current paths and so contribute to the damping. Bharali and Adkins [67] have pointed out that the complicated eddy-current distribution causes the subtransient reactance to diverge from that predicted on simple theory, and give a modification of the KD and KQ parameters for improved accuracy.

While the subtransient and transient inductances are well established in the literature, they do not have an identifiable existence. They are simplified concepts formulated to deal analytically with the exceedingly complex phenomena that occur in an electro-magnetic machine.

Time-constants. The subtransient time-constant is that of the damper or equivalent circuits operative immediately following the incidence of a sudden change of condition. Generally, the circuits concerned have a large resistance/inductance ratio and consequently the time-constants $\tau_d{}''$ and $\tau_q{}''$ are very short. The transient time-constants are up to 20 times as long. The effective resistance controlling a time-constant is a parameter derived from the loss due to the subtransient or transient current in a given part of the machine, and as the current path is not readily defined, its estimation is often rather difficult.

EXAMPLE 10.6: A 50 MVA, 13.5 kV, 50 Hz synchronous generator was run on open circuit and excited to give a terminal e.m.f. of 13.5 kV. A 3-phase terminal short circuit was applied at $t = 0$. The positive and negative amplitudes of the short-circuit current envelope, scaled from an oscillogram for one phase, are tabulated in columns p and n in the Table, for various times t. Estimate (i) the per-unit subtransient, transient and synchronous d-axis reactances, and (ii) the subtransient, transient and d.c. time-constants.

The current envelope has an asymmetry, like that in Fig. 10.26(b). If the d.c. component is extracted it can be plotted separately to a base of time and the time-constant obtained from the duration between the initial value and 0.368 of this value. The remaining symmetrical envelope is plotted as in the upper part of Fig. 10.27 and the rest of the constants obtained. For this purpose columns are added to the Table: $d = \frac{1}{2}(p + n)$ gives the d.c. component, and $a = \frac{1}{2}(p - n)$ yields the amplitude (above and below current zero) of the a.c. component. The results are:

t s	p kA	n kA	d kA	a kA	t s	p kA	n kA	d kA	a kA
0	50.8	−14.3	18.3	32.6	0.30	18.4	−14.1	2.2	16.2
0.02	43.7	−11.9	15.9	27.8	0.40	16.4	−14.3	1.1	15.4
0.04	38.6	−10.8	13.9	24.7	0.50	15.4	−13.8	0.8	14.6
0.06	34.3	−10.3	12.0	22.3	0.80	12.6	−12.6	0	12.6
0.10	29.1	−10.8	9.1	20.0	1.00	11.5	−11.5	0	11.5
0.14	25.3	−11.5	6.9	19.4	1.50	9.6	− 9.6	0	9.6
0.16	24.0	−11.9	6.1	18.0	...	...	...	...	...
0.20	21.9	−12.8	4.6	17.3	30.0	2.75	− 2.75	0	2.75

The rated current is 2.13 kA r.m.s. or 3.0 kA peak = 1.0 p.u. The initial peak of the a.c. component is 32.6 kA, whence $x_d{}'' = 3.0/32.6 = 0.09$ p.u. Extrapolating the plot gives the transient component peak as 22.4 kA, whence $x_d{}' = 3.0/22.4 = 0.13$ p.u. The steady-state short-circuit current is 2.75 kA peak, whence $x_{sd} = 3.0/2.75 = 1.1$ p.u.

The difference $32.6 - 22.4 = 10.2$ kA between the subtransient and transient currents is shown to fall to $10.2 \times 0.368 = 3.7$ kA in $\tau'' = 0.04$ s; similarly the difference $22.4 - 2.75 = 19.7$ kA between the transient and steady-state currents falls to 7.2 kA in $\tau' = 1.40$ s. A separate plot of the d.c. component, treated in a like manner, gives $\tau_a = 0.14$ s. The values required are thus:

(i) $x_d'' = 0.09$ p.u.; $x_d' = 0.13$ p.u.; $x_{sd} = 1.1$ p.u.
(ii) $\tau'' = 0.04$ s; $\tau' = 1.4$ s; $\tau_a = 0.14$ s.

Circuit theory

The short-circuit currents in Example 10.5 were evaluated on the assumption of zero resistance. If the field winding has a resistance r_f (all other circuits remaining resistance-free), the circuit analysis results in the appearance of two time-constants and corresponding exponential decays. Eq.(10.9) now becomes

$$i_d = \frac{\omega_1^2}{x_{sd}} \cdot \frac{1}{p^2 + \omega_1^2} \cdot \frac{1 + p\,\tau_{do}'}{1 + p\,\tau_d'} \cdot v_q \, 1 \qquad (10.11)$$

where $\tau_d' = \dfrac{11}{\omega_1 r_f} \cdot \dfrac{x_{ad}x}{x_{ad} + x} + x_f$ and $\tau_{do}' = \dfrac{1}{\omega_1 r_f}(x_{ad} + x_f)$

are the d-axis transient and open-circuit time-constants. As a result of the field resistance, i_d decays to v_q/x_{sd}, the steady-state short-circuit current limited by the d-axis synchronous reactance x_{sd}.

The more general case of five circuits (F, KD, KQ, D and Q), all with resistance, involves the solution of five simultaneous equations. As Adkins [65] shows, the task is formidable unless approximations are made and the speed is assumed invariant. The solution for phase A for a 3-phase short circuit on a machine initially excited, but unloaded, is

$$i_a = v_q \left[\frac{1}{x_{sd}} \cos(\omega_1 t + \theta_0) \right. \qquad \qquad \text{steady-state s.c. component}$$

$$+ \left(\frac{1}{x_d'}, -\frac{1}{x_d}\right) \exp\left(-\frac{t}{\tau_d'}\right) \cos(\omega_1 t + \theta_0) \qquad \text{normal-freq. transient}$$

$$+ \left(\frac{1}{x_d''} - \frac{1}{x_d'}\right) \exp\left(-\frac{t}{\tau_d''}\right) \cos(\omega_1 t + \theta_0) \qquad \text{normal-freq. subtransient}$$

$$- \frac{x_d'' - x_q''}{2x_d''x_q''} \exp\left(-\frac{t}{\tau_a}\right) \cos(2\omega_1 t + \theta_0) \qquad \text{double-freq. transient}$$

$$\left. - \frac{x_d'' + x_q''}{2x_d''x_q''} \exp\left(-\frac{t}{\tau_a}\right) \cos\theta_0 \right] \qquad \text{asymmetric (d.c.) transient}$$

$$(10.12)$$

The normal-frequency transient can be identified in Fig. 10.27 and the d.c. component in Fig. 10.26(b). The reactances and time-constants concerned are summarized below. They are based on the resistances $r_d = r_q = r$, r_f, r_{kd} and r_{kq}; the leakage inductances $L_d = L_q = L$, L_f, L_{kd} and L_{kq}; and the axis mutual inductances L_{ad} and L_{aq}, for the five circuits D, Q, F, KD and KQ. The reactances are formed from the inductances by multiplying each by ω_1. The various combinations that give the operative reactances and time-constants are as follows:

Reactances

Synchronous: d-axis

q-axis

$$x_{sd} = x_{ad} + x$$
$$x_{sq} = x_{aq} + x$$

Transient: d-axis

$$x_d{}' = \frac{x_{ad}x_f}{x_{ad} + x_f} + x$$

Subtransient: d-axis

$$x_d{}'' = \frac{x_{ad}x_f x_{kd}}{x_{ad}x_f + x_f x_{kd} + x_{kd}x_{ad}} + x$$

q-axis

$$x_q{}'' = \frac{x_{aq}x_{kq}}{x_{aq} + x_{kq}} + x$$

Time-constants

O.C. transient: d-axis

$$\tau_{do}{}' = \frac{1}{\omega_1 r_f} \left[x_{ad} + x_f \right]$$

O.C. subtransient: d-axis

$$\tau_{do}{}'' = \frac{1}{\omega_1 r_{kd}} \left[\frac{x_{ad} x_f}{x_{ad} + x_f} + x_{kd} \right]$$

q-axis

$$\tau_q{}'' = \frac{1}{\omega_1 r_{kq}} \left[x_{aq} + x_{kq} \right]$$

S.C. transient: d-axis

$$\tau_d{}' = \frac{1}{\omega_1 r_f} \left[\frac{x_{ad}x}{x_{ad} + x} + x_f \right]$$

S.C. subtransient: d-axis

$$\tau_d{}'' = \frac{1}{\omega_1 r_{kd}} \left[\frac{x_{ad}x_f x}{x_{ad}x_f + x_f x + x_{ad}x} + x_{kd} \right]$$

q-axis

$$\tau_q{}'' = \frac{1}{\omega_1 r_{kq}} \left[\frac{x_{aq}x}{x_{aq} + x} + x_{kq} \right]$$

S.C. armature (d.c.)

$$\tau_a = \frac{1}{\omega_1 r} \left[\frac{2x_d{}''x_q{}''}{x_d{}'' + x_q{}''} \right]$$

Speed

The simplifying assumption of constant speed is not strictly valid. Post-fault unbalance between prime-mover and electromagnetic torques must result in rotor acceleration, as discussed in Sect. 10.8.

Backswing. The *initial* rotor movement of a turbogenerator following a 3-ph short circuit is in fact a brief retardation ('backswing'). The stationary flux 'trapped' by the stator (represented by the d.c. stator-current components) induces currents in the rotor winding and surface. An interaction drag torque results, and a delay of 30–100 ms may occur before any net increase in the forward rotor angle begins. The magnitude of the backswing is dealt with by Harley and Adkins [134].

Asymmetric loading

Analyses are more complicated when the stator phases are unequally loaded. *Transient.* Graphed results for line-line and line-neutral faults are given by Saha and Basu [68], two-axis analyses by Ching and Adkins [78] and by Jones [135]. In solid-rotor surfaces the eddy currents have ill-defined paths affected by saturation, so that the parameters of equivalent KD and KQ windings are not constants. Phase co-ordinate methods and phase-sequence components may be more tractable. The n.p.s. currents give rise to a backward travelling-wave field having, relative to the rotor, twice normal synchronous speed and so acting alternately on the d- and q-axes; the mean n.p.s. reactance therefore approximates to $x_2 = \frac{1}{2}(x_d'' + x_q'')$. A stationary field only is generated by z.p.s. currents; the z.p.s. reactance x_0 depends on the winding geometry (particularly that of the end-windings) and can vary widely, e.g. 0.03–0.15 p.u. for salient-pole and 0.01–0.10 p.u. for cylindrical machines.

Steady-state. Line-line 1-ph loading of a 3-ph machine can be represented in terms of equal p.p.s. and n.p.s. component stator m.m.f.s., the line-neutral condition requiring the addition of a z.p.s. component of the same magnitude. The p.p.s. component provides armature reaction, while the n.p.s. imposes on the field system an m.m.f. fluctuation at double supply frequency. The pole-flux pulsation can in turn be resolved into a pair of oppositely rotating component fields, one of which has thrice synchronous speed relative to the stator, inducing in the phase windings a 3rd-harmonic rotational e.m.f. which appears in the line voltage. Any resulting 3rd-harmonic stator current generates a 4th-harmonic effect in the rotor. In general, even-order harmonics appear in the field system and odd-order in the stator windings. Jones[135] gives an 'exact' solution for the line-line steady-state condition in terms of the inductance parameters. In salient-pole machines the pole-face damper windings mitigate n.p.s. components, but in solid-rotor machines intense eddy currents are induced in the rotor body and may cause serious local heating.

A related problem, discussed by Bonwick [133], concerns rectifier loads, for which the stator e.m.f. waveform suffers distortion to a degree determined by the transient or subtransient reactance.

Reactance levels

The influence of the major reactances on performance is set out below.

Synchronous reactance. The static stability limit, discussed in Sect. 10.8, is given by the power $E_t V_1 / x_e$, where x_e is the total reactance of the machine, its transformer and the network up to the point having the constant voltage V_1. If the highest steady-state pull-out load or torque is wanted, the machine constituent x_{sd} of the total x_e should be small. However, as the same result can be obtained by use of a voltage regulator to raise E_t rapidly, and as dynamic stability sets a more stringent constraint, a reduction of x_{sd} for this reason is rarely sought. There is, however, the further consideration of capacitive loading by cables and long overhead lines at the time of low consumer load. With positive excitation the synchronous reactance x_{sd} must not be greater than the equivalent capacitive reactance x_c, and this may provide a limiting level for x_{sd} in a generator, though there are means for avoiding the limitation. In the range $x_{sq} < x_c < x_{sd}$ the required excitation is negative.

Transient reactance. When sudden disturbances occur, the rotor flux is maintained by a current induced in the field winding. The same power-limit expression as before holds, except that E_t is replaced by the effective e.m.f. behind the transient reactance and that x_e is now the transient reactance of the machine and system. It is desirable to keep the transient reactance small in those cases in which dynamic stability is critical. A cage form of damping winding may considerably raise the stability by countering the rotor swing by its induction torque. A low transient reactance tends to reduce voltage dip when faults occur.

Subtransient reactance. Let a generator be subjected to a phase-phase short circuit. If the third phase winding is open, the highest overvoltage in it is given by $[2(x_q'' / x_d'') - 1]$ p.u., with a large harmonic content. If the third phase is capacitively loaded by cables or lines, resonance may raise the overvoltage considerably. It is thus important to keep the ratio of q-axis to d-axis subtransient reactance as near as possible to unity. Complete cage dampers and solid pole-shoes give a ratio x_q'' / x_d'' typically between 1.3 and 1.5 and are always sufficient in this respect. The separate pole-grid type of damper has a ratio between 2 and 3 and is wholly unsuitable where resonating conditions are foreseen. In general the absolute value of the subtransient reactance is less important than the torque that the damper can exert.

10.8 *Stability*

The capability of a power system to remain in synchronism when disturbed by load-changes, faults and similar contingencies is called its *stability*.

Power systems normally comprise synchronous generators sited in large power stations and connected to loads that are dynamic (i.e. comprise motors and rotating mechanical attachments) or static (e.g., heating, lighting and certain industrial processes). The load has an important influence on stability, as it may affect a generator that has become unstable to increase its pole-slip or to allow it to re-synchronize after slipping one or more pole-pairs. Under steady load it is economically necessary that full use be made of active and reactive power capability, and at the same time there must be an adequate margin of transient stability against faults and outages for reliable, good-quality supply to the connected loads. Thus both *steady-state* and *transient* stability are of concern.

The consequences of instability can be severe, even disastrous, to system operation. Assessment of system behaviour is extremely complicated, for it involves the contribution of each item of generating plant including boilers, governors, exciters, voltage controllers, protective equipment and switchgear, all having interacting responses complicated by various degrees of nonlinearity.

Steady-state stability

A condition of steady-state stability exists when, after a small disturbance, the system returns to the initial operating state. For generators operating on an infinite busbar (strictly, an unrealized condition) it is shown in Fig. 10.13 that the maximum-power limit is reached for a load angle of 90 deg(e.) for a cylindrical machine, and somewhat less, for example 80 deg, for a salient machine. These load angles correspond to the maximum steady-state power that can be transferred through the synchronous impedance to a fixed system voltage.

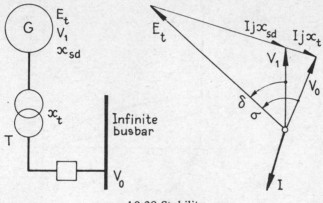

10.29 Stability.

Consider now the system of Fig. 10.29, in which a generator-transformer is connected between the machine and the inifinite busbar. The power/angle and voltage/angle curves now relate to the active power into the busbar through the impedance of the machine and its transformer, the terminal voltage V_1 of the machine, and the angle σ between the e.m.f. E_t and the busbar voltage V_0. So long as σ is less than the maximum-power angular divergence, the machine is stable to small disturbances; but if σ exceeds the maximum-power value, an increase in angle reduces the power, the mechanical drive torque exceeds the electromagnetic torque and the angle must increase still further. Thus regardless of damping the system is unstable. However, if the generator were provided with a continuously acting automatic voltage regulator, the tendency of the voltage V_1 to fall with increasing angle would invoke an increase in excitation and therefore in the active power developed. Provided that the excitation can be raised quickly enough, a positive synchronizing torque can be established to stabilize the machine. This is clarified by the electrical load diagram in Fig. 10.4, where the maximum power corresponding to a generator working at point S can still be developed after an increase of load angle at point T, provided that the excitation is raised (in the case considered) from 0.94 to 1.0 p.u.

The working region of operation at angles exceeding 90 deg is the region of *dynamic stability*. The factors of importance in determining the dynamic stability limit are the inertia of the machine and its prime mover, the transient reactance, damping and exciter response rate. Methods of determining limits are given by Gove [51], and a typical generator capability chart, comparable with that in Fig. 10.17, is shown in Fig. 10.30. The theoretical stability limit

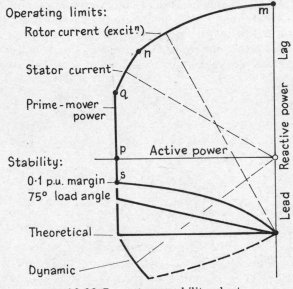

Operating limits:
  Rotor current (excitⁿ)
  Stator current
Prime-mover power
Stability:
  0·1 p.u. margin
  75° load angle
Theoretical
Dynamic

Lag
Reactive power
Lead
Active power

10.30 Generator capability chart.

line is based on a 90 deg load angle; the 0.1 p.u. margin is inserted; another limiting criterion is the 75 deg load angle; and the considerable addition to reactive power capability resulting from automatic excitation control shows the advantage to be gained by operation in the dynamic stability region.

Transient stability

The transient stability is the capability of a system to return to a steady-state operating condition following a large disturbance, such as a loss of a major load or of a large generating unit, or a severe fault. In systems loaded with heavy rotating plant, the mechanical reactions have caused generators to become unstable even after two or three safe swings. Instability may also occur at times of light system load, for with an extensive transmission network the demand for leading reactive power may require generators to be run at low excitation to provide it. Operation at leading power factor is critical in several ways: the permissible angular swing of the generator is less than in normal lagging operation for which the excitation level is high; the system voltage collapses more rapidly on fault and consequently reduces the stabilizing effect of connected loads; and the design of the generator excitation and governing has to be highly sophisticated if reliable equipment is to be achieved.

Typically, the fault condition against which transient stability is assessed is a 3-phase fault which, when cleared in 140 ms, results in the outage of a double transmission circuit. Trials have shown that to achieve adequate stability it may be necessary to provide protection equipment and switchgear capable of clearing such faults in less than 100 ms.

Swing curve. The equation of motion for a synchronous generator, based on eq.(7.5) of Sect. 7.3, can be expressed by equating the difference between the electromagnetic and prime-mover torques to the torque required to accelerate the inertia J and provide a pair of damping torques, one proportional to the velocity and one a constant, i.e.

$$\Delta M = J(d^2\delta/dt^2) + C_1(d\delta/dt) + C_2$$

It is more convenient to work in terms of power at the mean synchronous speed. Then $\omega_s \circ \Delta M$ becomes Δp, the difference between the electrical power p_e and the prime-mover input power p_m. The damping torques multiplied by ω_s become damping powers, and the product $J\omega_s$ can be written as the angular momentum m_o. Then

$$\Delta p = m_o(d^2\delta/dt^2) + k_d(d\delta/dt) + p_c \qquad (10.15)$$

The equation is not easy to solve for δ as a function of time. The power-difference term Δp is itself time-dependent. The electrical power varies not only with δ but also with voltage-regulator action, and the prime-mover power is affected by governor action. Again, during the swing the conditions may change rapidly because of the operation of circuit-breakers. The damping coefficient k_d can depend quite critically on the rate of change of load angle, particularly where significant damping is obtained by means of pole-face windings or induced eddy-current effects.

The momentum for a given machine is related by eq.(7.1) to the inertia constant H. This has a direct bearing on the frequency and amplitude of the rotor swing, for the larger the inertia of the generator-turbine combination, the more excess kinetic energy it can absorb for the same change in speed. Large modern turbo machines tend to have a rather low specific inertia because the physical dimensions do not increase in proportion to the rating, a consequence of the limitation on rotor diameter imposed by centrifugal force and the higher specific electric loadings that are obtained with advanced cooling methods. In the case of hydro machines a much larger inertia effect is essential because of the sluggish action of hydraluic governors made necessary by the great moving masses of water that they must control.

Various solutions for the dynamic stability equations have been given, such as that by Humpage and Saha [69], using various degrees of simplification in the boiler-turbine-generator system. The simplification acceptable depends largely on the number of rotor swings, and this in turn on the load. For static loads the first-swing stability criterion is adequate, but for highly dynamic loads a multi-swing investigation covering up to 20 s may be essential. The problem has some ramifications: for example, the auxiliary motors in a generating station must remain in step even if the main generators produce violent fluctuations of voltage and frequency at the motor terminals. In a computer study by Ramsden, Zorbas and Booth [71] the dynamic performance of induction motors is considered in relation to system stability. They show that the large cage induction motors connected to a supply system can be important in determining the behaviour of the system under fault conditions. Such motors generate large initial currents immediately subsequent to the fault occurrence, and these are additional to those contributed by the synchronous generators. Thereafter the latter machines swing to an increasing load angle, while the induction machines suffer a fall in speed. After the fault is cleared, the stability of the system depends upon whether all the synchronous machines remain in synchronism, and whether all the induction motors recover to normal slip.

Initial currents can be calculated on the assumption that the speed is constant; but for subsequent behaviour the effect of the appreciable speed-change is an essential factor, and the only method available to determine the variations in load angle or speed is a numerical step-by-step computation. The analysis is complicated by the difficulty in establishing a suitable reference frame. Kalsi and Adkins [81] discuss some analytical methods and compare the computed results with tests on micromachines; they conclude that the effect on the system of induction machines ought to be included in system-fault studies, but that simplified methods must be justified by comparison with measured values.

Damping

If the rotor speed departs from a steady synchronous value, any short-circuited path linking the gap flux or part thereof behaves with reference to the travelling-wave stator field as does the cage winding of an induction machine: slip-frequency currents are induced and the consequent I^2R loss dissipates some of the swing energy, developing a torque roughly proportional to the slip. At negative slip the damping torque acts to reduce the speed, and at positive slips to increase it. The estimation of the damping power $k_d(d\delta/dt)$ in eq.(10.15) is complicated, especially if the damping effect results from eddy-current loss in a solid rotor.

Damping circuit. In salient-pole machines, damping windings consist usually of solid copper bars through the pole-shoes. The ends of the bars can be connected together by short-circuiting straps to form a damping 'cage', as in Fig. 10.31. Sometimes interpolar connectors are omitted. An alternative method is to employ thick steel plate for the pole and shoe, or to use solid pole-shoes, in which the eddy-current damping effect is produced. It is possible theoretically to improve the damping by connecting the solid pole-shoes together by interpolar conducting strips, but the advantage gained is usually small. However, without interpolar connections, q-axis damping currents must flow in the pole-body and rim, and the ratio x_q''/x_d'' of the subtransient reactances is inher-

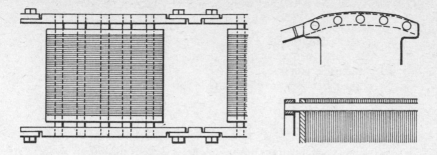

10.31 Damper windings for salient-pole machine.

ently greater than unity. The design of interpolar connections offers some difficulty in that they have to withstand tangential vibration between adjacent poles, and may require reinforcement against centrifugal force if the machine is subject to appreciable over-speed.

The rotors of turbo-generators are always of solid steel, and in addition there may be an effective damping cage effect provided by metallic slot wedges, with the end-caps acting as short-circuiting rings.

Negative-sequence damping. The constant-power term p_c in eq.(10.15) includes friction, which is normally comparatively small, and negative-sequence damping. Most stability studies involve fault conditions in which negative-sequence current components are concerned. These have a slip of approximately $s = 2$, setting up an asynchronous (induction) torque in a direction opposite to that of the rotor rotation; the effect is comparable with a friction torque.

Equation of motion

If eq.(10.15) can be solved, the motion of the rotor can be determined. Now Δp is a function of load angle δ, governor action and voltage-regulator response; damping powers vary with slip and fault conditions; and after some delay in the action of protective gear, circuit-breaker operation occurs to change the operating conditions. Only electronic computation methods with step-by-step calculation can properly cope with the evaluation of the equation of motion and make evident whether or not the machine concerned will remain in step.

The essential features of the calculation can be illustrated by use of the equal-area criterion (Sect. 7.3) and the following simplifying assumptions: (i) the prime-mover power remains constant, (ii) the generator power/angle relation remains unchanged for a given circuit condition, and (iii) damping is neglected so that the equation of motion reduces to $(d^2\delta/dt^2) = \Delta p/m_0$. These are employed in the following Example.

EXAMPLE 10.7: A 25 MVA, 33 kV, 50 Hz, 2-pole, 3-ph turbo-generator delivers a steady load of 20 MW over a double-circuit transmission line to an infinite-busbar system. A 3-ph fault occurs on one of the line circuits, and is cleared simultaneously at both ends (i) after 400 ms, (ii) after 650 ms. Obtain the swing curves and the critical switching time. The required data are:

Inertia constant $H = 2.3$ MJ/MVA
Generator e.m.f. (line) $E = 34.0$ kV
    transient reactance $x_d' = 13.0\ \Omega/\text{ph}$
Infinite busbar voltage $V_0 = 33.0$ kV
Reactances between generator and infinite busbar:
    prior to fault $x_1 = 17.4\ \Omega/\text{ph}$
    during fault $x_2 = 48.0\ \Omega/\text{ph}$
    after fault clearance $x_3 = 21.8\ \Omega/\text{ph}$

The power/load-angle relations are assumed to be sinusoids. From eq.(10.6), suitably modified for the three operating conditions and with output power taken for convenience as positive, the peaks of the three sinusoids are

$$P_1 = 33.0 \times 34.0/17.4 = 64.5 \text{ MW}; \quad P_2 = 23.4 \text{ MW}; \quad P_3 = 51.5 \text{ MW}$$

10.32 Transient stability: equal-area criterion.

Equal-area criterion. The three sinusoids are plotted in Fig. 10.32(*a*). The diagram (based on the treatment in Sect. 7.3) shows that the fault, if allowed to persist, is bound to lead to asynchronous running, for area A_2 is less than A_1. But if the fault is cleared before $\delta = 154°$ some additional compensating area is included. By trial it is found that clearance must take place by the time that δ reaches $138°$. For this *critical switching angle* the acceleration energy, proportional to the total area $A_1' + A_1''$, is just matched by the energy absorption proportional to $A_2' + A_2''$. The equal-area diagram does not give the *critical switch-*

ing time, to determine which it is necessary to find a solution to the equation of motion.

Equation of motion. The momentum is $m_0 = 2HS/\omega_s = 0.37$ MJ per rad/s = 0.0064 MJ per deg(e.)/s. The equation of motion with the simplifying assumptions made is given by $(d^2\delta/dt^2) = \Delta p/m_0$, whence the swing velocity and load angle are approximately

$$\omega = \Delta\delta/\Delta t = (\Delta p/m_0)\,\Delta t + \omega'$$

$$\delta = \delta' + \omega\Delta t$$

for a short-time-interval Δt, where ω' and δ' are the initial values of swing velocity and load angle respectively. Inserting numerical values with the time-interval $\Delta t = 50$ ms = 0.05 s gives

$$\omega = \omega' + 7.82\,\Delta p \text{ deg/s} \quad \text{and} \quad \delta = \delta' + 0.05\,\omega \text{ deg}$$

with Δp in megawatts. Prior to the fault $(t = 0-)$ the load angle is such that $64.5 \sin \delta = 20$ MW so that $\delta = 18.1°$. Immediately subsequent to the fault initiation $(t = 0+)$ the angle is the same but the power sinusoid changes to a peak of 23.4 MW for which the generator power is 7.25 MW: hence $\Delta p = 20 - 7.25 = 12.75$ MW available to accelerate the rotor. For $t = 0$ the *mean* $\Delta p = 12.7/2 = 6.37$ MW is taken, and the velocity at the end of the first interval Δt is

$$\omega = \omega' + 7.82\,\Delta p = 0 + 7.82 \times 6.37 = 50 \text{ deg/s}$$

The load angle at the end of the interval is

$$\delta = \delta' + 0.05\,\omega = 18.1 + 2.5 = 20.8 \text{ deg}$$

These become initial values for the second interval: $23.4 \sin 20.6° = 8.25$ MW so that $\Delta p = 20 - 8.25 = 11.75$ MW. The velocity and load angle are now

$$\omega = 50 + 7.82 \times 11.75 = 142 \text{ deg/s}, \delta = 20.6 + 142 \times 0.05 = 27.7 \text{ deg}$$

(i) Clearance at $t = 400$ ms: Tabulated values for 19 time-intervals are given below. Fault clearance occurs at the end of interval 8, where the power sinusoid changes to a peak of 51.5 MW, and the mean Δp is again taken as for $t = 0$. Graphs of swing velocity and load angle are given in Fig. 10.33. The machine swings over a wide angle but remains synchronized.

(ii) Clearance at $t = 650$ ms: The values up to interval 8 are the same as in (i), and continue under the same conditions up to interval 13, where a second transition occurs. The values are graphed in Fig. 10.33 (dotted line), showing that circuit-breaker operation is too late to avoid loss of synchronism.

Critical switching time. It is seen that the critical switching time exceeds 400 ms but is less than 650 ms. A guide can be obtained by a fairly simple adjustment of the switching angle in Fig. 10.32(c) to equalize the areas, giving the angle 138° as the critical angle. The dotted graph in Fig. 10.33 shows that this angle is reached in about 12 time-intervals, i.e. 600 ms, and this gives approximately the critical switching time.

Fault-clearance time, 400 ms

Interval No.	δ'	$\rho \sin \delta'$	Δp	ω'	$\Delta\omega$	ω	$\Delta\delta$	δ
0−	18.1	20.0	0					
0+	18.1	7.25	12.75	Initial, $\Delta p = 6.37$				
0	18.1	7.25	6.37	0	50	50	2.5	20.6
1	20.6	8.25	11.75	50	92	142	7.1	27.7
2	27.7	10.9	9.1	142	71	213	10.6	38.3
3	38.3	14.5	5.5	213	43	256	12.8	51.1
4	51.1	18.2	1.8	256	14	270	13.5	64.6
5	64.6	21.1	−1.2	270	−9	261	13.0	77.6
6	77.6	22.9	−2.9	261	−23	238	11.9	89.5
7	89.8	23.4	−3.4	238	−27	211	10.6	100.1
8−	100.1	23.0	−3.0					
8+	100.1	50.8	−30.8	Transition, $\Delta p = -16.9$				
8	100.1	50.8	−16.9	211	−132	79	4.0	104.1
9	104.1	50.0	−30.0	79	−234	−155	−7.7	96.4
10	96.4	51.3	−31.3	−155	−245	−400	−20.0	76.4
11	76.4	50.1	−30.1	−400	−235	−635	−31.8	44.6
12	44.6	36.3	−16.3	−635	−127	−762	−38.1	6.5
15	−42.1	−34.6	54.6	−319	427	108	5.4	−36.7
20	88.1	51.5	−31.6	524	−247	277	13.9	102.0
25	34.1	28.5	−8.5	−665	−66	−731	−36.5	−2.4

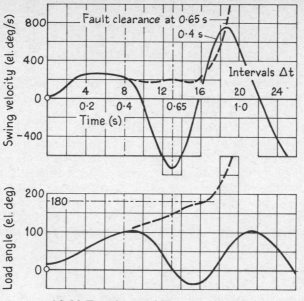

10.33 Transient stability: swing curves.

Improvement of stability

Governors contribute little to stability unless they have very fast responses to small changes of speed, e.g. full-throttle valve closure within 500 ms. Such a response demands special oil-pressure hydraulic servos. Automatic voltage regulators can act through the excitation system to change the generator power, but only by as much as the system voltage and rotor angle will allow. The most effective aid to stability is the fast-acting circuit-breaker. With rapid fault detection within 10 ms and air-blast breakers it is possible to attain full fault clearance in less than 100 ms. In this time-interval a large generator will not acquire much additional kinetic energy.

The complexities of system modelling have been discussed in Sect. 7.4. A generating set in particular is a mechanical-electrical conversion link between thermal or hydro plant at one end and an electrical supply network at the other. No single control element can be considered in isolation, for such elements interact and are usually nonlinear. However, the elements most immediately concerned in operation are the governing and the excitation, which must for stability be closely co-ordinated. Summarizing, the improvement of stability may be achieved as follows.

Steady state. Electrically, this is a matter of synchronizing and damping torques. Consider the system in Fig. 10.29. Let x_t be the reactance between the generator terminals and a point on the network that can be assumed to have infinite-busbar properties; and replace x_{sd} by x, the value most appropriate to the swing condition. Then adapting eq.(10.3), the synchronizing power coefficient is

$$p_s = [E_t V_0/(x + x_t)] \cos \sigma = [V_1 V_0/x_t] \cos (\sigma - \delta)$$

A strong synchronizing power depends on small reactances and divergence angles, and on the maintenance of system voltage and generator excitation (conditions not favourable if faults occur). The natural damping of the generator is raised if control secures that the machine supplies more active power when accelerating than when retarding during a swing.

Transient state. Improved steady-state stability benefits transient stability too, but there are further requirements. Network faults must be rapidly detected and cleared, and fast governor and valve control provided to reduce the mechanical drive. It may be possible to augment the inertia of a hydro-generator at the design stage, but the natural inertia of turbo-generator sets tends to be low.

Dynamic state. Control methods that cause torques to be produced in phase with the rate of change of load angle δ improve dynamic stability, in particular excitation control to increase the electrical output in phase with $+d\partial/dt$, and the prime-mover power in phase with $-d\partial/dt$, the former by excitation and the latter by governing. Many control quantities have been suggested, including $d\partial/dt$ and its higher derivatives, or the output power P_1 and its time-derivative and time-integral.

10.9 *Excitation and voltage regulation*

Small synchronous machines are sometimes built with permanent-magnet field excitation, but normally a d.c. excitation supply is necessary. Industrial motors may operate with a constant excitation. With generators the aims of the exciting system are: (i) to control the voltage for operation (if needed) near the steady-state stability limit; (ii) to regulate the voltage under fault conditions; and (iii) to facilitate reactive-power load-sharing between generators operating in parallel.

To obtain the greatest possible rating from a given mass of active material, turbo-generator design with forced cooling and high current densities gives unavoidably a high synchronous reactance compared with that normal in a low-speed machine. The synchronous reactance x_{sd} has as its main component the stator (or 'armature') reaction reactance x_a which depends on the electric loading. Thus typically with a leakage reactance $x = 0.15$ p.u. and a reaction reactance $x_a = 1.6$ p.u., the synchronous reactance has the high value $x_{sd} = 1.75$ p.u.

With reactances of this order, the rotor field excitation at full load and p.f. 0.8 lagging must be more than $2\frac{1}{2}$ times that for no load at the same terminal voltage. The exciter must therefore have a voltage and current rating to correspond. Allowing for temperature-rise and voltage margin, the exciter must have a a voltage range of the order of 4/1 over which it must possess a stable characteristic. Some tendency to instability occurs with weak fields when the reaction of the stator currents on the airgap flux is stronger than that of the exciting current. It is generally sufficient to have a gap length that makes strong excitation necessary and enables the exciter to achieve the wide range of voltage without undue saturation.

In large generators the excitation power is of the order of 0.005 p.u., approaching 0.01 p.u. during 'field forcing'. The power (which is little more than the field circuit I^2R) for the largest machines may be about 4 kW/MVA.

A compromise between large exciting currents at low voltage and thicker insulation for high voltage leads to a choice normally between 200 and 800 V.

Excitation systems

D.C. exciters. This is the traditional method. A d.c. generator is mounted on the main shaft and shunt- or separately excited, the output being fed to the rotor of the main generator through slip-rings and brushgear. Exciter construction follows normal d.c. generator practice, except that all or part of its field excitation may be provided by an auxiliary pilot exciter to improve the rate of response. Commutation difficulties at high speeds have sometimes led to geared or separately-driven exciters; low-speed hydro-generators may require exciters driven at some suitably higher speed by motors or by small water turbines.

Static excitation. Direct-current excitation can be obtained by means of a rectifier and a suitable supply of alternating voltage, and by this method the commutation limits inherent in the rotary d.c. exciter generator are avoided. The rectifier unit has no rotating parts, requires little maintenance and is immune to hazardous or dust-laden atmospheres. With associated control equipment it is accurate, reliable, compact, and fast-acting. There are several arrangements.

With an *a.c. exciter* machine mounted on the main shaft, its field fed from a pilot exciter whose field in turn is derived from a permanent-magnet generator, the exciter output is fed to the floor-standing rectifier, and the rectified current supplied to the main rotor winding through conventional slip-rings. The essential arrangement is shown in Fig. 10.34.

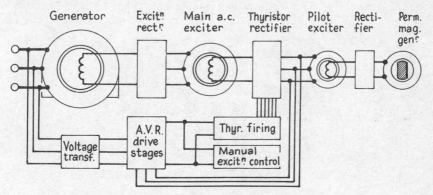

10.34 A.C. exciter/rectifier excitation for turbo-generator.

With *thyristors* in place of diodes a reliable excitation system can be achieved. A fully-controlled 3-phase bridge connection is most usual because of its advantages in securing lower voltage stresses on the semiconductors and greater utilization of the transformer capacity. With the high peak reverse voltage capability of of thyristors, two elements in series per branch are sufficient. Advances in thyristor technology, particularly in the on-state characteristics and in improved triggering, now permit the direct series connection without cascading.

Design of the rectifiers must be conservative, so that they can withstand the severe conditions imposed by system faults, pole-slip, asynchronous running and faulty synchronizing. Typical diode rectifier elements, rated at 300 A and 3 kV reverse peak voltage, can be connected in parallel banks, and are readily maintained.

Brushless excitation. The term 'brushless' is applied to a machine in which the conventional brushgear is eliminated. An exciter with a fixed field and rotating phase windings is mounted on the main shaft, the a.c. output being converted to d.c. by means of shaft-mounted rectifiers and fed directly to the main rotor winding, no slip-rings or brushgear being needed.

Most modern synchronous motors are equipped with brushless excitation. The requirements differ from those that obtain in large turbo-generators, for while the brushless system for generators is excellent from the viewpoint of general maintenance, to replace a faulty rectifier diode it is necessary to shut down the machine, involving a long and very costly outage. Further, it is not possible to use a field switch or to reduce the natural time-constant of the rotor field winding by externally connected resistors, so that damage of a more extensive nature may result in the event of an internal fault. Remote-indicating instrumentation without re-introducing slip-rings is more complicated.

The rotating rectifier should have the simplest mechanical arrangement to provide maximum reliability. Ideally the bridge arms should consist only of diodes in parallel since, with forward- and reverse-polarity diodes, this permits of the firm support of each unit in the metal housing of the rotating carrier. A construction suitable for a 500 MW generator is illustrated in Fig. 10.35: two T-section steel discs are provided, one for each polarity and insulated from

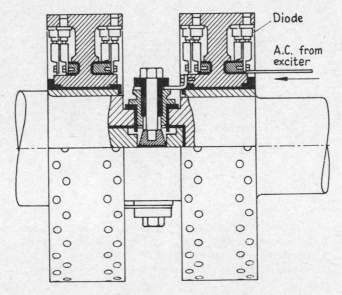

10.35 Rotating rectifier.

the shaft. The diodes are mounted on the rims with the inner flexible conductors joined to one of the a.c. phase terminals, each of which covers circumferentially somewhat less than 120° and is insulated from the T-section discs. Such a rectifier unit has an outside diameter of less than 1 m, and an axial length of about the same as would be needed for normal excitation slip-rings.

There are certain other ways of exciting a brushless machine, one being a combination of induction and rectification. In a machine operating on this principle, it is arranged that an auxiliary field is set up so as to induce an alternating e.m.f. in a winding carried on the rotor, and this is used with a rectifier to supply d.c. excitation to the main-field winding of the rotor. The auxiliary field may be stationary, pulsating or rotating relative to the stator, the only essential requirement being that it should alternate with respect to the rotor. For example, this field could be produced by space harmonics due to winding arrangement or distribution, or due to magnetic circuit geometry. It may have the same pole-number as the main field of the machine, in which case alternating voltages are induced into a single rotor winding which is closed through rectifiers. Alternatively the auxiliary field may have a different pole-number and induce e.m.f.s in a special pick-up winding connected through rectifiers to the main rotor exciting winding.

Self-excitation. Small generators, such as those for marine service, can be self-excited by means of a bridge-connected rectifier deriving its alternating voltage drive from a transformer with two primary windings per phase; one primary is excited from the generator terminal voltage, the other from the load current, as indicated in Fig. 10.36. As a result the excitation is responsive to both output voltage and load current, and is said to be compounded, or self-regulating. The advantages are rapid transient voltage response to suddenly applied loads together with the reliability of static equipment. As most generators have a residual rotor field too small to enable self-excitation to occur from a standstill start, a permanent-magnet pilot exciter is provided to estabilish an adequate generator terminal voltage for initiating self-excitation.

The method is applicable to synchronous motors, although here the initiation equipment is not needed. By proper design of the transformer and of the star-connected inductor used to terminate its 'voltage' primary windings, it is pos-

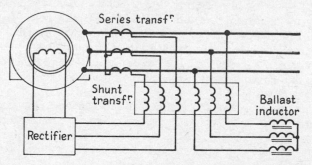

10.36 Compounded self-excitation.

sible to secure either constant power factor or constant reactive power for all motor loading conditions.

Double excitation

A major problem in the operating of generators connected to an extensive supply network is the supply of reactive power. A conventionally wound turbo-generator rotor can be made to furnish leading reactive power by underexcitation, but if stability is to be retained the limit of reactive power on zero active load is about 0.5 p.u. A fast-acting voltage regulator can make operation possible at load angles exceeding 90°, but dynamic stability is then endangered. Thus the turbo-generator with a conventionally-wound rotor (c.w.r.) operating at a leading power factor on an unloaded system works in its most unfavourable condition when the system voltage is most difficult to control.

To improve the conditions the double- or divided-winding rotor (d.w.r.) can be used. The d.w.r. makes it possible to shift the rotor m.m.f. axis with respect to the rotor body: it has the following features:

The d.w.r. has the same slotting as a c.w.r., but the slot conductors are separated by special end-connectors into two parts, the m.m.f. axes of which are mutually displaced (e.g. by an angle of 60°) to give *torque-winding* and *reactive-winding* m.m.f.s, individually controllable.

The m.m.f. axis of the *torque winding* is varied in proportion to the displacement of the rotor from a predetermined load angle, irrespective of whether that displacement results from a change of prime-mover power or from a system disturbance. For this purpose a control sensitive to load angle is required.

The m.m.f. of the *reactive winding* can, by the action of an automatic voltage regulator, be varied over a range of values from positive to negative as required by the divergence between the system voltage and that of a predetermined level of reference voltage.

The angle regulator controls the load angle to a value approximately equal to that between the torque-winding and reactive-winding m.m.f.s, although this condition is not essential and may, in fact, not be best for stability.

The essential concept can be conveyed by the m.m.f. diagrams in Fig. 10.37, drawn for constant generator full-load active power respectively at rated lagging and leading power factors. The rotor m.m.f. for the c.w.r. machine is F_2 and its resultant gap m.m.f. is F_0. The m.m.f. diagram for lagging power factor is based on Fig. 10.1 for the over-excited generator condition (b)(iii). The load angle δ is 40°, the machine being taken to have a short-circuit ratio $r_{sc} = 0.4$ and negligible saturation. If F_2 is reduced, the power factor rises and then becomes leading; and m.m.f. F_2 reduced by 33% results in a load angle of 98°, which lies in an unstable region.

Consider now the action in the d.w.r. machine, in which the reactive winding produces the m.m.f. F_{2r} along the gap-flux axis F_0, with the torque winding m.m.f. F_{2t} displaced from it by 60°. The lagging and leading power factor conditions are obtained by adjustment and reversal of F_{2r}, with the axis of F_{2t} unchanged.

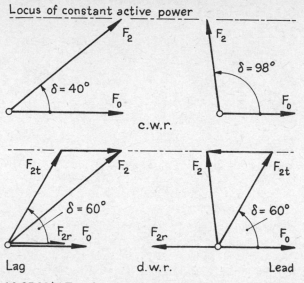

10.37 M.M.F.s of conventional and double-wound rotors.

There are several advantageous effects. The d.w.r. can control active and reactive outputs without altering the rotor position with respect to the gap flux, so increasing steady-state stability. The transient stability is improved, largely because of an increase in the q-axis time-constant. When angle- and voltage-sensing regulators are used independently, with the reactive winding held in line with the flux axis, it is possible to employ very high voltage-regulator gains, again improving the transient stability limits. The d.w.r. system has been analysed by several writers, including Harley and Adkins [83]. The success of the d.w.r. machine is dependent upon an advanced and sophisticated system of field control.

Excitation response

When a sudden load change on a generator occurs, demanding a change in excitation, the automatic regulating equipment must first operate on the exciter field; the exciter output voltage changes, and then the main rotor field current. Magnetic fluxes cannot build up instantaneously, so that the rapidity of response to a load change depends on a sequence of events each with a time delay: the voltage regulator, the exciter and the main field, together with closed circuits and eddy-current paths which tend to sustain the fluxes linked by them, all contribute some delay.

An indication of the rapidity of response of an exciter is the average rate of rise of the rotor terminal voltage following the opening of the exciter armature circuit. For turbo-generators a response of 0.5 may be adequate (i.e. a rise of 200 V/s for a nominal 400 V). The rate required for hydro-generators may be much higher as a consequence of the overspeed to which these machines may be subjected.

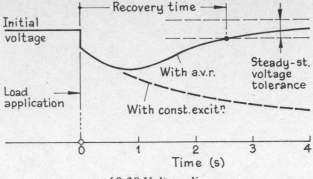

10.38 Voltage dip.

Fig. 10.38 shows stator terminal voltage variations for a generator on 0.4 p.u. load at p.f. 0.8 lagging, and undergoing at $t = 0$ a sudden increase of load to 0.8 p.u. The initial drop of about 4% is due to increased leakage reactance drop. Thereafter the voltage falls slowly as the gap flux alters to a value determined by the new armature reaction. After about 1 s the rising exciter current begins to restore the generator terminal voltage. A shunt-connected exciter gives the slowest restoration, separate excitation is more rapid, and if *field forcing* is employed (the application of e.g. twice normal field voltage) the delay may be reduced to little more than 2 s. Where the dropping of load is liable to cause overspeed, considerable voltage rises may initially occur. Fig. 10.39(*a*) shows conditions produced by suddenly unloading a 20 MVA hydro-generator working at a p.f. of 0.75 lagging and with a synchronous speed of 100 r/min. The water gate begins to close about 0.6 s after the load has been dropped. Meanwhile the speed rises, reaching 1.35 p.u. in 5 s. The voltage regulator reduces the field current but the rotor flux and overspeed combine to raise the generator voltage to 1.7 p.u. The curves in (*b*) are for the same conditions as in (*a*) except that an overvoltage overfrequency relay opens the field circuit; but there is still a voltage rise.

Field suppression. When a short circuit occurs in a machine in operation, not only should it be disconnected from the busbars but also its generated e.m.f. should be rapidly extinguished in order to avoid the aggravation of internal damage. The excitation must be interrupted by an automatic switch, by opening the main field or the exciter field circuit, or by reversing the field voltage, or by short-circuiting the main field winding and exciter.

The e.m.f. induced in a field winding disconnected from its exciter and closed on a resistor of resistance matching that of the field winding is twice normal voltage; but the field current may persist and cause the generator to feed into an internal fault, unless (as in turbo-generator) there are eddy-current damping paths in the rotor body to shorten the field persistence time.

Exciter reversal has little advantage over short-circuiting, but it reduces the field current more rapidly in the later stages. If the exciter is itself separately excited from a constant-voltage source, the generator field current can be

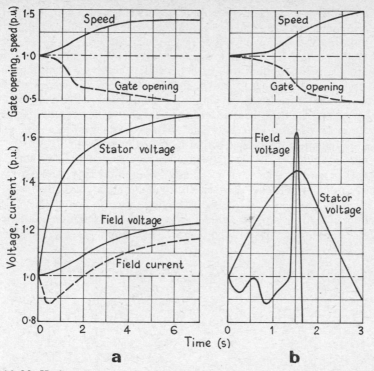

10.39 Hydro-generator: response to throwing off full load at p.f. 0.8 lagging.

rapidly reduced to zero. The main field switch should then be opened as soon as the stator fault current has been quenched, in order to prevent the build-up of a reversed main flux.

Automatic voltage regulators

Excitation control is required to maintain normal operating voltage, to vary the generation of reactive power, to maintain the voltage during external faults, and to increase the steady-state, dynamic and transient stability. Manual control of an exciter-field rheostat suffices for very small machines, but otherwise the use of automatic voltage regulators (a.v.r.) is universal.

Carbon-pile regulator. For small machines a carbon-pile form of regulator is common. It makes use of the resistance variation of a pile of carbon plates with variation of compressive force, the mechanical pressure being provided by an electromagnet energized from the supply to be controlled. The pile may operate directly on the main field circuit, but usually controls the exciter field.

Torque-motor regulator. This is a sensitive and quick-acting device comprising multi-tapped resistors connected into the exciter field circuit and brought out to closely spaced rows of silver contacts. Short-circuiting sectors roll over the contacts in accordance with the torque of a small split-phase disc motor against spring loading and eddy-current damping. Voltage setting is simple, and droop and compounding effects can be adjusted independently. The regulator has no parts subjected to wear and requires very little attention.

Vibrator regulator. The exciter field rheostat is rapidly switched in and out of circuit to correct deviations from nominal voltage setting. A control magnet is energized from the voltage to be regulated and its main contacts are closed by a spring. If the voltage rises, the pull of the spring is overcome and the contacts open, inserting the field rheostat into the field circuit. The resulting fall in voltage causes the contacts to reclose. The cyclic process takes place rapidly and repeatedly, holding the voltage within the limits prescribed.

High-speed regulator. The electromechanical regulators mentioned above are unsuitable for machines above about 75 MW owing to the large field currents required for the main exciter. High-speed regulators are required that can give rapid response rates with the largest generators. Such an automatic voltage regulator comprises error-detecting, follow-up, under-excitation limiter and under-voltage protection units, with appropriate amplifiers. Fig. 10.40 shows

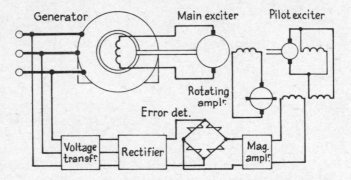

10.40 Automatic voltage-regulating system.

the essentials of a typical regulator. The error unit is a semiconductor bridge of elements with a voltage/current characteristic of the form $v = k\,i^\beta$, the exponent β being different in each pair of elements; the output voltage from this bridge is zero for the specified reference input voltage. A power of a few milliwatts from the bridge is amplified by magnetic and rotary amplifiers up to 40 kW or more for the field of the main exciter. The pilot exciter acts through the rotating amplifier, with a 'bucking' winding fed through the error-detecting circuit. The operating frequency of the magnetic amplifier is usually 50 Hz, but it may be given a faster response by use of a higher frequency, e.g. 400 Hz. Manual control of the system is provided for emergency, with automatic follow-up.

The limitation of the level of under-excitation can be achieved by monitoring the generator load angle, using a permanent-magnet generator on the main shaft to indicate the rotor position.

Continuous-acting a.v.r. are available for permitting stable operation in the region of dynamic stability, with load angles up to 120 deg(e.). In case of an a.v.r. failure the undervoltage protection boosts the excitation to a safe limit to preserve synchronous running.

10.10 *Generator control*

A generator/prime-mover set is a very complicated and expensive plant. Its starting and running require sophisticated operating and control equipments.

Starting

In starting a hydro-generator, the main considerations are those outlined in Sect. 7.2. The operation is more complicated for a steam turbo-generator.

Much of the process of starting a turbo-generator 'from cold' concerns the thermal problems of the turbine expansion and stressing, more particularly of its rotor. Thermal bends in rotors may develop during run-up and cause damaging vibration, through which the set must be 'nursed', It may take several hours for a turbo-generator set to become hot enough for loading, so that a common practice is to maintain some plant as 'spinning reserve' with the turbines taking steam largely for temperature maintenance, so that more rapid loading can be achieved. In large modern stations the starting and stopping processes are automated. The control engineer can preset the required active and reactive load and the rate of loading, and then initiate boiler starting, turbine, run-up, synchronization and loading by one switch operation.

Synchronizing

Almost all synchronous generators operate in parallel with others, and must be switched into operation by a synchronizing process. The conditions to be satisfied are these: (i) the speed of the machine to be connected to live busbars must be such that its frequency is nearly the same as that of the busbars and preferably 0.2–0.4 Hz high; (ii) the terminal voltage should be within 5% of that of the busbars, and co-phasal with it within about 5 deg(e.), The usual method of synchronizing a generator on to the system busbars is to apply rotor excitation with the coupling circuit-breaker open and to adjust the open-circuit voltage until it is in magnitude equal to that of the system. The phase displacement of the generator and system voltages is observed by synchroscope or lamps, and its rate of angular change adjusted by control of the generator speed. When the rate of change is slow, and the displacement within about 5 deg(e.) the circuit-breaker is closed. With manual synchronizing, allowance must be made by the operator for the delay, 200–400 ms, between initiating circuit-breaker action and the actual closure of the contacts. On large generators use is made of automatic synchronizing equipment with the appropriate voltage and angle sensors.

Another method ('rough synchronizing') consists in closing the coupling circuit-breaker with the rotor field unexcited, then raising the excitation and simultaneously controlling the speed until the generator self-synchronizes There are three successive processes concerned: (i) the magnetizing time following closure of the circuit-breaker, with a peak stator current occurring within the first two a.c. periods and rapidly decaying asymmetric and subtransient components; (ii) a duration of asynchronous running during which induction torque and governor action bring the speed close to normal synchronous speed; and (iii) the synchronizing time following closure of the d.c. field switch. Generally, the current peaks in (i) are smaller than those due to short-circuit faults, and unless a generator is operated on a daily two-shift duty the method has attractive simplifications, especially for automatic control.

Synchronizing currents. If at the instant of closure of the main circuit-breaker there are divergences of angle, frequency or voltage between the incoming machine and the busbars, there will be a subtransient synchronizing-current surge in the stator windings, with its effects reflected into the rotor. For a phase divergence α an active-power-producing current $i_\alpha = (v_1/x_d'')\sin\alpha$ flows, and if there is also a frequency divergence $s\omega_1$ due to a slip s, a further current of similar form (with $s\omega_1 t$ replacing α) occurs; in the limit, these in combination give a maximum of $2v_1/x_d''$, which may be very large. For a voltage difference Δv a switching current $i_v = \Delta v/x_d''$ will occur, demanding reactive power.

Resynchronizing. Restoring synchronous operation to a generator after it has lost synchronism following a system disturbance can be achieved without disconnecting it from the busbars, provided that certain limitations are observed. There is a maximum slip beyond which it is practically impossible to resynchronize, this maximum being determined by the ceiling voltage of the excitation system. The second most important requirement is that the excitation should be applied at a definite point in the slip cycle before the limiting swing angle is reached and with due regard to the field-circuit time-constant and the operating delay of the circuit-breaker. The conditions necessary can be established with the aid of a comparatively straightforward control loop. The conditions of asynchronous running are referred to later.

Steady-state operation

After synchronization, a generator is loaded by adjustment of the prime-mover governor to give the required output power, and of the excitation to cause the machine to take its assigned share of the reactive load. Normal operation of a generator involves maintaining a balance between the inputs to the various items of plant and their outputs. The required conditions are set by the operator, and thereafter control devices (such as the turbine governor and the boiler control loops) automatically maintain the condition. More elaborately, the sensing of the variables that determine the plant efficiency can be fed to a computer programmed to print out efficiencies and performance details on which the control engineer may take action. On-line computer control is a further step.

A factor of importance in the operation of large power-system generators is the maintenance of full loading for economic reasons. This involves steady-state stability in both the static and dynamic regions. Stable operations can be obtained by prime-mover governors and automatic voltage regulators in combination. A conventional speed governor by itself has little potential in maintaining stability, but substantial advantages can be obtained by the inclusion of acceleration sensing in governor-actuating mechanisms. The governor-a.v.r. combination forms a complex control system that can be expressed by an interconnection of transfer functions with feedback and feedforward, delay, dead-zone, differentiator and integrator elements.

In the chain modelling a steam turbo-generator, elements are required to represent the time response of the boiler, steam-flow control, turbine, generator, voltage regulator and system load:

Boiler. The transient response is slow, with a time-constant typically of 100 s, and for short-duration changes it may be regarded as an 'infinite steam busbar' of constant pressure and temperature.

Steam-flow control. The main element is usually a mechanical flyball governor backed by a two-stage hydraulic amplifier acting on the steam valve; it has characteristically a dead-zone of about 0.06% speed change, a fixed 'droop' which is higher for full-load high-efficiency sets and smaller for low-load sets, and a somewhat inflexible performance, but is reliable. The disadvantages are overcome by electrical governing, and a number of further controls can be added. It can be made to act on both high- and intermediate-pressure steam valves to give a faster mechanical response, provide acceleration feedback to aid system damping, and incorporate a load-control loop with integral action to zero the steady-state error for accurate setting. The most important control signals for dealing with load disturbances are the speed error and its derivative (i.e. rotor acceleration). For large disturbances the inherent mechanical limits on the rate of valve opening or closure mean that the acceleration signal has a dominant control function.

Turbine. This is an extremely complicated unit to model. It may be represented by a system of chambers interconnected by ducts. A chamber has a significant volume, in which a change in boundary conditions (such as mass-flow) produces time-dependent changes in the fluid conditions. A duct introduces modifications to the steam flow by friction and pressure-drop. Chamber conditions are found in large-diameter piping as well as in the main turbine cylinders and in the condenser, while pipes have also a partly ductlike nature, and valves can be modelled as pipes with 'nozzle' restrictions.

Generator. The generator model is based on the Park concept. For small-perturbation studies in which the current can be expressed as $i = i_0 + \Delta i$, the equations (Sect. 7.5) can be linearized to the following, stated in per-unit terms of voltage, power and flux.

Axis voltages:
$$\Delta v_d = \Delta v_d{}'' + x_q{}'' \Delta i_q$$
$$\Delta v_q = \Delta v_q{}'' - x_d{}'' \Delta i_d$$
$$p(\Delta v_d{}'') = [(x_q{}' - x_q{}'')\Delta i_q - \Delta v_d{}'']/\tau_{q0}{}''$$
$$p(\Delta v_q{}'') = [\Delta\psi - (x_d{}' - x_d{}'')\Delta i_d - \Delta v_q{}'']/\tau_{d0}{}''$$
$$p(\Delta\psi) = [\Delta v_f - (x_d - x_d{}')\Delta i_d - \Delta\psi]/\tau_{d0}{}'$$

Power:
$$\Delta P_e = v_{d0}\Delta i_d + i_{d0}\Delta v_d + v_{q0}\Delta i_q + i_{q0}\Delta v_q$$

Terminal voltage:
$$\Delta v_1 = [v_{d0}\Delta v_d + v_{q0}\Delta v_q]/v_0$$

Voltage regulator. The field voltage v_f from the exciter system is controlled by the automatic voltage regulator, which gives a control signal to an amplifier to provide the excitation required. Commonly the equipment comprises an a.c. exciter, rectifier, magnetic and/or rotary amplifiers and sensors. The terminal voltage of the turbo-generator provides the main control signal, which may include line-drop compensation and rotor-angle elements. The difference between the terminal and the reference voltage can be taken as acting through a single time-lag network with an exciter-stability feedback.

Load. The network influences the loading conditions in a complex manner. For the simplified case of a generator connected through a tie-line of reactance x_e to an infinite busbar of voltage V then the power flow is dependent upon the overall load angle, and changes of d- and q-axis generator current can be related to the d- and q-axis generator voltages, the load angle and the voltage of the infinite busbar.

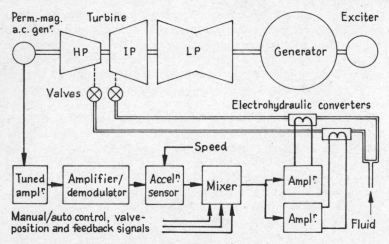

10.41 Electronic governing system for turbo-generator.

Fig. 10.41 shows a schematic diagram of an electronic governing system. A speed-sensing unit is driven from the turbine to give a frequency signal, which is then processed by means of a tuned circuit to provide a speed-error signal indicating the magnitude and sense of the deviation of the turbine from rated speed. To this signal is added a reference voltage indicating the load required from the set at nominal system frequency (e.g. 50.00 Hz) and the summed signal positions of the four electrohydraulically controlled high-pressure governor valves. Each valve-position servo includes a signal and a power amplifier. The output from the latter energizes the coil of an electrohydraulic valve to admit oil to the relay stage to position its governor or interceptor valve. Position transducers provide feedback signals to close the position-servo loops. Transient feedback terms provide appropriate damping of the fast response obtained. Acceleration sensors are included to improve control action and to limit speed rise resulting from load rejection.

Before synchronizing, the governor acts as a speed controller to stabilize the speed of the set. After synchronization, the system frequency is dependent upon the mean speed gain of all the working turbine governors. Variation of load with frequency depends on the gain of the speed-loop control (i.e. the 'droop'). With the electronic governing the speed change between no load and full load can be as low as 1%, compared with 4% for a mechanical governor. In purely mechanical governing the time-lags in the hydraulic relays are long enough to have a major effect on the generator speed which may swing to 5% above and then 20% below normal in the 50 s following a large disturbance. Such corresponding frequency swings could adversely affect the behaviour of station auxiliaries motor-driven from the generator transformer.

Asynchronous running

Turbo-generators may occasionally be required to operate asynchronously, as for example in the process of self-synchronizing. Further, with the rise in rating it becomes increasingly desirable, in the event of a machine losing excitation or desynchronizing for some other reason, that it should be able to run asynchronously at reduced output until normal running can be restored, in order to avoid the lengthy outage entailed by disconnection and re-loading. A basic problem in assessing the asynchronous performance is that of analysing the eddy-current phenomena in the saturable solid steel of the rotor body, for the effective torque depends on the total rotor I^2R loss, only part of which occurs in the field winding. Tests have shown that as the slip increases there is an increase of generated power (as an induction generator) up to a peak at typically 0.01 p.u. slip, but the peak torque depends on the stator voltage and the rotor-field connections, as might be expected. If, when operating at full load, the machine loses its d.c. excitation, it runs asynchronously with a slightly lower load but with a stator current much higher than rated value; however, the stator current level can be reduced to about normal if the load is quickly reduced to one-half normal value. Such operation should be possible for quite prolonged periods unless it gives rise to unacceptable voltage reduction on the system, as it may do if the rating of the machine is an appreciable fraction of the system rating.

The mean point of *sustained* asynchronous operation is determined by the trend of the mean torque/slip characteristics and the combined turbine-governor torque/speed characteristics. The point of intersection of these represents the mean operating point about which the speed and torque pulsate. With a rise in speed, a quantity Δw of kinetic energy is absorbed by the rotor which varies over the slip cycle and is given by the energies contributed by (i) the net induction torque, (ii) the synchronizing torque present if the rotor is d.c. excited, and a quantity dependent on the inertia J and given by $J(\frac{1}{2}s_m{}^2 - s_m)$ where s_m is the mean slip. A typical relation between these is given to a base of load angle δ in Fig. 10.42. If, at any instant, the change Δw in kinetic energy is zero, the rotor speed becomes equal to the synchronous speed and the machine attains a condition in which it may pull in to a synchronous condition. Malik and Cory [80] point out that the torque contribution (i) and the inertia are design parameters, but the synchronizing torque (ii) can be manipulated by means of excitation and voltage-regulator control. By applying a

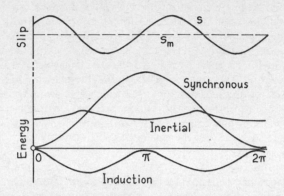

10.42 Energy and slip during asynchronous operation.

sufficiently large excitation at a suitable instant in the slip cycle, it is possible
to synchronize the machine successfully. Account must, however, be taken
of the presence of nonlinear factors of importance, in particular the response
of the excitation current to the exciter voltage, and the effect on the induction
and synchronous torques of the very rapid rate of change of speed over the
slip cycle.

Protection
The unscheduled shutdown of a generator has potentially serious economic
consequences. Its protective system, including that of the prime-mover, must
provide for early-warning signals, automatic remedial action, and tripping
only if essential. The regions and the possible contingencies associated with
them are listed below with particular reference to a large steam turbogenerator.
Mechanical. (Overspeed; bearing temperature; shaft distortion; thermal
expansion). A generator could accelerate to destruction on sudden load-
shedding were it not closely speed-controlled, e.g. by rapid-acting electrohydraulic governor referred to an accurate frequency measurement. The basic
speed function is supplemented by acceleration sensing. Rotor eccentricity
is monitored as it can produce destructive vibration.
Stator. (Over-current, -voltage, -temperature; over- and under-excitation;
winding faults; oversaturation, core heating). Under-excitation results in flux
leakage and end-plate core loss. After a short circuit, asynchronous operation
is tolerated provided that the current is not excessive and the slip is
monitored. As over-voltage protection does not respond to oversaturation at
subsynchronous running, a voltage/frequency limiter may be incorporated
in the a.v.r.

Rotor. (Over-current, -voltage, -temperature; winding faults; unbalanced load, n.p.s. current; stability limit). Over-current level and duration are limited by the short-term temperature-rise. An inter-turn fault leads to overheating, and computation of the current appropriate to the loading provides a check. A low-resistance earth fault is a hazard demanding shutdown. Excitation control adjusts the field current in accordance with the operating conditions (steady load, no load on open circuit or synchronized or after load-shedding, system fault, re-closure disturbance, asynchronous operation, re-synchronizing). The a.v.r. is provided with signals of the rotor speed and acceleration to improve transient performance. Rectifiers in the excitation system require special protection.

Coolants. (Over-temperature, under-flow, leakage, purity, pressure). Sensing the temperature, flow-rate and leakage for all coolant circuits (air, hydrogen and water) enables early remedial action to be taken. Leakages can be deleterious as they may cause erosion in coolant paths through hollow conductors.

Protection and control are increasingly becoming electronic. As a fail-safe condition is vital and the loss of a control signal must initiate operation of the prime-mover valves, the scheme incorporates 'redundancy' and duplication. Future protection practice is likely to abandon separate *ad hoc* 'dedicated' devices in favour of an integrated system for the complete turbogenerator set, based on the microprocessor. Some proposals are given in Ref. [136].

Pumped-storage plant

Where the terrain is suitable, spare generating capability on power systems at times of light load may be employed in driving hydraulic pumps to raise water to a high level and store it in a reservoir. At times of peak system load the water is used to drive hydro-generators to assist the base-load stations. Operational flexibility is an important feature. The plant can be started and fully loaded in the generating mode in a few minutes, and the output is readily adjusted to system demands. Similarly, a sudden reduction of system load releases surplus energy from steam and nuclear stations on the system which can be absorbed by pumping. The pumped-storage plant can also be employed for synchronous compensation.

In high-head installations it is possible to combine the hydro pump and turbine into a single unit, to which a synchronous machine can be attached to work in either generator or motor modes as required. There are special problems associated with starting the motor/pump mode. It is hydraulically necessary for the hydro runner to be below the outflow level to provide a suction head of 15–30 m. With the main inlet valve closed, the braking torque developed by a runner is high, but when dewatered it falls to less than 0.01 p.u. Hence it is necessary for starting to depress the water level temporarily, using compressed air. The static friction torque at breakaway is between 0.05

and 0.15 p.u. of normal, but this can be reduced by pumped high-pressure bearing lubrication. With the starting-torque demand so reduced, various starting methods (Sect. 10.11) such as asynchronous, synchronous, pony motor or pony turbine, or low-frequency rectifier/inverter systems are feasible, as discussed by Hammons and Loughran [84].

For most reversible hydro pump/turbine machines the maximum pumping head is greater than the head for turbine operation, and the optimum pumping speed is higher than for driving. Further, the utilization of pumped-storage water in some schemes involves a wide range of head, and the turbine/pump units would have an unacceptably low efficiency if they were connected to normal fixed-speed motor/generator machines. It is therefore desirable to adopt pole-change synchronous machines. There are several solutions to this problem. One is to use for the salient-pole rotor say 16 poles, uniformly spaced but with shoes arranged asymmetrically. When all are excited in successively NSN ... polarity, a 16-pole field is produced in the gap with strong harmonics; for a 12-pole system certain pole excitations are reversed and others are left dead to produce the required pole-number; again the gap field has strong harmonics. The stator winding is pole-changed in ways similar to those used in induction machines, with due regard to harmonic suppression.

Where a reversible hydro machine is impracticable, and the pump and turbine are separate, the turbine can be used to start the pump motor, and the electrical design is not affected. With a reversible hydro units and a single synchronous machine, the latter must be self-starting.

10.11   *Motor control*

The inherent advantages of a synchronous motor are its high efficiency, high or leading power factor, and ability to retain synchronous running during voltage depressions. The motor can be built for speeds from the highest (e.g. 3 000 r/min at 50 Hz) down to the lowest likely to be required in practice and for which the poor performance of induction motors makes them less attractive, quite apart from their uneconomically low power factor. Constructionally, synchronous motors may have cylindrical rotors with embedded exciting windings as in a turbo-generator, a form suitable for high-speed 2-pole machines; solid poles (4 or more) also suitable for the higher range of speeds; salient poles with pole-face bars inserted for starting and damping and either confined to the pole-shoes or extended between the poles to form a cage winding; or may be built as cylindrical- or salient-pole-rotor synchronous-induction machines. As a *purely* synchronous machine develops a torque only at synchronous speed, a chief development has been in methods of introducing means for self-starting.

The methods of excitation described in Sect. 10.9 are equally applicable to synchronous motors, but almost all modern motors are provided with brushless excitation. The application is, however, not so simple as it first appears, because of the behaviour of the field circuit during the self-starting process.

Starting

Although a synchronous motor may be run up by use of a convenient 'pony' motor attached to its shaft, self-starting is by far the most usual method: it employs the asynchronous (induction) torque developed in a cage winding, which may have deep bars, or sometimes only in solid salient-pole-shoes if the starting duty is light. The d.c. field winding could be closed to provide an effect comparable to that in a double-cage induction machine, but the stator current is increased thereby without much gain in starting torque; it is therefore usual to keep the field circuit open until the motor approaches synchronous speed. A characteristic feature of the starting torque/speed relation is a pronounced 'dip' at one-half synchronous speed, the severity of which is a function of the stator- and cage-winding resistances and of the saliency. The following brief descriptions list the most common starting methods.

Direct-switching. Motors with direct-on-line starting and with a limit imposed on the starting current must be designed to have an appropriate impedance at standstill as well as suitable acceleration and synchronizing characteristics. The limiting impedance may, however, be incorporated in the supply system, as when a motor is supplied from a step-down transformer. In generating stations, motors driving auxiliaries may be allowed starting peaks up to 6 or 7 times full-load current as, in spite of the negligible system impedance in such cases, the power level of the generating plant is very high.

Star-delta switching. This method is the same as for induction motors. It is suitable only for low-torque applications.

Special-connection switching. Various methods are available for reducing the starting current (at the expense of also reducing the starting torque). Machines with parallel stator windings may have these series-connected for starting; or the stator winding may be tapped so as to raise its impedance.

Reduced-voltage switching. This is effectively an increase of the impedance at starting by the inclusion of inductors or auto-transformers. The reactance of an auto-transformer has an important influence on synchronizing. Where a large machine is concerned and has a main transformer allocated to it, a standard tap-changer may be employed for voltage reduction at starting.

Rotor resistance. This method can be used with synchronous-induction motors. The rheostats are balanced to avoid a partial single-phasing effect which might cause the machine to 'lock' at one-half synchronous speed.

Synchronous. In particular installations, such as electrical ship-propulsion and pumped-storage stations, a motor may be connected to a stationary generator. If both generator and motor fields are excited and the prime-mover is started, the two machines pull into step at a low frequency and accelerate together up to normal running speed. Another method of achieving synchronous starting is to supply the motor from a static frequency converter, such as has been developed for variable-speed induction motors. The operating features of interest are the effects on the supply network introduced by harmonics and reactive power demand. The harmonics can be reduced by adopting a 12-pulse converter.

Pony-motor starting is suitable for easy duty against little or no mechanical load, and for cases where disturbance to the supply network must be strictly limited. The direct-switching method is simple and minimizes switchgear, but the starting current is large (e.g. 6 times full-load current) and of very low power factor. The increasing capability of supply networks and the usual tolerance of occasional voltage dips have made current limitation of less significance, with the result that synchronous motors are commonly switched direct-on-line.

Run-up

A salient-pole synchronous motor runs up in a manner basically similar to that in an induction motor, with the cage and field windings developing an induction torque, but the conditions are complicated by the magnetic and electric asymmetry introduced by the field winding. The 3-phase stator winding, when energized, produces a synchronous travelling-wave field. The rotor, running with a slip s varying from $s = 1$ at standstill to a value approaching zero immediately before it synchronizes, presents to the stator field a fluctuating gap reluctance and consequently a pulsating torque is developed. In deriving the torque/speed relation only the mean torque is of significance (apart from the possibility of interaction with some load-torque fluctuation as might occur in a reciprocating compressor).

Dynamic circuit analysis. Consider a 2-pole motor with a salient-pole member (rotor) carrying a field winding F defining the d-axis, and also an actual or equivalent cage forming separate short-circuited axis coils KD and KQ. The other member (stator) has a symmetrical 3-phase winding ABC energized by balanced voltages of angular frequency ω_1. In accordance with two-axis representation the voltages and currents of the phase windings are transformed into equivalents in axis coils D and Q, as discussed in Sect. 7.5.

In a synchronous machine in the synchronous steady-state condition the windings D and Q have direct applied voltages and carry constant direct currents. During run-up, however, the machine runs with a slip s varying from $s = 1$ to $s = 0$. It can be inferred that the voltages and currents in D and Q, and also those induced in KD, KQ and F (if it is closed) must alternate at slip frequency $s\omega_1$. These conditions can be inserted into the impedance matrix of Fig. 7.11. But we have the difficulty that the angular speed $\omega_r = \omega_1(1 - s)$ is a variable. It is usual to assume the motor to operate in the steady state at a series of constant slips and to assemble the results to give the torque/slip relation. Initial switching transients are ignored as too brief to have any significant effect. The instantaneous D and Q applied voltages, which are equal in magnitude and phase displaced by $\frac{1}{2}\pi$ rad in time, are therefore

$$v_d = v_m \cos s\omega t \quad \text{and} \quad v_q = -v_m \sin s\omega t$$

and ignoring stator resistance they can be equated to the d- and q-axis linkages by

$$v_d = p\,\psi_d + (1 - s)\,\omega_1\,\psi_q \quad \text{and} \quad v_q = p\,\psi_q - (1 - s)\,\omega_1\,\psi_d$$

and the instantaneous electromagnetic torque is given by

$$M = \tfrac{1}{2}\omega_1[\psi_q i_d - \psi_d i_q]$$

The two voltage equations taken simultaneously lead to the result that $\omega_1\psi_d = v_m \sin s\omega_1 t$ and $\omega_1\psi_q = v_m \cos s\omega_1 t$. As for the axis currents i_d and i_q, these are functions of the stator reactance and the reflected effects of rotor resistance and reactance. Let the resulting effective impedances of the D and Q coils be z_d and z_q having respective phase angles α and β. Then the two currents are

$$i_d = (v_m/z_d) \cos(s\omega_1 t - \alpha) \quad \text{and} \quad i_q = -(v_m/z_q) \sin(s\omega_1 t - \beta)$$

The torque can now be found: its instantaneous value is

$$M = \tfrac{1}{4} v_m^2 \left[(1/z_d) \left\{ \cos \alpha + \cos(2s \omega_1 t - \alpha) \right\} \right.$$
$$\left. + (1/z_q) \left\{ \cos \beta - \cos(2s \omega_1 t - \beta) \right\} \right]$$

indicating that the d- and q-axis torque contributions are separate, and that each has a pulsation at twice slip frequency. The physical explanation in simple terms is as follows. The stator develops a travelling-wave field with its N-S axis rotating at angular speed ω_1, which overtakes the more slowly rotating rotor at a relative speed $s\omega_1$. As a stator pole approaches a rotor salient pole 'from behind', it exerts a retarding reluctance torque; and as it overtakes and passes the rotor pole it exerts an accelerating reluctance torque. As this action takes place for each stator pole, the total torque fluctuates at frequency $2 s\omega_1$.

For most purposes it suffices to take the mean torque, for which the double-slip-frequency terms vanish to give

$$M = \tfrac{1}{2} V_1^2 \ \text{Re}[1/z_d) + (1/z_q)]$$

where the two axis impedances are here expressed in complex operator terms. The equation states, as might be expected, that the torque in terms of synchronous power is the same as the rotor power input, the machine operating as an asymmetrical induction machine. The complex impedances are expressed as functions of the axis synchronous reactances and time-constants:

$$z_d = j x_{sd} \frac{(1 + js\omega_1 \tau_d') (1 + js\omega_1 \tau_d'')}{(1 + js\omega_1 \tau_{d0}')(1 + js \omega_1 \tau_{d0}'')}$$

$$z_q = j x_{sq} \frac{(1 + js\omega_1 \tau_q'')}{(1 + js\omega_1 \tau_{q0}'')}$$

The d-axis time-constants must include the effects of the field circuit connection and external resistance. The stator resistance has been neglected for simplicity.

As in the case of the polyphase induction motor, the resistances and reactances of the rotor circuits play a major part in determining the asynchronous performance of a salient-pole synchronous motor. But the problem is more complicated in that there are three main rotor circuits concerned, namely the q-axis damping winding and the d-axis field and damping windings. Each of these may be represented to a reasonable approximation by resistance and inductance.

The two most common rotor constructional forms have (i) solid-steel poles (or pole-shoes), (ii) fully laminated poles with a cage winding in the pole-face.

In (i) the torque during starting results from the substantial eddy-current loss in the pole-shoes at slip frequencies, while in (ii) these losses are confined to the damper bars. In both cases the heat produced has to be effectively dissipated to avoid undue temperature rise. The pole-shoes or damper bars may be connected from pole to pole by interpolar links or end-rings, and this may have a substantial effect on the q-axis impedance and on the asynchronous performance.

Fig. 10.43 shows a typical set of component *mean* torques in a synchronous

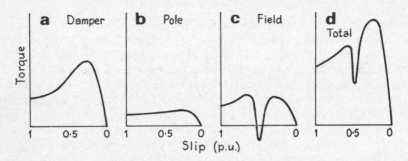

10.43 Motor starting torque components.

motor during run-up, with the field winding closed through a resistor. The damper-winding torque (*a*) is characteristic of a cage induction motor; the contribution of the solid pole-shoes and poles in (*b*) has a similar form, but the eddy-current paths are much more complex and are affected by local saturation; and the torque (*c*) due to the field winding shows a saliency effect strong enough to produce the characteristic torque-reversal at one-half speed. The total torque (*d*) shows that there is still a dip at half speed. The relative magnitude of the torque components is naturally dependent on the design of the machine, an intricate process. A method of determining starting performance from design data is given in detail by Widger and Adkins [86].

Brushless excitation. The rectifiers in the field circuit prevent bi-directional current flow. The pulsating unidirectional field current may produce sufficient 'generating' torque to make run-up impossible, and high inverse voltages appear across the rectifier bridge. These effects are alleviated by shunting a resistor across the field winding, but can be prevented only by switching out the rectifier during the starting process. Various methods have been devised for this purpose. A permanently shunted field winding may be disconnected from the rectifier by a mechanically operated switch, a reliable method but one which involves random synchronizing when the machine reaches a low slip. Alternatively the switch may take the form of a thyristor, retained in the off position until excitation is applied, in which case 'point-on-wave' synchronizing is possible. In more elaborate methods the rectifier bridge is short-circuited during run-up by reversed thyristors fired from the induced field e.m.f. through zener diodes. The behaviour of motors with field rectifiers has been investigated by Chalmers and Richardson [85].

Synchronizing

A large synchronous motor or compensator, started against inertia by a separate pony motor, can be synchronized as for a generator. Alternatively the pony- and main-motor stator windings may be connected in series for starting, the initial voltage division between them being typically 0.8 to 0.2 p.u.; when the speed of the set approaches the synchronous, the main motor excitation is applied, it absorbs most of the applied voltage, and pulls into step. Subsequently the pony motor (an induction motor with a very low rotor resistance, or with one pole-pair fewer than the main machine, or even with a solid unwound rotor) is short-circuited to neutral.

Generally a synchronous motor is self-starting and direct-switched. The induction torque for acceleration falls to zero at a small value of slip. If excitation is now applied to the rotor d.c. winding, a synchronizing torque is developed. The machine has to be designed so that the limit of induction torque is at a slip small enough for the synchronizing torque to pull the machine into step against the load torque; in some cases it may be necessary to reduce the slip by supplying the motor at overvoltage derived from a autotransformer to prevent pole-slipping and the consequent fluctuations of stator current.

The synchronizing conditions involve several torque components. The synchronous torque is a function of the load angle δ, and its growth following initiation of the d.c. excitation is exponential. The induction torque is proportional to the slip $s = (d\delta/dt)$, the inertial torque for a machine with p pole pairs and inertia J is $(J/p)(d^2\delta/dt^2)$, and the dead-load torque can be considered as constant. The equation of motion during the synchronizing process therefore has the form

$$K_1(d^2\delta/dt^2) + K_2(d\delta/dt) + K_3 \cdot f(t) \sin \delta = M_l$$

where M_l is the load torque and $f(t)$ expresses the growth of the field excitation. The solution is analytically difficult, but can be computed in numerical cases.

When the d.c. excitation is switched on, the pulsating synchronous torque is superimposed on the steady induction torque. The motor may pull into step almost immediately: it may continue to run as an induction motor but with speed and current fluctuations: it may revolve several times and then pull into synchronism. The required condition is to have the motor synchronize on its first swing through the steady-state operating position. The inertia of the rotor and its connected load is a very important factor. The angle δ obtaining at the instant of switching in the d.c. excitation also affects the pulling into step. If the angle is such that the machine generates (i.e. if δ is an angle corresponding to a forward position of the rotor) the speed decreases because initially the generated energy is abstracted from the kinetic energy of the rotating masses. The rotor swings into the motoring region, upon which an additional motoring torque is produced to speed it up again.

The field excitation is an important factor as determining the maximum synchronizing torque; the larger the field current, the more certain is the pulling into step.

Operation

Normal conditions. A synchronous motor is often run with constant excitation, in which case the power factor varies with the mechanical load and becomes low and leading on no load. Excitation compounding may be provided by the method shown in Fig. 10.36 if the uncontrolled power-factor condition has to be avoided.

Asynchronous running. Apart from loss of synchronous operation on fault or excessive load, synchronous motors are sometimes required to run asynchronously as part of the normal duty-cycle. In electric ship-propulsion, for example, they may be required to cover the speed range between full-ahead and full-astern, and deliver full-load torque to the screws during the reversal. Special design of the damping is necessary to yield a roughly constant ratio of torque to stator current. The design is intricate, as it is necessary to investigate the detailed linkage in symmetrical pairs of damper bars and to assess the heating of the bars and pole-shoes. A 'brushless' form of excitation is not likely to be practicable in these circumstances.

Two-speed Motors. A method of pole-changing for a pumped-storage generator/motor was described in Sect. 10.10. Rawcliffe and his co-workers [87] have devised a scheme for a two-speed salient-pole machine with a similar field-pole change but with the stator winding arranged for pole-amplitude modulation (Sect. 8.15), the single stator reconnection being comparatively simple.

Variable-speed Motors. Some of the frequency-changing methods used with induction motors (Sect. 8.16) can be applied to the synchronous motor, the rotor cage being replaced by a field winding with d.c. excitation — usually brushless.

Protection

Apart from the special requirements of solid-state devices where employed, the standard protective features for a synchronous motor comprise:

(i)   A thermal relay giving overcurrent and phase-unbalance protection, with high-set instantaneous overcurrent and earth-fault devices.
(ii) Undervoltage protection, which because of voltage maintenance by the machine itself if it has a large inertia may require reverse-power back-up or a suitable electronic sensor.
(iii) Overall differential (circulating-current) protection for phase-phase and phase-earth faults.
(iv) Field-failure protection.
(v)   Overpower relay for protection against mechanical overload or failure, in the case of some kinds of load.

Protection of large machines may be required to operate in case of pole-slipping by tripping the motor or opening the field circuit and initiating re-synchronization, using for example a power-factor relay based on electronic logic devices. Overheating of the field system may be detected by a thermal sensor; this may have a parallel-connected inductor to increase the current in the thermal element for protection during asynchronous running.

Where brushless excitation systems are employed, more elaborate rotor protective devices are necessary, particularly for the solid-state rectifier elements.

10.12   *Synchronous-induction motor*

If the rotor windings of a slip-ring induction motor are excited by a direct current, the m.m.f. distribution produces alternate N and S poles in the same way as does a 3-phase current, but the pole axes are fixed with respect to the rotor. In Fig. 10.44(a), the direct current I_d from an excitation system flows through

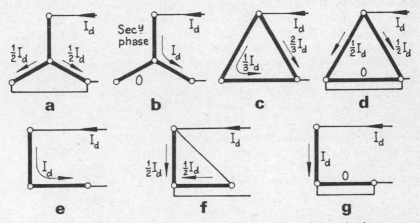

10.44 Synchronous-induction motor: connections for excitation.

one phase winding and then through the other two in parallel, so that the m.m.f. pattern is identical with that of a normal 3-phase excitation at the instant that the current in one phase is at its peak value. Other possible connections are also shown in Fig. 10.44. A cylindrical-rotor slip-ring induction motor can thus be readily synchronized. It is started on rotor rheostats through a change-over switch, and when a small slip has been reached, the switch is moved to provide d.c. excitation. The machine has then a starting torque that may be twice full-load torque or more, and a reasonable pull-out synchronous torque with the possibility of power-factor correction as an additional feature. If necessary the machine can be *inverted*, i.e. have the 3-phase supply connected by slip-rings to the rotor windings, with the starting rheostat and excitation system now readily connected to the stator. To cover both 'normal' and 'inverted' cases, the terms 'primary' and 'secondary' are used respectively for the 3-phase supply windings and the d.c.-excited windings.

The designer's object is to obtain the greatest d.c. m.m.f. with minimum excitation power, to distribute the I^2R loss evenly over the secondary windings, to obtain adequate damping for synchronous running, and to secure specified starting torque. The excitation m.m.f. depends mainly on the length of the air-gap: a long gap gives a stiffer machine with a larger overload capability, but the starting is impaired and more excitation power is required. A short gap gives opposite effects, and to limit power-factor variation it may be necessary to apply compounding.

As can be inferred from the current distributions in Fig. 10.44, unbalanced secondary heating is unavoidable with a 3-phase secondary, but a 2-phase arrangement, as in (e) or (f), equalizes the heating. However there is, in (e), no closed secondary loop apart from that through the exciter, so that the damping is not as good as in (f). A reduction in exciting current can be achieved by use of a multi-turn secondary, but as a consequence the induced secondary e.m.f. at standstill is high, introducing insulation problems.

Equivalent secondary current. Connection (a), with phase currents in the ratio $1:\frac{1}{2}:\frac{1}{2}$, has I_d corresponding to the peak i_m of the equivalent alternating current, so that in r.m.s. terms $I_a = I_d/\sqrt{2}$. The Table below compares in like manner the features of the seven connections. All the 3-ph cases require the same exciter power for a given m.m.f., resistance and turns per phase. All the 2-ph connections take $12\frac{1}{2}\%$ more power and the gap flux has a somewhat greater harmonic content. Because they include a closed circuit methods (a), (f), (d) and (g) − particularly the two latter − give good damping.

Secondary connections for synchronous-induction motors
(1) Relative d.c. excitation current for a given m.m.f.
(2) Secondary input resistance in terms of resistance per phase;
(3) Relative exciter voltage, (1) x (2).
(4) Relative exciter power, (1) x (3).
(5) Distribution of (4) between the secondary phases.
(6) Starting qualities.
(7) Damping qualities.
(8) Asynchronous operation on overload after pull-out.
(9) Notes on applications.

E = excellent; G = good; N = normal; P = poor; S = single-phase

Case	(1)	(2)	(3)	(4)	(5)	(6)	(7)	(8)	(9)
(a)	1.41	1.50	2.12	3.0	2/0.5/0.5	N	G	G	Small machines
(b)	1.23	2.00	2.46	3.0	1.5/1.5/0	P	P	S	Steady drives
(c)	2.12	0.67	1.41	3.0	2/0.5/0.5	N	P	S	Seldom used
(d)	2.45	0.50	1.22	3.0	1.5/1.5/0	N	E	G	Modified
(e)	1.06	3.00	3.18	3.4	1.7/1.7	N	P	S	Seldom used
(f)	2.12	0.75	1.59	3.4	1.7/1.7	N	G	G	Commonly used
(g)	1.50	1.50	2.25	3.4	3.4/0	N	E	N	Patented

Salient-pole motor
The disadvantage of the motor described above is that if the secondary winding has a moderate standstill induced e.m.f., the number of turns is such that a large d.c. excitation at low voltage is necessary. The restriction on slot conductor area is lifted by use instead of a salient-pole construction, so combining the high

efficiency and low excitation loss of the salient-pole synchronous motor with the starting characteristics of the induction motor. The d.c. field winding is normal; and independent starting winding connected to slip-rings occupies slots in the pole-shoes. Each winding can be designed almost independently to fulfil its allotted function, with consequent improvement in synchronous running and exciter-circuit conditions.

Current diagram
When the machine operates synchronously, the locus of the stator current for constant rotor excitation is a circle, Fig. 10.6. In Fig. 10.45 the only difference

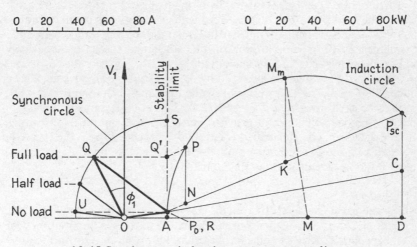

10.45 Synchronous-induction motor: current diagram.

is that OR is V_1/z_{sd} instead of V_1/x_{sd} as it is necessary to include primary resistance, and the angle V_1OR is $\arctan(x_{sd}/r)$. The impedance of an induction motor at synchronous speed is closely that of the magnetizing branches, z_m in Fig. 8.10, so that the no-load current phasor OP_0 fixes P_0 as the centre of the current circles for synchronous working. The intercept PN represents full-load induction-motor power output, and $Q'P_0$ is therefore a corresponding power for the synchronous condition. The theoretical limit of synchronous stability is the axis P_0S, which meets the voltage axis at a point remote $V_1/2r$ from 0; but as this distance for normal machines is very great, it is sufficient to make P_0S parallel to the voltage axis. The O-curves of constant power become the straight lines through Q' for full load and through P_0 for no load.

The secondary excitation is set off from P_0 as a circular locus with a radius corresponding to the a.c. equivalent of the exciting current. The primary currents are measured from 0. The maximum load for synchronous operation is P_0S to scale, the associated primary current being OS. On synchronous no-load operation the primary current is OU, at a low leading power factor.

EXAMPLE 10.8: A 30 kW, 500 V, 3-phase, 50 Hz, star-connected synchronous-induction motor has, as an induction machine, a short-circuit current of 200 A at p.f. 0.35 and a no-load current of 30 A at p.f. 0.15. The ratio of effective stator/rotor turns per phase is 1/1.3, the phase resistance ratio is 1/1.2, and the rotor excitation connection is (*a*) in Fig. 10.44. Find the starting, stalling and pull-out torques, the rotor excitation for full load (synchronous condition) at a p.f. of 0.90 leading, and the current and power factor at one half and no load. Fig. 10.45 is drawn to scale for the machine. The induction-motor circle is drawn as described in Sect. 8.7. To obtain the synchronous motor circle, draw OQ leading the voltage axis by arccos 0.90 = 26° and a horizontal no-load line P_0U; then $Q'Q$ parallel to P_0U is displaced vertically by a scale distance of 30 kW to fix Q. Then P_0Q is the radius of the required synchronous-motor circle. The full-load torque is $Q'P_0$ to scale; the synchronous pull-out torque is P_0S; the induction stalling torque is M_mK; and the starting torque with direct switching is $P_{sc}C$. In per-unit terms of full-load torque, the synchronous pull-out torque is 1.74 p.u. and the stalling torque is 2.17 p.u. The starting torque, inherently 1.11 p.u., can be increased by use of a starter resistance up to a limit of 2.17 p.u. The a.c. equivalent of the rotor d.c. excitation is $P_0Q = 60$ A, so that the exciting current is $I_d = 60\sqrt{2} = 85$ A for unity turns-ratio, and 85/1.3 = 65 A for the turns-ratio given. Stator currents and p.f.s. (all leading) from the diagram are: full-load, 43 A at 0.90; half-load, 35 A at 0.62; no-load, 30 A at 0.15. Closer estimates are obtained if the exciter loss and the stator and rotor I^2R losses are included in the no-load loss.

Performance

Three important characteristic torques are (i) starting in the induction mode, (ii) pull-in, i.e. synchronizing from the induction to the synchronous modes, and (iii) pull-out, corresponding to maximum torque in the synchronous mode. There is naturally a close relation between (iii) and the full-load p.f.; thus a motor excited to run at a p.f. of 0.9 leading when developing full-load torque might have a pull-out torque of 1.5 p.u., but if it were designed for unity p.f. on full load, the pull-out torque and the necessary exciter rating would both be considerably reduced. Fig. 10.46 shows current, reactive power and p.f. curves typical of a machine excited to unity p.f. on rated active-power output in (*a*), while (*b*) shows the same quantities for a motor with a strong excitation. In case (*b*) the machine has a large overload capability, but over the range of normal load its p.f. is reduced and the I^2R loss is higher. To accommodate a load that varies widely, the exciter may be compounded by means of a current transformer and rectifier to optimize the p.f. and loss automatically.

The synchronous-induction motor is rarely built for ratings below 25 kW, as the exciter system is an additional expense and the p.f. adjustment of small units is comparatively unimportant. However, if a given drive calls for a low speed and therefore a multi-pole machine, and for a high starting torque, a synchronous-induction motor offers advantage over both plain induction and plain synchronous machines.

A 'good' induction motor makes a 'bad' synchronous motor. The former has a short gap length and a high magnetizing reactance, typically 3.0 p.u. This in

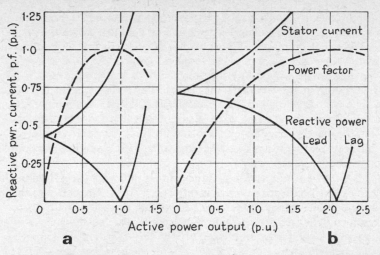

10.46 Synchronous-induction motor: performance characteristics.
   (*a*) Full load, p.f. unity.
   (*b*) Full load, p.f. 0.8 leading.

a synchronous machine corresponds to the synchronous reactance, and a value of this order would entail excessive change of p.f. with load and a limited pull-out torque. Reduction of the synchronous reactance implies a longer gap, and with this the induction machine is subject to pull-out torque limitation. The synchronous-induction motor in this respect demands compromise design.

10.13 *Synchronous compensator*
A synchronous compensator is designed to run without mechanical attachment but with its excitation so controlled as to provide a load taking a purely reactive power. Located at strategic points in a supply network, compensators can act as 3-ph inductors or capacitors. With over-excitation the machine is stable and capable of accepting leading reactive powers at almost zero p.f. When under-excited the input lags but the stability is reduced. Full rating at zero p.f. lagging can be obtained with excitation reversed, although such a mode is unusual. Compensators in large networks are equipped with automatic excitation control to maintain stable operation; the important requirements are related to the rapidity of response, which determines how quickly corrective measures can be applied to restore stability if lost. The control-loop time-lags and the reactances and time-constants of the compensator are crucial. The transient reactance should be lower and the subtransient reactance higher than those of other large synchronous machines on the system in order to mitigate voltage fluctuations caused by erratic lagging loads.

Hunting phenomena require the compensator to have a natural frequency appropriately related to its system, in respect of the natural frequency of other large machines, the position of the compensator in the network, and the voltage range over which it must work. The compensator is often arranged to maintain approximately constant the voltage at its own terminals, both at high-load times when the power factor is lagging and on light load when the system power factor is leading by reason of line-charging effects. The former requires the compensator to be over-excited, the latter under-excited.

Starting can be achieved by the same methods as those for motors. The conditions, however, are very different. The starting and synchronizing torques are low; but as a compensator, rated typically at 50 MVA, may represent a large fraction of the system rating, even small per-unit demands may be significant. Again, the machine may be required to start when the system is heavily loaded with lagging reactive power, or when the voltage is above or below normal, or even when the line is 'dead'.

The speed of a compensator is selected to give the cheapest machine that will fulfil the specification, which is influenced by the frequency of the system and the possible overspeeds and frequencies that may occur with hydro-generators. Cylindrical-rotor 2- and 4-pole designs are costly, and 6- or 8-pole designs may be preferred on this count. The former pair give a long-core small-diameter shape to the rotor, while with the latter the diameter is larger and the length smaller. Variation of dimensions, particularly of core length, affects the values obtainable of leakage, transient and subtransient reactance, and it also has an influence on the losses, more particularly those in windage and friction.

The service duty of a compensator is fairly easy as there is no torque output demanded. Machines can be built for mobile and out-door situation. Machines on a permanent site can be hydrogen-cooled, the absence of a shaft extension simplifying the achievement of gas-tightness. The external heat-exchangers can be integral with the frame.

Fig. 10.47 shows typical compensator characteristics. At (*a*) is the single V-curve for constant voltage operation. In this case 1.0 p.u. excitation gives 1.0

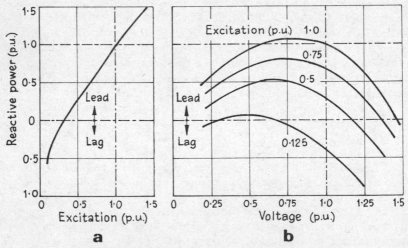

10.47 Synchronous compensator: characteristics.

p.u. of full rated reactive leading power. At about 0.3 p.u. excitation the current falls to a very small active value corresponding to losses other than those due to I^2R. At lower excitations the machine acts as a synchronous inductor. The maximum lagging reactive power may be limited to about 0.3 p.u. to avoid instability on low (or reversed) excitation. The curves (*b*) refer to performance at constant excitation but varying terminal voltage.

10.14   Surge voltages

The impact of a high-voltage surge on a synchronous machine has an effect comparable with that on a transformer; but a stator winding is structurally more complex, being an elaborate arrangement of coils insulated with material of high permittivity and embedded for about one-half of their length in iron-bounded slots. With the usual two-layer winding there will be at least two conductors, conductively separate but coupled both capacitively through their insulation and inductively by their mutual slot and overhang magnetic flux.

Consider a bar-type stator winding constructed with single-turn coils so that each slot has two conductors. Ignoring the discontinuity introduced by the end-windings the contents of one slot may be approximately represented by the equivalent circuit of Fig. 10.48(*a*). The self-inductances *L* of the bars have

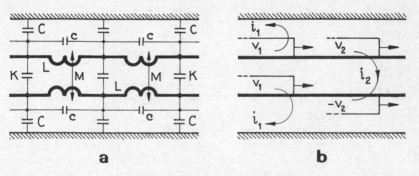

a   b

10.48 Surge voltage in stator winding.

mutual coupling *M*, and there are mutual capacitances *K* between conductors, *C* to earth, and *c* between coils in adjacent slots due to their proximity in the overhang. These parameters are expressed per unit length of conductor. From the differential equations relating voltages and currents in this network, Robinson [62] found that the necessary conditions are satisfied by two sets of travelling waves, Fig. 10.48(*b*). The first set comprises a pair of equal voltages v_1 accompanied by currents i_1 associated with the 'transmission-line' system formed by a conductor and the earthed slot-wall. The second set consists of a pair of equal but oppositely-polarized voltages v_2 with currents i_2. The respective currents and voltages are related by the surge impedances presented to them

$$Z_{01} = v_1/i_1 = \sqrt{[(L+M)/C]} \text{ and } Z_{02} = v_2/i_2 = \sqrt{[(L-M)/(C+2K)]}.$$

The velocities of propagation also differ. They are

$$u_1 = 1/\sqrt{[(L+M)C]} \quad \text{and} \quad u_2 = 1/\sqrt{[(L-M)(C+2K)]}.$$

There is an upper limit of the propagation of travelling waves, however, due to the series capacitance c which resonates with $(L+M)$ to form a rejector network. If, therefore, a step-function surge is analysed into a constant term plus a frequency spectrum, only those components below the frequency of rejector resonance can pass through the winding as travelling waves. (The effect of this limitation is substantially to slope the wavefront to a finite initial steepness, and to make the steepness progressively less as the waves advance into the winding by slowing down the higher-frequency components.) The upper-frequency components in the spectrum, which are rejected, form an additional standing voltage component distributed substantially in accordance with the capacitances of the equivalent network: this leads to a high voltage-gradient in the turns of the stator winding near the line terminal.

Where the winding has two or more turns per coil, high transient voltages occur across the comparatively weak insulation of the adjacent turns, intensified by the comparatively large ratio C/K.

With considerable simplification a synchronous machine may, from the surge viewpoint, be represented as a cable with the appropriate surge impedance and propagation velocity. The surge impedance varies from about 600 Ω down to 50 Ω for large turbo- and hydro-generators, and the surge velocity is of the order of 20 m/μs.

Generator-transformer units

Surge voltages may be impressed on the terminals of a generator by transfer through the associated transformer from the h.v. transmission system, though few generator insulation failures can be attributed to transferred surges, the magnitude of which is small compared with the line surge voltage for generators of rating below 500 MW. The transfer mechanism is considered by White [63] to be partly inductive and partly capacitive, and he gives analyses with the generator represented by an LC network on the assumption that the steepness of the h.v. line surge has been considerably reduced by its passage through the transformer. The *inductive* transfer is based on the concept that, when loaded by terminal LC circuits, the transformer l.v. winding may be regarded as a voltage source with internal leakage inductance, and the transferred surge voltage sets up an oscillation of amplitude shared by division between the transformer and generator inductances. The *capacitive* transfer is initially between the h.v. and l.v. sides of the transformer, appearing as a voltage pulse of the same polarity as the line-surge voltage and transferring to the generator terminals through the inductance of the l.v. winding. Two factors contribute to suppressing the capacitive transfer. First, with normal h.v./l.v. capacitance (typically 1—2 nF) feeding into a low generator surge impedance, the decay time of the initial transferred voltage pulse is of the order of 0.1 μs and therefore short compared with the line-surge rise time: thus the full theoretical maximum capacitive voltage (limited in any case by the ratio of h.v.—l.v. and l.v.-earth capacitance) can never be fully developed. Second, the l.v. winding inductance and resistance combine to delay the build-up of transferred current by a time-constant of about 10 μs, so that again the development of the theoretically possible voltage at the generator terminals is inhibited. Hickling and Winder [64] give the results of practical surge transfer tests, which demonstrate that the mechanism of surge-voltage transfer is predominantly inductive.

10.15 *Testing*

Small and medium-rated synchronous motors offer no difficulty in testing, but large generators (up to 1 000 MW) and compensators cannot be direct-loaded, and indirect methods are accepted. The primary use of works testing equipment is to prove that each machine tested has the required performance in respect of load characteristics, loss and temperature-rise, and excitation. Works testing never involves the supply and dissipation of the rated power of a large machine, as this is quite impracticable: even the losses alone may be several megawatts.

In a *back-to-back* test, two identical machines are coupled both mechanically and electrically, and by suitable angular displacement at the mechanical coupling it can be arranged that both machines work at rated voltage and current to provide a thermal test under load conditions. The opportunity for such a test is rare, but it may be set up for research purposes to confirm the reliability of the simpler routine methods.

In a *reactive power* test two machines are connected together electrically only; they may differ in every way except in working voltage. By adjustment of the field excitations, rated current at a power factor near zero can be circulated between the machines. Driving power can be supplied electrically or by a motor coupled to either machine. Inductive reactors can be used to make up any significant difference in the ratings of the tested machines.

Separate *open-* and *short-circuit* tests are made in most cases. In each, only part of the full-load loss occurs: on o.c. mainly the core loss, on s.c. mainly the I^2R loss. As these losses occur in different parts of the machine it becomes a matter of careful estimate backed by test experience to deduce the heating that will result on service load. Determination of excitation characteristics follows from the methods adopted for heating tests. For accuracy a zero-power-factor point for normal voltage and current is needed besides the data from the o.c.c. and s.c.c.

Turbo-generators

These machines do not readily develop adequate torque from an a.c. supply to start against bearing friction, and they are usually coupled to a driving motor (of rating up to 5 MW) and gear-box or other method of speed control to deal with overspeed. For tests within the voltage range of the loading machine, the turbo-generator is connected to a group of inductors and brought up to speed (unexcited). On applying field excitation, the terminal voltage and reactive current are built up, and the turbo-generator synchronized with a motor-generator for a.c. driving. The starting motor is then disconnected. The motor-generator makes up any deficiency in the reactive power taken by the inductors, and its power input provides the basis for the calculation of the turbo-generator loss.

Hydro-generators
Salient-pole machines can usually be started from a low-frequency supply. In horizontal-shaft machines the damper winding provides enough torque to over- come friction, but in vertical-shaft generators it is usually essential to jack the rotor to allow the bearings to flood. The generator is accelerated by increasing the frequency of the test-bed motor-generator, then synchronized on to a suitable normal-frequency supply, if necessary after an overspeed run. The driving set is then disconnected. As the generator now is driven electrically, its total loss is given by a terminal input power measurement, but care has to be taken in view of the very low power factor (which may typically be of the order of 0.02).

Synchronous motors
These are tested in the same way as hydro-generators. However, when rated at high power factors the reactive-load condition is different from the rated con- dition in such a way as to produce a deficiency in core loss and an excess in excitation loss. These divergences are calculable and the temperature-rise can be corrected accordingly.

Synchronous compensators
The reactive-power test is the normal working condition, with the loss rather less than the convention summation of o.c. and s.c. losses. A feature is the test on underexcitation for lagging reactive output, usually at some under- voltage level, to demonstrate the range and stability of the excitation equip- ment.

Test-schedule
The following characteristics and losses may be measured:

(*a*) Open-circuit characteristic and loss.
(*b*) Short-circuit characteristic and loss.
(*c*) Zero-power-factor characteristic and loss.
(*d*) Temperature-rise (i) by full-load z.p.f. over-excited run, or (ii) by equiva- lent heat run. The latter comprises a rise measurement on the stator with excita- tion but no stator current, followed by a test with rated stator current but mini- mum excitation. The total temperature-rise is then obtained by combining the results.
(*e*) Overspeed test, e.g. 0.25 p.u. for turbo- and 1.0 p.u. for hydro-generators, the latter in an enclosure; as the windage loss varies as the cube of the speed, considerable drive power may be demanded.
(*f*) High-voltage tests.
(*g*) Insulation-resistance tests, made before and after (*f*).
(*h*) Waveforms, interference, gap length, balance, vibration, bearing currents, magnetic symmetry.

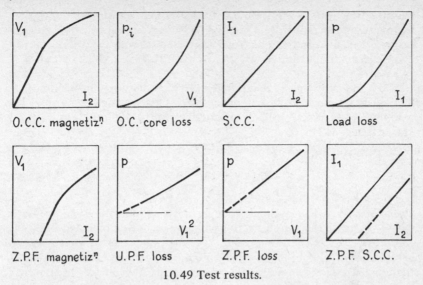

10.49 Test results.

Not all of these are necessarily made on each machine. The curves obtained from the major tests are shown in Fig. 10.49. For small machines the testing schedule is much simpler, and may be confined to (*a*), (*b*), (*c*), (*d*) and (*f*), apart from general mechanical constructional verification. Excitation and control systems are subject to full testing schedules before association with the machine itself.

The specified temperature-rises allowed are given in Sect. 4.7. Embedded temperature-detectors are called for in large machines. High-voltage tests are specified in BS 2613. Winding resistances are measured by a d.c. method, so that eddy-current and load-loss additions can be estimated.

Efficiency. The estimation by loss-summation is recognized in BS 269. The losses to be included are:

Excitation: field I^2R, rheostat, brushes, exciter system.
Fixed: core, bearings, brushes, windage and friction.
Direct Load: I^2R loss in stator windings.
Load (stray): conductors, iron parts, end windings, etc.

The load loss is due to (i) stray fields in solid parts near the stator, particularly coil-supports and end-covers, tooth-stiffeners and the outer packets of stator core laminations; (ii) field distortion increasing tooth loss; (iii) field pulsation arising from slotting and stator m.m.f. harmonics, setting up pole-face loss; (iv) eddy currents in stator slot conductors and end-connectors, increased by saturation in the teeth.

EXAMPLE 10.9: Evaluate the full-load efficiency of a synchronous machine of rating 23.4 MVA at p.f. 0.8 lagging from the following losses: Bearing friction 68 kW, measured by volume of lubricating oil circulation and temperature-rise; Windage 220 kW, from o.c. test unexcited, measuring cooling-air volume and temperature-rise; Core loss 165 kW, from o.c. test with excitation to give calculated e.m.f. at rated load, corrected for windage; Winding loss 200 kW, from full-load z.p.f. test; Stator d.c. I^2R loss 62 kW; Exciter loss 14 kW.

The load (stray) loss is the difference, 138 kW, between the winding loss and that calculated from the d.c. resistance. Then the losses sum to 763 kW. The output is 23 400 x 0.8.= 18 750 kW and the input is 18 750 + 736 = 19 513 kW. The efficiency is

$$18\ 750/19\ 513 = 0.961 \text{ p.u.}$$

Specific tests

Some details of the more important tests are given.

Open-circuit test. The machine, whether motor or generator or compensator, is driven at rated speed with the stator windings on open circuit and the field current varied, to give the o.c.c., Fig. 10.7. With the machine driven by a calibrated motor the input to the synchronous machine, representing the core, friction and windage losses, can be calculated and the core loss and mechanical loss separated, Fig. 10.50(a).

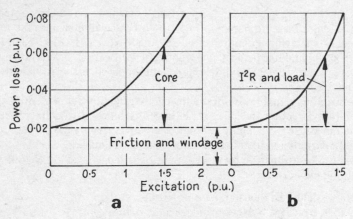

10.50 Power/excitation curves.
(a) Open circuit. (b) Short circuit.

Short-circuit test. The machine is driven at rated speed with the stator windings short-circuited. The field current is varied to give stator currents up to (or slightly over) full-load value. The relation between stator and field currents is substantially rectilinear, Fig. 10.7. Again the calculated input gives the loss, in this case the mechanical, stator I^2R and load losses summated, the core loss normally being negligible. The input power for the range of field currents is shown in Fig. 10.50(b), and the mechanical loss is readily separated.

Zero-power-factor test. A pseudo full-load test is carried out by loading the machine on static or synchronous inductors (or a combination thereof) so that it operates at normal voltage and current. The gap flux is 4–5% above normal full-load level and the field current is 20–30% high. The test serves as a heat-run, with temperature-rises a little above normal: but appropriate correction factors can be applied. The I^2R and load (stray) losses are close to the normal full-load values.

Axis reactances. The d- and q-axis synchronous reactances, x_{sd} and x_{sq}, may be found from a slip test. The machine is coupled to a driving motor, the field circuit is opened, and a normal positive-sequence voltage applied to the stator. The drive torque is adjusted so that the machine loses synchronism and runs at a low value of slip. Oscillograms, Fig. 10.51, are taken of (i) the e.m.f. induced

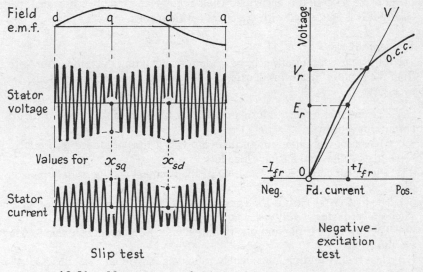

10.51   Measurement of axis synchronous reactances.

in the field winding, (ii) the stator applied voltage, and (iii) the stator current: then x_{sd} is the maximum and x_{sq} is the minimum ratio of stator voltage to stator current. The method has the advantage of direct measurement, but the field saturation conditions are not normal. The slip should be small enough to avoid undue eddy-current induction in damper windings and pole-shoes. The test can be made with the machine self-driven, in which case a closely control-lable supply voltage, of magnitude 0.3 p.u. or less and of normal frequency, must be available. The slip speed tends to fluctuate because of the variation of torque with load angle, with the result that x_{sd} tends to be under-estimated. For large machines the slip must be very small. In the alternative negative excitation test, Fig. 10.51, the machine is run on no load in parallel with a supply network. The field current is steadily reduced to zero, reversed, and then raised until the machine loses synchronism at a field current $-I_{fr}$ and a stator terminal voltage V_r. A straight line OV is drawn through the point on the o.c.c. corresponding to V_r, and the stator, e.m.f. E_r is determined from it for an excitation $+I_{fr}$. Then $x_{sq} = x_{sd} \cdot V_r/(V_r + E_r)$, where x_{sd} is the d-axis synchronous reactance as given by the line OV.

Ventilation

Machines of medium rating are often installed with cooling gas entering and leaving by ducts. The pressure-drop has to be reproduced on test by wire screens. so that the gas intake shall not exceed that in service. Larger machines with closed-circuit ventilation are tested with or without coolers, in the latter case again with provision for reproducing service conditions. Compensators with air coolers mounted on the sides of the stator frame can usually be tested under conditions approximating adequately to those in the intended service. With vertical-shaft hydro-generators the enclosure of the air circuit on site is formed by the concrete foundations, and for works test enclosure is omitted because the pressure-drop in it is negligible. The cooling-water circuit is provided, particularly if it is concerned with direct-cooled field coils.

Temperature-rise

The following assumptions may be made for calculating the temperature-rise from the result of heat-runs in cases for which close accuracy is not sought or necessary:

 (i) The airgap is a thermal insulator, so that the rises of stator and rotor are not interdependent.
 (ii) The stator rise is the sum of three individual rises produced respectively by windage, core, and I^2R plus load loss.
(iii) The stator core loss is a function of the terminal voltage as determined by the o.c.c. test.
(iv) The load (stray) loss is the same on short-circuit, zero-power-factor and rated power-factor conditions, and is independent of temperature.
 (v) The rotor rise is the sum of two individual rises, respectively due to windage and to field I^2R.
(vi) Heat re-distribution immediately following individual heat-runs is negligible.

Parameters

Evaluation of the several quantities required for the two-axis analysis of a synchronous machine is described in Sect. 12.6.

11 Synchronous Machines: Construction and Design

11.1 *Types of synchronous machine*

Synchronous machines for general power-supply networks (at 50, 60 and occasionally 25 Hz) may be classified as follows:

Turbo-generators. Driven by steam turbines at high speeds (up to 3 600 r/min) and in ratings up to 1 000 MW.

Hydro-generators. Driven by water turbines at speeds between 90 and 1 000 r/min and in ratings up to 750 MW.

Engine-driven generators. Various forms of prime-mover of the internal-combustion type, with speeds up to 1 500 r/min and ratings up to about 20 MW. Gas-turbine drives may be of higher speed and rating.

Compensators. Self-driven machines operating at speeds up to 3 000 or 3 600 r/min and in ratings up to about 100 Mvar.

Motors. The four main types are salient-pole, cylindrical, and salient or cylindrical synchronous-induction machines. The plain salient-pole synchronous machine has the highest efficiency and can be used for constant-speed industrial drives such as compressors, blowers, pumps and fans, in particular where the starting conditions are suitable or power-factor improvement is economically desirable.

Almost all synchronous generators are 3-phase star-connected machines, with the field windings carried on cylindrical or salient-pole rotors. The stator windings are of double-layer form so that harmonics can be reduced by chording. The phase-spread is almost invariably 60 deg(e). With large turbo-generators the armature (stator) m.m.f. may reach 300 kA-t/pole, and to avoid excessive demagnetization effects the field (rotor) m.m.f. must be of comparable magnitude as defined by the short-circuit ratio. As it is undesirable to have high core and tooth saturation levels, a long airgap is necessary, from 5 mm in a 1 MVA machine to 100 mm or more in a 500 MVA machine. Typical short-circuit ratios are 0.4 for machines with quick-acting excitation control, 0.7–1.0 for medium-speed machines, and 1.0–1.5 for hydro-generators and compensators.

11.2 *Turbo-generators*

Steam turbines run efficiently only at high speeds, so that 2-pole generator

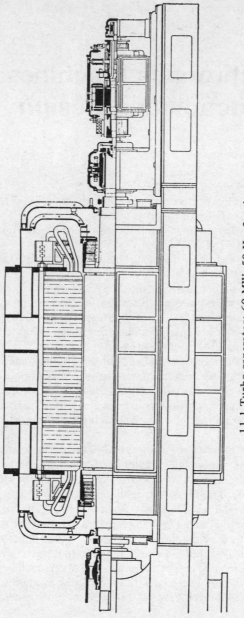

11.1 Turbo-generator: 60 MW, 50 Hz, 2-pole.

construction is common even at high ratings. The electric and magnetic loadings are exceptionally high. The rotor diameter is limited to about 1.2 m, and in 50 Hz machines the active core length must be of the order of 10 mm per MVA. A 500 MW generator may thus have a core length of 5 m, and a shaft length of 12 m. The outside diameter of the stator core may be about 3 m and of the outer casing about 4 m. Fig. 11.1 shows the general arrangement of a 60 MW air-cooled machine. Larger ratings make more sophisticated cooling essential.

Rotor

The I^2R loss per unit mass of conductor material in a 750 MW generator is of the order of 150 W/kg in the stator and 500 W/kg in the rotor. The rotor is therefore the limiting member. Within the diameter imposed by considerations of centrifugal force, deflection and critical speed, an excitation must be provided in accordance with the stator electric loading and the short-circuit ratio; and the exciting winding has to be contained in slots of such width as to leave teeth of adequate tensile and bending strength, able to carry the magnetic flux without excessive saturation. A cylindrical construction is necessary. Rotor bodies are normally machined from single-ingot forgings, and in the largest sizes may have a finished mass of 100 Mg. It may be necessary to use nickel-chrome-vanadium-molybdenum steel of ultimate strength up to 800 MN/m². Slots are milled out axially to receive the windings, typical profiles being shown in Figs. 3.11 and 4.14. The first critical speed is typically 1 200 r/min, the higher-order speeds occurring at 2.7, 3.0 . . . times this figure. One of these may be near to the normal running speed or the specified overspeed.

Windings. The rotor exciting winding is a major limiting feature. A concentric-coil winding is almost invariably adopted because of its simpler overhang geometry. Along the active length of the rotor the coils are retained in milled slots by wedges, but the coil ends outside the slots are contained within end-bells. To reduce the partial short-circuiting of the rotor flux through the end-bells, they are usually made from a non-magnetic austenitic steel (e.g., 18% Mn, 3% Cr, 0.5% C) of ultimate strength 1150 MN/m². A typical end-winding arrangement within an end-bell is illustrated in Fig. 11.2. In smaller machines with conventional cooling, the winding is formed from complete coils of silver-bearing copper strap. Larger machines have separate coil-sides placed into the slots and subsequently jointed at the ends.

Direct-cooled Windings. Windings with a high specific I^2R loss (e.g. 1 kW/kg) require direct cooling by hydrogen or water flow through hollow conductors. Joints at the non-drive end are more complicated, as insulating tubular connections to inlet and outlet manifolds are necessary to convey the coolant. For more even stressing of the teeth, it is usual in large machines to employ parallel-sided teeth. The result is a useful increase (e.g. of 30%) in the available slot area. But the slot shape is now such that each coil of the exciting winding has to be made to a shape and size corresponding to its position in the slot. The advantage of extra copper area is considered to outweigh the complication. The rotor winding of a 750 MVA generator, for example, may carry a current of 2.5 kA at 1 kV.

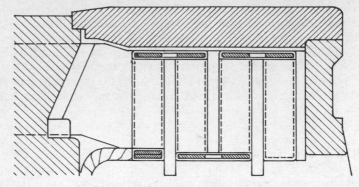

11.2 Retaining ring and rotor overhang.

Double Windings. The d.w.r. arrangement (Sect. 10.9) increases the leading reactive power capability and improves both steady-state and transient stability. Ideally the two individually controlled windings would be at an electrical angle of 90°, but in a 2-pole rotor this would involve slotting the polar horn where the magnetic flux density is greatest; in practice the angle is about 60°. Realization of a d.w.r. precludes a concentric field winding because of its complicated overhang geometry. There are thermal problems, too, for in the single-winding rotor all conductors carry the same current, whereas in the d.w.r. the two field windings carry different currents, their ratio changing with the active and reactive power levels.

Damper Windings. The solid cylindrical rotor body presents conducting (but comparatively resistive) ferromagnetic layers at its surface, providing somewhat indeterminate paths for axial eddy currents which damp magnetic fields that have a travelling-wave relation to the rotor. High surface losses therefore occur when the induced eddy currents dissipate non-synchronous field energy. This is of advantage in developing 'induction' torques, but is deleterious under certain 'normal' operating conditions, particularly those of 1-ph or of unbalanced 3-ph loading that give rise to strong negative-sequence fields. In large generators it may be found necessary to provide more definite arrangements for damping, (i) by the use of conductive metal rotor-slot wedges interconnected to form in effect a *damping cage,* or (ii) by the fitting of high-conductivity strip conductors beneath the wedges to form an intrinsic *damping winding.* Method (ii) can provide for continuous axial paths which bridge the flexibility slits that may be milled circumferentially around the rotor body to equalize the mechanical stiffness of the rotor in the direct (polar horn) and quadrature axes.

Stator
The core is built up from sectors such as that shown in Fig. 4.13. Although grain-oriented steels, arranged as in Fig. 2.19, (*a*), give lower core loss, the orientation is not everywhere favourable, and cold-rolled non-oriented steels containing 3% silicon are more usual. Mechanical problems arise from the double-frequency vibration set up by the rotation of the axis of the magnetic field.

Windings. For small machines the terminal voltage is usually fixed at some standard system level. For very large generators permanently connected to a transformer, the designer's choice is less restricted. A high voltage gives reduced current but entails more slot insulation; a typical compromise is 15 kV for generators of 100–200 MVA rating, and 25–30 kV for large machines. Even so, the current per phase may reach 20 kA, making essential a winding with parallel circuits, which must be electrically balanced in respect of e.m.f. and leakage reactance to avoid inter-circuit circulating currents. The 2-layer lap winding is universal because of its geometric symmetry, its lower load (stray) loss, and its simple construction with only two shapes of coil-side. Such a winding is readily chorded and parallelled. Copper is always used as a conductor material to make the best economic utilization of the available slot space. Current densities depend on the method of cooling: windings direct-cooled with circulating water may be worked at current densities of 12 A/mm^2, corresponding to a specific $I^2 R$ loss of over 300 W/kg. The conductors must be formed of strands, insulated and properly transposed to limit eddy-current loss. Coils are insulated with mica paper or glass-fibre tape, bonded with epoxy or polyester resin to give mechanical and electrical strength and permanence.

Airgap Windings. These are discussed in Sect. 3.12 and 12.7.

Electromechanical Forces. In a large turbogenerator, the electromagnetic force between conductors in a slot may reach 10 kN/m as a result of the attraction and repulsion between their currents. The forces pulsate at twice rated frequency, thus vibrating the conductors nearly 9 million times per day in a 50 Hz machine. The slot wedging must be such as to secure the slotted parts of the winding to prevent fatigue fracture of the conductors and fretting of their insulation. The end windings, however, present a more difficult problem, due not only to the complex static geometry of the overhang (which comprises coil ends of involute shape arranged on the frustrum of a cone), but also to the fact that the working condition is dynamic. The overhang structure is not rigid, and the force on any given short element of conductor depends on its instantaneous position, its mass and stiffness, and its complex interaction with neighbouring coil-ends in the assembly and with the bracing structure. The mechanics of the end-winding problem has been tackled by several authors [146]. Lawrenson found that the forces in general can be resolved into axial, radial and peripheral components, of which the axial appears to be the largest; that the forces on the two legs of the same coil-end differ considerably, as do those between different coils in the same phase group; and that small geometrical modification of the overhang assembly can greatly affect the force pattern and level. In practice, an end-winding supporting structure of brackets and rings of insulating material is common to all designs. Banding with glass fibre or with polyester cord which shrinks and tightens with rise in temperature is common. In some arrangements the assembly is clamped to its support by non-magnetic steel or non-metallic studs passing through the winding layers, but the space for these is very limited. The whole end-winding and its support is sometimes allowed to move axially to take up the effects of differential expansion.

Core-end Heating. Laminating the stator core limits the eddy-current loss in the steel resulting from the pulsation of fluxes in the *radial* plane. At the ends of the core, the high magnetic potential gradients in the core and particularly in the teeth (resulting from the necessarily high levels of electric and magnetic loadings) produce external flux components that enter and leave the core surface *axially*. The core lamination has no limiting effect on eddy-currents induced by axial fluxes, so that considerable eddy loss occurs in the end regions of the core and teeth. Further, the leakage fields may penetrate metallic core, tooth and overhang clamping structures and induce eddy loss therein. The core-end region can consequently develop an excessive temperature-rise. Control of the problem is maintained usually by the fitting of conductive screens of copper or aluminium over the core clamping plates, and in large machines these screens may have to be water-cooled. Shielding of the teeth is not feasible, so to minimize loss and improve cooling the teeth near to the end of the core may be slit radially. The important factors are the geometry of the stator and rotor cores, the stator end-packets and the conducting structural parts and screens, and the saturation levels. Contributions by several workers are listed in Ref. [147].

Two- and Four-pole Machines
In Europe the basic industrial supply frequency has been standardized at 50 Hz, and turbogenerators with 2-pole construction (3000 r/min) are most common. The main 60 Hz areas are in the Americas and in parts of Japan, requiring 2-pole speeds of 3600 r/min. When 4-pole generators have been used they have usually formed part of a low-pressure line of cross-compound sets, notably in the U.S.A. However, with the advent of nuclear-energy generation the 4-pole machine is more suitable, particularly with wet-steam systems in which the volume of steam per unit of power generated is very large. Turbine speeds of 1500 or 1800 r/min are then necessary for the higher ratings.

Cooling
Although in large machines the full-load loss may total no more than one or two per cent, this may amount to 10 MW or more. Removal of the heat demands a comprehensive cooling system using appropriate cooling fluids. Economic considerations usually determine the choice of coolant between air, hydrogen, oil and water. Hydrogen replaces air for 2-pole ratings of 50—60 MW. The point at which a change to direct hydrogen cooling is adopted has been progressively lowered, and the pressure raised to about 5 atm. For the highest contemplated ratings it becomes necessary to employ water as a coolant, employed for the stator (with hydrogen cooling for the rotor) or for both stator and rotor. An analysis of water cooling is given by Noser and Kranz [105].

In any winding, the thermal drop across the insulation is nearly two-thirds of the total conductor temperature-rise. The potential gain in passing the coolant through (or in near contact with) the conductor within the insulation is very great, making possible a substantial increase in rotor m.m.f. But there are several technical problems to be solved, such as corrosion, erosion, and electrolytic action; water connections, arrangement of manifolds, entry and exit water seals; flexure and vibration.

Large Turbogenerators

A possible future need for very large units of ratings up to 2000 MW raises the problems of their design, construction and transport. Since some Russian engineers [104] assessed the feasibility of 750 MW machines in 1963 there have been significant developments, doubling prospective ratings that can be achieved with refinements of the present constructional methods.

Mechanical Limitations. Equation (1.16) shows that the output is proportional to the product of the specific electric and magnetic loadings A and B, the rotor dimensions D and l, and the speed n (which is settled by the standard frequency). The product $D^2 l$ directly affects the rating, so that even a small increase in D is significant; but the rotor stress is proportional to $D^2 n^2$, and Dl is a factor in determining the critical speed. The present stress limit imposes a rotor diameter of about 1.2 m, and l is limited by flexural considerations to not more than about $5D$. Stress at the rotor surface should not exceed 50–60% of the yield stress for normal speed, and 75–85% at a 20% overspeed.

End Bells. Titanium alloy may be required for these, the stress in which is largely self-produced. The alloy has a yield point no greater than that of nickel-steel, but its density (4500 kg/m^3, compared with 7800 for steel) lowers the self-stress and permits it to cope more readily with the centrifugal force of the rotor overhang that it has to contain. It is non-magnetic.

Critical Speed. Refined calculation is essential, taking account of its relation to rated speed, flexibility slitting, thermal expansion of the body and its windings, and the characteristics of the generator-turbine shaft coupling and the bearings. The complete unit has several critical speeds related to the generator and turbine rotor sections. Vibration problems may be serious, for while stator deformation may be no more than 30 μm on the radius, the strong vibration at twice rotational frequency may be too near to the natural frequency of the stator and its frame, or to a rotor critical speed.

Transport. A potentially difficult design limit is that of permissible transport weight and profile. This applies particularly to the low-pressure turbine casing, which must be split into parts for subsequent assembly on site. The stator presents a more difficult problem, as the wound core cannot be split.

Electrical Limitations. The specific electric loading can be raised towards 300 kA/m with direct water cooling, but with consequent reaction on the leakage fields, electromechanical forces and operational reactances. The short-circuit ratio, however, remains at about 0.5 to maintain stability in light-load leading-p.f. conditions on the associated transmission network. Subtransient reactance should be high to limit fault levels and the stresses in the shaft and in the stator winding. The transient reactance should be low for stability. But these two reactances are not independent.

A 2000 MW generator might have a rotor of diameter 1.25 m, of length 7 m and of mass 80 tonne; an airgap of 150 mm radial; a 300 tonne stator; and a total full-load loss of 140 MW.

11.3 Hydrogenerators

Synchronous generators for hydraulic turbine drive are designed individually, as they must match the overspeed (typically twice normal) of the turbine as well as the short-circuit ratio, leading reactive power and transient reactance requirements of the connected power network. The speed may demand a multipolar structure and a large diameter, with provision for transport to site in sections. Turbine governing and transient stability determine the total inertia, most of which must be provided by the generator. Both horizontal- and vertical-shaft designs are employed.

Horizontal-shaft. These are for impulse turbines. The bearings must include a thrust type to resist the axial component of the turbine force. The construction has largely been superseded by the high-speed Francis turbine except for pumped-storage plants where the pump is mechanically decoupled when not in service.

Vertical-shaft. Sets are normally sunk into concrete pits to minimize crane lift. Typical arrangements are shown in Fig. 11.3. *High-speed* machines (*a*) and (*b*) have guide bearings above and below the generator rotor for dynamic stability and to preserve the airgap length. *Low-speed* multipolar machines have large diameters: (*c*) has a thrust bearing below the rotor, giving easier access for maintenance, possibly with the stator jacked up; (*d*) is 'overhung' with the thrust/guide bearing closer to the rotor mass-centre; (*e*) is a compact arrangement with the thrust bearing between the two rotors and a guide bearing immediately below the 'umbrella' generator rotor. The diagram in Fig. 11.4 shows the essential mechanical features of a Francis turbine set. Pneumatic brakes bring the set to rest when brought out of service, and are used hydraulically as jacks when the set is at rest to allow oil to flood the thrust bearing or permit maintenance of the pads.

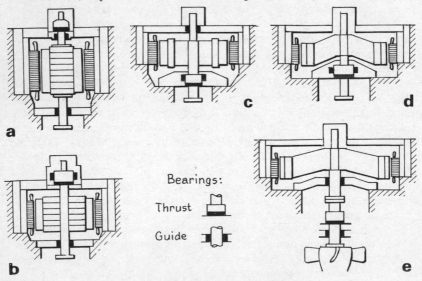

11.3 Vertical-axis hydro-generators.

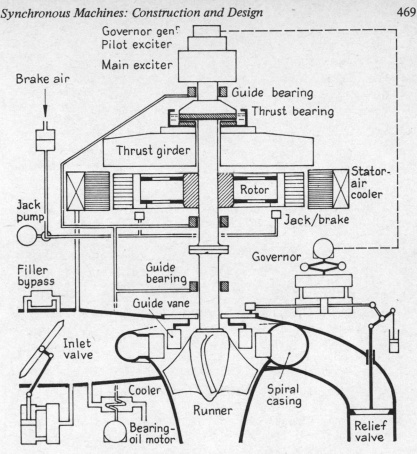

Governor gen^r
Pilot exciter
Main exciter
Brake air
Guide bearing
Thrust bearing
Thrust girder
Stator-air cooler
Rotor
Jack pump
Jack/brake
Filler bypass
Guide bearing
Governor
Guide vane
Inlet valve
Cooler
Spiral casing
Runner
Bearing-oil motor
Relief valve

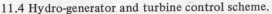

11.4 Hydro-generator and turbine control scheme.

Rotor

The salient poles are carried on a rotor body. For high-speed machines the body is (*a*) machined with its shaft from a forging; or (*b*) machined separately, with stub shafts attached thereafter; or (*c*) built up from discs shrunk on to a shaft. These methods are illustrated in Fig. 11.5. In the case of medium- and low-speed machines, method (*c*) may be used for diameters up to about 4 m, or (*d*) the body may be fabricated from a cast-steel spider mounted on the shaft and carrying a laminar ring of overlapping segmental plates tightly bolted to resist slipping under hoop stress. The rim rests on the spider arms and is driven from them by floating keys, thus relieving the spider of centrifugal force other than that due to its own mass. Up to a peripheral speed of about 80 m/s it is feasible to use a segmented spider. At higher peripheral speeds (up to 110 m/s normal and 220 m/s overspeed) a solid disc becomes essential.

Laminated poles are used on all medium- and low-speed constructions. They comprise thin steel plates clamped between heavy end-plates and secured

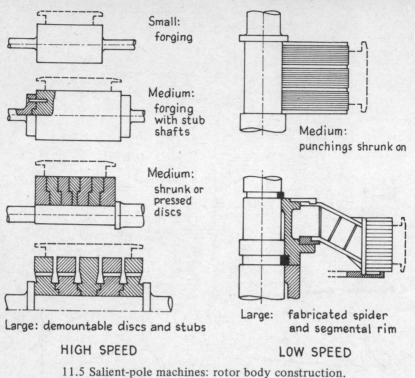

Small: forging

Medium: forging with stub shafts

Medium: shrunk or pressed discs

Large: demountable discs and stubs

HIGH SPEED

Medium: punchings shrunk on

Large: fabricated spider and segmental rim

LOW SPEED

11.5 Salient-pole machines: rotor body construction.

by studs or rivets. The poles may be bolted to the rotor body if the peripheral speed does not exceed about 25 m/s. For higher speed the poles are dovetailed, as shown in Fig. 11.6.

Windings. The field coils are formed from rectangular-section copper (or aluminium) strap, wound on edge, with inter-turn insulation cured and consolidated under a pressure exceeding that due to the centrifugal force on the coil at overspeed. The coils of machines with axially long poles are fabricated from punched straight lengths of copper, Fig. 11.6(*a*). Certain turns can be made of wider strap to produce additional cooling fins. The method also facilitates the construction of 'canted' field windings, (*b*), which better withstand centrifugal force.

Damper windings. With solid poles the eddy-current damping may be sufficient, but with laminated poles a damper winding is employed, comprising copper bars through the pole-shoes with end connections attached to the pole end-plates. Interpolar connection between the damper windings is not normally necessary.

Stator

The stator of a large machine is fabricated in 3, 4 or 6 parts, sometimes completely wound except for coil joints, so that the transport problem is relieved.

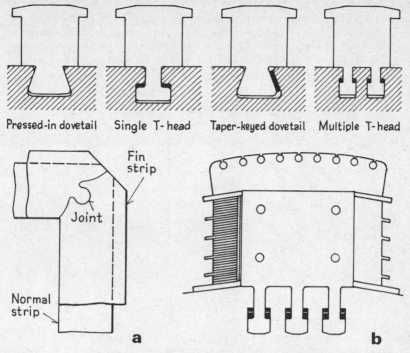

Pressed-in dovetail   Single T-head   Taper-keyed dovetail   Multiple T-head

11.6 Salient-pole fixings and field windings.

It is possible to employ grain-oriented magnetic sheet steel, Fig. 2.19(*b*), but the advantage is not great.

Windings. The two-layer diamond-coil form is normal, with single-turn bar or multiturn coils. Bar windings have Roebel or similar transposition arrangements. The insulation is generally of class B mica, with phenolic, epoxy or polyester bonding varnish.

Cooling

Closed-circuit air cooling is frequently employed, with air/water heat-exchangers mounted outside the stator frame. In very large machines the liquid cooling of both rotor and stator windings is considered to be feasible. Fig. 4.15 shows a water-cooled field winding.

11.4   *Industrial generators*

Special-purpose and standby synchronous generators are sometimes useful in industry. They are driven from diesel engines or gas turbines, sometimes as a transportable unit, in sizes typically between 50 and 1 000 kVA. The most usual design has a 4-pole salient rotor with brushless excitation, and a conventional stator winding. The performance guarantees include automatic voltage regulation, voltage dip and recovery time, voltage build-up when starting, and low harmonic content. A typical gas-turbine/generator set for 500 kW is fully enclosed on a single base, complete with turbine starting, cooling, fuel and oil systems, excitation control, switchgear and main cable connectors, all in a total mass of 6 Mg. With a top-mounted exhaust outlet, the plan dimensions are 4.5 m x 1.5 m, and the height is 2.75 m. Typical spider and pole punchings for a 6-pole generator are shown in Fig. 11.7.

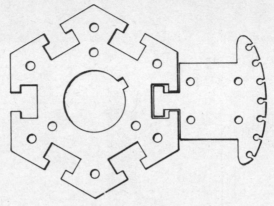

11.7 Spider and pole punchings.

11.5 *Compensators*
Synchronous compensators for the control of reactive-power demand in supply sytem networks may be permanently sited, or may require to be transported at intervals in accordance with the growth and changing pattern of system conditions. They are designed to be compact and almost self-contained, with the minimum of external connections. Machines of 50 Mvar upward are often built with hydrogen cooling. Four- or six-pole construction is usual, with salient rotor poles and a conventional stator winding. Apart from excitation, the losses are low, and a power factor of 0.02–0.03 p.u. can be achieved.

11.6 *Motors*
The constructional features of motors do not differ from those of generators of comparable rating and speed, except that the salient-pole design is the most common, and that (unlike the generator) the machine is almost always required to be self-starting, using the pole-face cage as a starting winding. There is no essential difference between the stators of polyphase synchronous and induction motors of comparable rating.

11.7 *Main dimensions*
The output S of a synchronous machine is a function of the air-gap diameter D, core length l, speed n, and the specific magnetic and electric loadings $\bar{B}$ and A. In machines of the highest rating the limiting feature is the peripheral speed $u = \pi Dn$: then from eq.(1.16)

$$S = 11K_w\bar{B}AD^2ln = 1.1\,K_w\bar{B}Alu^2/n$$

where the winding factor of the stator is of the order $K_w = 0.96$.

Salient-pole machine. With a mean gap density $\bar{B}$ = 0.65 T and an electrical loading A = 60 kA/m, and with dovetailed solid-pole construction to permit a peripheral speed u = 100 m/s, the output available per unit length of core is

$$S/l = (400/n) \times 10^6 \text{ VA/m}$$

A 20-pole 50 Hz machine has n = 5 r/s and D = 6.3 m, so that with the loadings stated it can develop S/l = 80 MW/m of core length. The main dimensions may be subject to several adjustments to meet specified values of short-circuit ratio, synchronous and transient reactance, overspeed, transport facilities and inertia constant. The per-unit transient reactance varies directly as the product $\bar{B}A$, and inversely as a function of the pole-pitch and the shape of the interpolar space. Hydro-generators located at the end of long transmission lines may have particularly low synchronous reactance (e.g. 0.55 p.u.) and transient reactance (e.g. 0.2 p.u.); exceptional demands for supply-system charging current may be met by raising the short-circuit ratio or even by the use of nega-

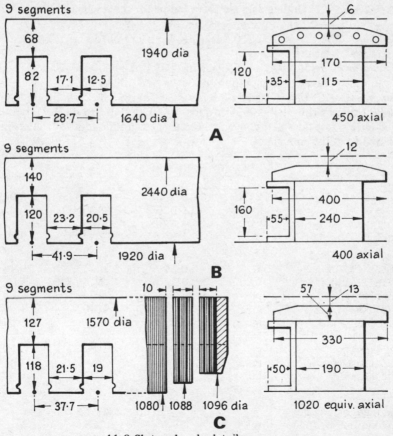

11.8 Slot and pole details.
Dimensions in mm; not to scale.

tive excitation. Large-diameter low-speed generators of the punched-segment rim construction can be given higher inertia by increasing the depth of the rim. In higher-speed units with disc rotors, a demand for higher inertia may require an increase in diameter (involving more elaborate pole fixings), or the provision of a flywheel, or the application of turbine runaway speed limiters. *Cylindrical machine.* Considering 2-pole, 50 Hz turbo-generators with a peripheral-speed limit of u = 195 m/s and therefore a diameter D = 1.25 m, the magnetic loading may be 0.7 T. Liquid cooling enables the electric loading to be of the order of 220 kA/m, using stator and rotor current densities that respectively approach 10 and 18 A/mm². The output per unit of core length is then

$$S/l = 125 \times 10^6 \text{ VA/m}$$

The rotor of a 750 MW generator with these loadings must have an active length of 6 m.

EXAMPLE 11.1: Outline data are given below for three salient-pole machines: the essential stator and rotor dimensions are set out in Fig. 11.8.
Generator A. 1 250 kVA, p.f. 0.8 lagging, 6.6 kV, 3-ph, 50 Hz, 20-pole, 300 r/min for engine drive.
Generator B. 3 750 kVA, p.f. 0.8 lagging, 10 kV, 3-ph, 50 Hz, 10-pole, 600 r/min for hydraulic drive requiring an inertia of 3 400 kg-m².
Compensator C. 5 MVA leading/2.5 MVA lagging, 11 kV, 3-ph, 50 Hz, 6-pole, 1 000 r/min machine with a full-load loss of 125 kW (0.025 p.u.). To reduce load (stray) loss the stator core ends are stepped and the tooth stiffeners insulated from the core plates.

		A	B	C
Rating				
Output	kVA	1 250	3 750	5 000
	kW	1 000	3 000	0
Line/phase voltage	kV	6.6/3.8	10/5.8	11/6.35
Line/phase current	A	109	217	263
Main Dimensions				
Magnetic loading	T	0.51	0.48	0.46
Electric loading	kA/m	38.2	51.5	55
Stator bore	m	1.64	1.92	1.08
Gross core length	m	0.45	0.40	1.00
Ducts, no./width	mm	6/10	5/10	14/10
Net core length	m	0.39	0.35	0.86
Iron length	m	0.351	0.315	0.775
Radial gap length	mm	6	12	13
Pole pitch	m	0.26	0.60	0.57
Stator				
Winding/connection	—	double-layer diamond/star		
No-load flux per pole	mWb	59	116	261
Turns per phase	—	300	240	120
Number of slots	—	180	144	90

		A	B	C
Conductors per slot	—	2 x 5	2 x 5	2 x 4
Coil pitch	—	9	12	12
Conductor size	mm	5 x 5	2(10 x 4)	2(8.5 x 4.5)
area	mm^2	25	80	77
Length of mean cond.	m	1.11	1.38	2.07
Resistance	mΩ/ph	560	174	136
Current density	A/mm^2	4.4	2.7	3.4
Total f.1. I^{2}R loss	kW	20	24.5	28
Copper mass	kg	475	1 520	1 060

Rotor

		A	B	C
Type of pole	—	laminated	laminated	solid
fixing	—	bolted	dove-tailed	screwed
Damper bars/pole	—	8	0	0
Turns/pole	—	71$\frac{1}{2}$	99$\frac{1}{2}$	99$\frac{1}{2}$
Conductor size	mm	35 x 1.4	55 x 1.3	50 x 1.75
area	mm^2	49	71	87
Full-load current	A	135	207	270
Current density	A/mm^2	2.75	2.9	3.1
Length of mean turn	m	1.49	1.41	2.52
Resistance	Ω	0.89	0.42	0.37
Volt drop	V	120	86.5	100
I^{2}R loss	kW	16.1	18	27
Copper mass	kg	950	630	1 300
Peripheral speed	m/s	25.5	60.5	11.2
Overspeed	p.u.	0	0.8	0

Magnetization

		A	B	C
Flux density, core	T	1.24	1.31	1.33
pole	T	1.20	1.23	1.40
teeth ($\frac{1}{3}$)	T	1.61	1.67	1.74
gap	T	0.77	0.725	0.79
M.M.F./pole, no-load	kA-t	7.5	8.95	10.35
full-load	kA-t	9.65	20.6	26.9
Armature m.m.f./pole	kA-t	4.91	15.1	16.0

Performance
Full-load losses:

		A	B	C
stator I^{2}R	kW	20.0	24.5	28.0
rotor I^{2}R	kW	16.1	18.0	27.0
core	kW	20.0	35.0	38.0
mechanical	kW	3.5	14.5	23.0
load (stray)	kW	7.4	12.0	5.0
total	kW	67.0	104.0	121.0
Output	kW	1 000	3 000	0
Input	kW	1 067	3 104	121
Efficiency	p.u.	0.937	0.967	0

11.8 *Design*

The general remarks in Sect. 1.3, and those specific to induction machines
in Sect. 9.3, may be taken as applying to the computer design of synchronous
machines: but the process, particularly for very large machines, is considerably
more complicated making synthetic design at present impossible.

The approach to design programming is greatly affected by the type and
size of the machine. In the case of small standard or semi-standard machines,
many of the dimensions and performance criteria are imposed by international
and manufacturers' standards, greatly reducing the number of variables; synthesis of the remainder of the design becomes practicable. The optimum
design is generally based on minimized production cost (occasionally mass),
and marginal improvements in performance are less important.

For large machines the number of design variables is so great that optimizing may vary with the application. The capitalized value of an optimized performance item (such as subtransient reactance) may outweigh the increase in
price that the customer is prepared to pay for it. Again, the machine may be
designed as part of an extensive system and its individual cost be of less consequence than its performance in the system.

A comprehensive synthesis program might take several man-years to develop, and could be both restrictive and obsolescent when completed. An
analysis-synthesis combination may be more practicable, with the synthesis
routines dealing e.g. with the adjustment of a few interrelated dimensions to
meet a particular item in the specification, and the analysis routines coping
with many of the design calculations (unbalanced magnetic pull, magnetization, inertia, stresses, resistance, leakage reactance and losses). Here the 'con-
versational mode' is valuable.

In the course of a design, coefficients may be used that are empirical functions of, say, two variables expressible as polynomials: techniques are available for interrelating them so that they can be stored in a main program concerned with the required coefficients. In all computer-aided design, the task
of forming interrelations between variables is vital.

No-load magnetization of large turbo-generators. An example of the problems that arise in turbo-generators is the predetermination of the no-load
magnetization characteristic. This is discussed by K. J. Binns. The magnetic
circuit concerned is shown in Fig. 11.9. The flux distribution at the centre
of the airgap is first assumed (it is approximately trapezoidal), the flux per
pole being estimated from the required stator e.m.f. The useful flux is divided into a number of paths, bounded by the centre lines of the rotor slots and
continuing in the rotor core to cross the q-axis at right-angles. For simplicity
the stator slotting is considered as replaced by an equivalent slot system
having angular alignment with the rotor slots but the same reluctance as the
actual slots. The flux quantity passing through each flux path is determined.
The rotor leakage flux is added to the rotor flux and then, because the
cross-sectional area of ferromagnetic material at each point on a path can be
found, the variation of the flux density along each path can be estimated.
From each value of B the corresponding H is obtained from the magnetization
curve of the steel. By taking a considerable number (e.g. 100) of points along

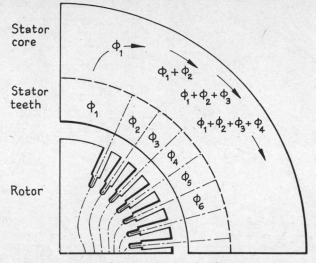

11.9 Calculation of magnetic circuit of turbo-generator.

each flux path it is possible to integrate H to obtain the drop of magnetic potential along the path. The m.m.f.s are used to determine the rotor slot current. If the assumed gap flux distribution is correct, each slot will require the same current; but if there is a discrepancy, the assumed distribution is modified until the error is minimized. The final flux distribution is employed in calculating the open-circuit e.m.f. of the machine. To program the calculation the independent variable is taken as $B,$ and the H/B relation fitted by a number of linear equations each valid for a part of the H/B curve. The variation of B with length x along a path is expressed as a polynomial in $x,$ and the magnetic potential difference along a path integrated by application of Simpson's cubic rule. Iteration enables the necessary equality of the rotor slot currents to be approximated, and finally a harmonic analysis of the gap-flux distribution is used to derive the r.m.s. value of stator e.m.f.

The saving of time compared with 'hand' calculation is important, and the accuracy is markedly improved because of the point-by-point rather than the region-by-region assessment. The 'hand' calculation of this problem requires a saturation curve to be determined for each part of the magnetic circuit on the assumption of no leakage. The curves are then corrected and combined to give a saturation curve for the whole circuit per pole with the exception of the rotor core. The core m.m.f. must be calculated from a knowledge of the field form, and as this m.m.f. does not primarily affect the rest of the circuit, it is added to the total m.m.f. at the end of the o.c.c. calculations. In dealing with the regions, the stator and rotor teeth are assessed graphically (Fig. 2.18), the polar horn as a large tooth and the stator core in terms of its average flux density. End-bell leakage is graphed, and the rotor slot leakage based on slot permeance. The saturation curves that emerge are shown in Fig. 11.10: they are

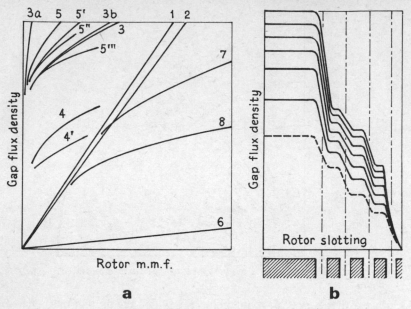

11.10 Magnetization curves and field form.

1. Air line for unwound part of the rotor.
2. Air line for wound part of the rotor.
3. Stator, combination of core (3a) and teeth (3b).
4. Wound part of rotor; 4' corrected for end-bell leakage.
5. Unwound part of rotor; 5' corrected for end-bell leakage;
   5" partly corrected for slot leakage; 5''' corrected for end-bell and slot
   leakage.
6. Slot leakage.
7. Magnetization curve for unwound part $(1 + 3 + 5''')$.
8. Magnetization curve for wound part $(2 + 4')$.

The distribution of gap flux density is now found for a series of rotor excitations, the results being shown in (*b*) of Fig. 11.10. These are *field-form* curves: the steps above the slots tend to slope more steeply at points in the gap more remote from the rotor surface, and the field-form curve tends toward a trapezoidal shape, which is the starting point for Binns' method.

12
Types and Applications

12.1 Power transformers
Transformers for power-supply networks were described in Sect. 5.19. Here the discussion is extended to a number of special types and specific applications.

Aluminium Windings
Most large transformers have square- or rectangular-stranded copper conductors to ease the coil-winding process and to limit eddy-current loss. Metal-market trends may make manufacture more economic with windings of aluminium, a soft and ductile metal liable to 'creep' under pressure, with a thermal expansion higher and a density and conductivity lower than that of copper. These differences react on the design. For wire windings the main

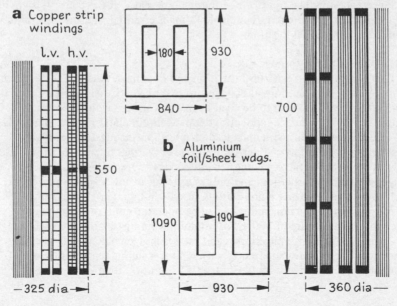

12.1 Cores and windings of 500 kVA 50 Hz 3-ph transformer with copper and aluminium coils (dimensions in mm).

problems concern creep, thermal expansion and the natural formation of a
high-resistance surface oxide skin. An alternative is *sheet* thickness 0.15 mm
upward) or *foil* (less than 0.15 mm). The interleaving insulant may be paper,
capacitor tissue or very thin polyester film. Figure 12.1 compares the cores
and windings of an oil-immersed 500 kVA 3-ph 50 Hz transformer with (*a*)
copper and (*b*) aluminium-foil windings. The active material and overall
dimensions for the same p.u. leakage reactance are:

Type	Core	Masses (kg) L.V.	H.V.	Total	Dimensions (m³)
(a) Copper strand	620	160	215	995	1.7 x 1.1 x 1.6
(b) Al foil/sheet	835	95	130	1060	2.0 x 1.1 x 1.8

The l.v. winding is of sheet, of width corresponding to the window height.
To avoid very thin foil, the h.v. winding has four shorter coils in series. As
indicated in Fig. 3.5, there is some distortion of current density over the
axial length, the additional $I^2 R$ loss being compensated by the lower core
flux density and loss.

Sealed Transformers
Sealing slows the ageing process by protecting the transformer from
contaminants in the ambient atmosphere. The change of oil volume with
temperature is accommodated by a gas cushion or a tank with flexible walls.
The latter can more readily withstand a sudden internal pressure rise due to
an internal fault. The transformer, completely sealed at the factory, is applied
for indoor sites where non-flammable filling is essential, and for mineral-oil-
filled units in critical locations.

Dry (Class C) Transformers
Oil-free transformers with air insulation and cooling have useful features for
protected sites in factories at voltages up to 11 kV. Virtual elimination of the
fire hazard enables them to be sited near load centres, avoiding long cable
runs. A protective barrier allowing adequate air circulation is necessary. Class
C insulating materials with enhanced chemical, mechanical and electrical
properties have influenced design; polyester coverings for winding conductors
give a better space factor, polyester-bound glass-fibre materials serve for interlayer and barrier insulation, and polyester resins of low viscosity can be used
without solvents to give a non-porous cured winding that is mechanically
strong and moisture resistant, and can be operated continuously with a
hot-spot temperature of 200°C. Aluminium windings are usually preferred.

A comparison of typical losses and masses is given below: C is for a dry
air-cooled Class C insulated range with a mean winding temperature of 170°C,
and O is for oil-immersed units with a 75°C limit on temperature rise.

Rating	(kVA)		300	500	1000	1500	2000
Core loss	(kW)	C	1.2	1.5	2.4	3.5	4.4
		O	0.7	1.0	1.7	2.4	3.0
$I^2 R$ loss	(kW)	C	6.2	8.0	13	18	25
		O	4.6	6.8	11	16	21
Mass of coils		C	1.1	1.5	2.4	3.2	3.8
and core	(t)	O	0.8	1.1	2.0	2.7	3.4

Sealed 'dry' transformers with nitrogen filling are sometimes used where site conditions demand. Sulphur hexafluoride (SF_6) can be alternative: at a pressure of 3 atm it has an electric strength approaching that of oil, and a heat transfer 2½ times as effective as for air. It is inert, non-flammable and non-toxic, but requires strengthened tank construction.

Moulded Windings

Winding assemblies placed in moulds or between resin/glass cylinders can be vacuum-impregnated with resin. Such *cast-in-resin* windings are safe from external contaminants and are mechanically strong enough to withstand radial short-circuit forces. The manufacture is simple: casting liquid is poured into the mould and hardened. The cast product is 'flame-retardant' and 'hard-to-ignite': even when subjected to flame from an external source, no toxic or aggressive vapours are emitted. Both h.v. and l.v. windings can be foil-wound to yield an impulse strength comparable to that of an oil-immersed unit because of the almost linear surge-voltage distribution and the low residual axial short-circuit force. Losses and noise levels are lower than for either liquid-filled or dry types. Thermal protection is provided by solid-state monitoring, with hot-spot detection by temperature sensors buried in the l.v. windings.

Rectifier Transformers

Seven basic transformer/rectifier connections for 3-ph equipments in Fig. 12.2 are classified below, (i) in accordance with the number of pulses (i.e., commutations per period), (ii) as single- or double-way (i.e., transformer secondary currents either uni- or bi-directional), (iii) primary and secondary winding connections, and (iv) the use of a bridge connection or an interphase inductor where relevant.

(*a*) Three-pulse; single-way; delta or star primary, zig-zag secondary.
(*b*) Six-pulse; single-way; delta or star primary, double-star secondary; interphase inductor.
(*c*) Six-pulse; single-way; star primary with isolated star-point, diametral secondary.
(*d*) Six-pulse; double-way; delta or star primary and secondary; bridge.
(*e*) Twelve-pulse; single-way; delta or star primary, quadruple zig-zag star secondary; three (or two) interphase inductors.

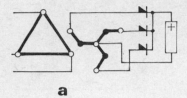

a

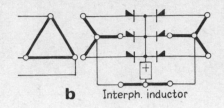

b Interph. inductor

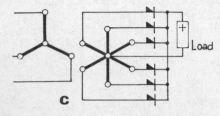

c Load

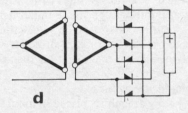

d

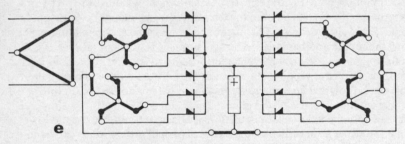

e

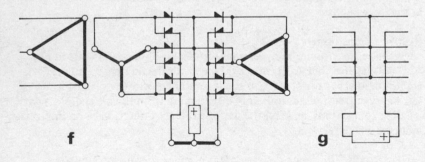

f

g

12.2 Transformer-rectifier connections.

(*f*) Twelve-pulse; double-way; delta or star primary, delta and star secondaries; two parallel bridges, interphase inductor.

(*g*) Twelve-pulse; double-way; windings as (*f*); two series bridges.

Waveform distortion of the transformer current increases its r.m.s. value and therefore the apparent-power rating for a given output. Large rectifier transformers distort the supply voltage waveform, so that for 100 kW upward it may be desirable to increase the number of pulses.

Mercury-arc rectifiers are normally connected in single-way forms (*a, b, c, e*), but all connections suit silicon diodes. Connection (*e*) is common for high-current low-voltage rectification. For higher voltages up to 2 kV, connection (*g*) can be used with two or more diodes in series in each path. In 6-pulse cases (*b, c, d*) the relative transformer ratings are 1.2, 1.2 and 1.0, with an inductor of rating 0.05 for (*b*). The relative ratings for 12-pulse arrangements (*e, f, g*) are 1.3, 1.0 and 1.0, with (*e*) and (*f*) requiring inductors respectively of rating 0.06 and 0.01. For large outputs where harmonic limitation is essential, a 24-pulse system can be furnished by two separate 12-pulse units with outputs in parallel, the 15° phase-shift being obtained by a 7½° shift on each primary resulting from auxiliary windings in the delta connection. For example, a typical transformer/rectifier unit for an electrolytic plant has a d.c. output of 30 MW (75 kA at 400 V). Two units with phase displacements of +7½° and −22½°, corresponding to a combined displacement of 30°, can together give 60 MW with a 12-pulse output. With ±7½° and ±22½° the output is 24-pulse with a rating of 120 MW.

Uneven firing of thyristor rectifiers causes a d.c. component of current to persist in the transformer secondary, distorting the flux excursions into oversaturation in alternate half-periods. The high and harmonic-burdened m.m.f. raises the leakage flux, making it necessary to provide eddy-current screening. Magnetizing-current offset is discussed by Yacamini and de Oliveira [120].

Railway-traction Transformers

A typical modern traction system has a h.v. 1-ph contact wire fed from 1-ph substation transformers, and locomotives with 1-ph transformers supplying a controlled direct voltage to d.c. series motors through a rectifier, or to 3-ph induction motors through a d.c. link inverter. For the British Rail electrification at 25 kV nominal, supply is taken from the 50 Hz main transmission network at voltages up to 275 kV.

Substation Transformers. The contact-line voltage has a range of 16.25−27.5 kV. Substation transformers are connected on the primary side across two lines of the feeding network and are fully insulated. They are subject to single or double lightning surges, and are normally provided with off-load tap-changers set to take account of the local supply-system voltage regulation and the traction loading. Rectifier control on the locomotives results in a contact-wire current demand of fluctuating magnitude and considerable harmonic content. The supply system can generally accept the inherent 3-ph unbalance and harmonic distortion, but occasionally it may be necessary to fit harmonic filters. Service conditions require short-circuit currents to be limited by adequate transformer leakage reactance (e.g., 0.1 p.u. for a 10 MVA unit), and 2 p.u. surge voltages to be withstood. Typical transformer design features include two- or three-limb 1-ph core types (Fig. 2.9), a normal flux density up to 1.65 T, and work-hardened or silver-bearing copper-strip windings to withstand short-circuit forces. Cooling methods (Sect. 5.16) are usually ON, but large units of 20 MVA and over have ONAN/OFAN/OFAF cooling by means of a separate heat-exchanger with air-blast fans and a pump in the oil pipework, the changeover being controlled by sensors responsive to hot-spot temperature.

Locomotive Transformers. These have to be dimensionally small and of light weight, and must not contain a potentially toxic fluid. A suitable form is that of Fig. 2.10, with a silicone liquid coolant.

Arc Furnace Transformers

Electric arc furnaces, often of large rating (e.g., 50 MVA) are employed in the production of tonnage steels and of special steels from scrap. Operating

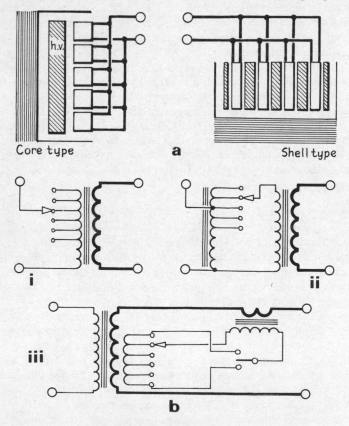

12.3 Arc-furnace transformers.

conditions are severe: short-circuits are frequent, justifying a conservative design to avoid damage, ageing and repeated 'coil hammering' resulting from axial forces. Characteristic features are the wide range of secondary voltage (down to 0.25 of the maximum) and the very large secondary currents, for which only solid and permanent connections are feasible. As regards construction, the *shell type* in (*a*) of Fig. 12.3 has a low leakage reactance, easier secondary connections, and the facility for tappings with an arbitrary number of turns. In the *core type* the h.v. winding has to be placed next to the core and be provided with tappings, but is normally preferred. Unavoidable unbalance of the total impedance per phase results in unequal arc-electrode voltages which affect the transformer core fluxes and sometimes call for the use of five limbs. The transformer leakage reactance is in series with the considerable inductive reactance of the connections to the electrodes. In large installations the total phase reactances must be both minimized and balanced. The load-voltage regulation problems determine the choice of transformer connections. Three schemes are shown in Fig. 12.3(*b*).

Flux Regulation (i). For minimum secondary output voltage, the primary is tapped up to the maximum number of active turns, resulting in a low core flux and e.m.f. per turn. Tapping down has the opposite effect, the core flux and e.m.f. per turn being raised and so increasing the secondary voltage level. However, in the idle part of the primary winding there are induced e.m.f.s which may become excessive should there be a supply-system voltage transient. As a consequence the method is uneconomic for primary voltages over 50 kV and for secondary regulation down to less than 0.4 of the maximum voltage.

Main and Intermediate Transformers (ii). A main tapped autotransformer feeds a separate step-down unit having a primary voltage in the range 20—40 kV. To increase reliability the two transformers are in a common tank. Large units have a 3-ph regulating transformer and three separate 1-ph step-down units sited symmetrically around the furnace.

Main and Booster Transformers (iii). A fixed-ratio main transformer provides a secondary voltage at the middle of the load voltage range. A variable voltage is tapped off a tertiary winding to feed an auxiliary unit with its secondary in series with the load, to raise or lower ('boost' or 'buck') the electrode voltage. The scheme is suitable for supply networks of 66 kV upward.

Low-voltage Transformers

Very large currents at low but adjustable voltage are required in *electrochemical* processes. The massive l.v. conductors and the problems of inductive reactance resemble those for arc-furnace transformers. A.C. *welding* transformers have secondary voltages between 70 and 100 V, with an adjustable series inductor to stabilize the arc and to improve restriking after current zeros. Part of the inductive reactance is built into the transformer. Capacitor correction of the inherent low power factor is provided.

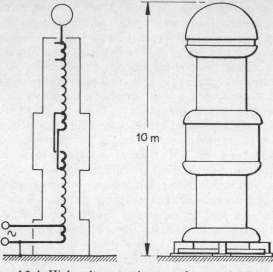

10 m

12.4 High-voltage testing transformer, 1.2 MV.

High-voltage Testing Transformers

Industrial-frequency h.v. testing transformers are used to impose a test voltage
on a workpiece (e.g., an insulator bushing), the current being very small
except at the instant of breakdown under electric stress in the workpiece. The
transformer windings must withstand the breakdown voltage transients, so
that much of the winding design centres on electric stress distribution.
For the secondary (h.v.) winding, concentric layers of length inversely
proportional to radius form approximately equal capacitors to give a linear
radial stress distribution. A single transformer unit can develop up to 600 kV
r.m.s.; for high voltages up to 1200 kV a vertical cascaded construction,
Fig. 12.4, can be adopted. The end turns of each unit are tapped and
insulated to act as the primary input of the earthed-end unit or to provide
excitation and load current to the next higher unit. The transformer in
Fig. 12.4 provides 1200 kV and a nominal primary current of 1 A in an
overall height of 10 m. Two 1200 kV assemblies arranged vertically can
develop about 2.2 MV in a height of 18 m. With the h.v. terminal isolated, the
primary may take a *leading* current because the self-capacitance effect
outweighs the magnetizing reactance.

Equipment testing for very high voltage systems may demand testing
transformers of such dimensions that they can be erected only outdoors, with
due provision for adverse weather conditions.

Mining Transformers

These work in onerous conditions, with restricted headroom and a
potentially explosive ambient atmosphere. The modern underground unit is
'dry', has a Class C insulant, and is enclosed in a flame-proof casing, the
surface temperature of which must not exceed 60 °C.

Small Transformers

Industrial Frequency. The open air-cooled construction has given place to cast-resin or hermetically sealed units in gas- or oil-filled containers. Core-plates are fitted into pre-wound coils. The *toroidal* core has magnetic advantages for 1-ph units, as it has no gaps and with continuous c.r.o.s. strip all the flux is in the preferred grain direction. Methods of winding toroidal cores are well established with either superposed or separated primary and secondary windings.

High Frequency. For mobile equipments and frequencies in the range 0.4–1.6 kHz, thin c.r.o.s. or nickel-iron core plates may be used with windings insulated by high-temperature resins. A fully sealed 30 VA 1.6 kHz unit may have a mass as low as 0.1 kg. Moulded ferrite cores with an E-shaped limb-yoke closed by an I-shaped top yoke have also been employed.

High-frequency Low-mass. Transformers rated up to 200 kVA 400 Hz may be required in large aircraft, where the non-payload mass must be limited. Designers employ high flux and current densities, and concentrate on heat removal by forced oil cooling. With a winding temperature of 150 °C, a power/mass ratio of 9 kVA/kg may be attained for a 400 Hz operating frequency, but with some sacrifice in losses and voltage regulation.

Transformers for Climatic Extremes

Cold. In such lands as Alaska and Siberia the winter ambient temperature may reach −60 °C, affecting metals and oils. Mild steel is subject to brittle fracture, so that tanks must be made of aluminium or low-alloy steel. Oils become highly viscous, inhibiting circulation and leading to maloperation of conservators and tap-changers. Protection demands a low-temperature lockout device, preheating of oil, and modification of the switches in Buchholz relays. Other problems concern freezing in oil/water heat-exchangers, oil-pump ratings and the possible deterioration of solid insulation.

Hot. Shade temperatures approaching +60 °C have been recorded in California and the Sahara. Some effects have been noted in Sect. 4.5. As there is a diurnal variation, the mean ambient temperature may often be taken as about 40 °C, but transformers must usually be de-rated. Tank paints must withstand humidity, rain and sandstorms, and contain a fungicide. Prevalence of lightning makes necessary the provision of surge diverters.

Superconducting Windings

Wilkinson [27], in a study of the possibility of operating transformers under cryogenic conditions, assumed that a superconducting metal film could be found capable of working at the required current density while immersed in the leakage magnetic field. He concluded that the core must be excluded from the refrigerated region, for apart from the unknown performance of magnetic steels at very low temperatures, the dissipation of the core loss might require a power several times greater than that extracted. Super-conducting windings would have to conduct in the presence of a leakage field of the order of 3.5 T and with very high linear current density. In return there would be a prospective saving of the entire I^2R loss. In a 570 MVA design with a 50 kW core-loss reduction effected by lowering the core mass from 100 to 65 tonne, the conventional I^2R loss of 3 MW would be eliminated. Against this must be set a refrigeration power of about 0.5 MW, leaving a net loss of about 2.7 MW on full load. Development and refrigeration-plant costs might absorb the capital equivalent of the saving in transformer loss. Wilkinson concluded that such a project is not yet economic even were the necessary materials available. However, *supercooled* forms are possible, using tank techniques developed for holding liquefied gases.

12.2. Solid-state control

The rate of technical progress in power electronics is such that new devices and more sophisticated control methods — such as the gate-turn-off thyristor which simplifies commutation — make a frequent appearance. Some details of semiconductor devices, further to those of Sect. 8.16, are given below.

Rectifier and Inverter

Operating modes are described for idealized conditions and without specific details of auxiliary commutation circuitry, for 3-ph bridge rectifiers (a.c./d.c.) and inverters (d.c./a.c.).

Consider an ideal sinusoidal 3-ph supply of fixed angular frequency ω, line terminals ABC, and line-to-neutral voltages of instantaneous value.

$$v_a = v_m \sin\omega t \qquad v_b = v_m \sin(\omega t - 2\pi/3) \qquad v_c = v_m \sin(\omega t - 4\pi/3)$$

applied to diode and thyristor bridges, Fig. 12.5.

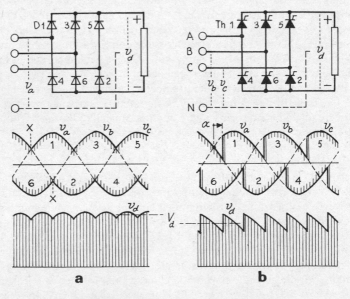

12.5 Diode and thyristor rectification.

Diode Bridge Rectifier. In (*a*), diodes D1, D3 and D5 have in common the positive rail of the load circuit, but at any instant only that diode backed by the most positive phase voltage will conduct, the other two being automatically reverse-biased. Similarly, of the diodes D2, D4 and D6 connected in common to the negative rail, only that with the most negative phase voltage can conduct. Natural or line commutation between successively conducting diodes takes place at the 'crossover' instants X where successive phase voltages are momentarily equal: the phase with a reducing voltage quenches, while that with an increasing voltage takes over the conduction. The voltage to neutral of the positive rail is therefore a succession of sinewave tops, and similarly (but displaced) for the negative rail. The resultant output voltage v_d is the intercept between the upper and lower waveforms, yielding the output voltage (shown hatched) of *mean* value V_d given by

$$V_d = (3\sqrt{3}/\pi)v_m = 1.65\ v_m = 2.34\ V_1$$

where $V_1 = v_m/\sqrt{2}$ is the r.m.s. phase voltage. On V_d is superimposed a ripple of frequency corresponding to $q = 6$ pulses per period. If the load current is assumed to be inductively smoothed to a constant value I_d, the phase currents are 'rectangular blocks' of duration one-third of a period, of magnitude I_d and centred about their respective phase-voltage peaks. The *fundamental* component of a phase current is thus cophasal with its phase voltage, so that it has a phase angle of zero. But the harmonic component currents result in the power/volt-ampere ratio being less than unity.

Thyristor Bridge Rectifier. If, as in Fig. 12.5(*b*), the diodes are replaced by thyristors triggered to conduct at the crossover points X, the conditions are the same as with the diode bridge in (*a*). But it is now possible to apply phase control of the rectifed voltage by imposing a time-delay angle a (measured from X). That thyristor connected to the phase having the highest instantaneous forward voltage will not conduct at X but will be inhibited until it receives a gate pulse. The mean output voltage is now

$$V_d = (3\sqrt{3}/\pi)\ v_m \cos a = (3\sqrt{6}/\pi)\ V_1\ \cos a$$

The output waveform has a greater harmonic content than for the diode bridge, but its mean level is adjustable. The phase current contains a fundamental displaced by a from the phase voltage, so giving an a.c. input power factor $\cos a$, besides harmonic distortion. Thus delayed firing controls the output voltage level with the disadvantage of a demand for a lagging reactive power input.

Thyristor Bridge Inverter. The arrangement of a basic form of thyristor bridge inverter is given in Fig. 8.46.

D.C.-link Converter

The term 'converter' refers to the combination of input rectifier and controlled-frequency inverter. Two basic forms are shown in Fig. 12.6. The *voltage-source* converter (a) has a thyristor rectifier, so controlled that the

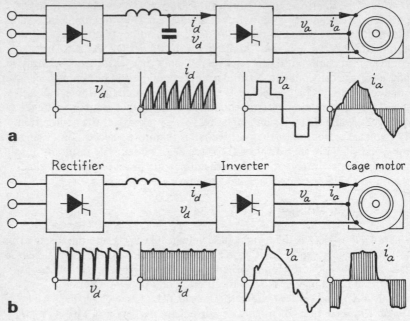

a

b

12.6 Voltage- and current-source d.c. link converters.

link voltage is an approximately linear function of the inverter switching frequency and so also of the motor speed. The motor voltage waveform is stepped, but gives rise to a roughly sinusoidal motor phase current. The link has a smoothing capacitor-inductor combination to provide a stable link voltage. The waveforms of the link voltage v_d, and current i_d, and of the motor voltage v_a and current i_a, are indicated. The *current-source* converter (*b*) has a rectifier controlled to give a nearly constant current i_d through a smoothing inductor. The inverter chops the current and allocates it to the motor phase windings at the command frequency. The rectifier is controlled with respect to the motor load torque, and the inverter frequency is related to the motor terminal voltage to maintain a proper voltage/frequency ratio. Some details of particular forms of the d.c.-link converter are given below, together with notes on other methods that exploit the versatility of semi-conductor devices.

Six-step Voltage-source Converter, Fig. 12.7(*a*). The 3-ph primary supply feeds a phase-controlled rectifier which delivers a filtered voltage to the d.c. link, whence an adjustable-frequency motor supply is developed by the inverter. The link voltage and the output frequency are controlled together by interrelated firing to achieve a constant motor airgap flux. Inverter thyristors Th1-Th6 are triggered at 60° intervals of the output frequency demanded, and each conducts for 180°. Forced-commutation circuits (not shown) turn each thyristor off at the end of its conduction period. Diodes

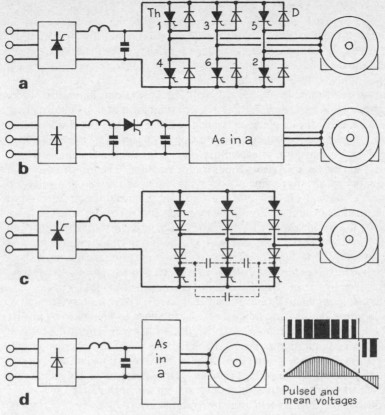

12.7 D.C.-link converters.

D1-D6, connected in inverse parallel, permit the flow of inductive load or regenerated current. When the inverter feeds a motor with star connection, the phase voltages have the six-step waveform v_a in Fig. 12.6(a), the harmonics of which have a magnitude inversely proportional to their order. The 5th and 7th harmonics are responsible for torque fluctuations at six times the fundamental output frequency, and at minimum frequency (e.g., 5 Hz) it may be desirable to apply feedback stabilization. When the motor is run into negative slip by an overhauling load, it regenerates back to the link, but energy can be returned to the primary supply only if the input rectifier can invert. The converter is efficient, reliable, and adaptable to standard motors operated at frequencies up to about 300 Hz. Although the large link capacitor limits the response rate and the rectifier phase control lowers the primary power factor, the system is in wide use for single- and multi-motor applications with or without closed-loop control.

Voltage-controlled Chopper Converter, Fig. 12.7(b). Here the input rectifier is a diode bridge, which provides the link with an approximately constant direct voltage through a filter, so avoiding the power-factor reduction

inherent in phase control. Voltage control of the inverter input is by a
chopper thyristor in the link, turned on and off to develop 'blocks' of
voltage of fixed magnitude but variable duration. After low-pass filtering, a
smoothed and adjustable direct voltage is passed to the inverter, from which a
quasi-square voltage output is delivered to the motor. In emergency the link
can be supplied from a standby battery.

Current-source Converter, Fig. 12.7(*c*). The overcurrent limitation needed by
a thyristor suggests the adoption of a constant-current d.c. link obtained by
feedback control of a phase-controlled rectifier with inductor smoothing. The
inverter has auto-sequential commutation by means of a diode in each arm
and a delta connection of capacitors (shown for the lower arms by dotted
lines), so that each thyristor conducts for $2\pi/3$ rad. The motor takes phase
currents that are approximately constant in magnitude, but at a voltage that
depends on the load: for a motor on no load the inverter output voltage is
near zero. Regeneration is possible by phase control of the input rectifier.
The basic system requires 12 thyristors, of modest turn-off capability but
able to withstand transient voltage spikes resulting from the current
commutation. The large link inductor gives some protection against short
circuit. The upper limit of output frequency (normally about 150 Hz) is
influenced by the motor leakage inductance, which should be low.

Pulse-width-modulated Converter, Fig. 12.7(*d*). The p.w.m. system has a
diode bridge rectifier providing a constant voltage to the link. The inverter
is basically the same as in (*a*), but requires a further six thyristors for the
complex commutation sequence. Each half-period of the inverter output
waveform comprises a series of voltage pulses of equal magnitude but of
duration such that the voltage-time integral builds up a roughly sinusoidal
waveform with a moderate harmonic content. As both voltage and
frequency controls are performed within the inverter, the system responds
rapidly to commanded changes. The technique is sophisticated and imposes
high dynamic voltage stresses on the thyristors; and as the generation of
high output frequencies demands very fast switching rates, the p.w.m.
converter is better adapted to low motor speeds unless provision is made for
automatic modification to six-step operation at the higher ranges of speed.
The converter is suitable for high-performance single- and multi-motor
drives, with regenerative and battery-standby facilities. Motors should have
a high leakage reactance to limit the effect of high-frequency switching
harmonics. Control reliability can be improved by the use of micro-
processors. For motors of up to 10 kW rating, thyristors can be replaced by
power transistors, their good switching capability permitting higher output
frequencies and simpler control.

Sinewave Synthesizer
This is a power conversion technique in which power transistors are used to
switch a d.c. source to generate a synthetic 1-ph waveform of variable voltage
and frequency. The process is basically digital. Using optimized algorithms,
digital synthesis is performed by microprocessor control, selecting voltage

steps of appropriate magnitude to give a 'synthetic' simulation of a sinusoidal output over a range of frequencies and with a high conversion efficiency.

Cycloconverter
The intergroup inductors shown in Fig. 8.44 prevent short circuits at various instants in the firing cycle. Hamad *et al* [124] describe a drive in which the split stator winding of the induction motor replaces these inductors, simplifying the converter control.

12.3 Induction motors: operating conditions
Sophisticated applications of induction motors, particularly those concerned with speed control, involve supply characteristics that differ from the 'normal' constant voltage and frequency, and modify the behaviour of the machine. Further, solid-state inverters introduce voltage and current harmonics that may result in fluctuations of the torque.

Supply Characteristics
The discussion is based on steady-state conditions with voltages and currents assumed to be *sinusoidal*, circuit parameters of resistance and inductance to be constant. A simplified equivalent circuit is employed, with the shunt magnetizing branch purely inductive and the complex factor c in eq.(8.10) taken as unity. Referred rotor quantities are written without primes.
Variable Voltage, Constant Frequency. This supply condition applies to

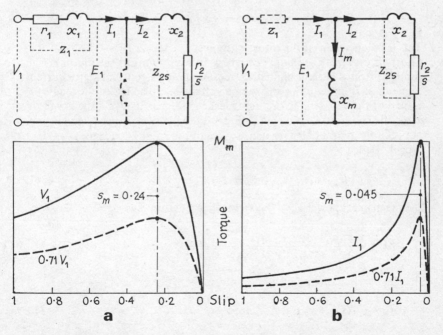

12.8 Induction motor: voltage- and current-controlled operation.

motors controlled by stator voltage reduction. The simplified equivalent circuit for a frequency f_1 is given in Fig. 12.8(a). Writing eq.(8.11) with $c = 1$ gives for the rotor current

$$I_2 = \frac{E_1}{z_{2s}} \cong \frac{V_1}{(r_1 + r_2/s) + j(x_1 + x_2)} \tag{12.1}$$

whence the torque per phase for a $2p$-pole machine with a synchronous speed $\omega_s = 2\pi f_1/p$ is

$$M = \frac{V_1{}^2}{\omega_s} \cdot \frac{r_2/s}{(r_1 + r_2/s)^2 + (x_1 + x_2)^2} \tag{12.2}$$

The stator and rotor currents, the gap e.m.f. E_1 and the torque are related to the slip as indicated typically in Fig. 8.11. For a given slip s the torque is proportional to $V_1{}^2$. The torque/slip curves in Fig. 12.8(a) are for voltages of V_1 and $0.71V_1$ in a motor with $r_1 = r_2$, $x_1 = x_2$ and $r_1/x_1 = r_2/x_2 = 0.50$ at the operating frequency f_1. The slip for maximum torque is

$$s_m = r_2/\sqrt{[r_1{}^2 + (x_1 + x_2)^2]} \tag{12.3}$$

The maximum torque is therefore

$$M_m = \frac{1}{2} \frac{V_1{}^2}{\omega_s} \cdot \frac{1}{r_1 + \sqrt{[r_1{}^2 + (x_1 + x_2)^2]}} \tag{12.4}$$

This is independent of the rotor resistance.

Constant Current, Constant Frequency. In some controls the motor is supplied from a current-limited source. Assuming a constant stator current I_1 then, apart from requiring the source voltage V_1 to adjust to the load by reason of the volt drop in the stator leakage impedance $I_1 z_1$, the effect of z_1 is nil. The equivalent circuit in Fig. 12.8(b) indicates that I_1 divides between the magnetizing reactance x_m and the rotor impedance $z_{2s} = [(r_2/s) + jx_2]$. Hence

$$I_2 = E_1/z_{2s} = I_1 \cdot jx_m/(jx_m + z_{2s}) \tag{12.5}$$

The mechanical power is $I_2{}^2(r_2/s)$, whence the torque is

$$M = \frac{I_1{}^2 x_m{}^2}{\omega_s} \cdot \frac{r_2/s}{(r_2/s)^2 + (x_m + x_2)^2} \tag{12.6}$$

The slip for maximum torque is found to be

$$s_m = r_2/(x_m + x_2) \tag{12.7}$$

which is much smaller than for the voltage-limited condition, as is shown in the torque/slip curves for currents I_1 and $0.71\,I_1$. The motor has the same leakage reactances as in (a), together with $x_m = 10x_1 = 10x_2$. The maximum torque per phase is

$$M_m = \tfrac{1}{2}\frac{I_1{}^2}{\omega_s} \cdot \frac{x_m{}^2}{x_m + x_2} \simeq \tfrac{1}{2}\frac{I_1{}^2\,x_m}{\omega_s} \qquad (12.8)$$

proportional to $I_1{}^2$ and again independent of the rotor resistance. The approximation is valid if x_2 is small compared with x_m; if this is so, then I_2 and I_m have for the maximum torque the same magnitude $I_1/\sqrt{2}$, the former leading and the latter lagging I_1 by 45 °. In practice, however, I_m is usually such as to saturate the magnetic circuit, which affects x_m. Comparing eq.(12.2) for voltage-source with eq.(12.6) for current-source operation at standstill ($s = 1$) it is seen that the starting torques are roughly proportional to

$$r_2/(x_1 + x_2)^2 \text{ and } r_2/(x_m + x_2)^2$$

and as x_m is normally much greater than x_2, the constant-current starting torque is likely to be low.

Variable Voltage and Frequency. Rated torque can, provided that the low-speed ventilation is adequate, be developed at all speeds when a predetermined normal gap-flux density is preserved by adjusting the supply voltage to suit the operating frequency. Consider the gross torque of a motor supplied (i) at a reference angular frequency ω_0, the gap e.m.f. being E_0, the slip s_0, and the rotor leakage impedance comprising the resistance r and the reactance x_0; and (ii) at a different frequency ω, slip s, gap e.m.f. E, rotor resistance r (assumed constant) and reactance $x = x_0(\omega/\omega_0)$. From eq.(8.5) the gross torques per phase are

$$\text{(i) } \frac{E_0{}^2}{\omega_0{}^2} \cdot \frac{\omega_0 s_0 r}{r^2 + (s_0 x_0)^2} \qquad \text{(ii) } \frac{E^2}{\omega^2} \cdot \frac{\omega s r}{r^2 + (s x)^2}$$

With $E/\omega = E_0/\omega_0$ the gap flux is the same in each case, and the torques are proportional to

$$\text{(i) } \frac{\omega_0 s_0 r}{r^2 + (s_0 x_0)^2} \qquad\qquad \text{(ii) } \frac{\omega s r}{r^2 + [s x_0(\omega/\omega_0)]^2}$$

Then for a slip such that $s_0\omega_0$ the rotor torque and current are the same for any operating frequency; and in relation to the individual synchronous speeds all the torque curves have the same basic shape, as indicated in Fig. 8.49: the same torque and current occur, for example, with a slip of 0.05 at an operating frequency of 50 Hz, with 0.025 at 100 Hz and with 0.10 at 25 Hz. In a practical machine, however, the useful torque differs by reason of the rise of effective resistances due to skin effect (for the stator at high operating frequencies, for the rotor at low speeds), the speed- and frequency-dependence of the core and friction losses, and the modification of the leakage and magnetizing reactances with the saturation level.

EXAMPLE 12.1: A 3-ph 4-pole motor is required to develop per phase a rated torque of $M = 100$ N-m and a pull-out torque $M_m = 200$ N-m. The equivalent-circuit parameters are $r_1 = r_2 = r$, $x_1 = x_2 = x$ and magnetizing reactance x_m. For a sinusoidal supply at a frequency $f_0 = 50$ Hz the equivalent-circuit ratios are $r/x_{m_0} = 0.05$, $r/x_0 = 0.50$ and $x_0/x_{m_0} = 0.10$. Deduce the gross-torque/speed relations for frequencies 150, 100, 50 and 5 Hz and sinusoidal supply conditions to give (1) constant airgap flux, and (2) constant stator current.
General expressions for the torque are first obtained.
(1) *Constant Gap Flux.* From eq. (12.6) with $I_2 = E_1/z_{2s}$, dividing numerator and denominator by x_m^2 and replacing ω_s by $2\pi f/p = \pi f$, then

$$M_1 = \frac{E_1^2 r}{\pi x_m^2} \cdot \frac{1}{sf} \cdot \frac{1}{(r/sx_m)^2 + (x/x_m)^2}$$

(2) *Constant Stator Current.* From eq. (12.6),

$$M_2 = \frac{I_1^2 r}{\pi} \cdot \frac{1}{sf} \cdot \frac{1}{(r/sx_m)^2 + (1 + x/x_m)^2}$$

The slips for maximum torque from eq. (8.5) and (12.7), evaluated for frequency $f = f_0 = 50$ Hz and the given circuit ratios, are

(1) $s_{m_0} = r/x_0 = 0.50$ (2) $s_{m_0} = r/(x_0 + x_{m_0}) = 0.045$

and for $M_{m1} = M_{m2} = 200$ N-m,

(1) $E_1^2 r/\pi x_{m_0}^2 = 100$ (2) $I_1^2 r/\pi = 1090$

For any other frequency $f = f_0/k$, the circuit reactances become $x = x_0/k$ and $x_m = x_{m_0}/k$. Hence $r/x = kr/x_0 = 0.5k$, $r/x_m = kr/x_{m_0} = 0.05k$ and $x/x_m = x_0/x_{m_0} = 0.1$. The torques for any frequency f are therefore given by

(1) $M_1 = 2.0\dfrac{k}{s} \cdot \dfrac{1}{(k/20s)^2 + 0.01}$ (2) $M_2 = 11\dfrac{k}{s} \cdot \dfrac{1}{(k/20s)^2 + 1.21}$

The factor 2.0 in M_1 applies to all frequencies because E_1 and x_m are both proportional to the frequency.
The torque/speed curves are drawn in Fig. 12.9. Rated torque at 50 Hz is obtained with a slip of 0.14 in (1) and 0.012 in (2). With constant gap flux (1) and a 5 Hz supply, the torque at starting is somewhat below the rated value.

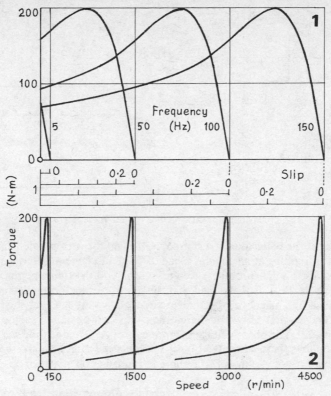

12.9 Example 12.1 : torque/speed relations
(1) Constant gap flux; (2) Constant stator current.

Non-sinusoidal Supply

In Sect. 8.11 the effect of *space* harmonics in the gap flux were considered, the supply voltage being assumed sinusoidal and symmetrical. A non-sinusoidal supply waveform impresses on the machine a *time*-dependent harmonic series of voltages: this, too, modifies the performance. The application of converters for speed control makes it necessary to examine the effects on the behaviour of a machine fed with voltages of complex waveform.

Let an induction motor be fed at a basic (fundamental) angular frequency ω from an inverter of direct link voltage V_d giving, for a phase A of a star connection of phases ABC, the six-step voltage v_a in Fig. 12.10(a). Fourier analysis of this idealized waveform gives a fundamental of frequency ω and amplitude $(2/\pi)V_d$, and an infinite series of harmonics of orders $n = 5, 7, 11, 13 \ldots (6y \pm 1)$ where y is an integer, and of amplitude $1/n$ of the fundamental. Thus

$$v_a = V_d(2/\pi)[\sin\omega t + \tfrac{1}{5}\sin5\omega t + \tfrac{1}{7}\sin7\omega t + \ldots]$$

Phases B and C have voltages given by substituting $(\omega t - 2\pi/3)$ and $(\omega t - 4\pi/3)$ respectively in the expression for v_a.

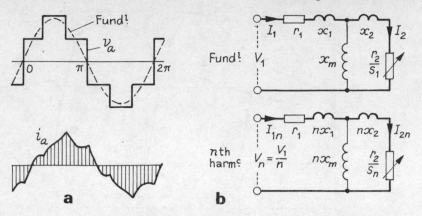

12.10 Motor operation on stepped voltage waveform.

Consider the behaviour of a star-connected motor with p pole-pairs and a cage rotor, running at an angular speed ω_r. The fundamental frequency corresponds to a synchronous speed $\omega_s = \omega/p$, and the fundamental slip is $s_1 = (\omega_s - \omega_r)/\omega_s$. It is assumed that the motor parameters (resistances, and leakage and magnetizing inductances) are constant and that the effects of the fundamental and of the harmonic components can be superposed.

Fundamental. The phase-voltage components of A, B and C produce corresponding phase currents of positive sequence (p.p.s.) which combine to develop a wave of gap m.m.f. travelling at fundamental synchronous speed ω_s. The slip is s_1.

Harmonic Order $n = (6y + 1)$. These develop corresponding phase currents and a p.p.s. travelling m.m.f. wave. The number of pole-pairs of the m.m.f. is still p because this condition is settled by the stator winding; but the m.m.f. wave moves at n times fundamental angular speed, and in the same p.p.s. direction. The slip is therefore $s_n = (n\omega_s - \omega_r)/n\omega_s$.

Harmonic Order $n = (6y - 1)$ These form an n.p.s. set of voltages, and corresponding currents developing an m.m.f. wave of p pole-pairs travelling at n times ω_s but in the opposite direction. The corresponding slip is $s_n = (n\omega_s + \omega_r)/n\omega_s$. Writing $\omega_r = \omega_s(1 - s_1)$, then the slip for a harmonic of order n becomes $s_n = [(n\mp1) \pm s_1]/n$. This is never very different from unity.

Equivalent circuits for fundamental and harmonic components are shown in Fig. 12.10(b), simplified by the omission of the shunt core-loss resistance. For a harmonic, the combined leakage reactance $n(x_1 + x_2)$ is large compared with the (assumed constant) resistance $(r_1 + r_2)$, and small compared with the magnetizing reactance nx_m, so that the magnitude of a harmonic current is determined mainly by the combined leakage reactance. The waveform of a typical phase current i_a is shown in (a).

Steady Harmonic Torques. Let the motor run on constant load and fundamental supply frequency. A given gap-flux component induces a corresponding rotor current component and interacts with it to develop a steady constant torque. The fundamental and harmonic torques are respectively

$$M_1 = I_2{}^2 r_2 / s_1 \omega_s \quad \text{and} \quad M_n = I_{2n}{}^2 r_2 / s_n \omega_s$$

Taking s_n as approximating to unity, the ratio of harmonic to fundamental torque is $M_n M_1 = (I_{2n}/I_2)^2 (s_1/n)$. At standstill ($s_1 = 1$), with the harmonic voltage only $1/n$ of the fundamental and with the leakage reactance n times as large, then $M_n/M_1 \cong 1/n^3$. This is -0.008 for the n.p.s. 5th and $+0.003$ for the p.p.s. 7th harmonic. The steady harmonic torques are thus negligible.
Pulsating Harmonic Torques. Interaction of a gap flux component with the rotor current induced by a flux of *different* harmonic order will produce a torque fluctuation. The effect is most prominent with 5th and 7th harmonic currents and the fundamental of the gap flux. In each case a torque fluctuation of six times fundamental frequency is developed, leading to a risk of vibration when the motor runs on a low-frequency supply.
Losses. Supply harmonics produce additional losses. Harmonic effects on core and stray losses vary with the saturation level and are not readily predictable. The 'normal' full-load $I^2 R$ loss is augmented by a factor typically between 1.04 and 1.2 depending on the leakage reactance and the skin effect in conductors at harmonic frequencies. As the harmonic $I^2 R$ loss is roughly independent of the load, its effect is more prominent in the no-load input power.

Winding Asymmetry

Stator winding asymmetry is sometimes unavoidable. For example, a change-pole winding in 48 stator slots can be made symmetrical for 4 poles, but reconnection for 6 poles introduces geometrical asymmetry, and m.m.f. unbalance is inevitable. With slip-ring rotors, asymmetric phase impedance may be produced by the external connections, or by some of the possible connections (Fig. 10.44) for d.c. excitation in the synchronous-induction motor. The problem is also raised by the use of thyristor chopping applied to the slip-rings in close-range speed control.

The phenomenon of rotor asymmetry was first noted by Goerges in 1896. He found that a polyphase induction motor, with a rotor winding connected to give effectively a single phase, could operate stably at two speeds, one near the synchronous and the other near one-half of that speed. The half-speed property of the 'Goerges effect' was explained by Lamme in 1915, using qualitatively the concept of two component travelling-wave fields in opposite directions to represent the effect of the 1-ph rotor winding.

When a balanced 3-ph voltage of positive phase sequence (p.p.s.) is applied to the *symmetrical* stator winding of an induction motor, the p.p.s. currents set up a travelling wave of m.m.f. moving with respect to the stator at synchronous speed ω_1 and at slip speed $s\omega_1 = \omega_1 - \omega_r$ with respect to the rotor (running at angular speed ω_r). The stator p.p.s. field induces e.m.f.s and currents of slip frequency $s\omega_1$ in the *asymmetrical* rotor winding. The unbalanced rotor m.m.f. can be represented by a combination of balanced p.p.s. and n.p.s. components. The p.p.s. rotor component moves at slip speed $s\omega_1$ relative to the rotor, and therefore at synchronous speed $\omega_1 = s\omega_1 + \omega_r$ relative to the stator; and a steady positive torque M_+ is developed by inter-action. The n.p.s. rotor m.m.f. moves at $-s\omega_1$ relative to the rotor, and therefore at $(-s\omega_1 + \omega_r) = [-s + (1 - s)]\omega_1 = (1 - 2s)\omega_1$ relative to the stator, inducing currents of this frequency in the stator windings and developing an n.p.s. interaction torque M_-. At small slips M_+ and M_- oppose; for $s = 0.5$ the n.p.s. torque vanishes; and for larger slips the p.p.s. and n.p.s. torques are additive. The resultant torque/slip relation therefore exhibits a dip near $\omega_r = \frac{1}{2}\omega_1$, and if a high degree of rotor asymmetry (such as a single rotor phase) is present, the machine may run stably near half synchronous speed as well as at 'normal' slips. A further outcome arises from the interaction of p.p.s. gap flux with n.p.s. currents, which results in the production of parasitic torques pulsating at the relative slip frequency of the components concerned. These have a time-average of zero, but can produce vibration except at one-half synchronous speed, for in this case the rotor n.p.s. component is stationary with respect to the stator and can induce no current therein. The p.p.s. and n.p.s. fluxes co-exist in the gap, going into and out of coincidence cyclically, causing high local saturation levels and (if half speed is to be a running condition) calling for voltage limitation or the choice of a larger frame size.

Analysis of machines with winding asymmetry is intricate. Many authors [122] have dealt with rotor asymmetry, but the case of unbalance in both stator and rotor has been given a generalized treatment on a rotating-field basis by Jha and Murthy [123].

Doubly-fed Motor

A polyphase induction motor with balanced windings on both stator and rotor can operate as a *synchronous* machine at a specified speed if each member is connected to a symmetrical polyphase supply. It can retain this speed over a range of load torque and develop a synchronizing torque. Such a machine could be used as a high-speed motor, or as a frequency-error detector with one member connected to a supply of reference frequency, the other to the source whose frequency deviation is to be detected.

Consider a 2-pole machine with stator and rotor fed respectively from supplies of angular frequency ω_1 and ω_2. Each member produces a travelling wave of m.m.f. around the gap with a direction corresponding to the phase sequence. The interaction torque pulsates at all rotor speeds ω_r except that for which $\omega_r + \omega_2 = \omega_1$, for only when the stator and rotor m.m.f. patterns move in synchronism can a sustained unidirectional torque be established, a condition given by $\omega_r = \omega_1 \pm \omega_2$, the sign depending on the relative stator and rotor phase sequences. If these are opposite, and if the two members are fed at the same supply frequency, the synchronous speed is $\omega_r = 2\omega_1$; thus a 2-pole 50 Hz motor runs stably at 6000 r/min. The excitation of the gap flux is shared by the stator and rotor m.m.f.s just as it is in the normal conventional synchronous machine in which $\omega_2 = 0$ (i.e., a d.c. supply) and $\omega_r = \omega_1$.

The maximum torque of an a.c. doubly-fed motor is always greater than that of a comparable singly-fed induction machine, and with its high speed represents a very compact power converter. But it suffers from (i) an inherent lack of starting torque (so requiring a pony motor for run-up to operating speed), and (ii) the presence of damping torques (sensitive to the leakage impedance parameters) that may be negative, tending therefore to amplify momentary divergencies from synchronism. Several authors [144] have dealt with the general analysis of the machine in the steady state, and with the origin and amelioration of instability.

12.4 Induction machines: types and applications

A brief survey of the operating characteristics of polyphase induction motors is given in Sect. 8.9. In the following, some general details are discussed and typical applications described.

Energy Economy

It has been estimated that the summated ratings of the twenty million motors in the U.K. approaches 100 GW, made up largely of induction machines rated below 150 kW and of *average* rating less than 5 kW. The choice of machines on the bases of efficiency and power factor is therefore significant in the context of energy conservation. The typical steady-state performance characteristics in Fig. 8.18 show that below 0.7 p.u. loading both efficiency and power factor fall, the latter more than the former. Customers sometimes call for 'oversize' machines on the valid grounds that they will run cooler, have a longer life, and can cope with unpredicted increase in loading: the result could, however, be that the machines rarely operate near optimum load. *Efficiency.* The two areas of loss are the motor itself and the cable and transformer plant concerned. Both are reduced by raising the motor efficiency, i.e. by adoption of more conservative electric and magnetic loadings. This means putting more metal into the machine, making it dimensionally larger and more costly. The economic motivation is the rising price of electrical energy, for the additional cost may well be recovered in a matter of months. But the choice of design is not simple, for economic claims can also be made for the use of high-temperature insulation to give a machine with *less* conductor material. The substitution, for example, of Class F (155 °C) for Class B (130 °C) insulation may give 10% more output from a given frame.

Variable-speed drives can economize energy in industrial process applications. About one-third of installed motor capacity in such industries is used to drive pumps. Pump output control by throttling is wasteful, whereas an efficient speed control can permit the drive to be adjusted to the pump demand. Similarly, the growing demand for heat-pump speed adjustment to follow the load is not economically satisfied by simple on/off switching. In many drives a precision control of speed becomes essential, as in machine-tool and paper-mill applications. Energy economy may outweigh the high capital cost of speed-control equipment.

Power Factor. A motor with a full-load p.f. of over 0.9 may have a no-load p.f. as low as 0.1, taking a substantial magnetizing current and reactive power, and perhaps incurring a tariff p.f. penalty. Some amelioration of the low-load running condition can be had by operating a delta-connected motor in star. More sophisticated methods have been developed, in which the start of each half-period of the input voltage is delayed in accordance with the load level; the p.f. and the efficiency are then considerably raised, together with a reduction in slip over the whole load range.

Single-speed Industrial Motors

The performance characteristics of an induction motor — full-load current, efficiency and slip, pull-out torque, starting torque and current — cannot be chosen independently. Nor is efficiency the sole criterion, for a high-efficiency motor may have insufficient margin to meet the inevitable departures from the specified loading conditions. One loss component, the rotor $I^2 R$ or 'slip' loss, is unique in that its minimum value is fixed almost entirely by the specified starting torque and current, for which typical values in per-unit of full-load ratings for a 50 Hz cage motor are:

Full-load rating (kW)	5	20	50	250	500	1000
Starting current (p.u.)	6–8	6–7	6–6.5	5–6	4.5–6	4.5–6
torque (p.u.)	2–2.5	1.5–2	1.2–1.6	1–1.2	1.0	0.75–1

Apart from small machines, the equivalent circuit per phase of a typical induction motor with a cage rotor approximates to that in Fig. 12.11 for p pole-pairs and a 50 Hz supply. This serves in the absence of specific parameters for a preliminary assessment of the performance. Further data can be incorporated. For example, if the starting current is to be limited to 4 p.u., the total per-unit leakage reactance $(x_1 + x_2')$ must be raised from 0.16 to 0.25; if the full-load slip is to be 0.01, then r_2' must be reduced to 0.01 p.u.; and for a higher efficiency the stator resistance must also be reduced. Consideration must often be given to the performance of the motor when started on a line voltage lower than normal, e.g. 0.8 p.u.

Transients. Switching and re-switching transients have been discussed in Sect. 8.21. The *peak instantaneous* starting current of a motor switched direct-on-line with the rotor at rest may be as much as 16 times the r.m.s.

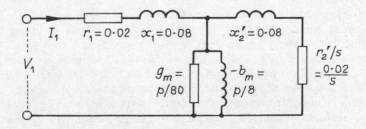

12.11 Typical per-unit equivalent circuit.

rated current, but there is not likely to be more than one such peak. When a motor running near synchronous speed is connected to the supply, the initial current transient may be as great as with the rotor at rest. The current decays in accordance with a time-constant given approximately by $(L_1 + L_2)/(r_1 + r_2')$ where L is the leakage inductance. Very high torques, of the order of 8 p.u., may occur when a motor is re-switched, the airgap flux thereafter decaying towards normal with a time-constant L_m/r_2', where L_m is the appropriate magnetizing inductance.

Current-displacement Rotors: The rotor bars of all large cage motors have profiles to exploit current-displacement effects, which markedly enhance the starting and run-up torques.

EXAMPLE 12.2: Preliminary design estimates for the minimum practicable full-load per-unit losses of an 11 kV 50 Hz 6-pole cage motor are: core, 0.0069; stator I^2R, 0.0048; rotor I^2R, 0.0054; friction and fan, 0.0054; load (stray), 0.005. The motor is to start and accelerate a pump up to a full-load slip of 0.01 with a terminal voltage not less than 8.8 kV (0.8 p.u.) and a starting current not more than 4.0 p.u. Figure 12.12 shows the torque/speed characteristic of the pump, and that of the motor at 8.8 kV for three copper rotor-bar sections A, B and C, all with the same total cross-section and full-load I^2R loss. Examine the feasibility of rotor bars for this duty.

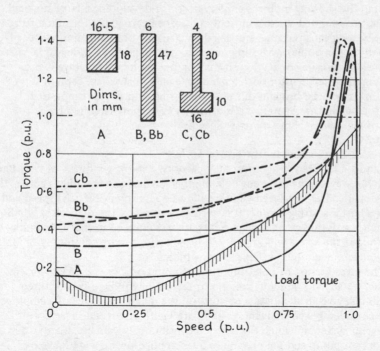

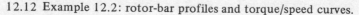

12.12 Example 12.2: rotor-bar profiles and torque/speed curves.

The full-load slip of 0.01 indicates a high-efficiency motor. The loss summation is 0.0275 p.u., i.e. a full-load efficiency of 0.973.

A,B: neither the rectangular nor the deep-bar rotor meets the requirements.

C: the T-bar just meets the torque demand, but at slips around 0.1 has insufficient margin for acceleration, unless the starting current is raised by reducing the stator leakage reactance.

Bb, Cb: if the chopper bars are replaced by bronze of like dimensions a satisfactory run-up is achieved, but the increase in rotor $I^2 R$ loss lowers the full-load efficiency (to about 0.96) and the full-load slip exceeds 0.01 p.u.

Thus the design cannot meet both efficiency and torque specifications simultaneously.

Double-cage Rotors. For a given starting current and full-load efficiency, high accelerating torque implies reduced pull-out torque and power factor, but the designer can manipulate the shape of the torque/speed relation. Most of the rotor starting loss occurs in the outer cage, and more conductor material is required than for the single-cage machine.

Slip-ring Motors. The cage motor is simple, reliable and robust. The slip-ring machine has collector rings and brushes that require detailed design, involve periodic maintenance and increase the cost, but in return a degree of control over the performance characteristics is made possible by external means. In general, the slip-ring motor is employed (i) for developing a large starting torque with limited starting current and rotor heating; (ii) for minimizing rotor heating on a repeated-starting duty; (iii) for obtaining limited speed reduction, sometimes with slip-power recovery; and (iv) for 'loss-free' speed control by cascade connection. It may on occasion be advantageous to combine in a single rotor both a main cage and an auxiliary 2-ph or 3-ph winding, the latter brought out not to slip-rings but to shaft-mounted resistors, which are operative at starting but are short-circuited by a centrifugal switch above a pre-set speed.

Two- and Three-speed Industrial Motors

An adjustable-speed cage motor may carry two stator windings, of the same pitch but developing different pole numbers; when one winding is active, the other is idle. The low-speed winding allows operation at reduced speed and gives a better starting torque. It is necessary to select the two stator windings in such a way that circulating currents are not induced in the idle winding by the flux of the active winding. A 2/1 speed ratio is more commonly obtained by the reconnection of a single stator winding.

The introduction of pole-amplitude modulation (Sect. 8.15) was a significant advance in induction-motor technology. Where two or three speeds can give an adequate load control, the p.a.m. motor can be employed with considerable cost advantage compared with an orthodox thyristor converter system in which the speed is continuously variable. Many p.a.m. motors for pump and fan drive have been applied over a wide range of outputs and with various pole combinations. The most usual machine for lower ratings have 4/6-pole 1½/1 speed-ratio design. At larger outputs, fans and pumps operate at low speed from 8/10-pole or 10/12-pole motors. A very large pumping system rated at 2240/500 kW has been built with a 28/46-pole p.a.m. stator winding.

Variable-speed Drives

A motor running in its normal low-slip condition can drive a load at any lower speed down to standstill by connecting the two through an *eddy-current coupling*, with the load speed regulated by adjustment of the coupling excitation. For suitable loads the advantages are that the motor is run up on no load, the load-starting torque can be as high as the motor pull-out torque, and the coupling excitation can be electronically controlled to give ramp acceleration and speed-holding.

Speed control of cage motors by *frequency variation* applies where the expensive and technically demanding converters give advantages that are economically justified. The motor, its control and its load should be 'matched', and provide maximized torque per unit of current. Normally the motor is best utilized if the airgap flux is maintained constant, although for high operating frequencies the necessary higher voltage amplitudes may not be economically obtainable.

The control equipment must be as reliable as the motor. The basic devices for a closed-loop system comprise the converter, gating circuits, control sensors and amplifiers, feedback devices, error amplifier, and reference sources for voltage and frequency. The error amplifier is responsible for comparing the actual operating condition with a demand condition and developing an output corresponding to the difference between them. The reference voltage may, for example, be a stabilized supply with a smoothed rectified output to a Zener diode, a voltage divider being provided for the demand setting. An oscillator gives an output from which the operating frequency is set. For speed indication an analogue tacho-generator may be mounted on the motor shaft, or a more precise digital system used.

Various operating characteristics can be achieved, for example (i) heavy-duty constant-torque, (ii) high-inertia, and (iii) variable-torque. Conditions (i) are found in ore-handling conveyors in which a smooth 'soft' start is required even when unloaded to avoid undue stress on pulleys and long conveyor belts. For (ii) the motor torque can be made proportional to the speed, enabling a smaller motor to be used and a higher efficiency reached. In (iii) the rate of increase of current can be controlled to limit inrush current at starting.

In most applications use is made of thyristor equipments. Phase-commutated static converters need relatively 'slow' components able to handle large switching powers. With forced-commutation components a balanced optimization of reverse-voltage level, recovery time and turn-on losses is necessary.

Small cage motors can be operated from 1-ph supplies, using a link current controlled by transistors and a 3-ph inverter output. Power transistors have fast-switching and inherent turn-off capabilities, and their application is likely to grow.

Motors. In adapting a standard 50 or 60 Hz induction motor for inverter speed control, the 'normal' performance must be re-assessed with reference to (i) the additional loss resulting from the harmonic content and its effect on temperature-rise; (ii) the leakage inductance required by the inverter mode and its augmentation where necessary by external inductors; (iii) the cooling at low speeds, which may demand a larger frame size or forced ventilation; (iv) the effect of centrifugal forces at maximum operating frequency; (v) the limit on maximum torque imposed by the inverter thyristor current rating; and (vi) the suppression in some cases of torque pulsation at low frequencies. Because of the high cost of control equipment, the motor torque/current ratio must be maximized.

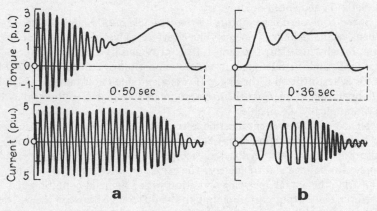

12.13 Starting transients.

Acceleration and Retardation. Figure 12.13 compares the no-load run-up torque and current for a small induction motor (*a*) switched direct-on-line at rated voltage and frequency, (*b*) accelerated by means of a linear rise of inverter output frequency at optimized rate. The latter gives a shorter run-up time, eliminates torque pulsations, and reduces the starting current because the motor operates throughout at a small slip and a high level of torque. In retardation, the condition for a high braking torque is that the negative slip should be small; this can be secured by a controlled rate of fall of the inverter frequency. On reversal, the machine operates at a small negative slip except at the instant just prior to standstill when the stator and rotor travelling-wave fields are momentarily opposed, and torque pulsations may appear. Such reversal is radically different from plugging (Sect. 8.18) in which the stored kinetic energy is lost as heat in the rotor winding.

Speed Holding. For some industrial applications, a cage motor has adequate speed-holding within its normal slip of 0.02–0.04 at full load. But compensation for slip can be provided by automatic adjustment of the inverter frequency to give a speed constant within ±½%. Closer speed-holding requires closed-loop control using a shaft-mounted tachometer, of analogue or digital form. The latter, which gives speed information in terms of the zero-crossing of the output waveform, has the higher accuracy. Control of the stator frequency is vital for precision speed control, and stable reference oscillators are employed; for the performance of the converter/motor system cannot be better than that of its references.

Instability. Otherwise stable motors on converter drive are prone to enter limit cycles (speed oscillations), more particularly on light load and low frequency. The phenomenon was investigated by Rodgers [126], who originated a study based on perturbation theory. Fallside and Wortley [127] have found that the limit of stability is two-valued and depends on the slip, the inertia and the motor parameters. The transfer function of the drive system has a dominant pair of complex roots (indicating oscillation) and a small damping. At low inverter output frequencies the oscillation is slow, resulting in a long settling time. Analysis shows that instability can be minimized by a reduction in rotor resistance and an increase in that of the stator, but such parametric changes may react unfavourably on the maximum-torque capability. Nanda and Mathew [128] suggest the use of external inductors to reduce low-frequency instability, the inductors being cut out when the motor operates in the higher frequency range.

Control. The control equipment may represent a major component of the total cost of a converter variable-speed drive. The basic elements comprise the rectifier, d.c. link inductors and capacitors, feedback circuits, inverter, gating circuits and logic, control sensors and amplifiers, reference devices for speed, voltage and frequency, and devices to give the appropriate voltage/ frequency relation. Optimization of run-up and speed-change performance may be included. The control 'package' is generally of some basically standardized form, with protective features to take account of thyristor ratings, voltage transients and thermal stability.

Pumps and Fans

Almost every process industry, whether the product is solid or fluid, needs pumps and fans. Drive-motor speed control replacing wasteful throttlers and dampers gives energy economy. For example, an induced-draught fan for gases at 230 °C required 200 kW for a gas flow of 85 m^3/s; for a throughput of 65 m^3/s and a fan torque proportional to the square of the speed, the useful power demand would be $200(65/85)^3 = 90$ kW. Thus a power reduction of about 100 kW could be obtained by fitting a speed-control unit between the motor and its switchgear, or by a pole-change motor.

In *waterworks,* pumps are commonly operated in parallel. All but one are constant-speed units, cut in or out to provide coarse control of the output well level. The remaining unit has a variable speed to act as a fine control. An advantage is that all motors are identical, any one being the speed-controlled unit. In *oilfields,* motors of rating up to 25 MW may be required to drive gas compressors, crude-oil and sea-water pumps, and may be sited in desert regions with a humid and dust-laden ambient air of temperature exceeding 60 °C. Refineries and chemical plants for plastic or fertilizer production also require large pump and compressor motors. The industry has preferred cage motors with direct-on-line starting. The operating speed of a large rotary compressor is from 5000 r/min upwards, normally driven through a gearbox from a 4-pole 50 or 60 Hz motor. Where the ambient air contains flammable gases, explosion-proof enclosures are necessary, or a construction in which the air (or nitrogen) pressure within the machine is maintained above the atmospheric. On some *oil tankers,* variable speed for the cargo pumps (driven by motors of rating typically 1 MW) is obtained by their direct connection to generators with variable-speed diesel-engine drive.

Pumping by *submersible motors* is a specialized but expanding field. Immersion eliminates the need for shaft seals because the motor runs in the working fluid. An example is the CO_2 coolant-gas circulator, Fig. 12.14, in a gas-cooled nuclear reactor. A 1200 kW cage motor drives the impeller fan at 2970 r/min, circulating gas at 300 °C and 20 atm pressure. The paramount requirements of maximum reliability are met by a 50 Hz constant-speed motor and a vane system of variable pitch. The motor design embodies a thermal barrier for the stator winding and a coupled pony motor fed from a 'safe' auxiliary supply in case of mains failure.

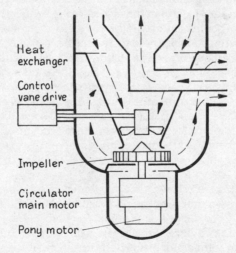

12.14 Heat-exchanger in gas-cooled nuclear reactor.

A typical aircraft *fuel-pump* drive is a cage motor, oil-cooled or provided with blast cooling in the tenuous atmosphere at high altitudes; the rating is 7½–11 kW, 200 V, 400 Hz, 3-ph; the stator has a star-connected winding in 24 slots and the rotor has 32 skewed slots carrying aluminium bars and end-rings; the active core length is 114 mm. Aircraft and military vehicles can carry only limited supplies of hydrocarbon fuel oil, which must be used with economy. Fuel pumping may take one-third of the total generated electrical power, and must cope with extremes of ambient temperature over which the fuel viscosity varies considerably.

Marine Auxiliaries

Many shipboard duties have been taken over by electric drives, particularly in large ships of high manoeuvrability. The economic voltages are 380—440 V for motor ratings up to 500 kW, 3.0—3.3 kV up to 1 MW and 6.0—6.6 kV for higher ratings. Where motors represent a major part of the system load, cage motors should have minimum starting current to permit of direct-on-line starting, and the generating plant should have a good transient recovery with limited voltage dip. Most drive requirements are met by single-speed cage motors with total enclosure and Class F insulation. Industrial-type control gear can be applied if it incorporates the special marine requirements: it is normally grouped at a control centre to facilitate sequence starting and maintenance.

Steel-bar Rolling Mills

The essential features of a steel-bar rolling mill, Fig. 12.15, include load-equalization by means of a flywheel. The to-and-fro passes of a billet impose on the motor a repeated shock loading of duration typically between 0.5 and 4 s, currents up to 3 p.u. and speed drops up to 30%. The conditions resemble those in Fig. 8.35. The slip-ring rotor has an external rheostat to give a limited range of adjustment to the duty cycle, and capacitors are often installed at the stator terminals to minimize the peak current demand and to delay the decay of the airgap flux. Binns *et al* [130] discuss the transient behaviour of a rolling-mill drive with particular reference to saturation effects nonlinearities and mass-elasticity oscillation frequencies.

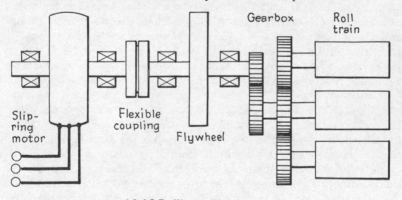

12.15 Rolling-mill drive.

Slip-ring Motors with Rotor-resistance Chopping

For variable speeds below the synchronous, slip-ring motors can be provided with external rheostats varied by thyristor choppers. The method is useful where efficiency is not a primary consideration, and where a robust system is needed (e.g., for heavy cranes) to give full-load torque over the whole speed range. Basic methods of electronic rotor control are: (i) burst firing, (ii) switching at rotor frequency by phase control, and (iii) high-frequency switching. In (i) the rotor power fluctuates at a comparatively low repetition rate and may be suitable only for drives with high inertia. Method (ii) is complicated, has a reduced power factor, generates large harmonics and has a slow response at small slips. Fast switching (iii) makes it possible by pulse-width modulation to maintain a substantially sinusoidal rotor current and a quick response.

The advantages of rotor chopping are that the rotor control is separated from the stator supply and is simpler than stator control. Chopping corresponds to an adjustable increase in the effective rotor-phase resistance, so modifying the torque/slip characteristics. High-frequency chopping (for example at 400 Hz) avoids the instability caused by the beating between the chopping and the stator supply frequencies. But the efficiency at low speeds is affected, the extracted slip energy being dissipated in harmonic power in the rotor, and induced in the stator from the rotor. In Ref. [145], Sen and Ma describe a method by which slip-ring currents are first rectified, then chopped by a thyristor operating with a controlled mark/space ratio; Holmes and Stephens deal with a system of asymmetrical chopping; and van Wyk gives an analysis of various methods of rotor chopper control, with comparisons, with the warning that for rapid response the analytical model of the system has to be refined. It is probable that transistor chopping could simplify speed control in ratings up to about 10 kW.

Cranes

Crane motors operate over a wide range of controlled speeds in either direction to raise and lower the hook. Lifting requires a relatively flat torque/speed characteristic to limit the change of speed between loaded- and light-hook conditions. Lowering a light hook sometimes needs a small motoring torque to overcome friction but a braking torque for loads. Thus a four-quadrant control is required. Regenerative braking is not possible at sub-synchronous speeds, so that alternative mechanical, magnetic, d.c. injection or plugging facility has to be provided for constant-frequency supplies.

Crowder and Smith [129] describe a sophisticated four-quadrant crane drive, Fig. 12.16, which combines phase control of the stator voltage with chopper control of the effective slip-ring rotor resistance. The stator thyristor control includes anti-parallel thyristors with additional pairs in two phases for reversal. The stator operates within the limit of rated current, and although it could control the motor at any set speed, the power loss at high slips would unduly restrict the performance. The rotor resistance is therefore adjusted to suit the demand speed by controlling the duty cycle of a chopper thyristor circuit across the external resistance R as a predetermined function λ of the rotor speed n. The function, which is different for motoring and braking operation, is such that maximum torque is obtainable with the rated current limit to all speeds. The speed is sensed by tachogenerator T. Braking by d.c.-injection is provided, but as this fails at low speeds, the control switches to plugging at a set speed n_1 for stopping.

The cost of a crane control system is not the sole criterion. The mounting of the equipment, the trolley space required for it, the safety, reliability and facility for maintenance are also significant.

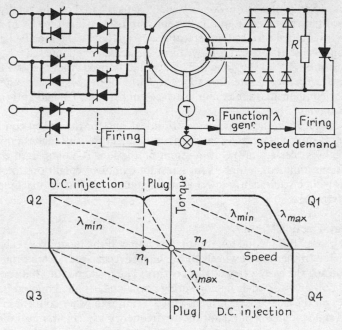

12.16 Crane control.

Motors for Hazardous Atmospheres

An ambient atmosphere becomes hazardous when a flammable gas, vapour or liquid forms with air a potentially explosive mixture that can be ignited by a flame, a spark or a hot surface. The significant ignition temperature in the presence of a flame, arc or spark is the *flashpoint* of the mixture, and by a hot surface is the *auto-ignition* temperature: these, for example, are -18 and $+540\,°C$ for acetone, -45 and $220\,°C$ for petrol. The auto-ignition temperature range is of first importance in a motor design in view of the surface temperature of its exposed parts, and has been classified as below:

Class	T1	T2	T3	T4	T5	T6
Max. surface temperature ($°C$)	450	300	200	135	100	85

The degree of hazard is identified by a 'zone number', defined in terms of the presence (during normal motor operation) of the explosive mixture continuously (Zone 0), likely (Zone 1) or unlikely and of short duration (Zone 2). The most hazardous condition is therefore 'Zone 0, T6'. For less demanding applications three basic designs are widely adopted for Zones 1 and 2. *Flameproof* (Ex c) for Zones 1 and 2. The construction must ensure that internal explosion is contained completely, and that explosion pressure build-up is withstood. The whole machine must be robust and have extra long spigots and flame paths on all mating joints (frame/endshields, shaft/bearing-caps, terminal-box/lid, etc.).

Increased Safety (Ex e) for Zones 1 and 2. This is based on the totally-enclosed fan-ventilated (t.e.f.c.) concept, but differs from the (Ex c) machine in that all internal surface temperatures are kept below the specified auto-ignition level so that explosions within the carcase are eliminated.
Non-sparking (Ex N) for Zone 2. The design is derived from the t.e.f.c. machine, with special attention to rotor/stator clearance, impact tested exterior components, a maximum surface temperature T3, and anti-tracking terminal board.

Most motors in the range are rotor-critical, and for the stalled condition the rotors must not exceed the specified temperature: this makes rapid disconnection necessary in the overcurrent protection. The temperature limitations imposed on motors for hazardous ambient conditions imply that ratings will be significantly lower than for standard industrial machines of the same frame size.

Railway Traction Motors

Three-phase 50 Hz induction motors, mounted in the locomotive cab and connected to the driving wheels through massive side-rods, were running in Italy during the early years of this century. The rigid coupling between the spring-borne motors and the non-spring-borne underframe imposed heavy stresses on the side-rods, causing frequent failure. In other countries the underframe mounting of d.c. or of low-frequency a.c. commutator motors proved more satisfactory, the motors being 'nose-suspended axle-hung' machines, as in Fig. 12.17, driving the wheels through a pinion and gear wheel. The advent of thyristor control has made it possible to substitute variable-speed 3-ph induction motors for the commutator machines, to give the advantages of lower mass, a higher limiting speed, and the elimination of commutator and brush maintenance. The robust cage rotor is invariably employed. However, besides the harmonic effects introduced by inverter feed, induction motors are more sensitive than are d.c. series motors to

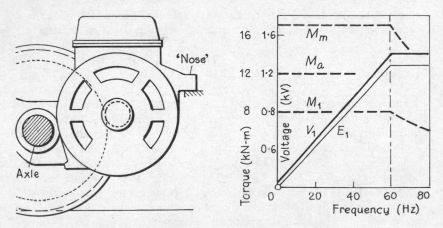

12.17 Nose-suspended axle-hung railway motor.

differential wear of the driving wheels, as a minor difference in the wheel diameters associated with a pair of motors can make a major difference between their slip speeds, resulting in torque and current unbalance. If a common rectifier output is provided to inverters individual to each motor of a multi-axle locomotive or of a rapid-transit multiple-unit train, the torque can be almost equalized with motor pairs in series, and compensation with motors in parallel, using sensors to feed back the respective slip speeds to the associated inverter gating control circuits.

Nose-suspended Axle-hung Motors. The general arrangement in Fig. 12.17 indicates the pinion and gear-wheel drive by dotted circles. The characteristic graphs apply to a constant-current motor of typical size. The rated torque is M_1, the pull-out torque is M_m and the torque demand for starting is M_a. These are maintained up to a speed corresponding to a 60 Hz inverter output frequency with the voltage V_1 such as to maintain the appropriate value of gap e.m.f. E_1 for constant gap flux. Above this frequency it is necessary to limit the voltage so that M_1 and M_m reduce as the frequency is raised (corresponding to field weakening in a d.c. series motor).

Inverted Motor. An inverted design, with the rotor rotating *external* to the stator, is described by Stokes *et al* [142]. One arrangement is shown in Fig. 12.18 with dimensions appropriate to a standard track gauge (1.435 m). The concept is novel in that the normal axle connecting a pair of driving wheels is replaced by a steel tube, with the rotor of the induction motor fixed to its inner wall. The stator, set within the rotor, is restrained from rotation by a torque tube through one of the main axle-box bearings, this tube serving also for the admission of cooling air at one end and of the stator phase connections at the other. The design has a number of difficult conditions to meet: the whole mass of the motor is non-spring-borne, the motor speed is low because there is no gearing, and a short airgap length of about 1½ mm has to be preserved. Dimensional restrictions impose the adoption of a 6-pole magnetic circuit and a rotor with a copper cage. By use of a d.c.-link converter for variable frequency and a series/parallel connection of the stator windings, the aim is to achieve a maximum speed of 200 km/h

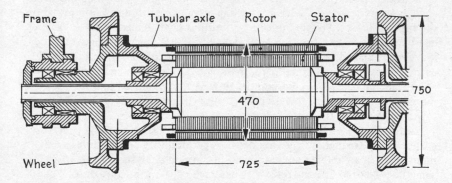

12.18 Tubular-axle railway motor (dimensions in mm).

(rotor speed 24 r/s), a maximum tractive effort up to 20 kN per motor, and regenerative braking down to 50 km/h. Further developments in the design relate to the elimination of the two inboard bearings by mounting the stator assembly on an I-beam. Besides the shock stresses imposed as the train passes over joints and crossings, the torque pulsations at low starting frequencies and the reduction of vibration of the mass-elastic motor structure to avoid damaging resonances have to be considered. The d.c.-link converter is intended to operate at 650 V from a d.c. contact-rail system, or from the output of a train-borne diesel-driven generator, or through a transformer from a 25 kV 50 Hz 1-ph overhead contact line.

A solid-rotor machine would be advantageous for the arduous conditions of railway traction, but it is not well suited to supply from a frequency-changing converter because of its sensitivity to weak-field operation at the upper end of the speed range. But a solid rotor with a change-pole stator could be used with a constant-frequency supply developed on the train by a d.c./3-ph or 1-ph/3-ph device fed from any contact line. The system would provide a high torque low-current start, a wide speed range controlled by a relatively small voltage variation, and regenerative braking. Besides its economic and robust design, an important advantage (not shared by inverter operation) would be the avoidance of interference with signalling circuits.

Solid-rotor Machines

For some applications an induction motor can have an unwound rotor, the torque being developed by eddy-current and hysteresis effects in an unlaminated 'solid' cylinder. Low-hysteresis materials (e.g. mild steel) have good starting and run-up torques, are robust and inexpensive. With a high-hysteresis material (e.g., cobalt steel) the rotor can self-synchronize. Small rotors can be formed from a ferrite. Notwithstanding the absence of a winding, the division of the rotor input power P_2 into the mechanical conversion power P_m and the rotor loss p_2 as given in eq. (8.3) is still valid. The normal full-load slip s is typically 0.2–0.4. The difficulty in predicting P_2 lies in the estimation of the parameters that determine it.

Consider a solid steel cylinder to comprise an assembly of thin concentric annular layers 1, 2, 3 . . . with a travelling wave of stator m.m.f. impressed on the airgap. Flux penetration into the outermost layer 1 raises it to a flux density approaching saturation level, develops in it a hysteresis loss, and induces an axial eddy current. Both effects develop torque. But as the eddy currents must have a circuital path, their closed loops have end regions of non-torque-producing circumferential flow. A flux modified in both magnitude and phase by the effects in layer 1 then enters layer 2, and so on towards the axis of the rotor cylinder.

Each volume element of active rotor material is subjected to a cyclic change at slip frequency: the flux density, permeability and hysteresis are interrelated in a complex and nonlinear fashion. In analysis the hysteresis is sometimes ignored and the resistivity assumed constant in spite of its dependence on the rotor temperature. Most solutions use Maxwell equations as a starting point, but express the outcome in terms of simplifying assumptions as below:

(i)   The rotor material has a constant permeability μ. Then from wave-propagation theory the rotor equivalent phase angle is $\phi_2 = 45°$.

(ii)   The rotor flux density has always the saturation value $\pm B_S$. Then $\phi_2 = 27°$ for a sine-distributed magnetizing force H at the rotor surface, and $23°$ with an H distribution distorted to conform with an assumed sinusoidal distribution of the airgap e.m.f. and flux density.

(iii)   The rotor B/H relation is constant up to B_S, after which the flux density is B_S for any higher value of H.

(iv)   The rotor flux density has the analytic form $B_1 = aH^b$ where a and b are constants and B_1 is the fundamental of the flux-density waveform for a sine-distributed H. Then ϕ_2 lies between $35°$ and $45°$ depending on the value chosen for b.

(v)   The rotor material is assigned a nonlinear complex permeability to include loss and hysteresis effects.

The diagrams in (a) of Fig. 12.19 give some idea of the distribution in a cylindrical solid rotor of the current and flux patterns with which an analysis

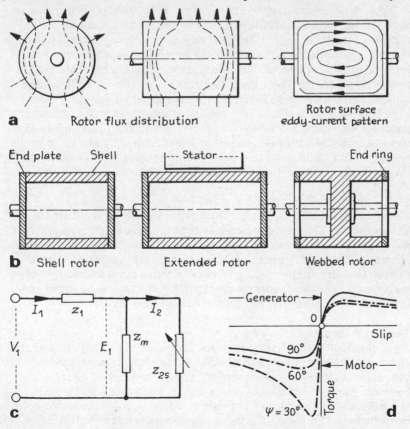

a   Rotor flux distribution   Rotor surface eddy-current pattern

b   Shell rotor   Extended rotor   Webbed rotor

c   **d**

12.19 Solid-rotor induction motor.

has to cope. The alternative constructions in (b) have the advantage of lower inertia, and show methods by which the eddy-current paths can be improved with end plates or copper and rings.

Chalmers *et al* [141] have shown that an equivalent circuit (c) can be set up for the machine; z_1 is the stator leakage impedance per phase of N_1 turns in series and a winding factor K_w, z_m is the magnetizing impedance for the stator and airgap (excluding the rotor), and z_{2s} is the effective impedance per stator phase introduced by a rotor of diameter D, length l, resistivity ρ and saturation flux density B_s. Then

$$z_{2s} = [A\rho l^2 (K_w N_1)^2 B_s / k_e D \Phi_s] \angle \phi_2$$

Account is taken of the means (if any) of improving the eddy-current paths in the end-factor k_e. The constants A and ϕ_2, respectively of the order of 4.6 and 27°, are affected by the chosen form of analysis (i–v) above. The variables are the slip s and the gap flux per pole Φ, and as Φ varies with the stator current an evaluation can be tedious.

Typical torque/speed relations in (d) for a *voltage-fed* machine relate to $\phi_2 = 30°$ and three values of the angle ψ in the complexor impedance $z_1 z_m / (z_1 + z_m)$. For $\psi = 90°$ (i.e., for negligible core and stator $I^2 R$ loss) the torque/slip relation is the same for generator and motor modes. For practical machines there is a marked asymmetry: with $\psi = 30°$ the peak torque for generating is four times that for the motoring mode. Chalmers [141] has also developed solutions for a *current-fed* machine.

Solid-rotor machines have been applied as short-time pony motors, machines with speed control by voltage reduction, control machines and instrument motors. The analysis is comprehensive enough to be applicable to the transient and asynchronous behaviour of turbo-generators (for which the slip may be 0.005 p.u.), and to the starting of smooth-rotor hysteresis motors.

12.5 Induction motors: protection and testing

Modern motors have a high power/mass ratio, low inertia and short thermal time-constants. The most common causes of failure are *environmental* (dirt, water, deleterious vapours etc.) inhibiting ventilation or degrading insulation; *mechanical* (bearings, lubrication, eccentricity, load jamming etc.); and *thermal* (winding failure, stalling, frequent starting duty) leading to over-temperature. The two former are avoided by proper enclosure, siting and maintenance, but thermal protection requires special devices.

Thermal Protection

Small cage motors of rating up to about 20 kW are 'stator-critical', i.e., the stator winding temperature exceeds that of the rotor. In consequence, protection is afforded by thermal overcurrent relays and by thermistors built into the stator winding. Above about 75 kW a cage machine is 'rotor-critical' and on delayed or frequent starts the rotor temperature-rise exceeds that of the stator. The stator overcurrent relays must then be set with due regard to the probable rotor temperature-rise. In slip-ring machines much of the rotor $I^2 R$ loss during starts occurs in the rotor external rheostat.

External thermal relays employ bimetal strips or equivalent devices, activated by heaters carrying a current proportional to the stator load current. In magnetic protection a pick-up coil coaxial with the rotor shaft senses a harmonic component of the airgap flux, which is characteristic of an unbalanced condition. Undervoltage relays protect against the effects of supply interruption, preventing restart and the voltage collapse that may follow should several motors attempt to restart simultaneously.

Electronic relays can provide closely controlled functions, and can be grouped to give sophisticated control by microprocessor.

Factory Testing

The basic tests in Sect. 8.22 are not easily applied to large motors when customers demand full-load testing. The following is a typical test procedure to verify performance on the works test-bed.

Process Testing. Preliminary checks are made on the conducting and insulating materials, core plates and mechanical parts liable to high stress. Built stator cores are ring-wound with flexible cable, excited at rated frequency to develop normal core-flux density to check core loss and reveal faults in interplate insulation. Windings are tested for insulation to earth and between subconductors to verify transposition, during the coil-fitting process. Built-in thermal detectors are checked.

Rotor Balance. At one or more stages in the rotor assembly the dynamic balance is adjusted. Two-pole machines of large length/diameter ratio are run on overspeed in the safety pit.

Characteristics. A variable-voltage supply of adequate capability is employed for no-load and rated-current locked-rotor tests, the latter including torque measurement for cage machines. Most large cage rotors have current-displacement conductors, and short-circuit tests at normal frequency give values not appropriate for pull-out conditions: an additional test at a frequency of about one-fifth of normal gives a better basis for estimating the pull-out torque. The test schedule provides data for determining the declared efficiency.

Full-load Testing. For large machines a back-to-back method is essential. It must have provision for accommodating the speed difference resulting from the positive and negative slips of the motor and generator machines, or (if they are mechanically direct-coupled) the difference in supply frequencies. Methods include: (i) a close-ratio gearbox, e.g., 1.05/1, when at least one of the machines has a slip-ring rotor to control its slip; (ii) a fluid coupling if both are cage machines; (iii) a differential gearbox resembling that in the drive of an automobile, the test machines replacing the road wheels, and the torque shaft, driven slowly by a geared auxiliary motor, depressing the speed of one machine (motor) and raising that of the other (generator). Thus the main supply of rated voltage and frequency provides the mechanical, core and stator $I^2 R$ losses and the magnetizing currents, while the torque shaft controls the drive-power exchange between the two test machines.

For a heat run on a single motor, a 'phantom load' method has been devised by Fong [137] using a superposed direct current, no mechanical coupling being necessary.

12.6 Synchronous machines: two-axis parameters

The reactance and resistance parameters of a synchronous machine with reference to its circuit model are essential in the analysis and prediction of its dynamic performance. The parameters are used in computer programs to evaluate behaviour under specified conditions of steady-state and transient operation, fault and disturbance. The equations for the basic 2-axis 5-circuit model, Fig. 7.13, are set out in the Table of Reactances and Time-constants in Sect. 10.7.

A test schedule for determining the relevant parameters for 3-ph machines is given in BS 4296:1968, *Methods of test for determining synchronous machine quantities,* based on IEC Recommendation 34-4. Of the various alternative methods, only the 'preferred' ones are given in the digest below. For convenience, the rotor is assumed to carry the field winding. In tests other than for d.c. resistance and q-axis subtransient reactance the machine runs (or is driven) at normal rated speed. The following abbreviations are used:

o.c.c. = normal-speed open-circuit (or 'saturation') characteristic;
s.c.c. = steady (sustained) 3-ph short-circuit characteristic;
t.s.c.c. = transient (sudden) 3-ph short-circuit characteristic;
z.p.f.c. = zero-power-factor (lagging) characteristic.

Impedances

x_{sd}	D-axis synchronous reactance: o.c.c. and s.c.c. (Fig. 10.7) based on air-line for unsaturated value x_{sdu}.
r_{sc}	Short-circuit ratio: o.c.c. and s.c.c. (Fig. 10.14).
x_{sq}	Q-axis synchronous reactance: negative-excitation or low-slip test (Fig. 10.51).
x_d', x_d''	D-axis transient and subtransient reactances: t.s.c.c. (Fig. 10.27).
x_q''	Q-axis subtransient reactance: phase-winding impedance test with rotor at rest and field winding short-circuited, measuring current and power with alternating voltage applied to each pair of stator terminals in turn; evaluation by formulae.
x_p	Potier reactance: o.c.c. and z.p.f.c. (Fig. 10.16).
r, r_f	Phase- and field-winding resistances: d.c. voltmeter-ammeter or conductance-bridge method.
r_2, x_2	Negative-sequence resistance and reactance: line-line sustained s.c.
r_0, x_0	Zero-sequence resistance and reactance: phase windings connected in series or parallel to 1-ph voltage source, with field winding short-circuited.
r_1	Positive-sequence phase-winding resistance: from known $I^2 R$ plus load (stray) loss for given current.

Time-constants

$\tau_{d_0}{'}$	D-axis transient o.c. time-constant: with normal no-load stator voltage, field winding suddenly s.c. and its current decay recorded.
$\tau_d{'}$	D-axis transient s.c. time-constant: with normal no-load stator voltage, sudden s.c. applied and stator current decay recorded (Fig. 10.27).
$\tau_d{''}$	D-axis subtransient time-constant: t.s.c.c. (Fig. 10.27).
τ_a	Armature (d.c.) time-constant: decay of a.c. field-current or of d.c. phase-current component from t.s.c.c.

Inertial Constants

H, T	Stored-energy coefficient, acceleration time: no-load retardation test, observing time for speed to fall from 1.1 (or 1.05) to 0.9 (or 0.95) p.u.

The 2-axis parameters based on these tests may differ from those derived from tests with different operational conditions, particularly for the q-axis. Shackshaft [138] proposes an extension of the test procedure to yield more realistic values. Retaining the 5-circuit 2-axis model and using the readily recordable stator and rotor voltages and currents, one quantity is maintained constant, a step change impressed on a second, and the voltage or current of the remaining two are observed. Modified analytical expressions are used to evaluate the parameters.

Difficulties in the test procedure for large turbogenerators has stimulated interest in variable-frequency measurement of the stator impedance over a range 0.001–200 Hz these data being analysed by Bode diagram. The stator is energized to give a pulsating flux, with the rotor stationary and aligned successively with the d- and q-axes. The method is discussed, and compared with the t.s.c. test, by Diggle and Dineley [138].

12.7 Synchronous generators: airgap and superconducting windings
Increase of turbogenerator ratings without undue extension of the physical dimensions has been actively considered by designers. Two methods have been suggested: (i) *airgap* windings with a slotless stator, or a fully slotless construction in which both rotor and stator are slotless and cylindrical; and (ii) *superconducting* rotor windings.

Airgap Windings
Use of the long airgap to accommodate a slotless stator winding, and its advantages, have been noted in Sect. 3.12, and possible arrangements for fixing the winding illustrated in Fig. 3.25. A comparison of (*a*) a fully slotless and (*b*) a conventional construction is given diagrammatically in Fig. 12.20. In Ref. [148], Davies gives examples of slotless stator winding arrangements, and Spooner analyses the fully slotless machine to show that substantial reductions in active length and total mass could be achieved with optimized magnetic and electric loadings.

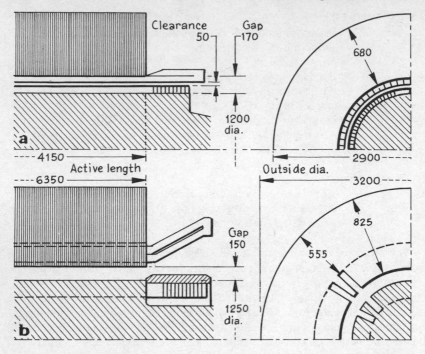

12.20 750 MW turbogenerators with airgap and conventional windings.

Stator. Elimination of teeth enables peak airgap flux densities to be raised to about 2 T (magnetic loading about 1.3 T), and the electric loading to 300 kA/m in a large machine, although it might then be more difficult to control axial leakage flux at the ends of the machine. There is much more freedom in the winding layout, which can be arranged for a more sinusoidal gap-flux distribution. The axial conductors are immersed in the main field and must be suitably transposed. Ideally, the whole winding including the end turns and the water-cooling connections would be assembled and encapsulated externally, then slid into the stator bore and firmly fixed thereto. The reactances would be inherently lower than in a conventional form of machine, improving transient stability but involving a greater fault current.

The whole torque developed on the stator winding falls directly on the conductors. In a 2-pole 750 MW generator the full-load torque is 2.4 MN-m, corresponding to a force of 3.2 MN at a radius of 0.75 m. Were it possible constructionally to spread this evenly over the surface of the stator bore, the specific stress would be only about 100 kN/m^2 (i.e., about the same as standard atmospheric pressure) and could be sustained by available polymer insulants.

Rotor. The cylindrical slotless-rotor forging could be of greater diameter than in a conventional machine and therefore stiffer and (if necessary) longer.

More of the periphery is available for winding. The winding depth is dependent on the radial gap between the stator bore and the rotor body, on the allocation of this dimension to the stator and rotor windings, and the clearance gap between them. The main problem is to retain the rotor winding against centrifugal forces by surrounding it with a retaining cylinder or by some equivalent means.

The data in the table below and in Fig. 12.20 compare a proposed slotless design with a conventional machine for 750 MW (833 MVA at p.f. 0.9), 50 Hz, 2-pole, with water cooling throughout.

Item		Slotless	Conventional
Magnetic loading	(T)	1.28	0.97
Elec. loading (stator/rotor)	(kA/m)	240/580	200/490
Current dens. (stator/rotor)	(A/mm^2)	5.1/12	5.1/12
Active length	(m)	4.15	6.35
Stator dia., outside	(m)	2.90	3.20
bore	(m)	1.54	1.55
Rotor diameter	(m)	1.20	1.25
Gap length, total	(mm)	170	150
clearance	(mm)	50	150
Reactance, synchronous	(p.u.)	0.79	1.87
transient	(p.u.)	0.06	0.35
subtransient	(p.u.)	—	0.22
leakage	(p.u.)	0.05	—
Losses: stator I^2R	(MW)	2.55	1.07
load (stray)	(MW)	0.75	1.65
core	(MW)	0.60	0.66
rotor I^2R	(MW)	2.84	3.44
mechanical	(MW)	0.75	1.00
total	(MW)	7.34	7.82
Efficiency	(%)	99.03	98.90

The relative active masses are in the ratio 1.0/1.7, as a result of the smaller length and diameter of the slotless machine. There is a significant reduction in the reactance parameters. It is here assumed that use is made of the concentric stator airgap winding described by Davies [148].

Superconducting Windings

Several metals and alloys, when cooled below a critical 'transition' temperature (which has a unique value for each and is within a few kelvin of thermodynamic zero), lose all electrical resistivity. Such materials are *superconductors*. But a superconducting wire can support a high current density only within strict limits of temperature and flux density: if either is exceeded, the material reverts to the resistive state. The material of engineering interest here is a niobium-titanium alloy, for which the limits are 9 K and 11 T. A typical conductor might comprise several hundred Nb-Ti wires, of diameter 0.025 mm, embedded in a copper matrix to provide thermal conductivity. Such a conductor could be worked at a current density of 120 A/mm².

Superconductors must not be subject to alternating magnetic fields and can therefore be used only for d.c. rotor excitation. A superconducting rotor winding is capable of developing very high magnetic flux densities (e.g., 5 T) with the virtual elimination of I^2R loss. At such densities the advantages of a ferromagnetic magnetic circuit are lost, but the rotor has to be provided with a steel forging for mechanical reasons. The stator winding, however, has neither core nor slots, and is held in a non-conducting and non-magnetic support structure. The very considerable stray field external to the stator is limited by a laminated outside cylinder of core-plate material or by an enclosing cover of highly conducting metal (the eddy currents in which serve the same purpose), providing an 'environmental' screen. An elementary cross-section of a superconducting-rotor turbogenerator then appears as in (*a*) of Fig. 12.21.

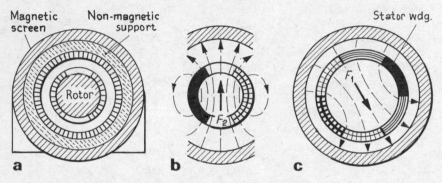

12.21 Fields in coreless machine.

In the conventional analysis of electrical machines, the flux has well-defined paths that carry either the working or the leakage flux components, a concept that arises naturally from the geometry of the magnetic circuit and the relatively short airgap. This approach is less satisfactory for machines with airgap windings, and fails altogether for substantially ironless machines. A different approach is now necessary to establish the flux patterns developed by the stator and rotor m.m.f.s F_1 and F_2, and to analyse performance in terms of self and mutual inductances, quantities determined by the effective winding radii. The flux pattern of a 2-pole rotor acting alone resembles that in (*b*), and for the stator winding that in (*c*), of Fig. 12.21. Hughes and Miller [149] show how the inductances and the rotor/stator mutual (coupling) inductance are determined. The coupling coefficient in an ironless machine is given by r^p, where r is the ratio of field-winding to stator-winding diameter, and p is the number of pole-pairs. Thus in a 2-pole superconducting machine with $r = 0.5$, only one-half of the total rotor flux is useful, whereas in a conventional machine the coefficient is about 0.9 because of the presence of ferromagnetic material and the shorter airgap. The power rating given by eq. (1.16) shows, however, that the air-cored machine can achieve a substantial increase in rating by reason of its much greater working flux density obtainable with a superconducting rotor winding, and by its inherently greater stator-winding diameter.

Rotor. The technological difficulties in the rotor construction are formidable. The field winding must lie in liquid helium at a temperature of about 5 K, and it must be thermally insulated from the external ambient temperature by gaseous helium. The rotor must be enclosed by thermal and magnetic screens, the latter to shield the winding from transient and negative-sequence stator fluxes. Some of the essential features are illustrated diagrammatically in Fig. 12.22.

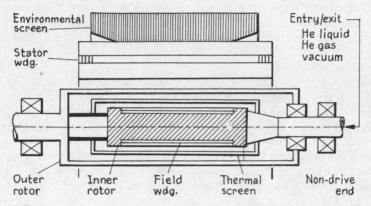

12.22 Features of turbogenerator with superconducting field system.

The *inner* rotor carries the superconducting field winding, immersed in liquid helium. The conductors must be so held as to prevent movement that might generate friction heat, and the coils housed in a strong slotted structure (e.g., of aluminium or stainless steel) to withstand the high magnetic and centrifugal forces. Heat from the stator penetrates into the rotor causing the liquid helium to vaporize: the temperature-rise of the vapour towards ambient temperature absorbs the heat, the gas being returned to the external refrigerator for recycling. The *outer* rotor cylinder has two main functions: it contains the vacuum necessary to maintain the low temperature of the inner rotor, and acts as a screen to prevent transient and negative-sequence fields from reaching the excitation winding. It is therefore a composite structure of magnetic and conducting layers, able to deal with the torques generated by conditions of unbalance. The exciting winding is connected to slip rings. The entry/exit channels for liquid and gaseous helium are contained in the hollow stub shaft at the non-drive end, with elaborate rotary sealing devices.

Development of practical superconducting turbogenerators is likely to result in a number of novel designs. A helical stator winding has been suggested, and an 'inertial' form of outer rotor to avoid passing fault torques

along the shell to the shaft. Lawrenson *et al* [150] have analysed the damping
and screening requirements of outer-rotor cylinders.

12.8 Synchronous generators: types and applications

Besides the large power-station machines discussed in Sect. 11.2 and 11.3,
synchronous generators are essential plant in ships and aircraft, are made in
fixed or transportable form for standby, are applied to local equipments such
as welders, provide the networks of small isolated and island communities,
and utilize process steam or burnable gases in factories. The range of ratings
and speeds is very wide. Prime-movers may be gas-, steam- or water-turbines,
diesel engines and even windmills. Engine-driven low-speed generators have
salient poles — up to 50 or 60 for a 120 r/min engine speed — with fully
interconnected damper windings to reduce the tendency to 'hunt'; solid poles
are used for high speeds. With low-pressure steam, turbogenerators have
4-pole cylindrical rotors.

Peak-load Provision

Peak-load plant may comprise steam, hydro, diesel or gas-turbine reserve
stations. Hydro- and gas-turbine sets can be run up and loaded in a few
minutes, but it takes some hours to start and load a steam turbogenerator
from cold, and it may be necessary to hold running machines as a *spinning
reserve*. An economic solution is some form of storage to absorb energy at
times of light system load and return it during peak-load periods, a method
desirable where the intermittent generation from tidal water, windmill and
wave machines is concerned. Storage methods are discussed by Davidson
et al [151].

Pumped-water Storage. Besides inland reservoirs, consideration has been
given to the use of the sea as a lower reservoir, and to deep underground
caverns with the pump, turbine and electrical machine sunk to the necessary
depth. The requirements are set out in Sect. 10.10.

Compressed-air Storage. If a very large 'pressure vessel', such as a disused
salt-mine, is available it can be used to store compressed air by coupling a
synchronous machine either to a compressor for storage or to a gas-turbine
for generation.

Other Methods. Thermal storage in hot water as part of the boiler plant of
a steam or nuclear station is an established method. The kinetic energy in a
large flywheel is effective if the windage loss can be minimized. Electrical
storage in batteries or in superconducting inductors makes some form of
d.c./a.c. inversion necessary.

Excitation

Three basic methods are shown in the simplified diagrams in Fig. 12.23,
suitable for both generators and motors.

(*a*) *Conventional*. A d.c. main exciter has its commutator rotor on the shaft
of the synchronous machine, its field being energized from a pilot exciter
with a permanent-magnetic field, and controlled through an automatic
voltage (i.e. excitation) regulator. The advantage of the permanent-magnet

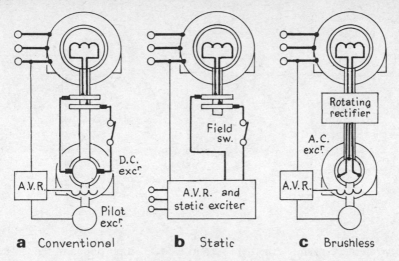

a Conventional   **b** Static   **c** Brushless

12.23 Excitation systems.

pilot exciter is that it does not have to rely on any auxiliary supply. The field of the synchronous machine is fed from the main exciter through slip-rings. Conventional field-suppression switchgear (not shown) limits damage in the event of an internal fault.

For small sets in which rapid response is not a criterion, the exciter system can be driven from the main shaft by V-belts, or replaced by a carbon-pile a.v.r.

(*b*) *Static*. The generator output voltage is controlled directly by a fast-response a.v.r. and a static thyristor rectifier unit, which takes an a.c. input from either the generator or an auxiliary supply. The rectified output is delivered to the main field through slip-rings. Field-suppression switching is normally provided. The system is accurate, compact and rapid-acting.

(*c*) *Brushless*. The main exciter has a 3-ph rotor winding, the output from which is rectified by a solid-state diode assembly, delivering field current to the generator directly without slip-rings. The a.c. exciter field is furnished by a p.m. pilot exciter through an a.v.r. Main-field switching is replaced by exciter-field suppression. The brushless system is the most common and its response approaches that of (*b*).

Compounded Self-excitation. The compounding scheme, Fig. 10.36, is useful for small isolated generating plant where, as in ships, the voltage dip and recovery time following the d.o.l. starting of cage motors is to be limited; and to synchronous motors required to run with a fixed p.f. over the whole load range or supply a pre-set leading reactive power for correcting other loads. One phase of the excitation circuit is represented by (*a*) in Fig. 12.24. Through the shunt and series transformers the stator phase voltage V_1 contributes $V = aV_1$ and the phase current I_1 contributes $I = bI_1$ to the excitation circuit. The rectifier and field winding are represented by a resistor, and X is the reactance of the ballast inductor. Converting (*a*) to (*b*) by the

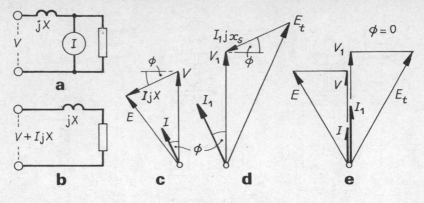

12.24 Compounded self-excitation.

Thevenin theorem shows the field current I_f to be proportioned to the scalar magnitude $|E|$ of the phasor sum $(V + I_jX)$. If I_1 leads V_1 (and I leads V by a phase angle ϕ as in (c)), the field current is

$$I_f = k_1 E = k_1 |V + IX\angle(\phi + \tfrac{1}{2}\pi)| = k_1 |aV + bI_1 X\angle(\phi + \tfrac{1}{2}\pi)|$$

where k_1 depends on the excitation-circuit parameters.

Phasor diagram (d) shows conditions in a stator phase of synchronous reactance x_s, with I_f generating the stator e.m.f. $E_t = I_f/k_2$ where k_2 is determined by the magnetic circuit. Then

$$I_f = k_2 E_t = k_2 |V_1 - I_1 x_s \angle(\phi - \tfrac{1}{2}\pi)|$$

Choosing the transformer turns-ratios so that $k_1 a = k_2$ and $k_1 bX = k_2 x_s$, the voltage triangles in (c) and (d) must be similar. This is possible only if $\phi = 0$, as in (e). Thus unity p.f. at all loads is the operating condition. Other constant p.f.s can be obtained by phase-shifting V or I by interchanging transformer phase connections or adding resistance to the ballast inductor and adjusting the parameters to give similar voltage triangles. Varying a by tap-changing makes possible a constant reactive power at all loads. These variants are described by Smith *et al* [152].

Industrial Generation
Generators separate from, or auxiliary to, a public main supply are employed in manufacturing and process industries where cheap fuel is a by-product that can be used in a gas turbine or for steam raising; where process steam at low pressure is required and is available from the exhaust end of a turbo set; where peak-lopping can reduce the maximum-demand component of the mains supply tariff; and where loads are too great or too impulsive for the supply system network. Industrial generating plant may not be permitted to run in parallel with the public supply, however, except by agreement on

'quality' (e.g., voltage waveform) and at times of system peak load.
Combined-cycle plant, incorporating a high-temperature feed to a gas-turbine-
driven generator and a separate steam turbogenerator using the rejected heat,
are increasingly applied to raise the overall thermal efficiency.

Standby Generation

A wide spectrum of effects, from minor domestic inconvenience to disastrous
breakdown of production processes, and even loss of life, may result from the
failure of a mains supply. Hospitals, food-processing plants, telecommunication installations, commercial office complexes and digital computers are
highly sensitive to interruption by mains failure. They are therefore provided
with standby plant arranged to give almost instantaneous back-up ('uninterruptible') or a start within the permissible delay.

Some standby schemes are shown in Fig. 12.25. In (a) the load is normally
provided by a motor-generator MG in parallel with a battery B, the two being
energized from the incoming mains through a rectifier R; on mains failure,
M is driven by the battery. Prolonged interruption requires a diesel-engine E
and a generator Gs as a mains substitute. In (b) an MG set has a flywheel F
and a clutch C that can be locked to a standby diesel engine E to start it
should there be a mains failure, using the kinetic energy of the flywheel F. If
a brief period of interruption can be tolerated, a mains failure initiates a
battery-fed starter motor to supply cranking torque to a diesel engine coupled
to an a.c. generator: operation is subject to proper engagement of the ring
gear on the engine flywheel. The engine speed, detected digitally from the
flywheel, controls the starter motor lock-out at about 30% of rated speed and
closure of the generator contactors at about 90%, together with shutdown on
some specified overspeed.

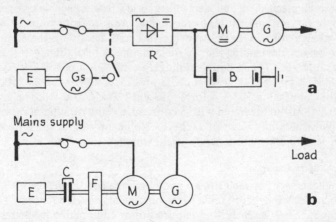

12.25 Standby schemes.

Uninterruptible standby for computers has typically to maintain a
frequency within ± 1% and a voltage within ± 5% of rated value. The engine/
generator set has to be liberally rated and be closely governed, as computer

loads have considerable variation. Standby supply for *small* computers, process controls and industrial data acquisition systems can, however, be derived from batteries and solid-state inverters.

Marine Generation

Plant of high capability is required for tugs, dredgers, tankers, large passenger and cargo ships, and off-shore oil rigs. Where several multi-speed a.c. motors with direct-on-line starting are employed (e.g., for winches), the transient behaviour of the generator is important: it may be necessary to select a machine of low reactance to minimize voltage dip, accepting the consequent high short-circuit levels and the need for automatic synchronizing. The plant rating depends on a close assessment of the load diversity. Marine generators are usually diesel-engine driven, but frequency-changing motor-generator sets are sometimes required for dockside use to connect ship and land-based mains supplies of different frequency and voltage.

Oil-rig Generation

A typical off-shore installation may require 15 MW gas-turbine/generator sets to feed motors driving 2.5 MW water-injection pumps and 9 MW compressors, besides general electrical services. Rig space is very limited and off-shore assembly is costly. Modules must be restricted in mass (e.g., 2000 tons) so that they can be floated out and lifted into place on the rig. Commissioning at a marine dock before float-out is usual, but complete testing and assembly at the maker's factory is more reliable.

Aircraft Generation

The a.c. generator for aircraft is a compact lightweight machine subject to high thermal, mechanical and electromagnetic stresses, capable of operating in the widely varying ambient temperature and pressure from grounded conditions to high-altitude flight. An overriding consideration is the minimum attainable mass, demanding the use of high-quality insulation, gap flux densities up to 0.7 T, and liquid cooling. Specific outputs of 0.7 kVA/kg can be attained. A typical 60 kVA (cont.) 3-ph 115 V 400 Hz 2-pole 24000 r/min generator can develop 90 kVA for 2 min and 120 kVA for 5 s. The generator is constructionally integrated with a constant-speed hydraulic prime-mover, a 3-ph permanent-magnet pilot exciter, a 3-ph main exciter and a rotating rectifier feeding the main generator field winding.

Wind-power Generation

The wind can vary between still-air and hurricane conditions, and is unpredictable. Wind-driven mills have consequently a wide range of operating speeds inherently, and require furling and braking mechanisms for protection. A mill coupled to a d.c. generator gives a simple wind-power system suitable for energy-storing loads such as resistance water-heaters. An a.c. generator/rectifier equipment can be used for battery charging, but for an a.c. output of controlled voltage and frequency a more sophisticated system is necessary.

Constant Speed. If the mill is controlled mechanically to secure a constant speed, a synchronous generator can provide an approximation to constant voltage and frequency. Alternatively a cage induction generator may be employed if a mains supply is available: the speed is not strictly constant but the system is not so difficult to control. Clearly the constant-speed system cannot make full use of the wind power available, and its low effectiveness makes it suitable only for low-power installations.

Variable Speed. The mill speed varies with the wind speed, improving utilization. Many methods are available to accommodate these conditions. Four are briefly described below.

1. An a.c. commutator machine with mains-frequency excitation of the stator develops a mains-frequency output from the rotor at all safe operating speeds.

2. A 3-ph generator feeds a d.c. link through a rectifier controlled by wind-speed and voltage sensors. The link output can be applied to d.c. loads directly, or to a.c. loads through an inverter.

3. A cycloconverter is used to develop a fixed-frequency output from the variable high-frequency voltage of a multipolar generator.

4. The output from a wind-driven generator is converted to a low (slip) frequency and fed to the rotor of a slip-ring induction machine, the stator of which is connected to a mains supply.

12.9 Synchronous motors: types and applications

Because small synchronous motors are more expensive than induction machines, their industrial application is restricted to drives for which their special characteristics are justified. For high ratings, however, the cost differential is reversed: synchronous machines have a higher efficiency and lower mass, particularly for low-speed applications such as reciprocating compressors, and power-factor control is advantageous. Excitation is normally brushless. Starting is discussed in Sect. 10.11.

Power-factor Correction. The use of synchronous motors for raising the p.f. of industrial loads is basically economic, depending on low-p.f. tariff penalty and reactive-power limitation. A few large synchronous motors can compensate for the reactive-power demand of smaller and more numerous induction motors. Synchronous machines (i) contribute to fault current when a short circuit occurs, and (ii) involve additional excitation loss if called upon to operate at a low leading p.f. The reactive power of capacitors (the most usual form of p.f. correction for induction motors) falls during a system voltage depression, but that of a synchronous motor with constant excitation will rise, providing a degree of voltage stabilization.

Two-speed Machines

For two speeds from a constant-frequency supply it is necessary to change the pole number for both rotor and stator. The number of stator poles is changed by applying the technique described by Fong, French and Rawcliffe [87] or by using two separate windings. For the rotor the problem can be solved in a number of ways: (i) a normal salient-pole design is adopted for the greater pole-number, while for the lower number certain poles are magnetically suppressed by short-circuiting the field coils, or pairs of poles at intervals around the rotor are excited to the same polarity and act as one; (ii) poles of uneven spacing and of dissimilar width are fitted, with pairs of narrow poles arranged to be excited to the same polarity; (iii) all pole bodies are the same and are evenly spaced, but the pole-shoes are shifted.

Fig. 12.26 shows the application of method (iii) for one quadrant of a 16/12-pole machine. (All four quadrants are similar.) The gap flux distributions are shown for the two modes, the excitations providing the marked n-s-n polarities. In each quadrant the central pole (o) is left unexcited for the 12-pole condition. The flux distribution clearly has considerable harmonic content for both modes, unlike method (i) in which the 16-pole condition would be normal and the whole distortion would occur for 12 poles. There is the advantage over method (ii) that the essential asymmetry applies only to the pole-shoes and leaves the pole cores spaced uniformly.

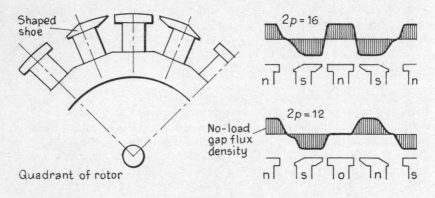

12.26 Pole arrangement for two-speed synchronous motor.

While pole-amplitude modulation windings may be used for the stator of a machine driving, say, a pump, large machines for use with reversible pump/turbine plant in pumped-storage schemes most usually have separate single-layer windings, which have better winding factors and are insensitive to the large subharmonic airgap flux distribution. The obvious drawback is that only one-half of the conductor material is usefully employed for each mode. A typical double winding has two bars per slot, one for each winding, adjoining bars of a winding being sited alternately at the top and the bottom of the slot. A design hazard is the differential expansion of the bars, as only one is energized.

In reversible pumped-storage machines a self-starting facility is necessary. Solid pole-shoes with copper interconnectors may be used to obtain the effect of a robust cage capable of dealing with thermal and mechanical stresses during the start. The stator windings may be current-limited by means of a series inductor, which is short-circuited before synchronizing.

Variable-speed Inverter-fed Motors

Some of the methods of speed control by variable frequency used with induction motors (Sect. 8.16) are applicable to the synchronous motor. A common arrangement is that in Fig. 12.7(c), except that the commutating diodes and their capacitor chains are no longer needed in the inverter.

The phase-controlled rectifier has natural line commutation. As the synchronous machine, unlike the induction motor, has e.m.f.s induced in the stator phases by the d.c.-excited rotating field system, the inverter is also line commutated, and by suitable triggering of its thyristors provides an adjustable motor-supply frequency. Speed control is virtually loss-free. Control equipment can be used to suppress hunting and loss of synchronism: the stator current-sheet pattern and its m.m.f. are maintained at a specified preset angle to the rotor m.m.f. axis (or to the e.m.f.) by deriving the thyristor turn-on instants in the inverter either from the rotor position or from the stator terminal voltage.

The speed, which depends only on the frequency of the inverter output, can if required be held within 0.05% of the set value, or even within 0.001% with digital control. The speed of a 2-pole machine with a cylindrical rotor can range up to 6000 r/min or more. Regenerative braking is possible by feeding energy back into the supply network without undue complication of the converter, keeping the link current constant and reversing the inverter and rectifier voltages. Reversal of rotation requires only a change in the inverter firing sequence. Standard thyristors can generally be employed, which with series and parallel connection can furnish voltages up to 30 kV and currents up to 1.5 kA, suitable for machines of large rating. To suit the load, the motor can be operated at constant torque (i.e., constant stator current with a fixed rotor excitation) and adjusted firing, or at constant power with constant motor voltage and control by field weakening.

Operation. The field of the synchronous machine must always be excited. The inverter is fed with unidirectional current from the phase-controlled rectifier through a smoothing inductor in the d.c. link. The d.c. pulses are routed by the inverter to the stator phases at appropriate instants that ensure (i) that the stator travelling-wave m.m.f. is in synchronism with the rotating field of the rotor, and (ii) that optimum interaction torque is maintained. The conditions represented by the simplified phasor diagram of Fig. 10.2(a) then apply.

The firing angle must not be advanced beyond the inversion limit, to ensure safe operation under weak-field or overload conditions. Provided always that there is a stator induced e.m.f. large enough to secure commutation, the machine runs at a leading power factor regardless of the excitation level. Controlled speed, current and torque are readily obtained electronically, and operation in all four quadrants is feasible. Williamson, Chalmers *et al* [153] have analysed the performance of an inverter-fed synchronous motor, and have proposed a method of induced brushless excitation, in which an auxiliary magnetic field set up in the airgap (e.g., by harmonics of the inverter waveform) induces in a special rotor winding an e.m.f. developing the main field current through a rectifier.

Starting. As the stator-winding e.m.f. is proportional to speed, it is zero at standstill, and too small at low speeds to permit of natural (machine) commutation of the inverter. For the speed range from zero to 5–10% of normal a 'switching' mode is necessary. The rectifier is phased back into a brief inversion mode to quench the link current for a few milliseconds before each change in the inverter firing pattern, giving current-free commutation. Firing is set with reference to the rotor position by a sensing device to give a 90° (elec.) angle between the stator and rotor m.m.f. axes, the ideal condition for maximum torque. As the motor accelerates, the more frequent quenching of the link current lowers its mean value, so that maximum torque can be developed only at standstill. Further, the low switching rate of the stator currents results in torque pulsation. Nevertheless, the combination of low-speed switching and higher-speed machine commutation can satisfy the starting, acceleration, speed-control and braking requirements.

Compared with the induction motor, the inverter-fed synchronous motor has the advantages of relative simplicity, lower cost, higher power/mass ratio and simpler speed control.

Cycloconverter-fed Motors
Large low-speed motors for grinding and tube mills, furnace fans, and generators or compensators that cannot otherwise be started and run up, may employ a frequency control based on the cycloconverter (Sect. 8.16), avoiding either speed-reduction gearing or an excessive number of poles. A typical gearless tubular grinding mill, with a normal speed of 10 r/min and a demand of several megawatts, has the rotor of the driving motor attached to the end plate of the mill near to the mill bearing. The surrounding stator has an airgap large enough to compensate for tube distortion and sinking of bearings. The stator is fed at frequencies in the range 0–6 Hz from a 50 Hz cycloconverter, with firing control from a rotor-position sensor. The cycloconverter may be operated to give a sinusoidal fundamental voltage as in Fig. 8.44, but the harmonic content and the reaction on the mains supply can be reduced by a 'trapezoidal' mode, in which the output voltage is varied only during the reversal of polarity, and then remains constant until the next reversal.

13
Special Machines

13.1 Range of special machines
Standard rotary 3-ph induction and synchronous machines of basically cylindrical geometry are of major industrial importance. However, their principles can be applied in many other forms and geometries. The progress from concept through experiment and inventive skill is admirably shown by Laithwaite [154].

Certain devices for voltage adjustment and regulation are based on the transformer and the induction machine. High-frequency 'inductor' generators have been used in eddy-current heating, though now challenged by power-electronic equipments. The induction-disc energy meter can be found at the terminals of factory and domestic electrical supplies. Commutator motors unrestrained by the supply frequency lift the speed restraint of more orthodox machines: the associated technological difficulties were overcome in the first decades of this century. The 1-ph commutator motor is still in wide use in fractional-kilowatt ratings, and the prevalence of 50 and 60 Hz supplies has led to demand for 1-ph induction motors to drive refrigerators, washing machines and domestic tools. Linear motors have been developed for railway traction, liquid-metal stirring and pumping, and duties concerned with locomotion or with restricted rectilinear motion.

Of the many novel and 'unorthodox' machines developed in recent years, it is possible to give here the main features of only some of them in outline.

13.2 Voltage regulators
Static equipments provide a variable voltage from a constant-voltage a.c. supply. They are based on the transformer and the induction machine.

Moving-coil Regulator
Wide and stepless voltage control can be obtained by varying the coupling between coils on a common magnetic circuit. Figure 13.1 shows the essentials of a 1-ph regulator: the 3-ph version comprises three 1-ph units or a 3-limbed core arrangement The core in (*a*) has windings *a* and *b* in opposition at the upper and lower regions of one limb. An isolated short-circuited coil *s* can be moved up and down the limb. The closer *s* is to *a*, the greater is their mutual inductance and the lower is the input impedance of *a*. If a voltage be applied

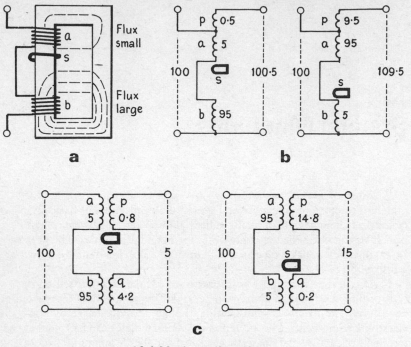

13.1 Moving-coil regulator.

to *a* and *b* in series, its distribution between them depends on their respective inductances, and the closer *s* is to one of them, the smaller is the proportion of the applied voltage across it. The voltages across *a* and *b* can in fact be varied between 5 and 95% of the total voltage applied by positioning *s*, which by its induced current varies the leakage flux and therefore the reactance.

There are several useful arrangements. In the auto connection (*b*) an auxiliary winding *p* is used, with one-tenth of the turns of *a* or *b*. Coils *a* and *p* act together as a normal transformer, so that with *s* at the top and 5% voltage across *a* there is a 0.5% voltage in *p* and an output of 100.5%. With *s* at the bottom, the voltage of *a* is 95%, of *p* is 9.5%, giving an output of 109.5%, with a smooth adjustment by varying the position of *s*. In (*c*) the turns-ratio of *a* and *p* is 6.43, and of *b* and *q* is 22.5, giving an output variable between 5 and 15%. This simple regulator, which introduces no phase displacement is discussed with further variants by Rawcliffe [102].

Sliding-contact Regulator

A two-winding or auto transformer, with windings on a toroidal core, has the outermost winding bared over a circular track swept by a contact brush or roller, so picking off an output voltage in steps of the voltage per turn. The device ('variac') is essentially 1-ph and has a rating limited by the current that can be satisfactorily collected by the sliding contact.

Induction Regulator

The construction resembles that of a slip-ring induction motor, but the movable member is locked at adjustable positions over a pole-pitch, and is often set vertically with its shaft connected to a worm/pinion device for adjusting the position and withstanding the torque.

Single-phase Regulator. Figure 13.2 shows the section through a 2-pole 1-ph regulator (*a*) and its connections (*b*). The constant-voltage primary supply is fed by flexible connectors to the inner (movable) member: its flux ϕ_m induces in the stator (secondary) winding an e.m.f. E_2 of magnitude dependent on the angle between the axes of the two windings. For the relative orientation (*c*), E_2 provides maximum boost to give an output voltage $V_2 = V_1 + E_2$. A 90° displacement (*d*) decouples primary and secondary so that $E_2 = 0$. A further 90° displacement (*e*) reverses the secondary e.m.f. to give the output voltage $V_2 = V_1 - E_2$.

The 1-ph regulator is always provided with a short-circuited q-axis compensating winding, shown black in (*a*), to reduce the reactance in the midway position and to stabilize the $I^2 R$ loss over the whole voltage range. Variation of primary and compensating m.m.f.s with displacement is shown in Fig. 13.3(*a*). If the two windings have the same equivalent resistance the combined $I^2 R$ loss is substantially independent of displacement. An output-voltage/displacement relation for a regulator with a ±20% regulation is shown in (*b*).

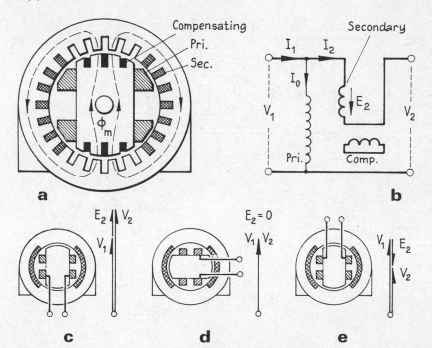

13.2 Single-phase induction regulator.

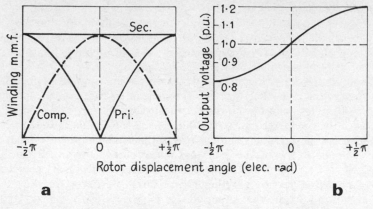

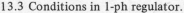

Rotor displacement angle (elec. rad)

a   b

13.3 Conditions in 1-ph regulator.

Three-phase Regulator. The electric and magnetic circuits resemble those of a
3-ph slip-ring induction motor, but with the primary supply to the movable
member fed through flexible connections, as in (*a*) of Fig. 13.4. The gap flux
is a nearly constant travelling wave. The secondary e.m.f. is therefore
constant in magnitude but its phase angle with respect to the primary voltage
varies with the position angle θ. The output voltage V_2 is the phasor sum of

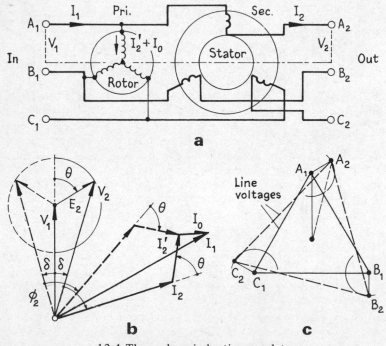

13.4 Three-phase induction regulator.

V_1 and E_2 given by the circular locus in (b). According to the sense of the angle θ, regulation follows either half (full-line or dotted arcs) of the locus. The change in magnitude and relative phase of a 3-ph output from $A_2 B_2 C_2$ with respect to the input voltage at $A_1 B_1 C_1$ is shown in (c).

Let the output per phase be I_2 at V_2 and phase angle ϕ_2. The primary current is the phasor sum of the magnetizing current I_0 and the secondary-current equivalent $I_2' = I_2 (E_2/V_1)$. Diagram (b) shows $(I_2' + I_0)$ to be smaller for the full-line locus, and the direction of the displacement θ is arranged for this condition. Further, V_2 is cophasal with V_1 only for the boost and buck maxima; for intermediate positions there is a phase displacement δ. This can be avoided by using a bank of 1-ph regulators, or by a 3-ph *double regulator* which comprises two mechanically coupled regulators with their secondaries in series and their primaries fed in opposing phase sequence.

A 3-ph regulator develops a torque $M = kV_1 I_2' \cos(\theta \pm \phi_2)$. For a unity p.f. load M is greatest at maximum boost and buck and is zero for $\theta = 90°$; for a purely reactive load it is zero at maximum boost and buck. In the double regulator the torques oppose to give a resultant proportional to $[\cos(\theta + \phi_2) - \cos(\theta - \phi_2)]$, which vanishes for a unity p.f. load. The rating depends on the maximum boost. A 3-ph regulator for 6.6 kV ± 10% and a current of 200 A/ph has the rating $\sqrt{3}$ x 6.6 x 0.1 x 200 = 230 kVA. Immersion of such a regulator in oil aids the insulation strength and reduces noise.

A general circuit analysis of the double regulator by Jha [88] includes a 'biased' form in which the E_2 locus is an ellipse.

13.3 Reluctance machines
These exploit the alignment principle (Sect. 1.5) for both generator and motor action, sometimes with hysteresis as a main or auxiliary effect.

Inductor Generators
Unlike conventional machines, these utilize the pulsation of a flux produced by cyclic variation of the magnetic-circuit reluctance. The *homopolar* form has a unidirectional gap flux excited by an annular field coil; the *heteropolar* form has alternate N and S poles around the gap periphery, with field coils wound in the plane of the armature coils. The machines are built for high-frequency outputs. *Guy* slotting has unwound 'inductors' on both sides of the airgap. *Lorenz* slotting comprises semiclosed armature slots and an otherwise smooth stator airgap surface between the field slots. If the Lorenz form has instead a dentated stator surface it is termed a *Guy-Lorenz* machine. The essential features are illustrated in Figs 13.5 and 13.6

Homopolar, Fig. 13.5. Two stator and two rotor cores form a single magnetic circuit, excited by an annular d.c. field winding on the stator. Rotor movement through one tooth-pitch produces one complete cyclic variation of reluctance and in consequence a pulsating flux. With S_2 rotor slots (or teeth) the frequency for a rotor speed n is $f = nS_2$, independent of the stator tooth arrangement, of which lower- and higher-frequency versions are illustrated.

Heteropolar, Fig. 13.6. This has single stator and rotor cores and a short gap

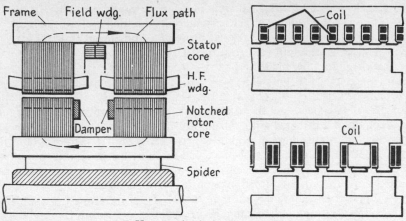

13.5 Homopolar inductor generator.

length, an important advantage in raising the output. The working flux is almost constant, so that no pulsating e.m.f. is induced in the d.c. field coil. To generate a frequency f with a rotor peripheral speed u the slot-pitch is u/f. This may be too small for the higher frequencies: it may then be necessary to adopt stator and rotor teeth of equal pitch with the stator coils lodged in larger slots of span equal to any suitable whole number of the smaller teeth.

Action. Detailed analysis, concerned as it is with variation of the reluctance of the gap with rotor position and the effects of h.f. currents and tooth saturation, is complicated. Bunea [89] for the Guy slotting, and Davies and Kay [90] for the Lorenz machine, give *ad hoc* analyses. The following is a simplified treatment.

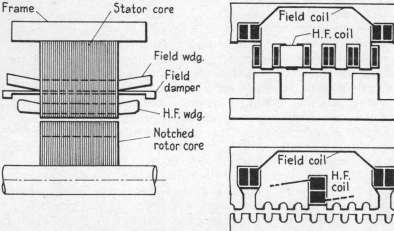

13.6 Heteropolar inductor generator.

The gap permeance fluctuates as the stator and rotor teeth move in and out of register. Let it be $\Lambda_g = \Lambda_0 + \Lambda_1 \cos\theta$, with θ an electrical angle of rotor displacement from the position of maximum gap permeance (i.e., minimum gap reluctance). Neglecting the ferromagnetic parts of the magnetic circuit, the no-load flux is $\Phi_0 = F_t\Lambda_g$ for a d.c excitation F_t. On load the stator current is $i_1 \sin(\theta - \phi - a)$, giving a corresponding reaction m.m.f. F_1, where ϕ is the (lagging) load phase angle and a is an additional lag caused by a shift in the flux axis consequent upon the reaction F_1. On load the flux is

$$\Phi = [F_t + F_1 \sin(\theta - \phi - a)] \,(\Lambda_0 + \Lambda_1 \cos\theta)$$

equivalent to a constant mean flux, a second harmonic (here neglected) and a fundamental-frequency alternating component

$$\Phi_m = \Phi_t + \Phi_a = F_t\Lambda_1 \cos\theta + F_1 \Lambda_0 \sin(\theta - \phi - a)$$

The phasor diagram, Fig. 13.7, shows e.m.f.s E_t and E_a corresponding to these flux components, and their resultant E_1. The machine can be represented by a source e.m.f. and a synchronous reactance $x_s = E_a/I_1$, with a load of impedance Z. The apparent-power output is a maximum for $Z = x_s$, and for this condition it is $S = (E_t^2/4x_s) \,[\cos\phi/(1 - \sin \phi)]$. However, saturation and core loss may have a large influence, difficult to assess with precision.

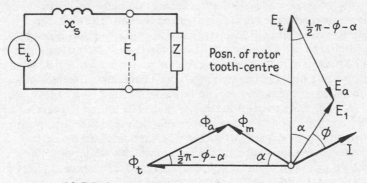

13.7 Inductor generator equivalent circuit.

Inductor generators have been built with ratings up to 2000 kVA for frequencies in the range $0.25 - 10$ kHz, e.g., for induction-melting processes.

Flux-switch Machine
For outputs between 0.1 and 5 kVA at frequencies $1 - 10$ kHz, flux-switch generators have been made, in which the flux reversal in a pair of stator coils is achieved by the rotation of a simple shaped rotor, Fig. 13.8, normally excited by permanent magnets. Flux-switch machines are designed for speeds in the range 150–850 r/s. Problems arise from high centrifugal force and high internal impedance.

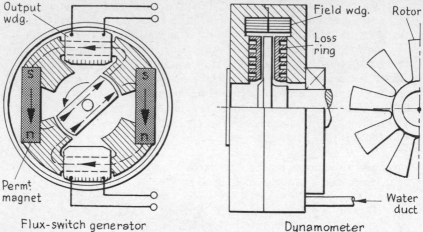

13.8 Inductor machines.

Dynamometer

This measures the torque and speed of a rotating machine and dissipates its power in heat. It is applied to the testing of engines, motors and turbines. Wide ranges of power and speed are required, combined with minimum inertia and low parasitic loss in windage and friction. The operating temperature-rise, typically 150 °C, must not produce undue dimensional change in the airgap. Chalmers and Dukes [143] describe a dynamometer, Fig. 13.8, based on the homopolar inductor generator. A d.c. field winding produces a flux between the loss rings, which have circumferential grooves for circulation of cooling water. The flux is modulated by a 9-blade steel rotor, inducing eddy currents in the loss rings. The torque is measured by a weigher arm and strain-gauge load cell, while the speed is derived from pick-up coils associated with a toothed wheel mounted on the shaft.

A variant is the eddy-current *coupling,* in which a salient-pole rotor drives a wrought-iron drum for torque transmission. Davies *et al* [142] give an experimental verification of the generalized theory.

Polyphase Reluctance Motors

Reluctance motors are cheap, robust and reliable. Despite low p.f. and specific output they have found application in such fields as the positioning of control rods in nuclear reactors and providing synchronized textile drives. The machine is adaptable to position and speed control by a variable-frequency supply. The limitations imposed by pull-in torque requirements are then avoided, the motor starting from rest and accelerating synchronously as the frequency is raised. Even for a simple motor at constant frequency, the pull-out torque is several times the pull-in torque. An additional advantage is that for steady synchronous load there is no rotor $I^2 R$ loss.

For variable-frequency operation the basic requirement of the rotor is a maximum difference between the gap reluctance of the d-axis (salient) and the q-axis. For fixed-frequency operation the further essential is adequate start, run-up and pull-in torques. Typical rotor constructions are shown in Fig. 13.9. The stator in each case carries a 3-ph winding. The simple and obvious saliency in (*a*) can be changed to (*b*) by interchanging the air and iron parts, the preferred flux paths now being essentially circumferential instead of radial, the non-preferred q-axis paths being of high reluctance on account of the large nonmagnetic spaces between the segment ends. The effective reluctance ratio is obtained without cutting away large sections of the rotor periphery and of the starting cage (which would be detrimental to asynchronous and pull-in performance). Rotor (*c*) is made by assembling laminations of salient form (or by machining a solid steel rotor body) and flanking it with copper. Form (*d*) is made by winding a grain-oriented steel ribbon into an annular shape, deforming it into a rough square, and cutting it across into four parts which are then assembled to give a 4-pole design. Figure 13.9 does not show the rotor cage

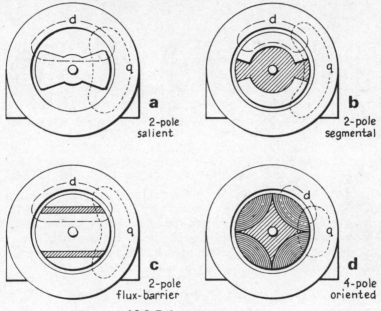

13.9 Reluctance motors.

Although usually small, reluctance motors can be built in ratings up to 100 kW or more, the larger machines having more advanced rotor designs embodying flux barriers to augment the reluctance ratio. Vernier motors have been made for subharmonic rotor speeds by slotting the rotor. Two-speed operation is also possible: a conventional 2-pole rotor still has some degree of saliency in a 4-pole field and can run stably at one-half normal speed. *Action.* The details of reluctance motor design present too great an asymmetry for the application of the generalized treatment in Sect. 7.5. Lawrenson [91] has greatly advanced the reluctance machine technology; he analyses the gap permeance distribution on which the reluctance ratio depends, deriving permeance and fringe factors in terms of the rotor geometry.

Starting. A constant-frequency machine has to start by means of the asynchronous (induction) torque. Fig. 13.10 shows at (*a*) a typical torque/ speed relation for a motor started from rest against constant full-load torque, and synchronizing successfully. The dotted line represents the 'steady-state' characteristic of the induction-torque component, which has a small negative value at zero slip. The small-slip region in (*b*) is for the same machine with an increased load inertia. In this case the motor reaches synchronous speed but is unable to hold it, with the result that it runs asynchronously in a fixed cycle (shown shaded) with large pulsations of torque and speed. Thus for a given motor the combination of load torque and inertia determines whether or not the machine can synchronize.

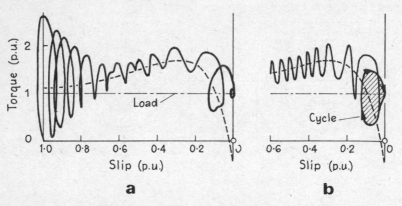

13.10 Reluctance motor: run-up torque/slip relation.

Performance. Good performance is a matter of reconciling conflicting factors. A high-reluctance ratio gives a high rating and synchronous pull-out torque, but impairs the pull-in torque and the stability. The following data (with the torques in terms of rated values) are typical for a 5 kW machine:

> reluctance ratio, 3–6; efficiency, 0.7–0.8; p.f., 0.6–0.8
> pull-out torque, 2–2.5 p.u.; pull-in torque, 0.9–1.2 p.u.

Hysteresis Motor

Whilst having some likeness to the reluctance machine, the hysteresis motor develops its polarization in a different way. In simplified terms, the travelling wave of stator m.m.f. F_1 magnetizes the rotor, establishing a distributed rotor flux of density B_2, caused by hysteresis to be displaced in angular position from F_1 by an angle δ, as in Fig. 13.11. At *subsynchronous* rotor speeds, the sine-distributed wave F_1 moves past a given element of the rotor volume, subjecting it to a change of magnetizing force *H*. The consequent value of B_2 is a function of the *cyclic* hysteresis loop of the rotor core material. The sequence of excitation through the values 0, 1, 2 . . . produces corresponding changes of B_2. There is a displacement angle δ between F_1 and B_2 for the volume element and therefore for the rotor as a whole, developing a torque in the direction of travel of F_1. The torque is produced without rotor current (except for a small eddy current), and as it depends only on δ it is independent of speed up to the synchronous provided that the cyclic hysteresis loop is not affected by slip. The motor thus has a starting and run-up torque. At *synchronous* speed the relation of F_1 and B_2 is roughly stabilized to a fixed polarization, resisting pull-out.

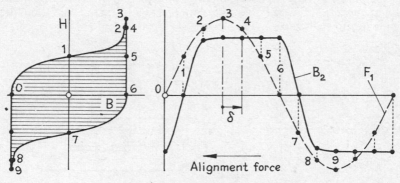

13.11 Hysteresis motor.

The output of a hysteresis motor is about one-quarter that of an induction motor of the same dimensions, making it suitable only for small ratings such as for driving gyros. The 3-ph (or 2-ph) stator winding is in closed or semi-closed slots to avoid minor parasitic hysteresis loops. The rotor is an unwound cylinder of magnetically hard chrome or cobalt alloy steel with a large hysteresis loop and a high resistivity to reduce eddy-current loss. For high-frequency supplies the rotor may have to be laminated, or sintered, or built up from alloy powder set in a synthetic resin. Alternatively the rotor is a number of separated discs with the flux entering the faces for better use of the active material, a construction that gives also a reduced inertia. *Hysteresis/reluctance Motor.* If two diametrically opposite slots are milled in the annular ring rotor of a 2-pole hysteresis motor, a degree of saliency is introduced so that, although the hysteretic torque is slightly impaired, a reluctance torque is developed. If the slots are used to accommodate a short-circuited copper cage or winding, a third (induction) torque is added to improve run-up.

13.4 Induction instruments
Essentially an induction-type measuring instrument employs two a.c. electromagnets acting on a conductive disc (usually of aluminium) able to rotate in the magnet airgaps. The magnet fluxes induce eddy currents in the

disc and are so set that the eddy current of one lies in the field of the other to produce interaction forces. A torque of unidirectional mean value is impressed on the disc, which will rotate provided (i) that the magnets are not sited at opposite ends of a disc diameter making the forces purely radial, and (ii) that there is a time-phase displacement between the fluxes. The induction torque on the disc is balanced by a spring for an indicating instrument, or by a brake for summation metering.

Ammeter

The basic arrangement of a 1-ph shaded-pole ammeter in Fig. 13.12(a) shows phase-displaced fluxes Φ_1 and Φ_S derived from a single magnet by encircling the right-hand half-pole by a *shading ring* comprising typically a single closed turn of copper. In this half-pole Φ_S is developed by the m.m.f. $I_1 N_1$ modified by the short-circuit current I_S induced in the shading ring. Thus Φ_S lags Φ_1 by

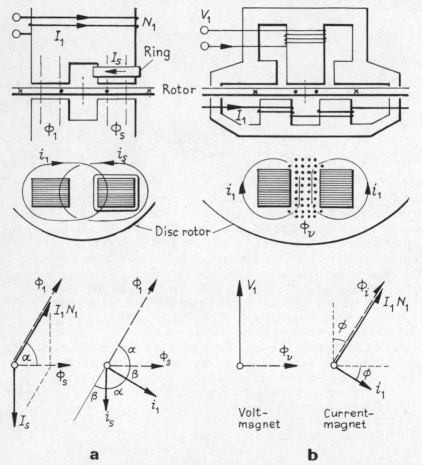

13.12 Induction instruments.

a *time*-phase angle a. In the disc, Φ_1 and Φ_S induce e.m.f.s. with a displacement a, and corresponding eddy currents i_1 and i_S, assuming both ring and disc to be purely resistive. Interaction of the current-flux pairs $i_1\Phi_S$ and $i_S\Phi_1$ gives torques proportional to

$$i_1\Phi_S\cos\beta = i_1\Phi_S\sin a \quad \text{and} \quad i_S\Phi_1\cos(\pi - \beta) = -i_S\Phi_1\sin a$$

because $(a + \beta) = \frac{1}{2}\pi$. The flux/current *space* orientations result in the component torques being additive to give a torque proportional to $(i_1\Phi_S + i_S\Phi_1)\sin a$. As both fluxes and eddy currents are proportional to I_1, the mean torque is $M = KI_1^2$, which can be equated to $c\theta$ when the disc movement is constrained to an angle θ by a spring of stiffness c. Modern forms of ammeter may use separate magnets with alternative methods to achieve the displacement a. In any case, the phasing results in a 'flux shift', a crude form of travelling wave.

Energy Meter
Here a two-magnet eddy-current device is used to give continuous summation of the active-power/time integral. An aluminium rotor disc, Fig. 13.12(b), on a vertical spindle with saphhire bearings or magnetic suspension, drives a revolution counter. The disc is driven by voltage- and current-excited magnets and its speed is controlled by a permanent-magnet brake. The flux Φ_v of the volt-magnet is in almost lagging phase-quadrature with the supply voltage V_1. For zero current in the current-magnet, Φ_v is symmetrically disposed, but for a load current I of phase angle ϕ the combination of Φ_v and Φ_i with the currents that they induce in the disc, develops a pulsating torque with a unidirectional mean proportional to the active power $V_1I_1\cos\phi$. The composite flux and eddy-current pattern is complex, but the eddy current induced by Φ_i may be taken to interact with Φ_v to produce torque. Practical meters must meet stringent specifications, and require (i) adjustment of the radial position of the brake magnet for full-load accuracy, (ii) a short-circuited 'quadrature' coil around the central limb to make the phase angle between Φ_v and V_1 precisely $90°$, and (iii) a 'low-load' adjustment which, by giving a slight asymmetry to the volt-magnet flux distribution, develops a no-load torque to balance the friction of the pivot and counter mechanism.

13.5 Permanent-magnet Machines
A small synchronous machine can be excited by permanent magnets, with the advantage of a compact design that eliminates field coils, a d.c. supply, excitation loss, and slip-rings and brushes. But the flux of such a machine is not readily controllable, and magnets are liable to suffer demagnetization. Typical properties for four classes of p.m. material are listed below, giving representative values of remanent flux density B_r, coercive magnetizing force H_c, maximum energy stored (BH)max and resistivity ρ.

Type		Material	B_r (T)	H_c (kA/m)	(BH)max (kJ/m^3)	ρ ($\mu\Omega$–m)
A:	Ceramic	Barium or Strontium ferrite	0.38	200	30	high
B:	Metallic	Alcomax	1.25	55	43	500
C:	Metallic	Hycomax	0.8	100	34	500
D:	Rare-earth	Samarium-cobalt	0.75	600	130	60

Type A material is cheap but bulky, readily mouldable with some tolerance, and suitable for low-rated production-run motors. Types B and C are hard, and so difficult to shape that very simple geometries are essential. Type B magnets must be long in the direction of magnetization, and type C are sensitive to demagnetization. Type D material, easily moulded and machined, can be be used for most machines, including those that are axially short, but is expensive. Thus a simple shape is necessary for B and C, while A and D are adaptable. The practical tendency is to employ A for low-cost machines, and polymer-bonded samarium-cobalt rare-earth magnets D where better magnetic and mechanical properties can be justified.

Magnetic-circuit design is a somewhat complicated process. The working flux density is lower than B_r because of leakage, reluctance and the effects of armature reaction. A demagnetizing effect is impressed on the p.m.s. when large currents flow in the 3-ph stator windings, (i) at starting (although the p.m. is partially shielded if the rotor has a cage), and (ii) at synchronous pull-in when a cage is inoperative.

Constructional Forms

These are influenced by the duties (such as self-starting) specified. Some forms are shown in Fig. 13.13, in which the direction of magnetization in the p.m.s is indicated by arrows. The basic 2-pole version (*a*) has a p.m. set within an annular rotor core carrying a starting cage. The core is split across a diameter by 'barrier' slots to prevent short-circuiting of the p.m. flux. Orthodox replacement of field windings by p.m.s of types B or C gives version (*b*), a salient-pole design, with the magnets and pole-shoes bolted on. Form (*c*) exploits the lateral magnetizability of type D materials, the poles at the rotor surface being of wedge shape to retain the p.m.s, an arrangement that resists demagnetization. The radical departure from the conventional form in (*d*) is constructed from rotor laminations carrying laterally-magnetized non-radial p.m.s, and the cage is augmented by metal inserts to give a cylindrical rotor. The balance between synchronous and asynchronous performance demands a careful selection of the design parameters. Annular shaped A or D p.m.s, axially magnetized and assembled with steel flux guides interleaved as in (*e*), provide a heteropolar flux at the gap surface: the use of flux guides shields the p.m. from demagnetizing m.m.f.s.

Several novel arrangments have been devised: one has a rotor of disc form presented as a 'faceplate' to a 3-ph winding with radial active conductors; the

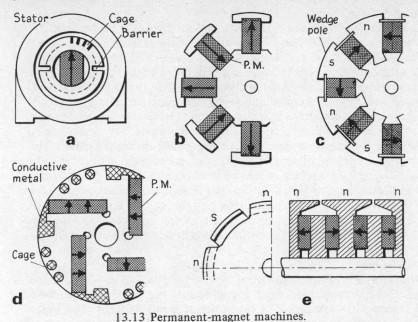

13.13 Permanent-magnet machines.

rotor disc may be of steel carrying circular- or sectorial-shaped p.m.s, or of type D material with a multipolar axial field impressed on it. These, and several other designs, have been developed by Binns and his co-workers [160].

Applications
Generator. The p.m. generator, which needs no starting cage, is efficient and robust. It has been adopted for the role of pilot exciter for the fields of large generators, and is of interest in the exploitation of 'alternative-energy' sources such as wind drive. Mainer [97] shows that the machine can be regarded as a self-regulating conventional generator in which the rotor m.m.f. is represented by the terminal m.m.f. of the magnets, with comparable expression for power and a similar phasor diagram.

Motor. Machines in ratings from 5 W to 50 kW have been used in process industries (e.g., synthetic fibres) where a precise speed is required for upwards of 200 associated motors, with overall speed control by a variable-frequency inverter. For speeds above 15000 r/min it is necessary to abandon the usual laminar rotor for a solid form. With die-cast rotors it is possible to insert the magnets before casting and magnetize them *in situ.* Efficiency and power factor are of importance where the starting current is limited by the inverter, and where high-inertia loads must be driven.

13.6 *Control machines*
A.C. machines for speed and position control systems include selsyns, synchros, 2-ph servo motors, tachometers and stepper motors.

Power Selsyns

The structure of a wide-span travelling crane is flexible in the plane of travel. Drive by a single induction motor with a cross-span shaft is not practicable. The use of a motor at each end of the span secures uniform travel, avoids skewing and improves adhesion, but it calls for an equivalent *electric shaft* or *electric tie*. Comparable requirements may occur in process industries (e.g., paper, steel, textile). Let two similar slip-ring induction motors have their stator windings connected to the same supply: when they run at the same speed, their rotor phase e.m.f.s are equal in magnitude, but are displaced if the rotor positions with respect to the airgap field are not the same. If the slip-rings are properly interconnected, an advanced position of rotor 1 causes a circulating current which retards 1 and advances 2, so that the two rotors tend to remain aligned. Machines so connected may be called 'power selsyns' if they are large enough to provide the load torque. The action tends to be lost at low rotor slips (i.e., running near synchronous speed) and for unequal torque demands. Under some operating conditions the speeds may fluctuate.

Pilot Selsyns

When two 3-ph induction machines are energized and interconnected as in Fig. 13.14, the two shafts retain a fixed relative angular position. Variation of the speed of the 'transmitter' T (by means of an auxiliary motor or, in the case of a crane, by one of the rail wheels) causes the speed of the 'receiver' R to be similarly varied. If R is loaded a small angular divergence is produced, the counterpart of the twist of a mechanical shaft. To change the speed of R without changing that of T requires a frequency change, which can be achieved with the aid of a separate motor and variable gear to drive machine D, which acts as an *electrical differential*.

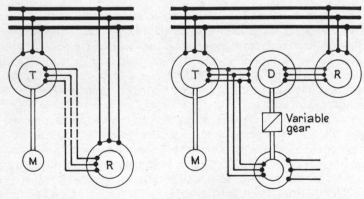

13.14 Selsyns.

Synchros

The term 'synchro' is generic for control machines having three stator and one or three rotor windings, normally operated at a frequency such as 400 Hz but otherwise resembling industrial machines in miniature. The winding axes

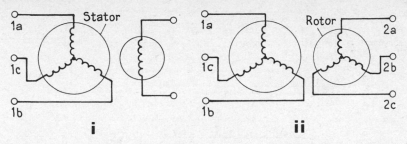

13.15 Synchros.

are shown conventionally in Fig. 13.15, and typical forms are listed below with reference to connections (i) and (ii) and the method of excitation. Despite appearances, the synchro is *not* a 3-ph machine: it has a single-phase input.

Transmitter (i). The rotor is salient and a.c. excited. The stator windings develop co-phasal e.m.f.s of magnitude depending on their angular position relative to that of the stationary rotor, with terminal voltages given by

$$v_{1a} = k \sin\theta \qquad v_{1b} = k \sin(\theta - 2\pi/3) \qquad v_{1c} = k \sin(\theta - 4\pi/3)$$

These uniquely define the rotor angular position θ.

Control Transformer (i). Stator input voltages v_{1a}, v_{1b} and v_{1c} define a position angle θ. The rotor induced e.m.f. is proportional to $\sin(\theta - a)$, where a is the angular position of the shaft of the rotor, which is cylindrical.

Receiver (i). The rotor input is an alternating voltage defining a, the stator input is a set of voltages defining θ. A torque proportional to $\sin(\theta - a)$ appears at the rotor shaft.

Differential Transmitter (ii). Both stator and rotor are cylindrical. One member is excited with voltages defining θ. The output from the other defines a difference angle $\beta = (\theta - a)$.

Differential Receiver (ii). When the voltages to one member define angle θ and those to the other an angle a, there is a torque on the rotor shaft proportional to $\sin(\theta - a)$.

The operations described are valid only for steady-state stationary conditions. When the rotor moves, rotational as well as pulsational e.m.f.s are generated in both stator and rotor.

Typical synchro systems are illustrated in Fig. 13.16. System (*a*) employs direct transmission of position data without feedback. The output synchro must develop a torque sufficient to position the load, and there will be a small load-dependent position error. In the closed-loop system (*b*) the operation is now determined by the error and the torque is derived from the amplifier-motor combination, minimizing the synchro torque demand and reducing the overall system error.

Ellipsing of stator and rotor, differences in coil resistance and leakage reactance, variation in the profile of salient members and rotor eccentricity all impair the position-angle accuracy and have to be minimized. Ellipsing and

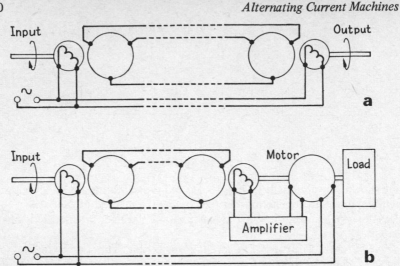

13.16 Synchro systems.

slotting introduce space harmonics. Residual error is indicated by a voltage output at the null position for which it should be zero, result from non-linearity and asymmetry in the ferromagnetic cores. It is usually necessary to 'match' synchro pairs with errors that correspond, so minimizing the resulting error of the combination. High-quality pairs may have a static error of less than 0.1 deg.

Control Motors
Some servo actuator requirements are met by the 2-ph induction motor having a cage (or equivalent) rotor and two stator windings set in space quadrature. One (main) stator winding is continuously excited to maximize the torque per unit of power in the other (control) winding. The characteristics are generally designed to give a stalling torque proportional to the control-winding voltage applied, and a torque that falls with rising speed for a given control voltage. With main and control voltages equal and in time-phase quadrature, the rotor effective resistance is such as to give maximum torque for a slip approaching 2 p.u., and therefore the maximum on sudden reversal (simply accomplished by reversing the control-winding voltage). Although this restricts the available load and stall torque levels, it provides an almost rectilinear fall of torque with rise in speed, giving an effective damping resembling that of viscous friction. Figure 13.17 shows typical steady-state performance characteristics to a base of output torque. In (*a*) the *input power* and *power factor* are plotted with rated voltage on both windings; the *output power* is roughly parabolic; the *control* curve is the variation of stalling torque with control-winding voltage. In (*b*) the speed/torque relation is shown for three control voltages. The efficiency is low, but this can be tolerated in a small machine.

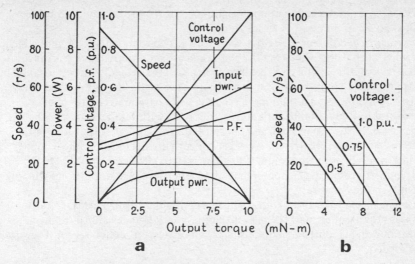

13.17 Two-phase servo motor characteristics.

Design criteria differ basically from those for industrial power drives. The basic requirement is an optimum performance in the servo system of which the 2-ph motor forms a part. An important criterion is that the torque be zero when there is no control-winding signal, a condition obtained by careful construction and a high rotor resistance. As the duty of the motor involves rapid speed change, its inertia must be small. Hesmondhalgh [93] deals with design for maximum stall torque and linearity for a given power and temperature-rise. The small dimensions of instrument 2-ph motors (e.g., rotor diameter 12.5 mm; stator outside diameter 40 mm; 8, 12 or 16 stator slots; number of poles up to 16) and the relatively strong second-order effects make it necessary to relate design to measurements on experimental prototypes.

Variable-frequency Control Motors. A variable-frequency supply has been suggested as a means for speed and position control with either cage induction or synchronous motors. The overall problem is that of determining the frequency response of the machine to variation of the 3-ph supply frequency to the stator winding, in terms of a transfer function relating output speed to demand speed; and achieving a specified rate of acceleration. The problems have been studied by West *et al* [155].

Tachometers

A tachometer is a small induction machine with a cage rotor and a stator with two windings set in space quadrature. One stator winding is supplied at constant voltage and frequency; an e.m.f. of the same frequency and proportional to the rotor speed is induced in the other. Errors arise from rotor asymmetry and imperfect quadrature setting of the stator winding axes. Resistance and leakage reactance reduce the effective flux as the speed is

raised, making the response nonlinear. Much of the error is reduced by use
of a 'drag cup' in place of a cage rotor.

Drag-cup Tachometer. The drive shaft carries a copper cylinder or cup, which
rotates in the airgap between the stator and a fixed internal core. As precision
depends on the stability of the cup, its material is selected with regard to grain
structure and stress-relief after forming, or even by boring out of solid stock.
To retain dimensional stability, all mating parts have matched thermal
expansion. Reduction of variation of resistance with temperature may be
achieved by the use of manganin, in spite of some loss of output that results
from the higher resistivity. Low-accuracy cups are made from drawn copper
tube, or even from aluminium if the inertia must be low. Errors due to
asymmetry result in the appearance of an e.m.f. in the stator pick-up winding
when the cup is at rest.

Permanent-magnet Tachometer. This is essentially a 1-ph generator. At low
speeds the frequency of the pick-up e.m.f. is very low. The output may be
rectified and smoothed, and the shaft then carries a direction sensor if reversal
is needed.

Stepper Motor

A stepper motor is employed for the direct control of angular position. Its
shaft is made to stop accurately in any one of a number of positions in a
complete revolution. A step may, according to the design, correspond to a
rotor angle between $180°$ and $1°$ (or less), by energizing the stator windings

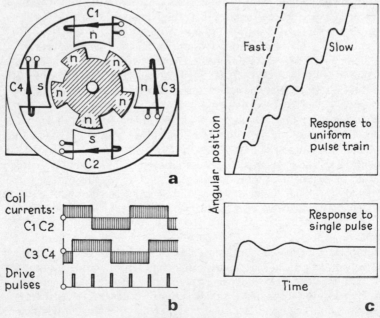

13.18 Stepper motor.

with unidirectional current pulses. By suitably timing the pulse sequence, the effective motor output can be in discrete steps, a nearly constant speed, or a specified rate of acceleration. The stepper is not strictly an a.c. machine, and only a brief outline is given here.

The operating feature is that the rotor moves into a 'preferred' magnetic position. In Fig. 13.18, with the pairs of stator coils C1C2 and C3C4 energized as in (a) to give the polarity indicated, the 5-pole homopolar permanent-magnet rotor is in the position of minimum reluctance. The angle between rotor poles is $72°$: hence if the current in either pair of stator poles is reversed, the least angle through which the rotor must move to regain a minimum-reluctance condition is $18°$, corresponding to 20 steps per revolution. Thus if the stator currents in the coil-pairs are reversed sequentially with the arrival of each control pulse in a pulse-train, as in (b), the rotor rotates in a series of defined steps. Digital logic circuits with transistor switching can meet such control demands economically.

The rotor response to a step may be oscillatory, with a frequency and damping depending on the inertial and frictional properties of the motor-load combination. Such a response is shown in (c), together with that for a rapid succession of pulses. If the pulse frequency is high, the rotor may be pulsed when it is in an under- or over-shoot position, introducing position error or stalling. For applications such as machine-tool control, when high slewing rates are necessary to start successive machining cycles, the error or stalling tendencies are eliminated by incorporating the stepper motor into a closed-loop system. [Ref 113.]

13.7 Commutator machines

The speed of both synchronous and induction machines is tied to the supply frequency. This restriction is lifted by use of a rotor carrying a closed-circuit winding fed through a commutator and a fixed brush system, because this combination has a *frequency-changing* effect, permitting a wide speed range with a fixed frequency supply.

D.C. Machine. A conventional d.c. (i.e., zero-frequency) machine has this capability. A direct current is fed to the commutator through the brushes, but the current in each rotor coil is reversed as the commutator sectors to which it is connected pass under a brush. Internally the rotor coil currents thus alternate at the speed frequency, yet at the brushes the currents have zero frequency. The commutator and brushes act as a frequency-changer. A similar conclusion holds for the generated e.m.f. between brushes: individual rotor coils have speed-proportional e.m.f.s of speed frequency ω_r which are rectified by the commutator/brush system. The action is independent of speed, accounting for the fact that stator and commutator-rotor windings can be connected in series to develop unidirectional torque. In brief, the frequency at the commutator brushes is the same as that of the field supply — in this case, zero.

In terms of the generalized theory, the basic conditions for a 2-pole d.c. machine are shown in (a) of Fig. 13.19. A field winding F sets up a d-axis flux, and a current I_q is fed to the rotor by diametral brushes enabling it to

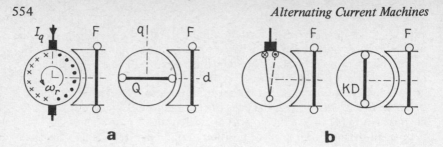

13.19 Single-phase and d.c. commutator motors.

develop a q-axis m.m.f. in the quasi-stationary equivalent coil Q representing the m.m.f. action of the rotor winding. Interaction of the two produces a unidirectional torque irrespective of the rotor speed.

Single-phase A.C. Machine. As the direction of rotation and of torque are unaffected by reversal of the terminal polarity of a d.c. series machine, it can work satisfactorily if the supply voltage alternates at angular frequency ω_1. The torque will, of course, fluctuate at a frequency $2\omega_1$ (because the current goes through zero twice per period) but will have a unidirectional mean value. Then across the rotor brushes the e.m.f. is proportional in magnitude to the speed ω_r while its frequency is ω_1, i.e., that of the field in which the rotor rotates. The a.c. series motor thus has the speed freedom associated with the d.c. machine. The technology of the 1-ph series motor is considered in Sect. 13.8.

Polyphase Machine. Consider now a 2-pole rotor carrying a closed winding with its coils connected to a commutator, and also symmetrically tapped to slip-rings into which balanced polyphase currents are fed. With the rotor *stationary* the polyphase currents set up a travelling wave field of angular speed ω_1 corresponding to their frequency, and induce e.m.f.s of that frequency in the winding. Thus slip-ring, winding-conductor and commutator-brush e.m.f.s and currents have all the same frequency ω_1. With the rotor *rotating* at angular speed ω_r, the speed of the travelling-wave field with respect to the winding conductors is still ω_1 because the tappings rotate with the rotor; the frequency of the conductor currents and e.m.f.s remains at ω_1. But relative to the fixed commutator brush axis the field travels at $(\omega_1 \pm \omega_r)$, and this is now the frequency in the external connections to the commutator brushes. Thus the commutator changes ω_1 in the conductors to $(\omega_1 \pm \omega_r)$ at the brushes. The sign depends on the direction of rotor rotation with respect to that of the travelling-wave field, and the resultant frequency is in fact that of the field with respect to the fixed brush axis, a fact basic to the operation of the polyphase commutator machine.

Commutation
When a rotor coil undergoes commutation, i.e., while it is short-circuited by a brush as in Fig. 13.19(*b*), the current reversal is opposed by the inductance arising from the leakage flux in slots and overhang. The commutation time

is lengthened and may lead to sparking. The effect, considered as a
'reactance e.m.f.' opposing current reversal, appears in d.c. machines, but is
augmented in a.c. machines by further rotational and pulsational e.m.f.s,
making commutation more difficult. The e.m.f.s concerned are:

1. *Reactance e.m.f.* This is directly proportional to the current being
commutated, the turns per coil and the number of coils commutated
simultaneously, and also to the speed, on which the time available for the
reversal process depends. It is an alternating e.m.f. E_{xc}, the greatest
amplitude of which occurs at peak current, rising and falling with the current.

2. *Rotational e.m.f.* This e.m.f., E_{rc}, is generated by rotation of the
commutating coils in any flux Φ_c that may exist in the commutating zone.
It has the frequency ω_1 of the alternating flux Φ_c in 1-ph machines; in
polyphase machines with a travelling-wave field Φ_1 of angular speed ω_1 it
is proportional to $(\omega_1 \pm \omega_r)$.

3. *Pulsational e.m.f.* Figure 13.19(*b*) shows that, in a machine with a fixed
main-flux axis, a short-circuited coil is mutually coupled to the field winding
F, and so has a 'transformer e.m.f.' E_{pc} induced in it if the main flux
alternates.

Summarizing, the e.m.f.s concerned in commutation are

$$E_{xc} = k_x I \omega_r \quad E_{rc} = k_r \omega_r \Phi_c \text{ or } k_r(\omega_1 \pm \omega_r)\Phi_1 \quad E_{pc} = k_p \omega_1 \Phi_1 \quad (13.1)$$

where k_x and k_r take account of coil turns and brush width, and $k_p = k_r \cos a$
with a the angle by which the brush axis diverges from the neutral (q-axis)
position. The resultant commutation e.m.f. E_c is the phasor sum of the
three components. Acting between brush heel and toe, it results in a
circulating current that loads the commutating coils, heats the brushes and
degrades the 'quality' of commutation.

Three-phase Commutator Motors

For speed ranges up to 4/1 a 3-ph commutator motor is cheaper overall than
alternatives using frequency control of speed. Simple control is often
adequate, the machine does not distort the supply waveform and the
efficiency is higher. Overload is not restricted by solid-state current limitation. The drawback is the need for commutator and brush maintenance.

Consider the phasor diagram (*a*) in Fig. 8.8. for a plain induction motor
with its stator supplied at frequency f_1 and its rotor running at a slip s. The
rotor phase e.m.f. sE_2 generates a phase current I_2 at slip frequency $f_2 = sf_1$,
with a lag ϕ because of the leakage reactance sx_2. The stator current I_1 has a
greater lag ϕ_1 by reason of its own leakage reactance x_1 and magnetizing
current component I_m. The 3-ph commutator motor is basically the plain
induction motor with means for injecting voltages of slip frequency into the
rotor for speed and power-factor control. As the injected voltages are derived
from the main supply of frequency f_1 the rotor winding must be connected
not to slip-rings but to a commutator, which converts the frequency to sf_1
internally.

Power Factor. Let a slip-frequency voltage V_j be injected into the rotor in leading phase quadrature to sE_2: then the phasor sum $(sE_2 + V_j)$ advances the phase of I_2 and therefore also of I_1, raising the p.f. at the stator terminals. *Speed.* The speed of a slip-ring induction motor can be lowered by absorbing part of sE_2 in an external impedance: for a given load current the slip must increase to maintain the torque, and the drop in speed is load-dependent. In the commutator motor, the slip is increased and the speed lowered by the injection of a voltage V_j in direct opposition to sE_2, but now the 'shunt' torque/speed characteristic is maintained. If $V_j = 0$, the machine operates in the same way as an induction motor. If V_j is reversed, the motor runs at a supersynchronous speed: the slip is negative, but with V_j greater than sE_2 the rotor current retains the direction that develops motor torque.

Stator-fed Shunt Motor

In this machine, Fig. 13.20, the stator windings are conventional, while the rotor has a commutator winding with six brushes per pole-pair. The injected voltage (and the corresponding power) is derived from a double induction regulator to provide speed control. Power-factor adjustment is obtained either by commutator brush-shift or by modification of the induction regulator. At the zero-voltage setting, the brushes are in effect short-circuited through the regulator windings and the machine runs as a plain induction motor with a rather higher rotor impedance. Regulator control gives a speed range typically of 0.5−1.5 times synchronous speed, with only a minor change with load. The motor can be built for ratings between 2 kW at 600 V and 2 MW at 11 kV. An example is an 11 kV motor for a generating-station draught-fan drive for 1.5/0.9 MW at 1000/850 r/min.

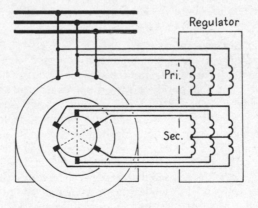

13.20 Stator-fed 3-ph shunt commutator motor.

Schrage Motor

This machine requires no separate injection device, but it is necessary to invert the stator and rotor functions. The rotor carries the primary 3-ph

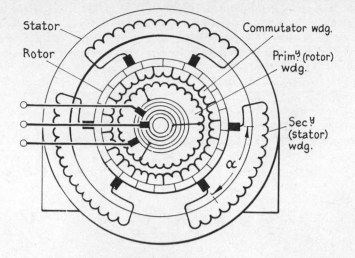

13.21 Schrage motor.

winding, fed from the supply through slip-rings; it also accommodates a low-voltage commutator winding. The stator has a 3-ph distributed winding, each phase terminated on a pair of commutator brushes, Fig. 13.21. The commutator brushgear comprises two movable rockers, geared together and fitted with three brushes per pole-pair. The rockers can be shifted in opposite directions by means of a handwheel or pilot motor. With the two brushes 'in-line', the stator phase winding terminating on them is short-circuited through the brushes and their common commutator sectors; an opposing shift of the rockers separates the brushes to include a required number of sectors into the stator phase circuit.

Consider a 2-pole machine, its primary (rotor) fed at angular frequency ω_1 and rotating at angular speed ω_r. The travelling-wave rotor field moves at ω_1 with respect to the rotor body and at $(\omega_1 - \omega_r) = s\omega_1$ with respect to the stator. The secondary (stator) e.m.f. is therefore of slip frequency, which is also that at the commutator brushes. As the coil e.m.f. in the commutator winding is proportional to the roughly constant gap flux, the voltage injected into a stator phase depends on the number of sectors between brushes. Thus with brushes 'in-line' and the stator phases short-circuited, the action is that of a plain induction motor. With brushes separated by an electrical angle a, the injected voltage enforces a rise (or fall) in speed. The no-load speed approximates to $\omega_r = \omega_1(1 \pm k \sin\frac{1}{2}a)$, where k depends on the turns ratio of the stator secondary and rotor commutator windings. The speed range attainable is typically 0.4–1.6 of synchronous speed.

Direct-on-line starting, with the brushes set in the minimum-speed position, gives 1.5 p.u. torque with 1.5–2 p.u. current. Commutation limits the rating to about 20 kW/pole, and the siting of the primary winding on the rotor restricts the supply voltage to about 600 V. The efficiency is a few percent lower than that of a plain induction motor. The power factor is reduced at low speeds, but in non-reversing motors the p.f. over the whole range can be improved by giving the brush rockers an unequal divergence from the 'in-line' position to change the phase angle of the injected voltage.

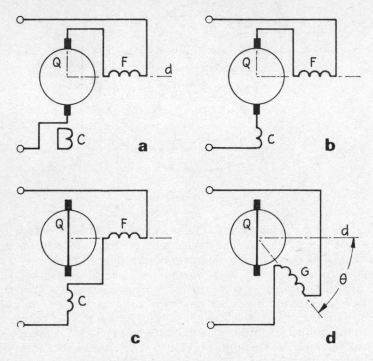

13.22 Single-phase commutator motors.

13.8 Single-phase commutator motors

Apart from series traction motors still in use on low-frequency railways in Europe, 1-ph commutator motors are uncommon in ratings over 5 kW. Fractional-kilowatt motors, however, are made in large numbers. Basic connections are shown in Fig. 13.22.

Series Motor (*a, b*). This resembles a d.c. series motor except that the whole magnetic circuit must be laminated to limit core loss. Small motors for drills and domestic machines can be run on both d.c. and 1-ph supplies.

Repulsion Motor (*c, d*). The machine has the rotor energized inductively: the rotor winding is designed for a low working voltage, and its brushes are joined by a short-circuit connection to provide a closed current path. The stator winding has either d-axis and q-axis sections (*c*), or is displaced by an angle θ to the d-axis (*d*).

Series Motor

The machine has a 'series' torque/speed characteristic. Its speed is readily controllable, typically 3000–10000 r/min, a rating up to about 500 W, and a shaft-mounted fan for cooling. To reduce winding reactance a 2- or 4-pole structure is employed without compole or neutralizing windings to ameliorate commutation, for in cheap machines (and cost is the competitive factor) with few rotor slots and sectors the reactance and pulsational ('transformer') voltages, eq. (13.1), may sum to 12–14 V. Brush pressure is high to restrict 'bounce', and brush positioning is crucial to satisfactory commutation.

The rotor winding, in slots skewed by one pitch, is conventional. The stator laminations have salient poles shaped to accept machine-wound field coils. Large core sections and short airgaps are employed to reduce saturation and exciting m.m.f. Production tolerance on rotor eccentricity must be such as to limit unbalanced magnetic pull, which may cause bearing 'hammer'. The

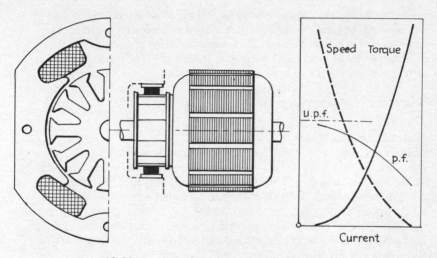

13.23 Small 1-ph series motor (full size).

cross-section of a 100 W 240 V 50 Hz motor is shown in Fig. 13.23, together with typical speed, torque and power-factor characteristics.

The behaviour equations of the machine can be set up directly from the matrix in Fig. 7.11. Reduced to its simplest terms, the motor has a d-axis stator winding F and a q-axis equivalent rotor winding Q. The respective voltages v_f and v_q sum to the applied voltage v_1 and the input current is $i_1 = i_f = i_q$: then

$$v_1 = [(r_f + r_q) + (L_f + L_q)\mathrm{p} + (L_{ad} + L_{aq})\mathrm{p} - L_{ad}\omega_r]i_1$$

For steady-state operation on a sinusoidal supply, writing $j\omega_1$ for p and with the total leakage impedance given by $r = r_f + r_q$ and $x = \omega_1(L_f + L_q)$, the r.m.s. current/voltage relation is

$$V_1 = [r + jx + j(x_{ad} + x_{aq}) - x_{ad}(\omega_r/\omega_1)]I_1$$

Apart from the volt drops in the leakage impedance, the voltage components are the d-axis (useful) and q-axis (non-useful) magnetizing reactances and the speed-dependent e.m.f. of rotation $E_r = L_{ad}\omega_r I_1$. The converted power is $E_r I_1 = M\omega_r$, whence the gross r.m.s. torque is $M = L_{ad}I_1^2$, the 'inverse-speed' or 'series' characteristic.

Figure 13.19(*b*) shows that in a practical machine there is an additional equivalent d-axis winding KD formed by the coils undergoing commutation and short-circuited by a brush. Its voltage equation is

$$0 = L_{ad} p i_1 + [r_{kd} + (L_{ad} + L_{kd})p] i_{kd}$$

which relates i_{kd} to i_1. A commutating coil becomes the short-circuited secondary to a primary F; to avoid sparking it is necessary to restrict the magnitude of the main flux and the number of turns per armature coil, and to raise the circuit resistance r_{kd} by the use of hard brushes. Further, the current reversals in the commutating coils develop the reactance voltage E_{xc} noted in eq. (13.1). Commutation in the 1-ph series motor is consequently difficult and strongly affects the design.

Neutralized Series Motor

The rotor q-axis flux adds nothing to the torque: it serves only to increase the series reactance of the machine and lower its power factor. The effect can be neutralized (or 'compensated') by the coil C in Fig. 13.22. The stator is provided with a winding distributed over the bore to simulate that of the rotor conductors. In (*a*), C acts as a transformer secondary of any convenient number of turns; in (*b*) it is in the main circuit and requires turns to match the rotor so that the m.m.f.s are equal and in opposition. In either case the q-axis flux is substantially reduced.

All large 1-ph series motors are neutralized, and commutation is assisted by (i) a multipolar structure to reduce the main flux per pole and the field-circuit reactance, (ii) narrow high-resistance brushes and single-turn rotor coils, demanding a commutator of large diameter, (iii) resistive connections between the rotor coil-ends and the commutator sectors, (iv) commutating poles set in the brush zone, and (v) where possible (as in supply to railway motors) a low supply frequency such as $16^{2/3}$ Hz.

Repulsion Motor

The functions of the neutralizing and rotor windings C and Q in Fig. 13.22(*a*) can be interchanged as in (*c*), with C a series winding and Q now short-circuited by connecting the brushes. F and C can further be combined into a single winding G with its axis displaced from the d-axis by an electrical angle θ as in (d). With $\theta = 0$, the stator and rotor m.m.f. axes are in quadrature and have no mutual inductance: no e.m.f. can be induced in the rotor at rest and no rotor current flows (except non-usefully in the coils undergoing commutation), and there is no torque. The stator takes mainly a magnetizing current and its input impedance is high. With $\theta = 90°$ the stator and rotor m.m.f. axes coincide: there is a large induced rotor current but no d-axis flux with which it can interact, so again there is no torque. The stator impedance is low as the machine acts like a transformer with a short-circuited secondary. At an intermediate value of θ (typically between $50°$ and $70°$) the stator winding is equivalent to separate windings C and F as in (c): there is a q-axis flux to induce rotor current and a d-axis flux for interaction torque production.

Assume an ideal 2-pole machine neglecting leakage impedance and with separate windings F and C on the stator. A stator current I_1 of angular frequency ω_1 in F develops the d-axis flux Φ_d in phase with I_1. The q-axis flux Φ_q results from the combination of the m.m.f.s of C and Q, so that Φ_q lags $90°$ on I_1. Let the rotor spin at angular speed ω_r: across its brushes is an e.m.f. $E_r = k\omega_r\Phi_d$ due to rotation in Φ_d and an e.m.f. $E_p = k\omega_1\Phi_q$ induced by Φ_q. Now E_r is in phase with Φ_d and E_p is in quadrature with Φ_q. They balance to give $E_r = E_p$, so that

$$\omega_r\Phi_d = \omega_1\Phi_q \quad \text{whence} \quad \Phi_q/\Phi_d = \omega_r/\omega_1$$

If $\omega_r = \omega_1$ the two fluxes are equal in magnitude, and as they are in both space and time quadrature they produce in the airgap a constant travelling-wave field. Below synchronous speed $\Phi_d > \Phi_q$, above it $\Phi_d < \Phi_q$; in each case the travelling wave is 'elliptical'.

From eq.(13.1) with appropriate change of subscripts, commutation can be considered with respect to $E_{rc} = k_r\omega_r\Phi_q$ and $E_{pc} = k_r\omega_1\Phi_d$, whence with the relation above and with due regard to phasing, the net e.m.f. in the commutating coils (apart from the reactance voltage) is $E_c = E_{pc}$ $[1 - (\omega_r/\omega_1)^2]$, which is zero at synchronous speed, for which the commutating conditions become the most favourable.

It can be shown that the equivalent circuit of the repulsion motor with a voltage V_1 applied to the stator winding G of N_g turns is representable by the *series* connection of a 'torque' winding F of $N_f = N_g \cos\theta$ turns to a rotor winding Q, with the short-circuit connection replaced by the magnetizing impedance z_m of the 'transformer' combination C-Q, where C has $N_c = N_g \sin\theta$ turns. Thus the repulsion motor is equivalent to a series motor with the impedance z_m connected across the rotor brushes.

The behaviour equations for an applied stator voltage V_1 and a stator current I_1, and a rotor current I_q, can be derived from the impedance matrix of Fig. 7.11. With a supply frequency ω_1 and a speed ω_r the voltage equations in the steady state are

$$\begin{bmatrix} V_1 \\ 0 \end{bmatrix} = \begin{bmatrix} r_g + \omega_1 L_g & \omega_1 L_{ag}\cos\theta \\ L_{ag}(\omega_1 \cos\theta - \omega_r \sin\theta) & r_q + \omega_1 L_q \end{bmatrix} \cdot \begin{bmatrix} I_1 \\ I_q \end{bmatrix}$$

The torque is $L_{aq}I_1I_q \sin\theta$, proportional approximately to $I_1{}^2$.

Repulsion motors are built in ratings up to 5 kW for industrial frequencies. The rotor winding is short-pitched to aid commutation, and as it is isolated it can be designed for optimum conditions of commutation in terms of turns and sectors. The non-salient stator winding has a layout to give an m.m.f. distribution roughly matching that of the rotor. The brushes are usually fixed, but adjustment of the torque/speed relation is possible if the brush-axis can be shifted. This facility would also permit reversal, but interchange of the stator winding terminals gives the same effect and is simpler.

13.9 Single-phase induction motors

A 3-ph induction motor can continue to run in a 'single-phasing' condition, Sect. 8.19, but has no starting torque. A 1-ph induction motor is therefore possible provided that some auxiliary means is available for starting: it is applied to many low-power drives requiring a roughly constant speed, where a 3-ph supply is not available and the power demand does not exceed a few kilowatts. Most such motors have ratings between 100 and 750 W for domestic refrigerators, fans and small tools. The rotor winding is almost invariably a simple skewed cage.

Plain Induction Motor

Consider first the basic machine with a single distributed stator winding and a uniform airgap. Ignoring harmonics, the stator winding can be assumed to develop a sine-distributed m.m.f. of peak value F_1, pulsating at the supply frequency ω_1 and centred about the d-axis. With the rotor at rest, the d-axis stator flux induces a pulsational e.m.f. in the rotor by transformer action, and the consequent rotor current develops in turn a sine-distributed m.m.f. F_2, centred also on the d-axis. As F_2 is coaxial with F_1, no interaction torque can be developed. If, however, the rotor is given an initial spin in either direction, it now has induced a rotational as well as a pulsational e.m.f., and F_2 is shifted from the d-axis, resulting in a torque in the direction of the initial spin to raise the speed if the load torque allows.

The behaviour of the machine can be based on (i) the *travelling-wave field* theory, or (ii) the *cross-field* (or 2-axis) concept.

Travelling-wave Theory. The fixed-axis alternating stator m.m.f. is regarded as the combination of a forward- and a backward-moving m.m.f. component, travelling in opposite directions around the airgap and exciting corresponding travelling-wave fluxes. With the rotor at standstill each component m.m.f. has the constant value $\frac{1}{2}F_1$, and each acts as in a polyphase machine to develop torque, but the torques are oppositely directed and balance out. When the rotor spins in the forward direction, the forward component develops the torque/slip relation so indicated in Fig. 13.24. The backward component gives a similar relation but with reference to a slip not of s but of $(2-s)$. The resultant torque is the difference of the forward and backward component torques, showing a zero at standstill for which $s = (2-s) = 1$, and another zero near to synchronous speed. These zeros are shown in Fig. 13.24,

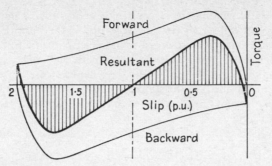

13.24 Single-phase induction motor: torque/slip relation.

but the intermediate values are approximate only, for while the stator forward and backward m.m.f.s $\frac{1}{2}F_1$ are equal, the gap flux distribution depends also on the rotor m.m.f. components that face them across the gap: these differ by reason of dissimilar impedances at all slips other than unity.

If started by external means, the rotor can run in either direction. The normal slip is greater than that of a comparable 3-ph machine, and the backward-moving field introduces additional core and $I^2 R$ loss. As in all 1-ph motors, the resultant torque fluctuates at frequency $2\omega_1$.

With the nomenclature of Sec. 8.2, the equivalent circuit (*a*) in Fig. 13.25

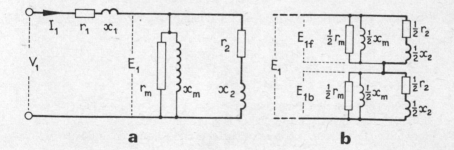

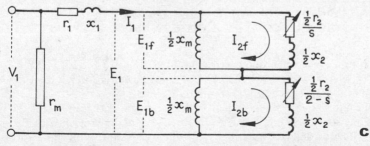

13.25 Single-phase induction motor: equivalent circuit.

can be set up for the machine at standstill. The two gap flux components induce in the stator winding the respective components E_{1f} and E_{1b}. These sum to the applied voltage V_1 (less the volt drop in the stator leakage impedance) and so can be represented as in (b) with one-half of the magnetizing and rotor impedances associated with each, in series. A simplifying approximation is to take the core and mechanical losses as represented by r_m across the terminals of the stator: then for a slip s in the forward direction the equivalent circuit becomes that in (c). In terms of synchronous power the net torque is

$$M = I_{2f}^2 \left[\tfrac{1}{2}r_2 / s \right] - I_{2b}^2 \left[\tfrac{1}{2}r_2 / (1 - s) \right]$$

giving the resultant torque/slip relation in Fig. 13.24.

EXAMPLE 13.1: A 200 W 240 V 50 Hz 4-pole 1-ph induction motor runs on rated load with a slip of 0.05 p.u. The parameters are: $r_1 = 11.4\Omega, x_1 = 14.5\Omega, \tfrac{1}{2}r_2 = 6.9\Omega, \tfrac{1}{2}x_2 = 7.2\Omega, \tfrac{1}{2}x_m = 135\Omega$, core and mechanical loss 32 W. Estimate the full-load performance.

Forward circuit. At $s = 0.05$, the rotor impedance is $(\tfrac{1}{2}r_2 / s) + \tfrac{1}{2}jx_2 = 138 + j7.2\Omega$. In parallel with $\tfrac{1}{2}jx_m = j135\Omega$ this gives

$$z_{2f} = 64 + j69 = 94\angle 47°\Omega$$

Backward circuit. Here $\tfrac{1}{2}r_2 / (2 - s) = 3.5\Omega$ and $\tfrac{1}{2}jx_2 = j7.2\Omega$. In parallel with $j\tfrac{1}{2}x_m$ this gives

$$z_{2b} = 3 + j7 = 7.6\angle 67°\Omega.$$

Output. The total series impedance is $z_1 + z_{2f} + z_{2b} = 78.4 + j90.5 = 120\angle 49°\Omega$, so that the power factor is approximately $\cos 49° = 0.66$. The forward and backward active rotor currents and torques are found to be

$$\begin{aligned}
E_{1f} &= 2.0 \times 94 &&= 188 \text{ V} & E_{1b} &= 2.0 \times 7.6 &&= 15.2 \text{ V} \\
I_{2f} &= 188/138 &&= 1.36 \text{ A} & I_{2b} &= 15.2/8.0 &&= 1.9 \text{ A} \\
M_f &= 1.36^2 \times 138 &&= 255 \text{ W} & M_b &= 1.9^2 \times 3.5 &&= 13 \text{ W}
\end{aligned}$$

The net torque (as a synchronous power) is $M = 255 - 13 - 32 = 210$ W, the shaft power is $210 \times 0.95 = 200$ W, the input power is $240 \times 2.0 \times 0.66 = 316$ W and the efficiency is $200/316 = 0.63$ p.u.

For the alternative *cross-field* theory, the treatment is that in Sect. 8.21 with the one stator winding 1D. The circuit equations are written with an applied voltage v_1 and stator current i_1. In the steady state with slip s, rotor speed $\omega_r = \omega_1(1 - s)$, and with ω_1 combined with the inductances to form reactances, the equations in r.m.s. terms are

1D: $\quad V_1 = [r_1 + j(x_m + x_1)]I_1 + jx_m I_2 d$

2D: $\quad 0 = [r_2 + j(x_m + x_2)]I_2 d + jx_m I_1 + (1 - s)(x_m + x_2)I_2 q$

2Q: $\quad 0 = [r_2 + j(x_m + x_2)]I_2 q - (1 - s)x m I_1 - (1 - s)(x_m + x_2)I_2 d$

From these the currents can be expressed in terms of V_1, s and the circuit parameters, and the torque derived. It is possible to relate $I_2 d$ and $I_2 q$ to the forward and backward currents of the travelling-wave theory by

$$I_2 d = [I_2 f + I_2 b]/\sqrt{2} \quad \text{and} \quad I_2 q = -j[I_2 f - I_2 b]/\sqrt{2}$$

These substitutions yield equations in I_1, $I_2 f$ and $I_2 b$ in the equivalent circuit of Fig. 13.25.

Starting

Single-phase cage motors are always built with some means for developing a torque at standstill to enable the machine to self-start. One method is to remove one pair of rotor cage bars (at opposite ends of a diameter in a 2-pole machine) giving the rotor a minor degree of magnetic saliency. Then, provided that the axis of saliency is out of alignment with the stator m.m.f. axis, an initial torque is produced at switch-on. The resulting movement of the rotor gives the necessary starting impulse, in a direction determined by the initial misalignment. However, the normal method is to employ the *split-phase* principle. A main (running) winding M and an auxiliary (starting) winding A, are so placed as to magnetize in equivalent space-quadrature. Then if the currents I_m and I_a can be made equal and in time-phase quadrature, a balanced 2-ph winding combination is achieved. The 1-ph motor has then a torque/slip characteristic resembling that of a normal 3-ph machine. However, only an approximation to such conditions can be achieved in practice. The arrangements of the stator windings commonly employed are shown in Fig. 13.26.

(a)*Resistance-start*. The starting winding A has a high resistance, inherent or provided by an external resistance R. At starting, the input current I_1 has the components I_m in the main winding and I_a in the auxiliary winding. I_m lags the supply voltage V_1 by typically $70°-80°$, I_a by $30°-50°$, giving a phase difference a between $30°$ and $50°$. The combination of the time-phase difference a and the space-phase difference $90°$ of the respective winding axes results in an asymmetric 2-ph field with a travelling but fluctuating wave of gap m.m.f. The rotor torque is proportional to $I_m I_a$ sin$a$. A starting torque of $1\frac{1}{2}-2$ times full-load value can be achieved. During run-up, winding A may be open-circuited by a switch or relay. On small machines with intermittent duty the switch may be omitted for cheapness to give a *resistance-start/resistance-run* mode; but if winding A is formed from high-resistance wire, it may overheat on delayed or too-frequent starts.

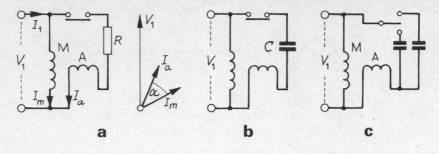

13.26 Split-phase motors: stator connections.

(*b*) *Capacitance-start/induction-run.* A greater phase-angle *a* and lower loss can be obtained by substituting a short-time-rated capacitor for the resistor *R* in (*a*) its value being typically between 30 and 100 μF. Winding A is open-circuited by a switch as the rotor approaches operating speed.

(*c*) *Capacitance-start/capacitance-run.* The capacitor in (*b*) can, if replaced by a properly rated unit, be left in circuit for normal running to raise the overall power factor and efficiency. But the optimum capacitance value for running is less than one-half of that for starting and the preferred arrangement (*c*) incorporates switching to change the capacitor value at a preset speed.

Comparative torque/speed curves for (*a*), (*b*) and (*c*) in Fig. 13.27 show the main (M) and auxiliary (A) torque contributions. Changeover is at 0.75 p.u. speed. In (*c*) As and Ar refer to the auxiliary winding with the starting and running capacitor values respectively.

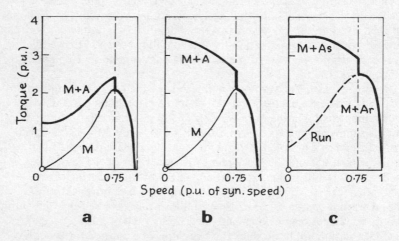

13.27 Split-phase motors: torque/speed curves
         (a) Resistance-start; (b) Capacitance-start; (c) Capacitance-start/run.

EXAMPLE 13.2: A 240 V 50 Hz 1-ph motor has main and auxiliary windings in space quadrature, and of respective impedance $z_m = 30\angle 75°$ and $z_a = 48 \angle 50°$ Ω at standstill. Estimate (*a*) the input current I_1 and torque M at starting and (*b*) the same quantities with a capacitor C of optimum value in series with the auxiliary winding.

(*a*) $I_m = 240/30 \angle 75° = 8.0\angle -75° = 2.1 - j7.7$ A
      $I_a = 240/48 \angle 50° = 5.0\angle -50° = 3.2 - j3.8$ A
      $I_1 = I_m + I_a = 5.3 - j11.5 = 12.7\angle -65°$ A

The starting torque is proportional to the product of I_m and that component of I_a in time-phase quadrature with it. Thus

$$M = k(8.0 \times 5.0 \sin 25°) = k(8.0 \times 2.1) = k16.8 \text{ units}$$

as shown in Fig. 13.28(*a*). The reactive component of z_a is +j36.8 Ω.

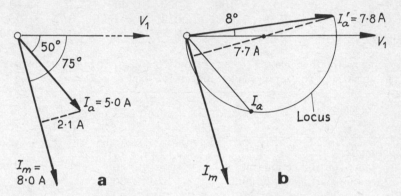

13.28 Example 13.1: capacitance-start.

(*b*) Draw the circular locus of I_a as the auxiliary-winding reactance is varied, Fig. 13.28(*b*). The maximum component of I_a in quadrature with I_m is found to be 7.7 A for a current $I_a = 7.8 \angle 8°$. Then $z_a = 240/7.8 \angle 8° = 30.7 - j4.3$ Ω; hence

$$I_a = 240/(30.7 - j4.3) = 7.7 + j1.1 \text{ A}$$
$$I_1 = 9.8 - j6.6 = 11.8\angle -42° \text{ A}$$
$$M = k(8.0 \times 7.7) = k61.6 \text{ units}$$

This is an almost fourfold increase in starting torque, with a slight reduction of the total current at starting but 2½ times the previous rate of heating in the auxiliary winding. The reactance component of z_a is −j4.3 Ω, so that the reactance of the capacitor is (36.8 + 4.3) = 41.1 Ω. Hence $C = 77$ μF.

Where the starting torque demand is greater than can be produced by a split-phase motor, a commutator winding may be incorporated in the rotor. *Repulsion-start*. The motor is built like a repulsion machine, but has a centrifugal device that acts to short-circuit all the commutator sectors together at about 75% of synchronous speed, converting the rotor winding effectively into a cage for normal-speed running.

Repulsion-induction. If, in addition to the commutator winding, the rotor carries an inner cage, the torques are additive. At standstill the cage develops no torque, but during run-up this rises, compensating for the inherent drop in the torque/speed characteristic of the commutator winding. At normal operating speeds the low cage impedance allows it to develop the major torque component. On no load the repulsion torque drives the motor slightly above synchronous speed, balancing the negative-slip counter-torque of the cage.

Performance

Typical values of efficiency and power factor for small 1-ph 50 Hz cage motors are:

Rating	W	40	100	200	400	750	1000
Efficiency	p.u.	0.38	0.50	0.60	0.68	0.72	0.75
Power factor	p.u.	0.45	0.55	0.60	0.65	0.67	0.70

Starting methods are chosen to suit the duty required. The table below gives starting currents and torques in per-unit of full-load values.

Machine	Rating W	Current p.u.	Torque p.u.
Split-phase			
Resistance-start/induction-run	10 – 40	6	1½
Capacitance-start/induction-run	100 – 800	5	3
Capacitance-start/run: one cap.	100 – 400	2½	½
two cap.	100 – 4000	5	3
Repulsion			
Repulsion-start/induction-run	100 – 4000	2½	3
Repulsion-induction	400 – 4000	3½	3

For capacitance-start, capacitors vary from 20–30 μF for a 100 W motor up to 60 – 100 μF for 750 W. Electrolytic capacitors are cheap, but their inherent loss may lead to breakdown if starts are too frequent.

For convenience of winding, the angle between the m.m.f. axes of main and auxiliary windings may diverge from ½π elec. rad, in which case the auxiliary winding may be represented by equivalent q- and d-axis windings in series. Leung and Szeto [95] set up the 2-axis impedance matrix for such cases, and show that it can be applied to most forms of 1-ph motor provided that the stator core is cylindrical. Performance can then be readily computed if the winding parameters are known with adequate accuracy.

Speed Control. This is usually achieved by control of the voltage. For small

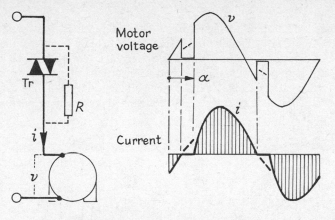

13.29 Voltage control for small 1-ph motor.

motors of output a few tens of watts, a single *triac* with phase control may be employed as in Fig. 13.29. A gating signal is delayed so that the triac Tr blocks for part of each half-period, the blocking mode beginning at a current zero, which occurs later than the zero of the applied voltage. The typical voltage and current waveforms show that variation of the delay angle a reduces their effective levels. The conduction angle is a nonlinear function of a and the inductance/resistance ratio of the motor. As the latter is speed-dependent, the time-harmonic content of the voltage applied to the motor terminals will vary with the parameters, with a and with the load. For cheapness the control has to be simple: gating signals can be derived from the voltage across the triac using a network of fractional-watt components. Continuous speed control can thus be achieved. Besides the double-supply-frequency pulsation of the torque, a series of odd-order voltage harmonics introduced by the control generates pulsations that are even multiples of the supply frequency. These make running noisy. Small fan motors used with ducted air-heating systems, convector heaters and window ventilation are particularly subject to noise production, a condition that may require attention to the vibration frequencies of the fan blades and cowling. The voltage and current harmonics can be reduced by shunting the triac with the resistor R shown in Fig. 13.29, to give the modified waveforms shown by dotted lines. With the triac blocking, a reduced voltage is maintained through R. The power loss in a small fan motor is not likely to exceed a few watts.

The principle of pole-amplitude modulation (Sect. 8.15) can be applied to a split-phase motor to obtain *two-speed* operation. It is shown by Broadway and Weatherley [94] that grading the number of turns in successive coils of the two stator windings provides the flexibility needed to design satisfactory p.a.m. motors that will, at each speed, give a performance comparable to that of the normal single-speed machine. There are strong harmonics in the space distribution of the airgap flux, but these are less objectionable in a small 1-ph motor than they would be in a larger 3-ph machine.

Construction

Single-phase induction motors have mean gap-flux densities in the range 0.35–0.65 T, and gap lengths of 0.2–0.3 mm to increase the magnetizing reactance. A concentric stator winding is preferred as, compared with a two-layer form, it requires fewer coils and end connections, the coils are readily wound by automatic machine, and the main and auxiliary windings can be accommodated in the same slots to give for each an approximately sine distribution of m.m.f.

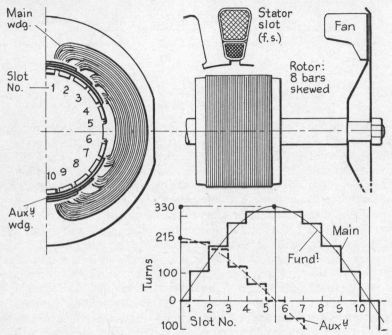

13.30 Resistance-start/run motor (one-half full size).

Figure 13.30 gives details of a resistance-start/run 240 V 50 Hz 1-ph motor of a few hundred watts rating for driving a domestic spin-dryer. The 2-pole machine has a main winding magnetizing on the horizontal axis and an auxiliary winding in space quadrature with it. The number of turns per slot in each half of the stator is:

Slot number	1	2	3	4	5	6	7	8	9	10
Cond./slot: main	102	94	78	42	0	0	42	76	94	102
auxy.	0	26	58	62	58	58	62	58	26	0

These give the m.m.f. distributions shown, each a rough sinusoid. The stator is clamped to an open frame with two ball bearings at the drive end. The rotor shaft is extended to carry a cooling fan.

The miniature 4-pole motor in Fig. 13.31 presents difficulty in winding

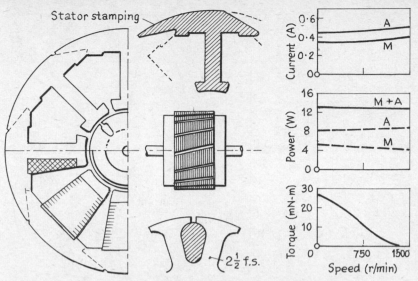

Stator stamping

13.31 Small capacitance-start/run motor (full size).

the stator, a problem solved here by assembling it from eight separate pole cores which can readily be wound individually. The main and auxiliary windings are thus identical, and a capacitor is connected in series with either to determine the direction of rotation in a capacitance-start/run mode. The motor is designed for 24 V 50 Hz to give a starting torque of about 25 mN-m using a 40 μF capacitor. The graphs show the contributions of the main (M) and auxiliary (A) windings to the current, power and torque.

For ratings above 500–750 W, the cost and performance of a 1-ph motor are such that a small standard 3-ph cage motor is a useful alternative where a 3-ph supply is available. For example, a 415 V 3-ph star-connected motor can be re-connected in delta for 240 V 1-ph operation. The supply is connected across one phase winding, and a capacitor across either of the others for split-phase starting and power-factor improvement.

Shaded-pole Motor

The shaded-pole motor is a split-phase machine, in which stator asymmetry is introduced to obtain self-starting. The induction ammeter, Fig. 13.12(*a*), is a *disc* form of shaded-pole machine: without spring restraint the rotation is continuous and can drive a small load. The normal shaded-pole motor of rating above about 20 W is *cylindrical,* with a salient-pole stator and a cage rotor. Figure 13.32(*a*) shows one-half of a typical 4-pole structure, with the main winding embracing the pole shank and the auxiliary winding (in the form of a 'shading ring') on one polar horn. The torque causes a point on the rotor to move from the unshaded to the shaded part of the stator pole. The magnetic conditions are complex: the performance is a function of the number of main-winding turns, the pole-arc/pole-pitch ratio, the extent of the shading and the shading-ring resistance, the rotor resistance and the skew.

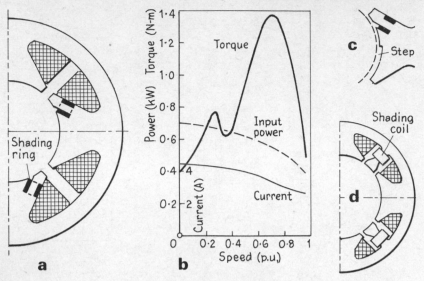

13.32 Shaded-pole motor.

The relations of torque, input power and current to speed for a 240 V 50 Hz 165 W 4-pole motor are given in Fig. 13.32(b). The important features are the starting, pull-out and 'dip' torques, and the temperature-rise. The torque/speed characteristic is like that of a 3-ph induction motor, apart from the dip at 0.4 p.u. of synchronous speed that results from the 3rd-harmonic in the gap-flux distribution. The dip thus depends on the parameters listed above, mainly the pole-arc/pole-pitch ratio and the extent of pole coverage by the shading ring. With a pole-arc greater than 0.9 of the pole-pitch, the 3rd-harmonic torque assists starting; and the shading ring should cover about 0.4 of the pole-pitch. The $I^2 R$ loss in the shading rings lowers the full-load efficiency to 0.2–0.6 p.u., although this is unimportant in the usual ratings of 15–250 W in which the machine is easily and cheaply manufactured. Such sizes have naturally considerable dimensional and performance tolerances. Analyses of the shaded-pole motor have been given by Butler and Wallace[114] and by Eastham and Williamson [115], but design rests heavily on tests to assess the motor parameters.

The rotor has a simple cage winding with a 60° (elec.) skew for optimum starting torque and minimized torque dip. The rotor resistance is greater than for any other 1-ph motor, to give starting and pull-out torques in the range 0.4–0.6 and 1.1–1.3 p.u. of rated full-load torque. Thick cage bars are used to suppress the higher-order space harmonics in the gap flux, the effective resistance being raised by reducing the end-ring section. The gap length is about 0.3 mm: too short a gap may result in torque pulsations at starting due to the rotor slotting.

Reluctance-augmented Motor. The starting torque may be raised by 'stepping' the airgap, as in Fig. 13.32(c). The cage-bar reactance and the mutual reactance in the two pole-surface regions differ, affecting the unshaded/shaded flux relation and the rotor-bar currents and phase angles. As a result, a small but helpful reluctance torque is developed.

Speed Control. Speed adjustment can be obtained by tapping the main winding to change the voltage per turn, or by some external voltage control. The method is suitable only for machines in which there is no undue dip in the torque/speed relation.

Reversal. Bi-directional machines are available in which two separate rotors are mounted on a common shaft, with one or other of the stators energized to give the appropriate rotation. The method is applied only for very small outputs where simplicity is a prime requirement. An alternative, (*d*) in Fig. 13.32, has separate shading poles and coils. In this 'ring-shift' method the shaded poles can be switched to associate magnetically with the main poles to either right or left for reversal.

Applications. Inspection of the torque/speed curve in Fig. 13.32(*b*) shows that shaded-pole motors are suited to loads of the 'fan' type having a characteristically square-law torque/speed and a cube-law power/speed demand. The starting torque is always limited. The simple and inexpensive construction possible in a small shaded-pole motor is exemplified by Fig. 13.33. The 13-bar rotor carries a fan at each end of the shaft to draw air into the body of a blower type of room-heater. For mechanical reasons the stator pole tips are connected by a narrow isthmus of stampings. Each shading ring comprises a single turn of 2.5 mm copper wire, the loop ends being soldered.

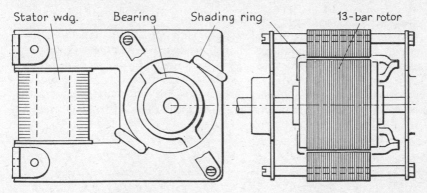

13.33 Small shaded-pole motor (full size).

Single-phase Reluctance Motor

Motors that can start, run-up and synchronize can provide for such loads as clocks, record-players and small computer peripherals. The rotor is structurally polarized, the space between the arms being occupied by conductors that form an equivalent cage. To develop a starting torque, the stator carries a split-phase winding with a capacitor in the auxiliary part. Initial synchronizing and subsequent synchronous running are, as might be expected, dependent on the d- and q-axis reactances, expressed in terms of the 2-axis concept. Intricate relations concern the dependence of the torque on the saliency, reactance ratio, cage design and resistance. The double-frequency torque pulsation at synchronous speed introduces small speed perturbations. Pull-in capability can, with proper design, be adequate for synchronizing inertial loads.

13.10   Linear machines

Although linear motion can be derived from a rotary machine by use of a simple mechanism, it can be more directly produced by a linear motor in which a translational force instead of a torque is developed. Three main operating duties, which influence the design and construction, can be classified:

Power (Drive) Machine. This is concerned with transport of masses, as applied to conveyors, haulers, electromagnetic pumps, travelling cranes and high-speed railway traction, in each case with an acceptable *power* efficiency.
Energy (Accelerator) Machine. The duty is to accelerate a mass from rest to a high speed within a specified time and distance, as for rope and car-crash test rigs, and the launching of aircraft and missiles. The criterion is the *energy* efficiency, the ratio of useful energy to the total primary electrical energy input, a figure that cannot exceed 0.5.
Force (Actuator) Machine. This develops thrust at rest, or a low speed over a short stroke, as in the operation of stop-valves, impact metal forming, door-closers, the stirring of molten steel, and small thrustor applications. The criterion is the ratio of *force* produced to input power.

Linear Machine Topology

Axial-flux Machines. These are derived directly from conventional rotary machines by a logical modification of shape that preserves the essential directional relation between magnetic flux, current and motion. The active gap region of a 2-pole 3-ph rotary machine of conventional *cylinder* geometry is shown diagrammatically in (*a*) of Fig. 13.34, the outer surface representing the stator. The main flux enters the rotor radially. The active axial conductors

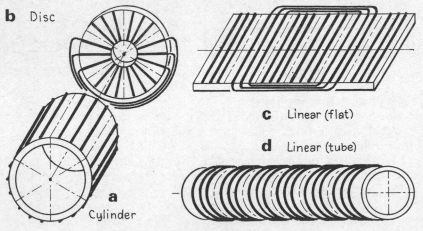

13.34 Heteropolar machine topology.

are indicated, with the phase bands separated by chain-dotted lines: for simplicity the overhang connectors are omitted. The primary 3-ph supply develops a travelling-wave field, with flux, conductor current and interaction force mutually at right-angles at any point.

If the front diameter of the cylinder is contracted and the rear expanded, the cylinder becomes a *disc* (*b*) which, though still rotary, is useful where the machine has to be accommodated in a restricted axial space. The active flux is now axial and the currents radial, producing a torque. The overhang connectors for forming one of the phases are shown. Alternatively a reluctance-motor action or a homopolar arrangement is possible, such machines being discussed by Lawrenson, Evans and Eastham [156].

Returning to the cylindrical form (*a*), let it be cut axially and 'opened out' as in (*c*): the gap region becomes typical of a *flat linear* machine. If (*c*) is re-rolled around a lengthwise axis it becomes a *tubular linear* machine (*d*), and as each originally straight active conductor is now circular, the overhang is eliminated apart from the links between associated phase bands.

Elementary realizations of the flat and tubular structures are shown in Fig. 13.35. Of the two basic members, one is the *primary* which produces the working flux across the flat or annular airgap, the other is the *secondary* which carries currents so disposed as to develop, with the primary flux, an axially directed interaction force. Any normal rotary machine has a linear counterpart: thus the primary could be a permanent-magnet or d.c.-excited field system, with a secondary carrying d.c. (as a shunt or series d.c. motor) or a.c. (as a synchronous motor). Alternatively the primary could be a.c.-excited and employ induction or reluctance operation.

Axial-flux and Transverse-flux Machines. In general terms, the gap-flux and secondary-current orientations so far described are respectively FG and AB in Fig. 13.36(*a*). A feature that limits the linear pole-pitch *Y* (and therefore the speed of a synchronous or induction machine, as is shown later) is the

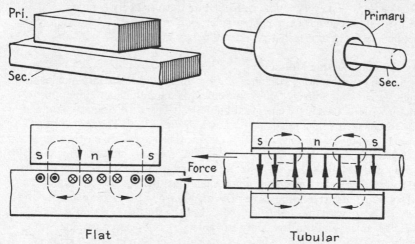

13.35 Flat and tubular linear machines.

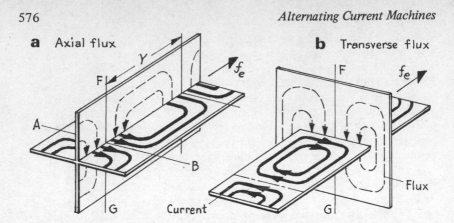

13.36 Axial- and transverse-flux relations.

excessive length of (i) non-useful axial currents in the overhang paths, and (ii) the yokes, which carry the flux between poles. For a high-speed machine it may be advantageous to adopt the *transverse-flux* orientation in Fig. 13.36(*b*). The magnetic path is shortened, reducing the mass of core steel needed for a narrow machine of long pole-pitch. The novel conditions that confront the designer of a transverse-flux machine are discussed by Laithwaite [99].

Special Features. The linear machine presents problems (*a*) of a linear movement between primary and secondary that, if continued, must cause one member to part company with the other, (*b*) of magnetic asymmetry, and (*c*) of unbalanced magnetic attraction between members. To cope with (*a*), one of the members must span the full operating distance. Magnetic asymmetry (*b*) arises from the fact that the machine has physical 'ends' (unlike a rotary machine in which every S pole is flanked on both sides by N poles), and further that it can have any pole number, odd or even. The attraction (*c*) is not balanced out as it is in a cylindrical rotary machine.

Axial-field Induction Motor

In d.c. and synchronous machines, both primary and secondary members must be fed conductively. More convenient is the polyphase induction motor with its secondary energized inductively from the primary. Some of the arrangements then possible are shown in Fig. 13.37. The *short-primary* form (*a*) is applied to long operating distances as it avoids a costly and uneconomic full-length primary winding. The *short-secondary* (*b*) is suitable for limited distances. In (*a*) and (*b*) the secondary member is a cage of bars in a slotted core, and mechanical provision must be made to withstand the strong force of attraction between the members. The coreless *plate* or *sheet secondary* (*c*) is simple and less massive, but the magnetic circuit is 'open' and ill-defined, and the secondary currents are no longer constrained to optimum transverse paths. The *double-primary* construction (*d*) gives a more definite magnetic circuit. For adequate clearance the airgap in all cases

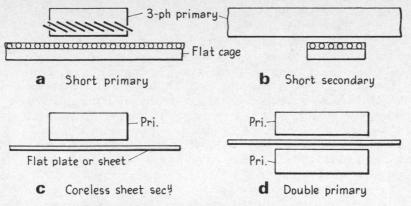

13.37 Flat linear 3-ph induction motors.

has usually to be substantially wider than in a conventional rotary machine.

To suit the application, either primary or secondary can act as the fixed *stator*, the other (moving) member being the *runner*. With a primary runner, appropriate means for feeding it electrically have to be provided.

Ideal Machine
The double pole-pitch of the idealized machine in Fig. 13.38 is considered as sufficiently remote from the ends to permit the assumption of magnetic symmetry. For a pole-pitch Y, the double pole-pitch gives the *wavelength* corresponding to a 2-pole unit in a rotary case.

Speed. With a primary supply frequency f_1 (angular frequency $\omega_1 = 2\pi f_1$) the travelling-wave field has the linear synchronous speed $u_s = 2Y f_1 = \omega_1 (Y/\pi) = \omega_1 /\beta$, where $\beta = \pi/Y$. For a slip s the speed of the runner with respect to the stator is $u = u_s(1 - s)$. Thus with f_1 = 50 Hz, Y = 50 mm and s = 0.2, the field travels at 5 m/s and the runner at 4 m/s. The synchronous speed for a 50 Hz supply and a pole-pitch of 1 m is 100 m/s = 6 km/min = 360 km/h, so that high linear speeds can be attained; but the long pole-pitch

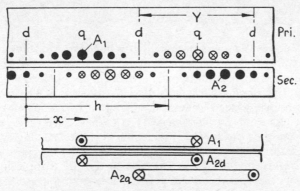

13.38 Ideal flat linear induction motor.

has magnetic disadvantages and it may be preferable to use a shorter pole-pitch with a higher supply frequency.

Thrust. Primary and secondary slot currents are represented in Fig. 13.38 by sine-distributed current sheets of peak density A_1 and A_2 (expressed in A/m). These produce corresponding peak m.m.f.s per pole $F_1 = A_1(Y/\pi) = A_1/\beta$ and $F_2 = A_2/\beta$. With ideal cores, a gap length l_g and a transverse width l, the conditions in a wavelength are (apart from the now linear geometry) those discussed for the 2-pole prototype rotary machine in Sect. 1.11 and illustrated in Figs 1.10 and 1.11: the angle θ corresponds to the linear distance x, and the torque angle λ to the force displacement h. In trigonometric terms, θ is replaced by $(\pi/Y)x = \beta x$, and λ by βh. Modifying eq. (1.10), the thrust per wavelength becomes

$$f_e = -Yl(\mu_0/l_g)F_1 F_2 \sin(\beta h)$$

in a direction to reduce h. The current sheet A_2 can be resolved into d- and q-axis sheets of peak values $A_{2d} = A_2 \cos(\beta h)$ and $A_{2q} = A_2 \sin(\beta h)$. The latter accounts for the interaction force, and the former is concerned with the 'transformer' action of power transfer from primary to secondary.

Power. Consider a flat plate secondary of width l, surface resistivity ρ and negligible leakage inductance. The runner speed is $u = u_s(1 - s)$ with respect to the stator, and su_s with respect to the travelling-wave field. An axial plate element dx at point x; where the gap flux density is B_{1x}, has an induced e.m.f. $e = B_{1x}l(su_s)$ which produces in the element a transverse current $A_{2x} \cdot dx = (e/\rho l) \cdot dx$. Interacting with B_{1x} this current develops an axial force $df = B_{1x}lA_{2x} \cdot dx$ At speed u the mechanical power is $dP_m = u \cdot df$, i.e.,

$$dP_m = (lu_s^2/\rho)B_{1x}^2 s(1 - s) = ksB_{1x}^2(1 - s) \cdot dx$$

where $k = (lu_s^2/\rho)$. The $I^2 R$ loss in the element is $dp = eA_{2x} \cdot dx = ksB_{1x}^2 s$, so that the power input to the element is $dP_2 = dP_m + dp$, which sums to $ksB_{1x}^2 \cdot dx$. Integrated over a wavelength the term $B_{1x}^2 \cdot dx$ becomes YB_1^2, whence

$$P_2 = YksB_1^2 \qquad P_m = P_2(1 - s) \qquad p = P_2 s$$

a secondary-power relation corresponding to that in eq. (8.3) for a rotary induction motor.

Goodness Factor. An electromagnetic machine is essentially a magnetic circuit producing a primary magnetic flux Φ_1 by means of a current I_1, linked to a secondary electric circuit producing an interaction current I_2 from an induced e.m.f. E_2. Laithwaite [158] coined the term *goodness factor* to express the capability of a machine to convert airgap power between the electrical form $E_2 I_2$ and the mechanical form $(f_e u)$ at a speed related to the supply angular frequency ω_1. The flux Φ_1 obtainable for a current I_1 is determined by the magnetic-circuit permeance, proportional to the effective permeability μ_m and an area/length a_m/l_m. The secondary current I_2 for an e.m.f. E_2 depends on the conductivity σ or the resistivity ρ of the conductors and an area/length ratio a_2/l_2. Then the goodness factor is given by the proportionality

$$G \propto [\mu_m(a_m/l_m)]\ [\sigma(a_2/l_2)]\ \omega_1 \propto \mu_m\sigma(a_m a_2/l_m l_2)\omega_1$$

Thus G, the power-conversion capability, depends on the physical properties of the active materials and the dimensions. If all the linear dimensions were doubled, G would be quadrupled.

An alternative viewpoint is to regard Φ_1/I_1 as a magnetizing inductance L_m. For a secondary resistance r_2 the goodness factor is proportional to $(L_m/r_2)\omega_1 = x_m/r_2$, the ratio of the magnetizing reactance to the secondary resistance (if the secondary leakage reactance is negligible).

The goodness factor, fully discussed by Laithwaite [158], is a valuable criterion for unconventional machines. In a polyphase axial-flux linear induction motor with ideal cores, the magnetic-circuit reluctance lies wholly in the airgap: then μ_m becomes μ_0 and l_m reduces to the effective gap length l_g. The secondary conductor material (e.g., copper, aluminium) settles the conductivity σ or the resistivity ρ; and if the secondary takes the form of a reaction plate of thickness d, its conductance is σd or d/ρ. The term $(a_m a_2/l_g l_2)$ contains the basic dimensions of the machine, and ω_1 is a measure of the speed of the travelling-wave field. The goodness factor for a double-primary 3-ph motor with a plate secondary of thickness d and a primary fed at angular frequency ω_1 to give a travelling-wave speed $u_s = 2Yf_1 = Y/\pi$ with a pole-pitch Y is

$$G = (\mu_0/\rho)(d/l_g)(u_s{}^2/\omega_1) \tag{13.2}$$

which can be used in determining the performance characteristics.

Practical Machine

Conditions in a real linear induction motor are more complex than those in the ideal machine. The long airgap demands more magnetization, there is a large primary leakage flux, and the discontinuous nature of the structure gives rise to important effects.

Short Primary, Cage Secondary. With relative motion, electrically 'dead' cage bars successively enter the energized primary region. Transient bar currents are induced in opposition to the primary m.m.f. in accordance with the theorem of constant linkage. Sect. 10.7, the working flux in a bar building up as it passes into the airgap, and decaying as it leaves. As a result, *entry* and *exit losses* occur. The rotor entry currents move with the cage into and out of phase with the primary travelling-wave field, giving rise to compli-cated flux patterns which are not uniform and vary with the slip and the primary pole number. Some flux-current combinations develop counter forces. The primary coils of a phase should be connected in series, for were they in parallel and connected to a constant-voltage source, the flux in all pole-pitches (including those at the ends of the primary core where the reluctance is much higher) would be constrained to be the same. Turns at the entry end would draw excessive currents, to develop the flux and to counterbalance the secondary current-reaction m.m.f. in that region.

Short Cage Secondary. Here the roles of current and flux are interchanged. With series primary phase connection the flux density is high just outside the secondary ends while in parallel connection the cage induced e.m.f.s are substantially the same as in a slice of a conventional rotary machine. However, as the secondary currents must flow in closed paths, some partly outside the primary core, there is a greater $I^2 R$ loss.

Plate or Sheet Secondary. The linear motor develops thrust by interaction, but the secondary has no distinct bars and end-rings so that the current pattern is less defined. The axial end-currents flow mainly but not entirely near the sheet edges: their effect is dependent on the width of the sheet relative to that of the primary core. In a practical machine the width of the sheet may be greater or less than that of the core, and may further be small in comparison with the pole-pitch in a high-speed motor. The sheet currents tend to flow in elliptical paths as indicated in Fig. 13.36(a), so that much of the current while contributing to the $I^2 R$ loss, develops no useful axial force. Moreover, the transverse distribution of the gap flux is distorted; it is greatest at the sides (where the primary and secondary currents are roughly at right-angles) giving a *transverse-edge effect.* Some typical flux distributions have been given by Preston and Reece [98]. In the case of a double-primary motor with a very long plate secondary (as in railway traction) the secondary plate may, to reduce its cost, be narrower than the primary cores. Its effectiveness is thereby reduced by the restriction on the active current path and by the comparatively greater length of the non-useful paths along the side edges. Cross-slitting of the secondary plate may be used in motors of limited travel to constrain currents to useful paths and so improve force production. In long-travel applications (e.g., rail traction) the cost would be prohibitive.

Performance

Design and performance assessments for 'power', 'energy' and 'force' machines differ radically, but a limited use of an equivalent circuit can be made for the two former types.

Equivalent Circuit. The leakage inductance of the secondary, particularly if it is a coreless plate, is usually ignored, for although the rated slip may be 0.05 or considerably more, the ratio of r_2/s to x_2 is likely to exceed 2. Design parameters are evaluated by methods such as those proposed by Laithwaite [82,158], or from standstill and no-load (synchronous speed) tests on an existing machine. Let the measured impedance of a primary phase with the runner fixed be $Z_1 = r_1 + (R_1 + jX_1)$, where r_1 is the primary ohmic resistance, and the term in brackets is the effective series impedance of the parallel connection of r_2 and x_m for $s = 1$ in (a) of Fig. 13.39. A reactance $X_0 = x_1 + x_m$ is obtained from a no-load test at $s = 0$. Then with $R_1/(X_0 - X_1) = k$, the equivalent-circuit parameters are r_1 together with

$$r_2 = R_1(1 + k^2) \quad \text{and} \quad x_m = (R_1/k)(1 + k^2)$$

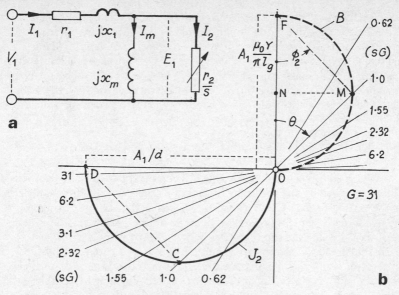

13.39 Equivalent circuit and flux/current diagram.

The gross thrust per phase for a *voltage-fed* machine is given by modifying eq. (12.2) to

$$f_e = \frac{V_1^2}{u_s} \cdot \frac{r_2/s}{(r_1 + r_2/s)^2 + x_1^2}$$

A similar modification of eq. (12.6) for a *current-fed* machine gives

$$f_e = \frac{I_1^2 x_m^2}{u_s} \cdot \frac{r_2/s}{(r_2/s)^2 + x_m^2}$$

The gross thrusts are reduced by frequency- and speed-dependent losses and by primary core loss, assessed (with some approximation) by formulae published in the literature: some are quoted by Boldea and Nasar [157]. They can be applied to 'power' machines in the steady state, but in 'energy' machines there is, strictly, no such state as the application involves continuous acceleration.

Power Drives

The large airgap necessary, particularly for high-speed machines, and the consequently lower magnetizing reactance and higher primary leakage than for conventional rotary machines, makes the simple circle diagram no longer valid, even with the modification in Fig. 8.12, although this does indeed bring out the changes with load of both the magnitude and phase of the magnetizing current and gap flux. Design and performance characteristics are normally obtained from a computer program, but a graphical presentation

is informative as it shows more clearly the effects of adjusted parameters.
Flux/Current Diagram. For a double-primary plate-secondary current-fed
linear motor, the diagram is based on the complexors of gap flux density B
and secondary current density J_2 (in A/m^2). These are expressed in terms of
the primary peak linear current density A_1 (in A/m) produced by the constant
primary current I_1, the pole-pitch Y, the effective gap length l_g, the secondary
plate thickness d and the secondary goodness factor G. Then for a slip s

$$B = A_1 \frac{\mu_0 Y}{\pi l_g} \left[\frac{j1}{1 + jsG} \right] \quad \text{and} \quad J_2 = - A_1 \frac{1}{d} \left[\frac{jsG}{1 + jsG} \right] \tag{13.3}$$

For varying slip, each term in brackets is a complexor with a circular locus on
a diameter given by the preceding term in A_1. The locus of B is a semicircle
on the vertical (quadrature or 'imaginary') axis of the complex plane in
Fig. 13.39(b). The locus of J_2 is a semicircle on the horizontal (datum or
'real') axis. For a given slip s, corresponding pairs of complexors B and J_2
(such as OM and OC, drawn for a slip such that $sG = 1$) are given by the
intersections with the loci of a line drawn through the origin 0 at an angle
$\theta = \arctan(sG)$. If the secondary leakage reactance is negligible and the
primary cores are ideal, OC represents a purely resistive secondary current
I_2 and CD the magnetizing current I_m, the two phasors summing to OD to give
the primary current I_1 that develops A_1. On the B-locus for the same (sG),
OM is the flux density and so is proportional to I_m: hence to suitable scales
MF represents I_2 and E_1, and the angle $\angle$OFM is the secondary angle of lag
$\phi_2 = \frac{1}{2}\pi - \theta$, from which the secondary p.f. can be found. It is clear that the
B- and J_2-loci are similar, and a single semicircle could serve for both phasor
quantities. Corresponding loci are described by Freeman and Lowther [157]
for single-sided motors having secondaries with and without core-steel backing.

The thrust (in N) for a machine with p pole-pairs is

$$f_e = A_1(pYl_1) \times \text{Re}|B^*| \tag{13.4}$$

where B^* is the conjugate of B. Re$|B^*|$ is obtained from the horizontal
intercept between the vertical axis and the B-locus. Its maximum value is
given by NM in Fig. 13.39(b) for $sG = 1$ and $\theta = 45°$: this maximum density is
$\frac{1}{2}A_1(\mu_0 Y/\pi l_g)$ and for it the thrust is a maximum and the secondary p.f.
is 0.71.

The secondary I^2R loss for a plate of resistivity ρ is

$$p_2 = J_2{}^2 \rho Y dl p \tag{13.5}$$

where l is the appropriate equivalent path length per pole. As the gross
output power is $P_m = f_e u$, the secondary efficiency is

$$\eta_2 = f_e u/(f_e u + p_2) \tag{13.6}$$

These formulae are somewhat idealized. The predicted performance is impaired by secondary end, edge and skin effects, and the primary discontinuity. The primary input is not considered: it has to supply the secondary input and the primary core and $I^2 R$ losses, and its p.f. is reduced by the large inherent primary leakage.

EXAMPLE 13.3: From the following data and the dimensions in Fig. 13.40(a), estimate the secondary characteristics of a large high-speed double-primary plate-secondary axial-flux induction motor supplied at a frequency f_1 = 173 Hz to give a synchronous speed u_S = 121 m/s.

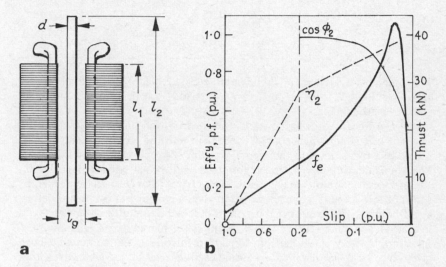

a   b

13.40 Example 13.3: performance of linear motor.

Primary. Total current-sheet density (both primaries together) A_1 = 225 kA/m, core-width l_1 = 0.25 m, pole-pitch Y = 0.35 m, number of pole-pairs p = 5, gap length l_g = 37.5 mm.
Secondary. Plate width l_2 = 0.50 m, thickness d = 6.25 mm, resistivity p = 0.091 $\mu\Omega$-m.
From eq. (13.2) the goodness factor is G = 31. The diameters of the B- and J_2-circles are respectively

$$A_1(\mu_0 Y/\pi l_g) = 0.84 \text{ T} \quad A_1/d = 36 \text{ MA/m}^2 = 36 \text{ A/mm}^2$$

The loci for G = 31 are as indicated in Fig. 13.39(b), giving pairs of values of Re$|B^*|$ and J_2 for a number of values of sG. Then eq. (13.4–6) are applied. For the secondary $I^2 R$ loss, the effective current path length per pole is taken as the mean of l_1 and l_2, giving l = 0.375 m. The thrust, secondary efficiency and power factor are plotted to a base of slip in Fig. 13.40 from the following table.

Slip s	(p.u.)	0.02	0.032	0.05	0.075	0.10	0.20	1.0
sG	(−)	0.62	1.00	1.55	2.32	3.10	6.20	31
θ	(deg)	32	45	57	67	72	81	88
ϕ_2	(deg)	58	45	33	23	18	9	2
B	(T)	0.71	0.59	0.46	0.33	0.26	0.13	0.03
$\mathrm{Re}\lvert B^*\rvert$	(T)	0.38	0.42	0.38	0.30	0.25	0.13	0.03
J_2	(MA/m^2)	19.6	25.4	30.2	33.1	34.2	35.6	36.0
f_e	(kN)	37.4	41.3	37.8	29.5	24.2	12.7	2.6
u	(m/s)	119	117	115	112	109	97	0
$f_e u$	(MW)	4.45	4.83	4.35	3.30	2.64	1.23	0
p_2	(MW)	0.14	0.24	0.35	0.41	0.44	0.46	0.48
	(p.u.)	0.97	0.95	0.93	0.89	0.86	0.73	0
$\cos\phi_2$	(−)	0.62	0.71	0.84	0.92	0.95	0.99	1.0

The same operating speed could be obtained by extending the pole-pitch to 1.0 m for a 60 Hz supply, or to 1.2 m for 50 Hz, at the cost of considerably greater length and mass.

Control. A linear motor has to start, stop, reverse, and run at constant normal or creep speeds, for which control equipment must be provided using methods corresponding to those employed for conventional rotary machines. Pole-changing, voltage variation or frequency adjustment are usable methods. Voltage variation results in increased secondary $I^2 R$ loss, especially for large thrust at low speed. Frequency variation gives a greater stator $I^2 R$ but better speed-holding. The speed signal may be a d.c. tachogenerator or (for very low speeds) an incremental encoder. Braking is achieved by plugging or d.c. injection. Feed by phase-controlled inverters introduces harmonics which can give rise to large thrust variations during each inverter period when the slip is small, but do not seriously affect the mean thrust. Connection, reconnection and plugging also produce thrust pulsations: induced currents of sub-harmonic frequencies generate travelling waves of gap flux with a speed lower than that of the secondary, and consequently develop reverse thrust, but they last only a few milliseconds.

Power-drive Applications

Industrial Drives. Linear machines have been found useful for cranes, conveyors, trolleys, turntables, door-operating units, small goods lifts (with a primary runner as part of the counterweight), foil tensioning and car-assembly platforms. A typical range of motors has a pole-pitch of 60 mm giving a synchronous speed of 6 m/s on a 50 Hz supply, and runner speeds from standstill up to 4.5−5 m/s. Primary windings are encapsulated. Secondaries for *single-sided* motors are aluminium plates backed by mild-steel sections, or simply steel plates. An optimum thrust/size ratio is obtained with a secondary of thickness about 3 mm, a width $l_2 = l_1 + 2Y/\pi$, and a clearance gap of about 2 mm. Backed secondaries develop an attraction or repulsion force that may be ten times the thrust: the forces depend in magnitude and sense on the slip and the terminal conditions − particularly those in pole-changing, frequency-changing and plug braking. In *double-sided* units the thickness of the secondary plate has to meet mechanical as well as electrical specifications. The clearance gap inevitably varies, but the mean clearance is usually about 2 mm. The apparent-power demand is about 6 VA/N of thrust, and in large units the primary cooling may need forced ventilation. The current and p.f. do not vary much with slip, and the thrust is normally a maximum at standstill, drooping smoothly to zero close to synchronous speed. In suitable applications the linear motor has the advantages of eliminating gears, avoiding contact with the work-piece (such as aluminium foil), high acceleration and retardation rates, easy adjustment of the thrust demand by the use of two or more separate units, and adaptability to disc drives in which double-primary motors exert tangential force on an annular secondary plate for slow rotational speeds.

Electromagnetic Pump. This has been developed specifically for circulating high-temperature liquid metals such as sodium, sodium-potassium alloy or bismuth as heat-transfer media in nuclear reactor plants. The flat linear pump, Fig. 13.41, has a double-primary core, with a 3-ph winding in slots. The 'airgap' comprises a rectangular-section tube which carries the liquid metal. Side bars are required to encourage an appropriate direction of current flow in the liquid. The force developed on the liquid produces a pressure differential which, multiplied by the flow rate, gives the mechanical power output. The power efficiency is low: the ideal value, $(1 - s)$, is substantially reduced by secondary effects. The liquid-metal channel, about 2 mm thick, shunts both the working flux and the secondary current; there is a contact volt drop at the side bars; fluid friction and turbulence reduce the effective thrust; and the large gap between the two primaries involves high levels of magnetizing current and lowers the p.f. The high-temperature conditions make thermal insulation necessary between primaries and the secondary.

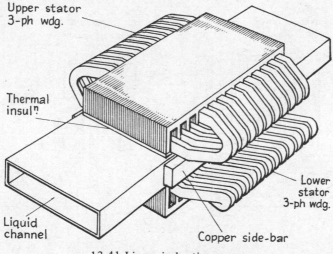

13.41 Linear induction pump.

Higher efficiency may be obtained with a tubular instead of a flat channel, or by combining axial with circumferential flow by use of an annular channel with a helical barrier surrounded by a primary resembling the stator of a rotary machine.

Stirrer. Stirring the molten charge in a metal furnace is possible by means of a short-primary linear structure clamped to the exterior surface, provided that the shell is nonmagnetic and the operating frequency is low. The charge acts as a secondary, the stirring of which is readily controlled.

Crane. Overhead crane speeds do not normally exceed about 1.5 m/s, with a duty requiring starting, acceleration, retardation, stopping and reversal. The travelling gantry carries single-sided primaries at each end of the span, the secondaries being provided by the tracks or joists on which the crane travels, Fig. 13.42(*a*), either untreated or provided with thin aluminium strips. Although operation on a standard mains supply is simple, the apparent-power demand can be reduced by providing a lower-frequency supply, e.g., from 15 VA/N at 50 Hz to 5 VA/N at 15 Hz. Linear motor drive dispenses with gears, clutches and coupling shafts, and is easy to control.

Medium-speed Traction. Conventional railway traction depends on the transfer of tractive effort through the wheel/rail contact, up to a limit imposed by friction and the adhesive weight. The linear induction motor is an alternative drive, independent of adhesion, with excellent regenerative braking capability and a high power/mass ratio. The secondary, which must extend the full length of the route, must be reduced to the simplest and cheapest form. The train must therefore carry the primary member, with power supplied through contact lines or an on-board prime-mover/generator. An assessment of the traction problem is given by Laithwaite [100]. The system in Fig. 13.42(*b*) shows a double-primary flanking a T-section aluminium-plate secondary. Mechanical clearance together with the plate thickness results in a considerable gap-length. As the synchronous speed is

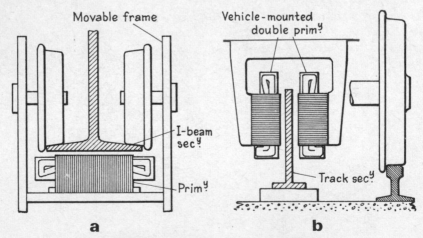

13.42 Linear induction drives for traction
(a) Crane; (b) Railway.

$u_S = 2Yf_1$ and the pole-pitch Y is fixed, the supply frequency f_1 should be proportional to the vehicle speed demanded. For short runs at low speed, however, a constant-frequency supply is adequate. On a conventional railway system the accommodation of a vertical secondary plate involves constructional difficulties: the change to a single-sided primary with a horizontal gap surface, acting on a thin horizontal aluminium sheet backed with a ferromagnetic core and bolted firmly down to the track, gives a more practicable design.

Accelerator Drives

These drives are concerned with imparting kinetic energy to a mass. An example is a car-crash test rig in which vehicles of mass in the range 0.5–2 tonne are driven into a concrete block, their speed being raised from zero to 50 km/h by a short double-primary 3-ph 50 Hz 3.3 kV 400 kW motor; the accelerating period is about 3 s and the energy imparted is about 4 MJ (1 kWh). Various realizations are possible to optimize the interrelations between the synchronous speed, the slip and the accelerating thrust at all points in the travel. With a primary covering the full length of the run, the pole-pitch could be made progressively longer to match the runner speed to the local synchronous speed, using a fixed-frequency supply; alternatively a constant pole-pitch primary could be supplied at a progressively rising frequency, a method that could also be used with short primaries. The design problem is very complex, for it concerns a combination of 'sustained transients' that persist without ever reaching a steady state. Onuki and Laithwaite [101] suggest that a machine with a few constant synchronous speeds is likely to be more practicable than one attempting to give a continuous rise throughout the launching process. Compared with alternative methods for mass acceleration, the linear motor has a lower plant cost and can achieve rates up to $1200g$ (12000 m/s^2) and final speeds up to 500 m/s.

Actuator Drives

Actuators and thrustors are most commonly pneumatic or hydraulic in action and require appropriate compressor plant. A solenoid-plunger device

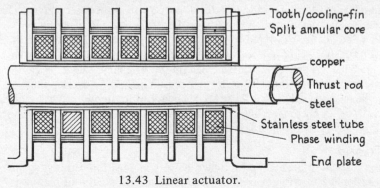

13.43 Linear actuator.

has a limited stroke. Long strokes with a thrust independent of the position
of the thrust bar can be obtained by means of a 3-ph linear machine:
Fig. 13.43 shows a typical construction of tubular form. The stator comprises
a stack of circular coils, separated by steel plates (which provide the 'teeth').
The runner is a steel bar, cylindrical or tubular, with a copper sheath and
either an outer coating of self-lubricating plastic (p.t.f.e.) or a pair of linear
bearings. Such actuators develop about 10 kN/m^2 of active pole surface and
an apparent-power/thrust ratio of about 15 VA/N with continuous rating
and forced cooling. For short-time ratings these values are considerably
higher. Representative data are:

Thrust	(N):	50	150	250
Stator length	(mm):	110	200	200
diameter	(mm):	130	130	150
Runner diameter	(mm):	20	30	50

Stator blocks can be added in line to give greater thrust to a common thrust
bar. Linear-motor actuators are robust, require no standby plant, and have a
single moving part with minimum wear. They are used to operate stop-
valves, hopper and sluice gates, and sliding doors, with control to move the
runner slowly, rapidly or in timed steps in either direction. The split annular
cores may in some designs be omitted, as the flux path outside the stator
coils, though long, has a large effective area of air path. An advantage is that
the stator coils are in direct contact with cooling air.

Single-phase Travelling-wave Linear Motor

As any rotary machine has a linear counterpart, a 1-ph linear motor could
be arranged as a split-phase machine. A development described by Watson
[159] adapts the artificial transmission line to generate a travelling wave.
Basically, this machine comprises series-connected primary windings with a
shunt capacitor at each junction, a 1-ph supply of angular frequency ω_1
applied at the input end, and a resistor of value equal to the characteristic
impedance to terminate the inductance-capacitance chain. In simplified
terms, with a coil inductance L and a shunt capacitance C per unit length
of primary taken as uniformly distributed, the gap flux wave travels away
from the input terminals and can develop thrust on a sheet secondary. The
'synchronous' speed of the wave is $u_S = 1/\sqrt{(LC)}$: it is *independent* of ω_1
and can be raised or lowered by reducing or increasing C. The terminating
resistance is $Z_0 = \sqrt{(L/C)}$. Raising the speed will reduce the thrust, so that
the machine has approximately a constant-power characteristic. Experimentally it is found that the performance is improved by overlapping
coils, so that the value of L must take account of intercoil mutual inductance.
The efficiency is comparable with that of a corresponding 3-ph linear motor,
and the power factor approximates to unity.

Levitation

A body is said to be levitated if, when its vertical position is raised against
gravitational attraction, it is held in stable equilibrium without material

contact with its environment. Practical methods are: (i) centrifugal balancing force, as in a telecommunication satellite; (ii) reaction force of a stream of particles or liquid or gas ejected downward from the body, as in a hovercraft or a vertical-take-off aircraft; and (iii) magnetic or electromagnetic forces. Only (iii) concerns us here.

Magnetostatic Repulsion. Strong forces of repulsion are produced between permanent magnets, one on a fixed surface and the other above it on the body to be levitated. The system is not laterally stable as the smallest departure from vertical symmetry produces a sideways destabilizing force. One p.m. (or both) could be replaced by a d.c.-excited coil.

Magnetostatic Attraction. Suitably oriented, two p.m.s develop an attractive force; and a body fitted with a p.m. system can be levitated by a second p.m. system fixed above it. The attractive force increases rapidly with decrease of the distance of separation, giving vertical instability. A d.c.-excited coil could be substituted for one of the p.m.s, stable levitation being achieved by feedback control of the exciting current using a form of proximity sensor.

Electromagnetic Force. Stable levitation is possible by means of currents induced in a conducting but nonmagnetic body by the field of an a.c.-excited electromagnet. Some features of the method are illustrated in Fig. 13.44:

(*a*) A vertically laminated circular core with a 1-ph primary exciting winding will support a coaxial secondary ring. The action resembles that of a shaded-pole motor. The secondary induced currents so modify the primary flux pattern as to provide an upward-shifting field, developing lift.

(*b*) With the ring in (*a*) replaced by a disc, the flux shift now produces both upward and outward-radial forces. As eccentricity increases the radial force in the direction of displacement, the disc is unstable and is expelled sideways.

(*c*) The expulsion force in (*b*) is countered by a radial component developed by 2-ph excitation. The magnitude and phasing of the exciting-coil currents

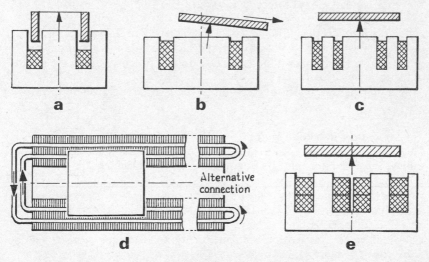

13.44 Electromagnetic levitation.

are adjusted to stabilize the secondary disc within certain limits of shape and levitation distance. True levitation is then obtained in accordance with the foregoing definition.

(*d*) The problem of laminating a circular core is greatly relieved by adopting one or other of the rectangular forms. The secondary plate is centred on the longitudinal axis over most of the core length, and if a small horizontal force is momentarily applied to it axially, it will continue to move by an action related to the behaviour of a plain 1-ph rotary induction motor.

(*e*) The 4-limbed ('XI') core arrangement is a practical form giving lift and guidance. If it is provided with a 3-ph in place of a 1-ph winding it will also produce horizontal thrust as a transverse-flux machine.

Electromagnetic levitation is a complex phenomenon, sensitive to the relative dimensions and geometry of primary and secondary, and to the material of the conducting secondary (whether or not it is ferromagnetic). The behaviour of a given system can be inferred if a map of the constant-phase contours of the flux can be drawn with due regard to the influence of induced currents: in the directions in which the resultant flux increasingly lags, the mechanically free parts will tend to move.

Tests made with a rectangular form of the Xi-core having the proportions in Fig. 13.44(*e*) gave the results tabulated below. The primary core was 180 mm wide and 450 mm long, levitating nonmagnetic secondary plates of thickness 6.3 mm and various widths. The figures relate to a lift of 20 mm and are given in terms of the mass of the secondary plate:

Plate width	(mm)	100	125	150
Primary apparent power	(kVA/kg)	4.6	2.6	1.9
active power	(kW/kg)	0.6	0.4	0.3
power factor	(p.u.)	0.13	0.14	0.15
Secondary power loss	(kW/kg)	0.18	0.15	0.12

Plates of width less than 100 mm could not be levitated. Larger widths showed a considerable stability to lateral displacement up to about one-half of the width. The levitation force was found to be proportional to the square of the exciting current.

A comprehensive review of electromagnetic levitation is given by Jayawant [161].

Transverse-field Induction Motor

For linear machines to operate at high speed on a normal industrial-frequency supply, the pole-pitch must be lengthy. In axial-flux machines this involves thick and massive 'yokes' to carry the axial flux. The transverse-flux form, in contrast, has an unrestricted pole-pitch as no flux has to be conveyed axially. Elementary realizations of the transverse-flux machine are shown in Fig. 13.45: (*a*) is for a 'flat' and (*b*) for a tubular topology. Individual phase cores and windings ABC are assembled in line, the flux in each being almost wholly transverse. The more sophisticated single-sided arrangement (*c*) gives lateral stability to the secondary, which is shown with steel backing plates to reduce

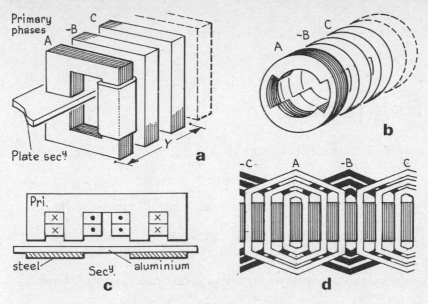

Primary phases

Plate sec^y

Pri.

steel   Sec^y.   aluminium

-C-   A   -B   C

13.45 Transverse-flux machines.

the magnetic-circuit reluctance. However, these elementary arrangements are equivalent to a single slot per pole and phase, and successive cores must have a minimum separation, as otherwise the axial thrust is impaired by the strongly harmonic axial flux distribution: expressed more simply, the secondary current induced by one core interacts less effectively with the flux of the next. To increase the pole-pitch and to reduce the harmonic content of the gap flux, successive cores are interlinked, the example in (*d*) showing a 3-ph two-layer concentric winding.

High-speed Ground Traction. Conventional wheel-on-rail traction systems provide all three essentials: *propulsion* through the wheel/rail contact, *guidance* by the rails and wheel flanges, and *support* through the axle bearings. At high speeds (e.g., 400 km/h or 110 m/s) the system fails because of rail and wheel wear, very high quality of track and maintenance, excessive noise, and insufficient adhesion to transmit the tractive effort. A contact-free support, with novel means for propulsion and guidance, is therefore needed. Methods of support by levitation have been described above. Magnetic attraction requires feedback control to stabilize the necessary gap of 10–20 mm. Magnetic repulsion between a line of track magnets and an inboard line of coils (perhaps superconducting) has been tried. A promising method would appear to be a suitably designed transverse-flux 3-ph linear motor with a single-sided primary, Fig. 13.43(*c*), carried below the train to react on a secondary plate firmly bolted to the track. Besides propulsion, this would provide guidance, and with an unbacked secondary might also develop adequate lift for support.

References

1. Say, M. G.: *Introduction to the unified theory of electromagnetic machines* (Pitman, 1971).
2. Binns, K. J.: Calculation of some basic flux quantities in induction and other doubly-slotted electrical machines, *Proc. IEE*, **111**, 1847 (1964).
3. Hague, B.: *Electromagnetic problems in electrical engineering* (Dover, 1965).
4. Billig, E.: Calculation of the magnetic field of rectangular conductors in a closed slot, *Proc. IEE*, **98** (IV), 55 (1952).
5. Mamak, R.S. and Laithwaite, E.R.: Numerical evaluation of inductance and a.c. resistance, *Proc. IEE*, **108**(C), 252 (1961).
6. Bruges, W. E.: Evaluation of certain ladder-type networks, *Proc. Roy. Soc. Edin.*, **62** (II), 175 (1946).
7. Agarwal, P.D. and Alger, P.L.: Saturation factors for the leakage reactance of induction motors, *Trans. AIEE*, **80**, 1037 (1961).
8. Chalmers, B. J. and Dodgson, R.: Saturated leakage reactance of cage induction motors, *Proc. IEE*, **116**, 1395 (1969).
9. Alger, P. L.: Calculation of the armature reactance of synchronous machines, *Trans. AIEE*, **47**, 493 (1928).
10. Reece, A.J.B. and Pramanik, A.: Calculation of the end-region field of a.c. machines, *Proc. IEE*, **112**, 2083 (1965).
11. Hawley, H., Edwards, I. M., Heaton, J. M. and Stoll, R. L.: Turbo-generator end-region magnetic fields, *Proc. IEE*, **114**, 1107 (1967).
12. Carpenter, C. J.: Application of the method of images to machine end-winding fields, *Proc. IEE*, **107** (A), 487 (1960).
13. Lawrenson, P. J.: Calculation of machine end-winding inductances with special reference to turbogenerators, *Proc. IEE*, **117**, 1129 (1970).
14. Field, A. B.: *Trans. AIEE*, **24** (1905); Field, M. B.: *Journ. IEE*, **37**, 83 (1906).
15. Swann, S. A. and Salmon, J. W.: Effective resistance and reactance of a rectangular conductor placed in a semi-closed slot, *Proc. IEE*, **110**, 1656 (1963).
16. Mullineux, N., Reed, J. R. and Whyte, I. J.: Current distribution in sheet- and foil-wound transformers, *Proc. IEE*, **116**, 127 (1969).

17. Stoll, R. L. and Hammond, P.: Approximate determination of the field and losses associated with eddy currents in conducting surfaces, *Proc. IEE,* **112,** 2083 (1965).

18. Mullineux, N. and Reed, J. R.: Eddy-current shielding of transformer tanks, *Proc. IEE,* **113,** 815 (1966).

19. Greig, J. and Freeman, E. M.: Simplified presentation of the eddy-current loss equation for laminated pole-shoes, *Proc. IEE,* **110,** 1255 (1963).

20. Chalmers, B. J. and Richardson, J.: Investigation of high-frequency no-load losses in induction motors with open stator slots, *Proc. IEE,* **113,** 1597 (1966).

21. Alger, P. L.: Induced high-frequency currents in cage windings, *Trans. AIEE,* **76,** 724 (1957).

22. Bates, J. J. and Tustin, A.: Temperature-rises in electrical machines, *Proc. IEE,* **103** (A), 471 (1956).

23. Morris, D. G. O.: Transformer equivalent circuit, *Proc. IEE,* **97** (II), 17, 735 (1950).

24. Henshell, R. D., Bennett, P. J., McCallion, H. and Milner, M.: Natural frequencies and mode shapes of vibration of transformer cores, *Proc. IEE,* **112,** 2133 (1965).

25. Curruthers, M. G. and Norris, E. T.: Thermal rating of transformers, *Proc. IEE,* **116,** 1564 (1969).

26. BS 171:1970, *Power transformers*; BS 2613:1970, *Electrical performance of rotating electrical machines*.

27. Wilkinson, K. J. R.: Superconductive windings in power transformers, *Proc. IEE,* **110,** 2271 (1963).

28. Lewis, T. J.: Transient behaviour of ladder networks representing transformer and machine windings, *Proc. IEE,* **101,** 541 (1954).

29. Dent, B. M., Hartill, E. R. and Miles, J. G.: Analysis of transformer impulse voltage distribution, *Proc. IEE,* **105**A, 445 (1958).

30. Adamson, C. and Mansour, E. A. A.: Nonlinear transformer-winding synthesis for improved surge-voltage distribution, *Proc. IEE,* **115,** 1821 (1968).

31. Schleich, A.: Behaviour of partially interleaved transformer windings subject to impulse voltages, *Bull. Oerl.*, No. 389/390, 41 (1969).

32. Alger, P. L.: *Induction machines* (Gordon & Breach, 1970).

33. Binns, K. J.: Cogging torques in induction machines, *Proc. IEE,* **115,** 1783 (1968).

34. Ellison, A. J. and Moore, C. J.: Acoustic noise and vibration of rotating electrical machines, *Proc. IEE,* **115,** 1633 (1968).

35. Binns, K. J. and Dye, M.: Effects of slot skew and iron saturation on cogging torques in induction machines, *Proc. IEE,* **117,** 1249 (1970).

36. Christofides, N.: Origins of load losses in induction motors with cast aluminium rotors, *Proc. IEE,* **112,** 2317 (1965).

37. Morgan, T., Brown, W. E. and Schumer, A. J.: Reverse-rotation tests for the determination of stray-load loss in induction motors, *Trans. AIEE,* **58,** 319 (1939).

38. Chalmers, B. J. and Williamson, A. C.: Stray losses in cage induction motors, *Proc. IEE*, **110**, 1773 (1963).
39. Schwarz, K. K.: Survey of basic stray losses in cage induction motors, *Proc. IEE*, **111**, 1565 (1964).
40. Chalmers, B. J. and Mulki, A. S.: Design synthesis of double-cage induction motors, *Proc. IEE*, **117**, 1257 (1970).
41. Elicki, M. S., Ben Uri, J. and Wallach, Y.: Switching drive of induction motors, *Proc. IEE*, **110**, 1441 (1963).
42. Rawcliffe, G. H. and Jayawant, B. V.: Development of a new 3:1 pole-changing motor, *Proc. IEE*, **103A**, 306 (1956).
43. Rawcliffe, G. H., Burbidge, R. F. and Fong, W.: Induction motor speed changing by pole-amplitude modulation. *Proc. IEE*, **105A**, 411 (1958).
44. Rawcliffe, G. H. and Fong, W.: Speed-changing induction motors, further developments in p.a.m., *Proc. IEE*, **107A**, 513 (1960); Speed-changing induction motors, reduction of pole number by sinusoidal p.a.m., *Ibid*, **108A**, 357 (1961); Close-ratio two-speed single-winding induction motors, *Ibid*, **110**, 916 (1963); Three-speed single-winding squirrel-cage induction motors, *Ibid*, **110**, 1649 (1968).
45. Fong, W.: Wide-ratio two-speed single-winding induction motors, *Proc. IEE*, **112**, 1335 (1965).
46. Chalmers, B. J. and Sarkar, B. R.: Induction-motor losses due to non-sinusoidal supply waveforms, *Proc. IEE*, **115**, 1777 (1968).
47. Largiader, H.: Design aspects of induction motors for traction with supply through static frequency-changers, *Brown Boveri Rev.*, **57**, 152 (1970).
48. Lawrenson, P. J. and Stevenson, J. M.: Induction-machine performance with a variable-frequency supply, *Proc. IEE*, **113**, 1617 (1966).
49. Ward, E. E., Kazi, A. and Farkas, R.: Time-domain analysis of the inverter-fed induction motor, *Proc. IEE*, **114**, 716 (1967).
50. Park, R. H.: Two-reaction theory of synchronous machines, *Trans. AIEE*, **48**, 716 (1929).
51. Gove, R. M.: Geometric construction of stability limits, *Proc. IEE*, **112**, 977 (1965).
52. Krishnamurthy, M. R.: D.C. dynamic braking for direct determination of rotor impedance and p.f. of induction machines, *Proc. IEE*, **111**, 1299 (1964).
53. Butler, O. I.: Effect of magnetic saturation on the d.c. dynamic braking characteristics of a.c. motors, *Proc. IEE*, **103C**, 185 (1956).
54. Shepherd, W.: Voltage and impedance conditions for the rapid plugging of induction motors including saturistor control, *Proc. IEE*, **111**, 1713 (1964).
55. Dubey, G.K. and De, G.C.: D.C. dynamic braking of inductance motor with a saturistor in its rotor circuit, *Proc. IEE*, **118**, 1585 (1971).
56. Slater, R. D. and Wood, W. S.: Constant-speed solutions applied to the evaluation of induction-motor transient torque peaks, *Proc. IEE*, **114**, 1429 (1967); Transient negative torques in induction motors due to rapid connection of the supply, *Ibid.*, **116**, 2009 (1969).

57. Smith, I. R. and Sriharan, S.: Transient performance of the induction motor, *Proc. IEE,* **113**, 1173 (1966); Induction motor reswitching transients, *Ibid,* **114**, 503 (1967); Transients in induction motors with terminal capacitors, *Ibid,* **115**, 519 (1968).
58. Enslin, N. C., Kaplan, W. M. and Davies, J. L.: Influence of transient switching currents and fluxes on the torque developed by a squirrel-cage induction motor, *Proc. IEE,* **113**, 1035 (1966).
59. A.S.A. C50.20 (1954): *Test code for polyphase induction motors and generators.*
60. Chalmers, B. J. and Bennington, B. J.: Digital-computer program for design synthesis of large squirrel-cage induction motors, *Proc. IEE,* **114**, 261 (1967).
61. Walker, J. H.: Operating characteristics of salient-pole machines, *Proc. IEE,* **100** (II), 13 (1953).
62. Robinson, B. C.: Penetration of surge voltages through a transformer coupled to an alternator, *Proc. IEE,* **100** (II), 453 (1953); **101** (II), 335 (1954); **103**A, 341, 355, 370 (1956).
63. White, E. L.: Surge transference characteristics of generator-transformer installations, *Proc. IEE,* **116**, 481 (1969).
64. Hickling, G. H. and Winder, A. L.: Surge transfer through transformer-generator units, *Proc. IEE,* **116**, 653 (1969).
65. Adkins, B.: *General theory of electrical machines* (Chapman & Hall, 1957).
66. Jones, C. V.: *Unified theory of electrical machines* (Butterworth, 1967).
67. Bharali, P. and Adkins, B.: Operational impedances of turbo-generators with solid rotors, *Proc. IEE,* **110**, 2185 (1963).
68. Saha, T. N. and Basu, T. K.: Analysis of asymmetrical faults in synchronous generators by the d-q-O frame of reference, *Proc. IEE,* **119**, 587 (1972).
69. Humpage, W. D. and Saha, T. N.: Digital computer methods in the dynamic response analysis of turbo-generator units, *Proc. IEE,* **114**, 1115 (1967).
70. McLean, G. W., Nix, G. F. and Alwash, S. R.: Performance and design of induction motors with square-wave excitation, *Proc. IEE,* **116**, 1405 (1969).
71. Ramsden, V. S., Zorbas, N. and Booth, R. R.: Prediction of induction-motor dynamic performance in power systems, *Proc. IEE,* **115**, 511 (1968).
72. Stephen, D. D.: Connecting large machines to power systems, *Proc. IEE,* **110**, 1425 (1963); Effect of system-voltage depression on large a.c. motors, *Ibid.,* **113**, 500 (1966).
73. Laithwaite, E. R.: Magnetic equivalent circuits for electrical machines, *Proc. IEE,* **114**, 1805 (1967); Carpenter, C. J.: Magnetic equivalent circuits, *Ibid.,* **115**, 1503 (1968).
74. Shackshaft, G.: General-purpose turbo-alternator model, *Proc. IEE,* **110**, 703 (1963).

75. Kron, G.: Generalised theory of electrical machinery, *Trans. AIEE,* **49**, 666 (1930); Applications of tensors to the analysis of rotating electrical machinery, *GE Rev.* (1942); *Tensors for circuits* (Dover, 1959).
76. Nicholson, H.: Dynamic optimisation of a boiler-turboalternator model, *Proc. IEE,* **113**, 385 (1966); Dynamic optimisation of a multimachine power system, *Ibid.* **113**, 881 (1966); Integrated control of a nonlinear turboalternator model under fault conditions, *Ibid,* **114**, 834 (1967).
77. Gibbs, W. J.: *Electric machine analysis using matrices* (Pitman, 1962).
78. Ching, Y. K. and Adkins, B.: Transient theory of synchronous generators under unbalanced conditions, *IEE Monograph* No. 85 (1953).
79. Hammons, T. J. and Parsons, A. J.: Design of microalternator for power-system stability investigations, *Proc. IEE,* **115**, 1421 (1971).
80. Malik, O. P. and Cory, B. J.: Study of asynchronous operation and re-synchronisation of synchronous machines by mathematical models, *Proc. IEE,* **113**, 1977 (1966).
81. Kalsi, S. S. and Adkins, B.: Transient stability of power systems containing both synchronous and induction machines, *Proc. IEE,* **118**, 1467 (1971).
82. Laithwaite, E. R.: *Induction machines for special purposes* (Newnes, 1966).
83. Harley, R. G. and Adkins, B.: Stability of a synchronous machine with a divided-winding rotor, *Proc. IEE,* **117**, 933 (1970).
84. Hammons, T. J. and Loughran, J.: Starting methods for generator/motor units employed in pumped-storage stations, *Proc. IEE,* **117**, 1829 (1970); Investigation into damping-torque coefficients and stability of a two-synchronous-machine system employed in synchronous starting, *Ibid.,* **117**, 2249 (1970).
85. Chalmers, B. J. and Richardson, J.: Steady-state asynchronous characteristics of salient-pole motors with rectifiers in the field circuit, *Proc. IEE,* **115**, 987 (1968).
86. Widger, G. F. T. and Adkins, B.: Starting performance of synchronous motors with solid salient poles, *Proc. IEE,* **115**, 1471 (1968).
87. Fong, W., French, J. R. and Rawcliffe, G. H.: Two-speed single-winding salient-pole synchronous machines, *Proc. IEE,* **112**, 351 (1965).
88. Jha, C. S.: Theory and equivalent circuits of the double induction regulator, *IEE Monograph* No.197U (1956).
89. Bunea, V.: Theory of medium-frequency pulsating-field machines, *Proc. IEE,* **111**, 1324 (1964).
90. Davies, E. J. and Lay, R. K.: Stator flux distributions in Lorenz-type medium-frequency inductor alternators, *Proc. IEE,* **113**, 2023 (1966). Rotor-surface flux distributions, *Ibid,* **114**, 1251 (1967); Performance of Lorenz-type medium-frequency inductor alternators, *Ibid.,* **115**, 1791 (1968).
91. Lawrenson, P. J. and Agu, L. A.: Theory and performance of polyphase reluctance machines, *Proc. IEE,* **111**, 1435 (1964); Lawrenson, P. J. and Bowes, S. R.: Stability of reluctance machines, *Ibid.,* **118**, 356

(1971); Lawrenson, P. J. and Gupta, S. K.: Fringe and permeance factors for segmental-rotor reluctance machines, *Ibid.,* 118, 669 (1971); Lawrenson, P. J., Mathur, R. M. and Stephenson, J. M.: Transient performance of reluctance machines, *Ibid.,* 118, 777 (1971).

92. Mainer, O. E.: Theoretical treatment of permanent-magnet alternators, *Int. Journ. El. Eng. Educ.,* 6, 417 (1968).

93. Hesmondhalgh, D. E. and Laithwaite, E. R.: Method of analysing the properties of two-phase servo motors and a.c. tachometers, *Proc. IEE,* 110, 2039 (1963); Hesmondhalgh, D. E. and Driver, K. L.: Design of the a.c. servomotor, *Ibid.,* 113, 1657 (1966).

94. Broadway, A. R. W. and Weatherly, M. R.: Pole-amplitude-modulated graded concentric windings for 2-speed 1-phase induction motors, *Proc. IEE,* 116, 1169 (1969).

95. Leung, W. S. and Szeto, W.: Generalised treatment of 1-phase motors with the aid of a digital computer, *Proc. IEE,* 116, 769 (1969).

96. Laithwaite, E. R.: Some aspects of electrical machines with open magnetic circuits, *Proc. IEE,* 115, 1275 (1968).

97. Fiennes, J.: New approach to general theory of electrical machines using magnetic equivalent circuits, *Proc. IEE,* 120, 94 (1973).

98. Preston, T. W. and Reece, A. B. J.: Transverse edge effects in linear induction motors, *Proc. IEE,* 116, 973 (1969).

99. Laithwaite, E. R., Eastham, J. F., Bolton, H. R. and Fellows, T. C.: Linear motors with transverse flux. *Proc. IEE,* 118, 1761 (1969).

100. Laithwaite, E. R. and Barwell, F. T.: Application of linear induction motors to high-speed transport systems, *Proc. IEE,* 116, 713 (1969).

101. Onuki, T. and Laithwaite, E. R.: Optimised design of linear induction-motor accelerators, *Proc. IEE,* 118, 349 (1971).

102. Rawcliffe, G. H., Broadway, A. R. W. and McLellan, P. R.: Moving-coil voltage regulator, some aspects of a generalised transformer, *Proc. IEE,* 114, 1692 (1967).

103. Davies, E. J.: Airgap windings for large turbo-generators, *Proc. IEE,* 118, 529 (1971).

104. Anempodistov, V. P., Kasharskii, E. G. and Urusov, I. D.: *Problems in the design of 750 MW turbo-generators* (Pergamon, 1963).

105. Moser, R. and Kranz, R. D.: Turbo-alternator with liquid-cooled rotor (*CIGRE Paper* 111 (1966)).

106. Lorch, H. O.: Feasibility of turbo-generator with super-conducting rotor and conventional stator, *Proc. IEE,* 120, 221 (1973).

107. *IEC Publication* 354: 1972; *BS Code of Practice* 1010: 1975.

108. Macfadyen, W. K., Simpson, R. R. S., Slater, R. D. and Wood, W. S.: Method of predicting transient-current patterns in transformers, *Proc. IEE,* 120, 1393 (1973).

109. Middlemiss, J. J.: Current pulsation of induction motor driving a reciprocating compressor, *Proc. IEE,* 121, 1399 (1974).

110. Binns, K. J. and Dye, M.: Identification of the principal factors causing unbalanced magnetic pull in cage induction motors, *Proc. IEE,* 120, 349 (1973).

111. Binns, K. J. and Schmit, E.: Some concepts involved in the analysis of the magnetic field in cage induction motors, *Proc. IEE*, 122, 169 (1975).

112. Fong, W.: Polyphase symmetrisation, *Proc. IEE*, 115, 1123 (1968); Broadway, A. R. W.: Part symmetrisation of 3-ph windings, *Proc. IEE*, 122, 145 (1975).

113. *Proc. Internat. Conference on Stepping Motors and Systems* (University of Leeds, 1974).

114. Butler, O. I. and Wallace, A. K.: Generalised theory of induction motors with asymmetrical primary windings, *Proc. IEE*, 115, 686 (1968).

115. Eastham, J. F. and Williamson, S.: Generalized theory of induction motors with asymmetrical airgaps and primary windings, *Proc. IEE*, 120, 767 (1973); Design and analysis of close-ratio two-speed shaded-pole induction motors, *Ibid.*, 120, 1243 (1973).

116. Jones, R. E.: Discharge detection in high-voltage power transformers, *Proc. IEE*, 117, 1352 (1970).

117. Valkovic, Z.: Calculation of the losses in 3-ph transformer tanks, *Proc. IEE*, 127(C), 20 (1980).

118. Allan, D. J., Mullineux, N. and Read, J. R.: Some effects of eddy currents in aluminium transformer tanks, *Proc. IEE*, 120, 681 (1973).

119. Hemmings, R. F. and Wales, G. D.: Heating in transformer cores due to radial leakage flux, *Proc. IEE*, 124, 1062 (1977). Carpenter, C. J., Sharples, K. O. and Djurovic, M.: Heating in transformer cores due to radial leakage flux, *Ibid.*, 124, 1181 (1977).

120. Yacomini, R. and de Oliveira, J. G.: Harmonics produced by direct current in converter transformers, *Proc. IEE*, 125, 873 (1978).

121. Brown, J. E. and Butler, O. I.: Zero-sequence parameters and performance of 3-ph induction motors, *Proc. IEE* Monograph 92U (1954).

122. Barton, T. H. and Doxey, B. C.: Operation of 3-ph induction motors with unsymmetrical impedance in the secondary circuit, *Proc. IEE*, 102, 71 (1955). Leung, W. S. and Ma, W. F.: Investigation of induction motors with unbalanced secondary winding connections, *Ibid.*, 114, 974 (1967). Butler, O. I. and Wallace, A. K.: Generalised theory of induction motors with asymmetrical primary windings, *Ibid.*, 115, 685 (1968). Guru, B. S.: Revolving-field analysis of asymmetric 3-ph machines and its extension to 1- and 2-ph machines, *IEE Elec. Pwr Applic.*, 2, 37 (1979).

123. Jha, G. S. and Murthy, S. S.: Generalised rotating-field theory of wound-rotor induction machines having asymmetry in stator and/or rotor windings, *Proc. IEE*, 120, 867 (1973).

124. Hamad, A. K. S., Holmes, P. G. and Stephens, R. G.: Phase-controlled circulating-current cycloconverter-induction-motor drive using rotating intergroup reactors, *Proc. IEE*, 124, 865 (1977).

125. *Electrical variable-speed drives*, IEE Conf. Pubn No. 179 (1979).

126. Rodgers, G.: Linearised analysis of induction-motor transients, *Proc. IEE*, 112, 1917 (1965).

127. Fallside, F. and Wortley, A. T.: Steady-state oscillation and stabilisation of variable-frequency inverter-fed induction-motor drives, *Proc. IEE*, **116**, 991 (1969).

128. Nanda, I. and Mathew, M. A.: Stability study of induction motors under variable-frequency modes in the presence of external inductors, *Proc. IEE*, **125**, 836 (1978).

129. Crowder, R. M. and Smith, G. A.: Induction motor for crane applications, *IEE Elec. Pwr Applic.*, **2**, 194 (1979).

130. Binns, K. S., Smith J. R., Buckley, G. W. and Lewis, M.: Predetermination of current and torque requirements of an induction-motor-driven steel-bar rolling mill, *Proc. IEE*, **124**, 1019 (1977).

131. Stokes, R. W., Lilley, M. J., Lockwood, M., Rash, N. M. and Spooner, E.: Tubular-axle induction motors for railway traction, *Proc. IEE*, **125**, 959 (1978).

132. De Buck, F. G. G.: Design adaptation of inverter-supplied induction motors, *IEE Elec. Pwr Applic.*, **1**, 54 (1978).

133. Bonwick, W. J.: Characteristics of a diode-bridge-loaded synchronous generator without damper windings, *Proc. IEE*, **122**, 637 (1975); Voltage waveform distortion in synchronous generators with rectifier loading, *Ibid.*, **127**, 13 (1980).

134. Harley, R. G. and Adkins, B.: Calculation of the angular backswing following a short circuit of a loaded alternator, *Proc. IEE*, **117**, 377 (1970).

135. Jones, C. V.: Steady-state line-line loading of the synchronous machine, *IEE Elec. Pwr Applic.*, **2**, 205 (1979), Gopal Reddy, I. and Jones C. V.: Line-line short circuit of a synchronous machine (computer-aided analysis), *Proc. IEE*, **118**, 161 (1971).

136. Daniels, A. R., Lee, Y. B. and Pal, M. K.: Nonlinear power-system optimisation using dynamic sensitivity analysis, *Proc. IEE*, **123**, 365 (1976). Fenwick, D. R. and Wright, W. F.: Review of trends in excitation systems and possible future developments, *Ibid.*, **123**, 413 (1976). Limebeer, D. J. N., Harley, R. G. and Nattrass, H. L.: Agile computer control of a turbo-alternator, *Ibid.*, **126**, 385 (1979). Moya, O. E. O. and Cory, B. J.: On-line control of generator transient stability by minicomputer, *Ibid.*, **124**, 252 (1972).

137. Fong, W.: New temperature test for polyphase induction motors by phantom loading, *Proc. IEE*, **119**, 883 (1972).

138. Shackshaft, G.: New approach to the determination of synchronous-machine parameters from tests, *Proc. IEE*, **121**, 1385 (1974). Shackshaft, G. and Poray, A. T.: Implementation of new approach to determination of synchronous-machine parameters from tests, *Ibid.*, **124**, 1170 (1977). Diggle, R. and Dineley, J. L.: Generator works testing, *Ibid.*, **128**(C), 177 (1981).

139. Wutsdorf, P.: Contribution to the discussion concerning methods of balancing flexible rotors, *Brown Boveri Rev.*, **5-74**, 228 (1974).

140. Bonwick, W. J. and Jones, V. H.: Performance of a synchronous generator with a bridge rectifier, *Proc. IEE*, **119**, 1338 (1972).

141. Chalmers, B. J. and Wooley, I.: General theory of solid-rotor induction motors, *Proc. IEE,* **119**, 1301 (1972); End effects in unlaminated-rotor induction machines, *Ibid.,* **120**, 641 (1973); Internal design of unlaminated-rotor induction machines, *Ibid.,* **121**, 197 (1974). Chalmers, B. J. and El Attar, M. M. K.: General theory of induction machines with unlaminated secondary and constant-current excitation, *Ibid.,* **125**, 666 (1978). Chalmers, B. J., Spooner, E. and Abdel-Hamid, R. H.: Parameters of solid-rotor induction machine with unbalanced supply, *Ibid.,* **127**(B), 174 (1980). Sarma, M. S. and Soni, G.: Experimental study of solid- and composite-rotor induction motors, *IEEE Trans.,* PAS-91, 1812 (1972).

142. Davies, E. J., James, B. and Wright, M. T.: Experimental verification of the generalised theory of eddy-current couplings, *Proc. IEE,* **122**, 67 (1975).

143. Chalmers, B. J. and Dukes, B. J.: High-performance eddy-current dynamometers, *Proc. IEE,* **127**(B), 20 (1980).

144. Bird, B. M. and Burbidge, R. F.: Analysis of doubly-fed slip-ring machines, *Proc. IEE,* **113**, 1016 (1966). Prescott, J. C. and Raju, B. P.: Inherent instability of induction motors under conditions of double supply, *Ibid.,* **105**(C), 319 (1958). Mansell, A. D. and Power, H. M.: Stabilisation of doubly-fed slip-ring machines using the datum-shift method, *Ibid.,* **127**(B), 293 (1980).

145. Sen, P. G. and Ma, K. H. G.: Rotor chopper control for induction-motor drive, *IEEE Trans.,* IA-11, 43 (1975). Holmes, P. G. and Stephens, R. G.: Wide speed-range operation of a chopper-controlled induction motor using asymmetrical chopping, *IEE Elec. Pwr Applic.,* **1**, 123 (1978). Van Wyk, J. D.: Variable-speed a.c. drives with slip-ring induction machines and a resistively loaded force-commutated rotor chopper, *Ibid.,* **2**, 149 (1979).

146. Lawrenson, P. J.: Forces on turbogenerator end-windings, *Proc. IEE,* **112**, 1144 (1968). Myerscough, C. J.: Calculation of magnetic fields in the end regions of turbogenerators, *Ibid.,* **121**, 653 (1974). Young, J. B. and Tompsett, D. H.: Short-circuit forces on turbo-alternator end-windings, *Ibid.,* **102**A, 101 (1955).

147. Tavner, P. J.: Measurements of the influence of core geometry, eddy currents and permeability on the axial flux distribution in laminated steel cores, *Proc. IEE,* **127**B, 57 (1980). Tavner, P. J., Penman, J., Stoll, R. L. and Lorch, H. O.: Influence of winding design on the axial flux in laminated stator cores, *Ibid.,* **125**, 948 (1978). Jacobs, D. A. H., Minors, R. H., Myerscough, C. J., Rollason, M. L. J. and Steel, J. G.: Calculation of losses in the end region of turbogenerators, *Ibid.,* **124**, 356 (1977). Howe, D. and Hammond, P.: Distribution of axial flux on the stator end-surfaces of turbo-generators, *Ibid.,* **121**, 980 (1974).

148. Davies, E. J.: Airgap windings for large turbogenerators, *Proc. IEE,* **118**, 529 (1971). Spooner, E.: Fully slotless turbo-generators, *Ibid.,* **120**, 1507 (1973).

149. Miller, T. J. E. and Hughes, A.: Analysis of fields and inductances in air-cored and iron-cored synchronous machines, *Proc. IEE,* **124**, 121 (1977); Comparative design and performance analysis of air-cored and iron-cored synchronous machines, *Ibid.,* **124**, 127 (1977).

150. Miller, T. J. E. and Lawrenson, P. J.: Penetration of transient magnetic fields through conducting cylindrical structures with particular reference to superconducting a.c. machines, *Proc. IEE,* **123**, 437 (1976). Lawrenson, P. J. Miller, T. J. E. and Stephenson, J. M.: Damping and screening in the synchronous superconducting generator, *Ibid.,* **123**, 787 (1976).

151. Davidson, B. J. *et al*: Large-scale electrical energy storage, *Proc. IEE,* **127**(A), 345 (1980).

152. Smith, I. R. and Garrido, M. S.: Current-compounded self excitation of synchronous motors, *Proc. IEE,* **114**, 269 (1967). Smith, I. R. and Nisar, P. A.: Brushless and self-excited 3-ph synchronous machine, *Ibid.,* **115**, 1655 (1968).

153. Williamson, A. C., Isaar, N. A. H. and Makky, A. R. A. M.: Variable-speed inverter-fed synchronous motor employing natural commutation, *Proc. IEE,* **125**, 133 (1978). Chalmers, B. J., Mohamadein, A. L. and Williamson, A. C.: Inverter-fed synchronous motors with induced excitation, *Ibid.,* **121**, 1505 (1974). Williamson, A. C. and Chalmers, B. J.: New form of inverter-fed synchronous motor with induced excitation, *Ibid.,* **124**, 213 (1977). Chalmers, B. J., Mwenechanye, I. M. and Pacey, K.: Brushless excitation system for a class of variable-speed synchronous motors, *Ibid.,* **125**, 754 (1978).

154. Laithwaite, E. R.: *All things are possible* (IPC Electrical-Electronic Press, 1976).

155. West, J. C., Jayawant, B. V. and Williams, G.: Analysis of dynamic performance of induction motors in control systems, *Proc. IEE,* **111**, 1468 (1964). Jayawant, B. V., Kapur, R. K. and Williams, G.: Dynamic performance of synchronous machines in control systems, *Ibid.,* **117**, 609 (1970).

156. Lawrenson, P. J. and Gupta, S. K.: Developments in the performance and theory of segmental-rotor reluctance motors, *Proc. IEE,* **114**, 645 (1967); Fringe and permeance factors for segmented-rotor reluctance machines, *Ibid.,* **118**, 669 (1971). Evans, P. D. and Eastham, J. F.: Segmented-rotor disc motor, *IEE Elec. Pwr Applic.,* **1**, 7 (1978); Disc-geometry homopolar synchronous machine, *Proc. IEE,* **127**(B), 299 (1980).

157. Boldea, I. and Nasar, S. A.: Optimum goodness criterion for linear-induction-motor design, *Proc. IEE,* **123**, 89 (1976). Shanmugasumdaram, A. and Parthasarathy, T. B.: Circle diagram for a compensated linear induction motor, *Ibid.,* **127**B, 197 (1980). Freeman, E. M. and Lowther, D. A.: Form of circle diagram for the linear induction motor, *Ibid.,* **124**, 1053 (1977).

158. Laithwaite, E. R.: *Induction machines for special purposes* (Newnes, 1966).

159. Watson, D. B.: Variable-speed single-phase induction motor, *Proc. IEE*, **119**, 188 (1975); Some characteristics of the single-phase travelling-wave machine, *Ibid.*, **124**, 771 (1977); Improved travelling-wave machine, *Ibid.*, **126**, 1162 (1979).

160. Binns, K. J., Barnard, W. F. and Jabbar, M. A.: Hybrid permanent-magnet synchronous motors, *Proc. IEE*, **125**, 203 (1978). Binns K. J. and Kurdali, A.: Permanent-magnet a.c. generators, *Ibid.*, **126**, 690 (1979); Binns, K. J. and Jabbar, M. A.: High-field self-starting permanent-magnet synchronous motor, *Ibid.*, **128**(B), 157 (1981).

161. Jayawant, V. B.: Electromagnetic suspension and levitation, *Proc. IEE*, **120**(A), 549 (1982).

Problems

1.1 A straight conductor of length 100 mm and carrying a current $i(t)$ lies at right-angles to a uniformly distributed flux of density $B(t)$. Given that, from $t = 0$, $i(t) = 5 \exp(-0.75\ t)$ and $B(t) = 1.5\ [1 - \exp(-1.25\ t)]$, calculate the maximum interaction force on the conductor and the mean force during the interval $t = 0$ to $t = 3.0$ s.

1.2 A cylindrical 2-pole a.c. machine, Fig. 1.10, has a rotor of active length 1.00 m and diameter 0.600 m within a stator of bore diameter 0.612 m. The m.m.f.s F_1 and F_2 are sinusoidally distributed. For a given load condition the peak gap flux density is 1.0 T. Calculate, per pole, the airgap reluctance, the peak airgap m.m.f. F_0, the flux and the gap energy.

1.3 A cylindrical synchronous machine has a sine-distributed flux of peak density 0.90 T in the airgap of radial length 20 mm. Given that the rotor m.m.f. F_1 is 1.5 times the gap m.m.f. F_0, and that the machine operates as a generator at a power factor 0.8 lagging, evaluate the stator m.m.f. F_2, the rotor angle δ and the torque angle λ.

1.4 An ideal 2-pole 50 Hz machine, Fig. 1.14(a) has the main dimensions $D = 1.2$ m, $l = 2.0$ m and $l_g = 15$ mm. The peak rotor m.m.f. of 10 kA-t/pole leads the peak stator m.m.f. of 15 kA-t/pole by an angle $135°$ in the direction of rotation at synchronous speed, both m.m.f.s being sinusoidally distributed. Determine (i) the resultant peak gap m.m.f. F_0, (ii) the peak gap flux density, (iii) the total gap energy, (iv) the electromagnetic torque, and (v) the converted power.

2.1 Evaluate the ratio of the gap reluctance of a salient-pole machine with a slotted armature to that with a smooth unslotted armature, using the relevant dimensions (Fig. 2.15) $y_s = 55$ mm, $w_s = w_0 = 20$ mm, $l_g = 4$ mm.

2.2 Semiclosed slots, Fig. 2.28(c), have the dimensions (i) $w_s = 12$ mm, total depth 35 mm, conductor 28 x 10 mm^2; (ii) correspondingly 6 mm, 77 mm, 70 x 4 mm^2. The conductors have the same area and d.c. resistance: compare their slot-leakage inductances. Comment on the sources of error.

2.3 The symmetrically disposed h.v. and l.v. coils on one limb of a core-type transformer have an axial length $L_c = 1.1$ m. The radial dimensions, Fig. 2.24(a), are a = 30 mm, b_1 = b_2 = 50 mm, 2x = 260 mm. The l.v. winding of 200 turns carries a current of 5 kA peak. Estimate (i) the radial force between the h.v. and l.v. coils per metre of periphery, (ii) the axial force developed if 10% of the h.v. turns at one end are tapped out.

2.4   The electromagnetic force between two rectangular-section conductors, set 1 above 2 in a parallel-sided slot and carrying equal alternating currents i in the same direction, is

$$f_e = \tfrac{1}{2}i^2 \left[(\mathrm{d}L_{11}/\mathrm{d}x) + (\mathrm{d}L_{22}/\mathrm{d}x) \right] + i^2 (\mathrm{d}L_{12}/\mathrm{d}x)$$

Estimate the peak mechanical pressure between the conductors for i = 4 kA r.m.s. and a slot-width w_s = 17.5 mm.

2.5   From basic principles evaluate the specific core loss [in W/kg] in a specimen of alloy steel subjected to a maximum flux density of 1.0 T at 50 Hz, using 0.5 mm plates; resistivity 0.25 $\mu\Omega$-m, density 7800 kg/m^3, hysteresis loss 175 J/(m^3-Hz).

3.1   A slot 20 mm wide has four layers of solid copper conductors each 14 mm wide and 8 mm deep. Assuming adequate transposition, find for an operating frequency of 50 Hz (i) the eddy-current loss-ratio in each layer within the slot, and the mean value; (ii) the overall loss-ratio if 40% of each turn lies in the slots and the eddy effects in the overhang are negligible.

3.2   A 20-pole stator has 180 slots with single-layer full-pitch coils, 6 conductors per slot, and all coils per phase connected in series. The flux per pole is 25 mWb. Compare the 50 Hz phase e.m.f.s (i) in a 1-ph winding with 5 adjacent slots per pole wound (the others being empty), and (ii) in a 3-ph star-connected winding with all slots wound.

3.3   A 4-pole 50 Hz 3-ph generator has a single-layer winding with 21 slots/ pole and 1 conductor/slot. The fundamental gap-flux component is 1.3 Wb, and there are a 10% third harmonic and a tooth ripple of maximum amplitude 6%. Calculate the r.m.s. components of the phase e.m.f. and the resultant phase e.m.f.

3.4   Work out a suitable arrangement for a 2-layer stator winding for a 3-ph 10-pole machine with 108 slots.

3.5   From the expression in Sect. 3.14 for the m.m.f. per phase of a full-pitch phase winding of N turns and spread σ, obtain an expression for the fundamental and harmonic components of the travelling-wave m.m.f. of a 3-ph winding with $\sigma = \pi/6$ rad, arranged symmetrically and carrying balanced 3-ph currents. Deduce the magnitude and speed of the first three components of the m.m.f.

3.6   An ideal 2-pole cylindrical machine has a uniform airgap of length l_g = 5 mm. The stator carries a full-pitch symmetrical 2-ph winding of 32 turns/ph, with respective phase currents i_a = 50 $\cos\omega t$ and i_b = 50 $\sin\omega t$, where ω = 314 rad/s. Draw developed diagrams of the gap m.m.f. distribution for t = 0, 2.5 and 5 ms, and calculate the peak gap-flux density. Deduce the effect of the flux pattern.

3.7   Using eq. (3.22) for the m.m.f. distribution of a full-pitch phase winding of spread σ, derive an expression for the 7th space harmonic in the resultant m.m.f. distribution of a 3-ph winding carrying balanced sinusoidal currents, including the effects of spread and chording. Discuss the effect of the 7th harmonic on the torque/speed characteristic of an induction motor.

3.8   A 3-ph 4-pole induction motor has a symmetrical 2-layer stator winding in 48 slots, with 10-turn coils short-pitched by 30° (elec.). Draw waveform diagrams of the m.m.f. distribution and find the maximum m.m.f. per pole when the phases carry symmetrical currents of 10 A r.m.s., at the instants (i) of zero and (ii) of one-half peak current, in phase A.

3.9   Briefly explain the effect of negative-sequence currents on the behaviour of (i) an induction motor, (ii) a turbogenerator. A single-layer winding has 6 slots/pole and 4 conductors/slot and carries unbalanced 3-ph currents. The r.m.s. sequence component currents for phase A are $I_{a+} = 100\angle30°$, $I_{a-} = 50\angle180°$ A. Plot the m.m.f. distribution over one double pole-pitch for the instant of peak current in phase A, and obtain the resultant peak m.m.f. per pole.

4.1   An enclosed conductor of cross-sectional area 0.04 m^2, initially cool, has to carry a short-circuit current of 18 kA for 50 ms. Estimate its temperature-rise. Data for copper: resistivity 0.021 $\mu\Omega$-m, specific heat capacity 400 J/(kg-K), density 8900 kg/m^3.

4.2   Derive an expression for the temperature-rise/time relation for an electrical machine, indicating the assumptions made. Discuss the factors that determine the rate of rise and the final steady temperature-rise on constant load. Show that for successive equal intervals of time, the temperature increments follow a geometric progression.

4.3   A motor has a constant core loss and an I^2R loss proportional to $(load)^2$. On 1.0 p.u. load its temperature-rise above ambient temperature is 31.5 °C after 15 min and 50.6 °C after 30 min, the I^2R loss being 2.5 times the core loss. Given that the heating and cooling time-constants are the same, estimate the final temperature-rise after a consecutive loading of 30 min at 1.0 p.u. load, 10 min at zero load and 25 min at 1.25 p.u. load.

4.4   A motor has a continuous rating of 30 kW with a final temperature-rise of 65 °C. Its heating time-constant is 50 min, and its cooling time-constant when shut down is 80 min. On a sustained duty-cycle of a load P for 10 min followed by shutdown for 20 min the maximum temperature-rise is 70 °C. Estimate P, stating the assumptions made.

4.5   The temperature-rise of a 100 kVA transformer with ONAN cooling (Sect. 5.16) is 12 °C after 0.5 h and 20 °C after 1.0 h on full load, for which the I^2R loss is twice the core loss. With fans fitted to give ONAF cooling the final full-load temperature-rise is 24 °C. For what rating will the final rise be equal to that for ONAN operation?

4.6   A 100 kW totally-enclosed motor has a finned outer casing of length
0.75 m and mean diameter 0.50 m, of emissivity 45 W/(m²-K).
The casing and core have a mass (assumed homogeneous) of 150 kg
with a specific heat capacity of 700 J/(kg-K). (i) Estimate the final
steady surface temperature-rise of the motor and its thermal time-
constant, when operating on full load with an efficiency of 0.95.
(ii) If an external fan is provided to raise the emissivity to 60 W/(m²-K),
what can now be the rating for the same efficiency and temperature-
rise as in (i)?

4.7   Justify the ratio (mean-square-current/mean speed) as a measure of
the temperature-rise of a motor working on a duty cycle.

5.1   Discuss the experimental methods used to determine the parameters
of the equivalent circuit, Fig. 5.1(*a*), of a 1-ph power transformer.
Such a transformer has its primary terminals connected to a 250 V 50 Hz
supply; its equivalent circuit has the values $r_m = 200\ \Omega$, $x_m = 75\ \Omega$,
$R_1 = r_1 + r_2' = 0.20\ \Omega$, $X_1 = x_1 + x_2' = 0.60\ \Omega$, and $N_1/N_2 = 3.0$.
For a load of resistance 2.0 Ω and inductance 4.0 mH connected
across the secondary terminals, calculate the primary current and the
secondary terminal voltage.

5.2   A 1.0 MVA 3-ph 50 Hz 6.6/0.433 kV transformer has a core loss of
3.2 kW. At 75 °C and full load, the $I^2 R$ and stray loss is 12.6 kW and
the leakage reactance is 0.088 p.u. Calculate the total loss and the
voltage regulation (i) for 1.25 p.u. load at p.f. 0.8 lagging, and (ii) for
0.75 p.u. load at p.f. 0.6 leading.

5.3   The magnetizing current of a 1-ph transformer approximates to a
fundamental and a 3rd harmonic. Three such transformers are connected
to form a 3-ph star/star bank, with the primary windings energized from
a symmetrical 3-ph 4-wire supply of sinusoidal waveform. What
secondary terminal voltage waveforms are to be expected (*a*) between
line and neutral, (*b*) between lines, in each case with the primary star-
point (i) connected to, (ii) disconnected from, the supply neutral?
Give reasons.

5.4   In a 1-ph transformer with the secondary open-circuited and the
154-turn primary connected to a 240 V 50 Hz sinusoidal supply, the
core-flux/magnetizing-current relation (in mWb and A) is

$$\Phi = 2i \quad (0 \leqslant i \leqslant 3.0\ \text{A}) \qquad \Phi = 6 + 0.2i \quad (i \leqslant 3.0\ \text{A})$$

Sketch the waveforms (giving significant values) of the inrush current
when the primary is suddenly energized (i) at a voltage peak, (ii) at
a voltage zero, and (iii) as (ii) but with a remanent flux of 2 mWb in
the core. Disregard the primary leakage impedance.

5.5   The h.v winding of a transformer with earthed neutral, assumed to be
uniformly distributed, has a capacitance to earth of $C = 180$ pF and
a total series capacitance of $c = 5$ pF between its extremities. The
transformer is subjected to a step-function surge of 200 kV. Estimate
the surge voltage that appears momentarily across the most highly

stressed 10% of the winding. Describe the subsequent changes of stress distribution. Explain the effect of (i) a stress shield and (ii) interleaved h.v. coils.

5.6 Two transformers in Scott connection are employed to give a 230 V 2-ph output from a 400 V 3-ph supply. The leading secondary phase has an output current of 100 A at p.f. 0.50 (lead), and a load impedance of $(1.73 + j1.0)$ Ω is connected to the lagging phase. Determine the currents in the 3-ph supply.

5.7 A 400/100 V 1-ph transformer has a thermal rating of 20 kVA. Estimate its rating when auto-connected for 500/400 V. When may such an arrangement be inadmissible?

5.8 A 1-ph 240/200 V auto-transformer, Fig. 5.21(a), has $N_1/N_2 = k$. Disregarding magnetizing current, show that

$$V_1 = kV_2 + I_1 \left[z_1 + (k-1)^2 z_2 \right]$$

where $z_1 = r_1 + jx_1$ and $z_2 = r_2 + jx_2$. An auto-transformer has $k = 1.2$; with the secondary short-circuited, the primary takes 40 A at 6 V and a p.f. of 0.3. Estimate V_2 for $I_2 = 120$ A at unity p.f., when $V_1 = 240$ V.

5.9 Using the equivalent star circuit, Fig. 5.23, calculate the primary (h.v.) per-unit terminal voltage required in a 3-ph 132/11/33 kV transformer such that a balanced 20 MVA load at p.f. 0.8 lagging can be taken from the l.v. secondary terminals at 11.0 kV. The m.v. tertiary is permanently connected to a balanced star load of 50 Ω/ ph inductors. The measured leakage reactances to a 40 MVA base are: primary with secondary s.c., 0.15 p.u.; secondary with tertiary s.c., 0.08 p.u.; tertiary with primary s.c., 0.09 p.u.

5.10 A 66/11 kV core-type transformer with connection Dy1 (Sect. 5.9) is tested on the 11 kV star-connected side to determine its phase-sequence impedances Z_+, Z_- and Z_0. The test connections to a supply ABC, and the measurements observed, are:
P.P.S.: Supply ABC respectively to $a_2 b_2 c_2$ with $A_2 B_2 C_2$ short-circuited: 1.3 kV (line), 1.5 kA (line), total power 1.4 MW.
N.P.S.: As for p.p.s. but with b_2 and c_2 interchanged: 1.3 kV (line), 1.5 kA (line), total power 1.4 MW.
Z.P.S.: Supply A to $a_2 b_2 c_2$ connected together, N to yn: 50 V, 1.5 kA, 7.5 kW.
Evaluate the phase-sequence impedances, and explain why $Z_+ = Z_-$, and why Z_0 differs from both.

5.11 Two 1-ph transformers have each a turns ratio 750/11. Their ratings and leakage reactances are: T1, 1000 kVA, $(0.08 + j\, 0.6)$ p.u.; T2, 200 kVA, $(0.02 + j\, 0.06)$ p.u. They are connected in parallel to supply a load of impedance $(0.8 + j\, 0.6)$ Ω at 1.1 kV. Estimate the loading of each transformer and the common primary voltage. Is the arrangement satisfactory?

5.12 Two 3-ph star/star connected transformers each have a leakage impedance of $(0.2 + j0.8)$ Ω/ph referred to the secondary. For the same rated primary voltage, the secondary open-circuit voltages are $245 \angle 0°$ and $240 \angle 0°$ V/ph. Estimate the common secondary voltage and the secondary currents when they operate in parallel to feed a 3-ph star load of $(10 + j0)$ Ω/ph. What circulating current will flow on no load?

5.13 Two transformers of the same rating are to supply in parallel a common load of $(0.8 - j\,0.6)$ p.u. Their respective leakage impedances are $z_1 = (0.05 + j0.20)$ and $Z_2 = (0.05 + j0.10)$ p.u. Estimate the percent tap-change required on transformer 2 in order to share the load approximately equally. Assume that the load and its p.f. are unaffected by the tap-change.

6.1 A new transformer design is based on an existing unit having a full-load $I^2 R$/core loss ratio of 2.0. The linear dimensions are increased by a factor k_1 and the conductor current density by a factor k_2 : the core flux density, number of turns and temperature-rise remain unchanged. Compare the new and existing designs in respect of (i) voltage, current and output per unit mass, (ii) losses and effective tank dissipation surface area, (iii) per-unit leakage reactance. Evaluate these for $k_1 = 1.2, k_2 = 1.3$.

6.2 The core loss at rated voltage of a 400 MVA 1-ph generator transformer is guaranteed at 800 kW. On test it was found to be $900 \pm 2\%$ kW, but the test voltage had a considerable harmonic content. The test was repeated with accurate voltmeters, one reading *r.m.s.* and the other *mean* values. The test voltage, set to rated level using the mean value, gave the core loss as 815 kW; the ratio r.m.s./mean voltage was 1.10. The makers of the core steel certify that the hysteresis/eddy-loss ratio is 3.0. Argue the case for accepting the transformer as meeting the guarantee.

7.1 A 100 MW 50 Hz 3000 r/min turbogenerator, running at rated load, is tripped by a busbar fault. Given that the inertia coefficient is $H = 4.0$ MJ/MVA, how long will the machine take to reach the overspeed trip value of 3400 r/min, the speed governor being inactive?

7.2 Develop the electrical (current/voltage) and mechanical (torque/current and speed/torque) performance equations for the 2-pole primitive 3-winding machine, Fig. 7.7(c). The airgap is uniform, the d-axis winding F is sinusoidally distributed, and the rotor has commutator brushes to give equivalent windings D and Q. The rotor shaft has a purely inertial mechanical load.

7.3 A 6-pole 3-ph induction machine has a uniform airgap and a sinusoidal gap-flux distribution. Using the 2-axis model, Fig. 7.14, write the general voltage/current equations, and derive an expression for the electromagnetic torque in terms of the currents in the four windings and the axis mutual inductances.

8.1 Test results for a 415 V 50 Hz 4-pole 3-ph delta-connected cage induction motor are: no-load, 415 V (line), 4.30 A (line), 261 W; locked-rotor, 110 V (line), 5.80 A (line), 303 W; stator resistance,

4.7 Ω/ph; stator/rotor leakage reactance ratio, 1.49. Obtain the component values for the equivalent circuit per phase, Fig. 8.7(*b*).

8.2 Estimate the shaft power of the machine in Q. 8.1 for a slip of 0.05 p.u.

8.3 The stator and rotor leakage impedances of a 440 V 50 Hz 4-pole star-connected induction motor at standstill are $z_1 = (0.40 + j1.0)$ and $z_2' = (0.235 + j0.39)$ Ω/ph respectively. The effective stator/rotor phase turns ratio is 1.60. Neglecting the magnetizing admittance, calculate the gross mechanical output power at 1440 r/min. What is the slip (i) for maximum output power, (ii) for maximum torque?

8.4 A 440 V 50 Hz 3-ph 6-pole star-connected slipring induction motor has the following parameters (in Ω/ph at standstill):

$$z_1 = (0.5 + j2.5) \quad z_2' = (1.0 + j3.0) \quad z_m = j55$$

Calculate the referred value of an external resistor to be connected into each rotor phase to achieve maximum torque at starting.

8.5 A 3-ph induction motor develops maximum torque M_m at slip s_m. Neglecting stator resistance, show that the torque at slip s is

$$M = M_m \cdot 2s_m s/(s_m^2 + s^2)$$

8.6 Using the expression in Q. 8.5, estimate the time to accelerate a 50 Hz 4-pole motor from rest to 1400 r/min with the shaft unloaded but having an inertia of 0.20 kg-m^2. The output power at M_m is 10 kW, and the ratio maximum/starting torque is 4.0.

8.7 A 400 V 50 Hz 4-pole 3-ph star-connected slipring induction motor has the following parameters referred to the stator: $r_1 = 0.20$ Ω/ph, $r_2' = 0.10$ Ω/ph, $X_1 = x_1 + x_2' = 0.75$ Ω/ph, rotor/stator e.m.f. ratio at standstill 0.5. The motor is used on a crane to start against a load torque of 300 N-m. Estimate the resistance to be added to each rotor phase. Which of the two possible values would be chosen?

8.8 A 220 V 50 Hz 3-ph delta-connected 6-pole induction motor has the referred standstill leakage impedances $z_1 = (1.0 + j1.9)$ Ω/ph, $z_2' = (1.24 + j1.9)$ Ω/ph. Neglecting the magnetizing impedance, determine (i) the gross output power at slip 0.05, (ii) the standstill torque when started in star, (iii) the change in line current when the terminals are switched from star to delta at a slip of 0.40.

8.9 A 3-ph cage motor, started on load, fails to accelerate beyond 1/7 of synchronous speed. Describe the physical phenomenon most likely to account for this.

8.10 Explain why, at standstill and at low speeds, the current in a deep-bar cage conductor is not uniformly distributed. Draw diagrams to show the distribution in magnitude and phase.

8.11 The standstill referred leakage impedances (in Ω/ph) of a 415 V 50 Hz 20 kW 4-pole delta-connected induction motor are: $z_m = (10 + j50)$, $z_1 = (0.5 + j1.0)$, outer cage $z_2 = (2.0 + j2.0)$, inner cage

$z_3 = (0.5 + j5.0)$, outer-inner cage mutual $x_{23} = j1.0$. The motor is switched direct-on-line: determine its starting torque, and the proportion thereof provided by the outer cage.

8.12   A 415 V 50 Hz 3-ph induction motor has, in a given operating condition, an equivalent star leakage impedance of $(3.5 + j7.2)$ Ω/ph. Neglecting core loss and magnetizing current, calculate the per-unit torque when the motor is fed through a 3-ph inverse-parallel thyristor converter having a delay angle $a = \pi/2$ rad. Assume the torque to be proportional to the square of the fundamental component of the motor input voltage.

8.13   Speed control by slip is applied to a 12-pole 50 Hz induction motor driving a load requiring 600 kW at 500 r/min and having a torque proportional to (speed)^2. Estimate the maximum slip-power that must be dealt with by the rotor circuit over the range from full speed to standstill.

8.14   The equivalent circuit per phase of a 420 V 50 Hz 6-pole 3-ph star-connected induction motor has the parameters: $z_m = j50$ Ω, $z_1 = (0.4 + j0.8)$ Ω, $z_2' = (0.5 + j1.0)$ Ω. Estimate the line current, power factor and torque when the machine is supplied at 45 V (line), 50 Hz, and the rotor speed is 1.0 r/s. Compare the no-load gap fluxes at the two frequencies.

8.15   Capacitors costing £9 per kvar are to be used to raise the power factor of a motor taking a maximum demand of 500 kW at p.f. 0.80 lagging. It is required to repay in 15 months the cost of the capacitors by the resulting reduction in the maximum-demand charge of £15 per year per kVA. What should be the rating of the capacitors? What other factors should be considered?

8.16   A motor attached to a load and a flywheel has a total inertia of 1000 kg-m^2. The load-torque/speed characteristic is rectilinear from zero torque at 600 r/min to 130 N-m at 570 r/min. With the motor developing a steady torque of 80 N-m, an additional torque of 300 N-m is applied for an 8 s period. Obtain an expression for the motor torque during the 8 s interval, and calculate its value at the end of the period.

8.17   Tests on a 400 V 50 Hz 3-ph delta-connected 4-pole induction motor gave the data: no-load, 400 V (line), 5.0 A (line), 440 W; locked-rotor, 100 V (line), 12.5 A (line), 660 W. Derive the approximate equivalent circuit. Estimate the initial braking torque when the motor is 'plugged' by interchanging the motor supply connections when it runs with a slip of 0.05.

8.18   A 420 V 50 Hz 6-pole 3-ph star-connected induction motor has $z_1 = z_2' = (1.0 + j2.0)$ Ω/ph and an inertia $J = 2.0$ kg-m^2. When running on no load at 955 r/min it is brought to rest by 'plugging'. Estimate the time to stop the motor.

8.19   Each phase of a 415 V 50 Hz 3-ph star-connected 4-pole induction motor has, at standstill, the leakage impedances $z_1 = (1.0 + j3.0)$ and $z_2' = (0.5 + j2.0)$ Ω. The magnetizing admittance is small enough to be negligible. The stator/rotor phase turns-ratio is 2.0. When the machine is running with a slip of 0.05, the stator is suddenly re-connected for d.c. braking and a 2.0 Ω resistor inserted in each rotor phase. Assuming the magnitude of the airgap flux to be unchanged, estimate the initial braking torque.

8.20 A 20 kW 3-ph 50 Hz 4-pole induction motor drives a load of constant torque 200 N-m; the system inertia is 60 kg-m^2. Disregarding transient effects, calculate the maximum duration of a supply interruption that will permit the motor to accelerate after the supply is restored. The motor torque/slip relation is

Slip	(p.u.):	1.0	0.8	0.6	0.4	0.2	0.06	0.04
Torque	(p.u.):	0.8	0.96	1.3	2.0	2.7	1.5	1.0

8.21 Use the theorem of constant linkage to explain, in terms of the stator and rotor currents and magnetic fields in a 3-ph cage induction motor, the following. (*a*) The transient currents and torque (i) when the motor at rest is switched direct-on-line, (ii) when the motor at normal speed is disconnected momentarily and then reconnected. (*b*) The momentary speed overshoot when the unloaded motor is rapidly accelerated from standstill.

8.22 A 400 V 50 Hz 3-ph star-connected 6-pole induction machine on rated voltage takes 8.5 A at p.f. 0.205 on no load, and 100 A at p.f. 0.375 at starting. The stator resistance is 0.46 Ω/ph. Draw a complete circle diagram per phase, and indicate the regions of motor, generator and brake modes. From the diagram determine the slip, shaft input power, electrical output power and power factor in the generator mode when the output current is 50 A.

9.1 Find suitable main dimensions D and l for a 7.5 kW 220 V 50 Hz 3-ph induction motor to run near 1500 r/min. Assume a full-load efficiency 0.86 and power factor 0.87, and specific loadings $B = 0.4$ T, $A = 22$ kA/m.

9.2 Obtain values for the rotor diameter and axial length for a 350 kW 50 Hz 10-pole induction motor from the following data: Specific loadings, $B = 0.60$ T, $A = 35$ kA/m, stator winding factor 0.92, axial length = 1.25 pole-pitch, efficiency 0.93, power factor 0.85.

9.3 The standstill equivalent circuit per phase of an induction motor has $z_1 = (0.8 + j1.9)$ Ω, $z_2' = (1.2 + j0.5)$ Ω, $r_m = 450$ Ω, $x_m = 36$ Ω. Estimate the values for a motor having every linear dimension doubled.

10.1 For rated terminal voltage V_1, an ideal 2-pole cylindrical synchronous machine requires a resultant gap m.m.f. $F_0 = 10.0$ kA-t/pole. Draw diagrams (Fig. 10.1) for the following stator and rotor m.m.f.s (in kA-t/pole), and state the operating mode, the rotor angle δ, the torque angle λ and the power factor: (i) 2.0 and 12.0 (ii) 8.0 and 6.0, (iii) 6.0 and 14.0.

10.2 A 2.2 kV 3-ph star-connected synchronous motor has a synchronous impedance $(1.0 + j10)$ Ω/ph. Determine the internal e.m.f. and the rotor angle when the input is 150 kW at p.f. 0.8 leading.

10.3 A generator with a synchronous impedance $(0 + j1.2)$ p.u. delivers full-load current to infinite busbars at a p.f. 0.8 lagging. Determine the per-unit current and power factor at the point of maximum power, the excitation being unchanged.

10.4 A 12 MVA 11 kV 50 Hz 3-ph 4-pole star-connected cylindrical-rotor synchronous generator supplies rated output at p.f. 0.9 lagging to an infinite busbar. Its synchronous reactance is 0.40 p.u. Determine the synchronizing torque for a shaft displacement of $0.35°$ (mech.). Neglect losses and saturation.

10.5 The machine in Q.10.4 has a system inertia of 20000 kg-m^2. Find its natural frequency of oscillation when operating on full load at unity p.f.

10.6 Show that a salient-pole synchronous generator connected to an infinite busbar has a synchronizing power coefficient per phase given by eq. (10.8). A 100 MVA 50 Hz 40-pole 3-ph star-connected hydro-generator with $x_{sd} = 1.3$ p.u. and $x_{sq} = 0.8$ p.u. has 2.0 p.u. excitation and a load angle of $30°$ when delivering 60 MW to an infinite busbar. The turbine-generator system inertia is 5×10^4 kg-m^2. (i) Determine the undamped natural frequency of oscillation for this loading. (ii) Estimate the runaway speed following a sudden loss of load and a simultaneous governor failure, and the time to reach a speed of 25 rad/s: the loss torque of the set is 2.5 ω_r^2 kN-m.

10.7 A 250 kVA 1100 V 3-ph star-connected synchronous generator has a resistance of 0.35 Ω/ph. Its open- and short-circuit test results are:

Field current	(A):	10	20	30	40	50	60
O.C.: line voltage	(V):	400	750	1000	1190	1320	1420
S.C.: current	(A):	75	150	225	–	–	–

From a plot obtain (i) the synchronous-impedance/field-current characteristic, and (ii) the full-load voltage regulation for p.f.s 0.8 lagging and leading.

10.8 Calculate the synchronizing power and torque coefficients per angular degree (mech.) at full load, for a 6.6 kV 1.0 MVA 0.8 p.f. (lag) 50 Hz 8-pole star-connected cylindrical-rotor synchronous generator having a synchronous impedance of $(0 + j0.6)$ p.u.

10.9 A 3-ph synchronous generator, operating at normal voltage and connected to a large power system, delivers 0.9 p.u. power at p.f. 0.8 lagging. Given that $x_{sd} = 1.1$ and $x_{sq} = 0.7$ p.u., (i) draw the phasor diagram and determine the load angle, (ii) develop an expression relating the output power to the load angle, and (iii) discuss the behaviour of the machine subsequent to a loss of excitation.

10.10 Tests on a 5 MVA 6.6 kV 3-ph star-connected cylindrical-rotor synchronous machine gave the following data:

O.C.C.: Line e.m.f. (kV)   3   5   6   7   8   8.8

Field current (A)   25   44   57   78 117 181

S.C.C.: Field current 62 A at rated armature current.

Z.P.F.: Field current 210 A at rated armature current and voltage.

Find the field current for rated load at a p.f. 0.80 lagging.

10.11 A 25 kVA 50 Hz 4-pole synchronous generator working on a busbar of rated voltage has per-unit d- and q-axis synchronous reactances $x_{sd} = 1.0$ and $x_{sq} = 0.6$. Estimate the maximum available torque for zero excitation.

10.12 A generator has a power/angle characteristic $P = (100 \sin\delta)$ MW. When supplying a load of 50 MW the power is suddenly increased to 87 MW. Show that the machine remains in step, and estimate the upper limit of the load angle.

10.13 Neglecting saturation and winding resistance, show that for constant speed and excitation, the terminal-voltage/current characteristics of an isolated synchronous generator are linear for zero p.f. loads and elliptical for unity p.f. loads.

10.14 A salient-pole synchronous generator has $V_1 = 1.0$ p.u., $E_t = 1.7$ p.u., $x_{sd} = 0.8$ p.u. and $x_{sq} = 0.5$ p.u. Estimate (i) the load angle for an output $S = (0.6 - j0.8)$ p.u., and (ii) the maximum load angle for steady-state stability, and the corresponding per-unit torque.

10.15 State the factors that limit the operating range of a synchronous generator in the steady state. An 11 kV 500 MW 0.9 p.f. (lag) generator has a synchronous reactance of 1.5 p.u. (i) What is the maximum leading reactive power permissible? (ii) With the machine loaded to 500 MW, 150 Mvar, the active power is reduced to 400 MW without change in excitation: use an operating chart to find the consequent change in the reactive power. (iii) If it is now required to absorb 150 Mvar, what is the change necessary in the excitation?

10.16 A 750 MVA 0.8 p.f. generator with $x_s = 1.4$ p.u. is connected to an infinite busbar through an external reactance $x = 0.6$ p.u. The machine has a continuously acting automatic voltage regulator. Allowing 20% of the power rating for the stability margin, find the active and reactive powers at the operating condition for which the MVA and practical stability limits coincide.

10.17 A 130 MVA generator, of synchronous impedance $(0 + j0.5)$ Ω/ph, supplies a load of 100 MW and 50 Mvar through an 11/132 kV transformer and a 132 kV transmission circuit. To a base of 100 MVA the transformer and line impedances are $(0 + j0.09)$ and $(0.02 + j0.15)$ p.u. respectively. Calculate (i) the generator internal e.m.f. required to hold the load voltage at 132 kV, (ii) the generator reactive-power loading.

10.18 Two 10 MVA 11 kV 3-ph star-connected synchronous machines, each with a synchronous reactance of 1.0 p.u., are paralleled. One is coupled to a prime mover, the other to a mechanical load, and both are excited to give 11 kV (line) on zero load. What maximum steady-state power can be exchanged between the two machines, and what are the corresponding terminal voltage, line current and power factor?

10.19   A 40 MVA 50 Hz 3-ph star-connected cylindrical-rotor synchronous generator has a synchronous reactance 0.50 p.u., transient reactance 0.10 p.u., d.c. time-constant 110 ms, and a.c. time-constant 600 ms. With the machine on no load and a terminal voltage of 11 kV (line), a symmetrical short circuit occurs at the terminals at an instant $t = 0$ when the voltage in phase A is zero. Sketch the waveform of the short-circuit current in phase A, and determine its instantaneous value at $t = 110$ ms. Why does an a.c. component appear in the field current and a 2nd harmonic component in the phase current? Apply the theorem of constant flux linkage.

10.20   A 50 MVA 13.5 kV generator operating at rated voltage on open circuit is subjected to a symmetrical 3-ph short circuit: the table gives phase-current values taken from the envelope of an oscillogram. Determine the per-unit subtransient, transient and synchronous reactances, and the subtransient, transient and armature time-constants.

Time (ms)		0	20	40	60	100	140	160
Envelope (kA)	pos.	51	44	38	34	29	25	24
	neg.	14	12	11	10	11	11½	12

Time (s)		0.2	0.3	0.4	0.5	0.8	1.0	30+
Envelope (kA)	pos.	22	18	16	15	13	11	2.7
	neg.	13	14	14	14	13	11	2.7

10.21   A 50 Hz synchronous generator has the reactances (p.u.) and time-constants (sec): $x_{sd} = 1.8$, $x_d' = 0.3$, $x_d'' = x_q'' = 0.2$; $T_d' = 1.1$, $T_d'' = 0.06$, $T_a = 0.25$. Calculate the per-unit current components in phase A, five periods after a symmetrical short circuit is applied at an instant that gives maximum current asymmetry.

10.22   A synchronous generator is represented by a 1.08 p.u. voltage source and a transient reactance of 0.15 p.u. The machine is connected by two parallel transmission circuits, each of reactance 0.25 p.u., to a large power network represented by 1.0 p.u. voltage in series with a reactance of 0.20 p.u. A 3-ph short circuit occurs on one of the transmission circuits at the generator end, and is subsequently cleared by switching out the faulted line. Using the equal-area criterion, estimate the critical clearing angle for the generator when, at the instant of fault, it is delivering 1.0 p.u. power.

10.23   Write down the general equation of motion for a synchronous machine with pole-face damper windings, connected to an infinite busbar and subjected to a small load-torque perturbation $M_d \sin \omega_d t$ (Sect. 7.3). Deduce the amplitude of the cyclic rotor oscillation. Given that the ratio (damping torque per rad/s)/(moment of inertia) is 0.6, show how the oscillation amplitude varies as the forcing frequency ω_d passes over 0.5–1.5 times the natural frequency of the rotor.

10.24 A synchronous motor drive, with an inertia of 420 kg-m^2 and a friction torque of 135 N-m (independent of speed), runs on no load at 500 r/min. The system is dynamically braked to standstill by isolating the stator terminals and then connecting them across resistors. If the field current is held constant at the value giving an initial braking torque of 5.4 kN-m, estimate the time for the set to come to rest, and the number of revolutions of the rotor in that time.

10.25 A 4-pole synchronous-induction motor has a symmetrical 3-ph full-pitch slipring rotor winding uniformly distributed in 48 slots. The synchronous-mode connection is (*b*) of Fig. 10.44, the excluded phase being short-circuited to form a damping winding. Sketch the m.m.f. distribution over two pole-pitches in the steady state, and show that $I_d/I_a = 1.23$.

10.26 On no load in the induction mode, the magnetizing current of a 440 V 50 Hz 45 kW 3-ph synchronous-induction motor is 30 A (line). The stator resistance is negligible; its leakage reactance is 1/10 of the synchronous reactance. Estimate, for synchronous operation at full load 0.90 p.f. leading, (i) the line current, (ii) the ratio of the synchronous to the induction magnetizing m.m.f., (iii) the pull-out torque.

11.1 Describe the main features of the design of a large low-speed vertical-shaft hydrogenerator. Discuss the effect on the design of the basic performance criteria.

11.2 A 100 MVA 150 r/min 50 Hz 3-ph vertical-shaft generator has a normal rotor peripheral speed of 75 m/s. Determine the rotor diameter and core length for specific loadings 0.7 T and 60 kA/m.

12.1 A 3-ph fully-controlled bridge rectifier is connected through a transformer to an a.c. supply. The leakage inductance per phase of the transformer is 2.0 mH, and the bridge is supplied at 400 V (line), 50 Hz. For a d.c. load current of 15 A and for ideal thyristors, calculate (i) the reduction in the d.c. output voltage by internal reactance volt-drop, (ii) the firing angle when the output voltage is 200 V d.c., and (iii) the overlap angle for the conditions in (ii).

12.2 Show that the mean output voltage of a 3-ph controlled rectifier is proportional to the cosine of the delay angle α. Derive an equivalent circuit for the system, including the overlap, when the load inductance is large enough to give a constant-current output.

12.3 A 3.3 kV 50 Hz 10-pole 3-ph star-connected induction motor is supplied from 3.3 kV busbars having a short-circuit level of 30 MVA. The core loss is 30 kW and the magnetizing current is 40 A. The leakage impedances are $z_1 = (0.18 + j1.6)$ Ω/ph, $z_2' = 0.40 + j1.6$ Ω/ph. Justifying appropriate assumptions, estimate (i) the torque, output and power factor of the motor running at a slip of 0.03, and (ii) the starting torque with direct-on-line switching.

12.4 A 415 V 4-pole 3-ph star-connected induction motor has $r_1 = 0$, $r_2' = 0.2$ Ω/ph, $x_1 = x_2' = 0.8$ Ω/ph and $z_m = j40$ Ω/ph. Estimate the torque at a slip $s = 0.05$, and the slip s_m for maximum torque,

(i) for a constant voltage V_1 = 240 V/ph, (ii) for a constant current I_1 = 50 A/ph.

12.5   The induction motor of a motor-generator-flywheel colliery winder is a 6-pole 50 Hz machine. The repeated load-power duty cycle is: uniform rise from zero to 1000 kW in 20 s; constant at 500 kW for 40 s; retardation in 10 s with power regenerated falling uniformly from 330 kW to zero; decking (winder at standstill) for 20 s. Estimate for the 90 s duration of the load cycle (i) a suitable continuous rating for the drive motor based on r.m.s. power, (ii) the mean power supplied to the load, (iii) the required flywheel inertia, given that the slip of the motor varies between 0.005 and 0.080 p.u.

13.1   The magnetic-circuit reluctance of an ideal 1-ph 2-pole reluctance motor varies, as the rotor axis turns through an angle θ, from a minimum S_n at $\theta = 0$ (when the rotor axis coincides with the axis of the stator flux Φ) to a maximum S_m when $\theta = \pi/2$ rad. Develop simple expressions (i) for the reluctance S in terms of θ, and (ii) for the mean torque developed when the rotor revolves at synchronous speed with a torque angle λ.

13.2   A 240 V 50 Hz 2-ph reluctance motor has a stator bore diameter D = 80 mm and an axial length l = 50 mm. The radial airgap length is l_g = 1.0 mm. Each stator winding has 675 turns, a winding factor 0.80 and negligible resistance. The gap reluctance S varies between 100 and 200 kA-t/Wb. Obtain an expression for the mean torque on the rotor at synchronous speed. Plot, for one phase to a time-angle ωt, the flux Φ, the reluctance S, also Φ^2 and $dS/d\theta$, and derive the instantaneous torque/angle relation for synchronous speed at a torque angle $\pi/4$ rad.

13.3   A mechanical load L1 is rotated by a prime-mover. It is required to drive a similar but remote load L2 in synchronism with L1. Show how this can be achieved by a pair of power selsyns, and indicate why there is only a limited range of speed over which synchronism can be maintained.

13.4   A hard ferrite material has a remanence B_r = 0.37 T, a coercive force H_c = 280 kA/m and a B/H characteristic approximating to a straight line between B_r and H_c. It is required to establish a flux of density 0.25 T in the airgap of a small 2-pole motor using ferrite blocks and flux-concentrating pole-shoes. The essential data are: stator bore diameter 30 mm, active length 30 mm, gap length 3 mm, pole-arc/pole-pitch 0.70, total/useful flux 2.0. Estimate (i) the minimum volume of ferrite required, (ii) the torque for a rotor m.m.f. of 200 A-t.

13.5   A 2-ph servo motor has two identical sinusoidally distributed stator windings disposed in space quadrature. The impedance of one winding when alone excited from a 400 Hz supply is (100 + j50) Ω. The winding resistance is 20 Ω. Calculate the standstill torque when the currents in the two windings are respectively (0.3 + j0) A and (0 + j0.15) A.

13.6 A commutator winding with three symmetrically placed commutator brushes rotates in a travelling-wave field. Describe the factors that determine the frequency, magnitude and phase of the e.m.f.s between brushes.

13.7 A 50 Hz 10-pole Schrage motor with rated voltage on the primary winding and with the connection between the tertiary and secondary windings open-circuited, gave 76 V between the commutator brushes set 180° (elec.) apart. The stator windings can be arranged to give a standstill e.m.f. of 120 V/ph. Estimate the total speed range of the machine.

13.8 A 1-ph series traction motor takes rated current I_1 at a p.f. 0.90 at a speed n_1 when supplied at 230 V 50 Hz. It has sometimes to operate on a $16^2/_3$ Hz system: estimate the voltage and power factor for the same current I_1 and speed n_1.

13.9 For a current of 2.0 A, the d.c. test on a 'universal' series motor gave 15 V across the field winding, 12 V across the stationary armature, and a ratio E_r/ω_r of 0.22 V per rad/s; the corresponding a.c. test values were 100 V, 24 V and 0.20. Estimate the speed and gross torque for d.c. and a.c. operation with the motor taking 2.0 A at 200 V.

13.10 A small 1-ph series motor with a resistance 25 Ω and an inductance 0.20 H runs at 1400 r/min and takes 1.5 A from a 240 V 50 Hz supply on a given load. For the same torque, sketch the appropriate current, voltage and speed waveforms with the motor supplied through a thyristor with a delay angle of 30°, and a flywheel diode.

13.11 A 1-ph repulsion motor develops a mean torque of 10 N-m at synchronous speed with the brushes displaced 30° (elec.) from the stator m.m.f. axis. Neglecting all losses and the leakage reactance, estimate the starting torque with a 45° brush displacement.

13.12 The equivalent circuit, Fig. 13.25(c), of a 115 V 50 Hz 1-ph induction motor has $r_1 = 1.0$ Ω, $x_1 = 2.0$ Ω, $x_m = 40$ Ω, $r_2 = 2.8$ Ω, $x_2 = 1.6$ Ω. Estimate the gross torque, output power, efficiency and power factor at a slip $s = 0.05$, taking the core and mechanical loss to be 25 W.

13.13 At standstill, a 200 V 50 Hz 1-ph 4-pole induction motor takes a line current of $50 \angle-70°$ A. Ignoring the magnetizing current, stator impedance and mechanical loss, estimate the torque at a slip of 0.05 p.u.

13.14 A capacitor-start 50 Hz 1-ph 4-pole induction motor has the speed/torque relation tabulated below, the auxiliary winding and its series capacitor being disconnected by a sensor at 1000 r/min. The load torque is constant at 2.5 N-m and the system inertia is 0.15 kg-m². The motor is switched on at $t = 0$: find the time t_0 at which the auxiliary winding is disconnected, and the time t_1 of run-up to 1400 r/min.

Speed	(r/min):	0	200	400	600	800	1000
Torque	(N-m):	6.0	7.8	8.5	8.7	8.6	7.5

Speed	(r/min):	1000	1100	1200	1300	1400
Torque	(N-m):	6.0	5.7	5.2	4.5	3.0

Answers

1.2 5.1 kA-t/Wb, 4.8 kA-t, 0.60 Wb, 1.13 kJ.

1.3 9.6 kA-t, $-21°$, $-148°$.

1.4 (i) 10.6 kA-t/pole; (ii) 0.88 T; (iii) 17.5 kJ; (iv) 36.3 kN-m
(v) 11.4 MW.

2.1 1.22.   **2.2** 1/2.73.

2.3 (i) 570 kN/m; (ii) 240 kN.   **2.5** $0.53 + 1.12 = 1.65$ W/kg.

3.1 (i) 1.02, 1.15, 1.42, 1.83, mean 1.36; (ii) 1.14.

3.2 (i) 1.47 kV; (ii) 0.96 kV.

3.3 $E_1 = 3860$, $E_3 = 260$, $E_{41} = E_{43} = 116$, $E_{ph} = 3875$ V.

3.4 $g' = 3$, requiring 4, 4, 4, 3, 3 (= 18) slots/ph in each 5-pole unit with coil-span 9 slots.

3.6 1600, 2260, 1600 A-t/pole; 0.40, 0.56, 0.40 T.

3.8 (i) 980; (ii) 990 A-t/pole.

3.9 1150 A-t.   **4.1** 145 °C.

4.3 84 °C.   **4.4** 44 kW.

4.5 132 kVA.   **4.6** (i) 100 °C, 33 min; (ii) 135 kW.

5.2 (i) 26.8 kW, 0.077 p.u.; (ii) 10.3 kW, -0.047 p.u.

5.5 90 kV.   **5.6** 66.3, 127, 60.3 A.

5.8 197 V.   **5.9** 1.09 p.u.

5.10 $Z_+ = Z_- = (0.21 + j0.46)$ p.u., $Z_0 = (0.01 + j0.10)$ p.u.

5.11 968, 246 kVA; no.   **5.12** 13.0 A, 11.5 A, 240 V; 3.0 A.

5.13 5.6%.

6.1 (i) 1.44, 1.87, 1.56; (ii) 2.92, 1.73, 2.63; (iii) 1.56.

6.2 Within guarantee.   **7.1** 1.27 s.

8.1 $r_1 = 4.7$, $x_1 = 10.0$, $r_2' = 4.3$, $x_2' = 6.7$, $r_m = 1980$, $x_m = 167$ Ω.

8.2 5 kW.   **8.3** 11.6 kW; (i) 0.21; (ii) 0.29.

8.6 0.92 s.

8.8 (i) 5.0 kW; (ii) 62 N-m; (iii) 22.7 to 68.2 A (line).

8.11 187 N-m, 93%.   **8.12** 0.46 p.u.

8.13 87.5 kW.

8.14 16.2 A, 0.96, 86 N-m; 1.0/1.11.

8.15 196 kvar.   **8.16** 122 N-m.

8.17 19 N-m.   **8.18** 3.4 s.

8.19 83 N-m.   **8.20** 22 s.

8.22 −0.09, 34.7 kW, 27.8 kW (34.6 kVA), 0.79 lead.
9.1 0.18 m, 0.135 m.   **9.2** 0.81 m, 0.31 m.
10.1 (i) Overexcited compensator, 0°, 180°, 0.0; (ii) Underexcited motor, −53°, −90°, 0.60 lag, or underexcited generator, 53°, 90°, 0.60 lead; (iii) Overexcited motor, −22°, −131°, 0.87 lead, or overexcited generator, 22°, 131°, 0.87 lag.
10.2 1580 V/ph, 15.6°.   **10.3** 1.85 p.u., 0.89 lead.
10.4 2.75 kN-m.   **10.5** 0.69 Hz.
10.6 (i) 1.05 Hz; (ii) 30 rad/s, 21 s.
10.7 (ii) 18%, −13%.   **10.8** 159 kW/deg, 2.02 kN-m/deg.
10.10 150 A.   **10.11** 53 N-m.
10.12 110° approx.   **10.14** (i) 12°; (ii) 54.5°, 2.08 p.u.
10.15 (i) 370 Mvar; (ii) 234 Mvar (lag); (iii) 37% reduction.
10.16 390 MW, 645 Mvar.   **10.17** (i) 16.4 kV; (ii) 80 Mvar.
10.18 4.5 kV/ph, 370 A, 1.0.   **10.19** 36 kA.
10.20 0.09, 0.16, 1.11 p.u.; 0.05, 1.65, 0.14 s.
10.21 3.4 p.u. (a.c.); 4.3 p.u. (d.c.).
10.22 70°.   **10.24** 15 s, 31.
10.26 (i) 65 A; (ii) 1.4; (iii) 63 syn W.
11.2 9.5 m, 1.0 m.   **12.1** (i) 9 V; (ii) 67°; (iii) 2.1°.
12.3 (i) 12.4 kN-m, 750 kW, 0.89; (ii) 5.4 kN-m.
12.4 (i) 237 N-m, 0.125 p.u.; (ii) 182 N-m, 0.005 p.u.
13.2 (0.1 sin 2λ) N-m.
13.4 (i) 23 cm³; (ii) 32 mN-m.   **13.7** 23−103 rad/s.
13.8 210 V, 0.99.
13.9 785 rad/s, 0.44 N-m; 670 rad/s, 0.40 N-m.
13.11 11.5 N-m.   **13.12** 304 syn W, 265 W, 0.69, 0.56.
13.13 16 N-m.   **13.14** 2.8 s, 5.8 s.

List of Symbols

All quantities are expressed in SI units. Unit symbols are those listed in 'Symbols and Abbreviations for Electrical and Electronic Engineering' (Institution of Electrical Engineers, 1976).

CAPITALS

A area
    integer
    linear current density
    specific electric loading
B magnetic flux density
$\bar{B}$ specific magnetic loading
C capacitance
    constant
    damping coefficient
    number of coils
D diameter
    limb spacing
E electromotive force
F magnetomotive force
G mass
    resistance/reactance ratio
H inertia constant
    magnetic field intensity
    overall height
I current
J current density
    moment of inertia
K coefficient, factor
    compliance

L inductance
    overall length
L   transference
M torque
N number of turns
P active power
Q charge
    reactive power
R radius
    resistance
S apparent power
    number of slots
    rating
    reluctance
    surface area
T absolute temperature
V voltage
W energy
    overall width
X reactance
Y admittance
    pole-pitch
Z impedance
    number of conductors

SMALL LETTERS

a acceleration
    conductor area

a constant
    dimension

a	number of paths	*m*	number of phases
b	breadth	*m*₀	momentum
	component flux density	*n*	harmonic order
	dimension		number
	gas pressure		speed of rotation
	pole-arc	*p*	active power
c	capacitance		number of pole-pairs
	cooling coefficient		specific loss
	damping ratio	*p*	d/d*t*
	dimension	*q*	charge
d	diameter		reactive power
	thickness		volume flow-rate
d	differential	*r*	integer
e	electromotive force		radius
f	force		ratio
	frequency		resistance
f	function	*s*	slip
g	gravitational constant	*s*	complex frequency
	slots per pole	*t*	time
g'	slots per pole and phase	*u*	coil-sides per slot
h	height, depth		linear speed
	load displacement	*v*	voltage
i	current	*w*	energy
j	90° operator		width
k	coefficient	*x*	displacement
	compliance		fraction
	constant		length variable
	factor		reactance
	fractional load	*x'*	transient reactance
	thermal conductivity	*x"*	subtransient reactance
l	gross core length	*y*	admittance
	length		dimension
m	integer		slot-pitch
	mass	*z*	impedance
	numbers of layers		conductors per slot

GREEK LETTERS

α	angle	β	root of capacitance ratio
	angular acceleration	γ	angle
	eddy-loss parameter		angular divergence
	load angle		slot-pitch angle
	resistance/reactance ratio		specific cost
	root of capacitance ratio	Δ	finite difference
	temperature coefficient	δ	density
β	angle		load angle
	phase coefficient	ε	chording angle
	phase-displacement angle		eccentricity

δ	per-unit impedance		μ_0	magnetic space constant
	radiation factor		ρ	resistance section
	regulation			resistivity
	shaft deflection		σ	angle
η	efficiency			angular phase-spread
θ	position angle		τ	time-constant
	temperature		τ'	transient time-constant
	temperature-rise		τ''	subtransient time-constant
Λ	permeance		Φ	magnetic flux
λ	permeance coefficient		ϕ	phase angle
	thermal emissivity		ψ	magnetic linkage
	torque angle		ω	angular frequency
	wavelength			angular speed
μ	absolute permeability			

SUBSCRIPTS

a	absorption		d	eddy-current
	acceleration		e	chording, coil-span
	active			effective
	air			electrical
	alternating-current			excitation
	armature-reaction		er	end-ring
	duct-width		f	field
	phase A			fluid
ad	d-axis mutual			friction
aq	q-axis mutual		g	gap
au	armature, unsaturated		h	harmonic
av	average			hot-spot
b	boundary			hysteresis
	braking		i	inlet
	duct-width			iron
	phase B			net-iron
c	cage		kd	d-axis winding
	capacitive		kq	q-axis winding
	coil		l	leakage
	conduction			line
	conductor			load
	control			loss
	critical		m	magnetizing
	damping			maximum
	phase C			mean
d	direct-axis			mechanical
	direct-current		mc	mean-conductor
	distribution		mt	mean-turn
	disturbing		n	harmonic-order
	duct			winding

o	opening	*t*	test
	outlet		tooth
	outside		torque
	overall		total
	overhang		transient
oc	open-circuit		turn
p	pole	$t\frac{1}{3}$	at one-third tooth level
	pulsation	*u*	tube
	transient-peak		unsaturated
ph	phase	*v*	convection
pl	pole-leakage		voltage
q	quadrature-axis	*w*	width
r	radial		winding
	radiation		window
	reactive	*x*	at position *x*
	relative		reactive
	remanent	*y*	yoke
	resistance	*z*	impedance
	ring		zigzag
	rotational, motional		
s	mechanical-storage		
	pole-shoe	0	characteristic
	slot		free-space
	space		natural
	starting		no-load
	stray-loss		normal
	synchronizing		resultant
	synchronous		zero-phase-sequence
	system	1	fundamental
sc	short-circuit		input
sci	ideal short-circuit		primary
sd	d-axis synchronous		stator
sdu	d-axis synchronous,		supply
	unsaturated	2	outer-cage
sl	shoe-leakage		rotor
sq	q-axis synchronous		secondary
t	at time *t*	3	inner-cage
	tank		tertiary

Index